**Charge Storage, Charge Transport
and Electrostatics
with their Applications**

STUDIES IN ELECTRICAL AND ELECTRONIC ENGINEERING

Vol. 1 Solar Energy Converson: The Solar Cell (Neville)
Vol. 2 Charge Strage, Charge Transport and Electrostatics with their Applications (Wada, Perlman and Kokado)

STUDIES IN ELECTRICAL AND
ELECTRONIC ENGINEERING 2

Charge Storage, Charge Transport and Electrostatics with their Applications

Edited by

Yasaku WADA
Faculty of Engineering, University of Tokyo, Tokyo, Japan

M. M. PERLMAN
Collège Militaire Royal de Saint-Jean, Quebec, Canada

Hiroshi KOKADO
Faculty of Engineering, Tokyo Institute of Technology, Tokyo, Japan

KODANSHA LTD.
Tokyo

ELSEVIER SCIENTIFIC PUBLISHING COMPANY
Amsterdam – Oxford – New York

1979

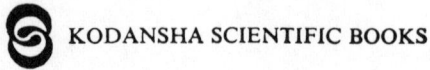 KODANSHA SCIENTIFIC BOOKS

Copyright © 1979 by Kodansha Ltd.

All rights reserved
No part of this book may be reproduced in any form, by photostat, microfilm, retrieval system, or any other means, without the written permission of Kodansha Ltd. (except in the case of brief quotation for criticism or review)

Library of Congress Cataloging in Publication Data

International Workshop on Electric Charges in Dielectrics, Kyoto, 1978.
   Charge storage, charge transport, and electrostatics with their applications.

   (Studies in electrical and electronic engineering ; v. 2) (Kodansha scientific books)
   Bibliography: p.
   Includes index.
   1. Charge transfer--Congresses. 2. Electrostatics --Congresses. I. Wada, Yasaku. II. Perlman, Martin M. III. Kokado, H. IV. Title. V. Series.
QC173.I556 1978   537.2   79-20991
ISBN 0-444-99769-5

Copublished by
KODANSHA LTD.
12–21 Otowa 2-chome, Bunkyo-ku, Tokyo 112
and
ELSEVIER SCIENTIFIC PUBLISHING COMPANY
335 Jan van Galenstraat, P.O. Box 211, 1000AE Amsterdam, The Netherlands

ELSEVIER NORTH-HOLLAND, INC.
52 Vanderbilt Avenue, New York, N.Y. 10017

Printed in Japan

EDITORS

Edited by

Yasaku WADA
Faculty of Engineering, University of Tokyo, Tokyo, Japan

M. M. PERLMAN
Collège Militaire Royal de Saint-Jean, Quebec, Canada

Hiroshi KOKADO
Faculty of Engineering, Tokyo Institute of Technology, Tokyo, Japan

CONTENTS

Preface ix

I ELECTROSTATICS

The Effect of Corona Generated Excited Molecules on Surface Potential Decay in Polyethylene 3
Charge Absorption and Charge Transport in Polymers Wetted by an Insulating Liquid and Subjected to an Electric Field 10
New Method for Charge Density Measurements in Dielectric Liquids 19
On the Method to Solve Poisson's Equation by the Substitute Charge Method 24
Temperature and Friction Speed Dependence of Frictional Electrification between Metal and Polymers. Contribution of Molecular Motion of Polymers to Frictional Electrification 30
Corona Charging of a Spherical Particle Having an Extremely High Resistivity 35
Maximum Charge of a Spherical Particle Imparted by Pulse Charging 40
Corona-Charging Properties of Polymer Powders 45
Surface Potential Decay in CdS Powder Layers 50
Electrification of Liquids in Laminar Flow through a Small Diameter Metal Pipe 55
Electrokinetic Phenomena in Electrical Insulating Oil/Impregnated Cellulosic Insulating Paper Systems 60
Measurement of Space Charge Quantity in Dielectric Film by Maxwell Stress Method 65
Surface Potential of SiO_2 during Ion Implantation 70
Triboelectrical Properties of Toner and Carrier 75
A New Idea for Improved Electrical Precipitator Design 79
Impact Electrification of Oil Jet 84
Behaviour of Fluids under Electric Field—Promotion of Combustion, Vaporization and Heat Transfer— 89

II ELECTRETS

Bioelectrets and Bound Water in Biological Molecules 97
Charge Distribution and Transport in Polymers 103
Spectroscopic Study of the Internal Potentials Topography for Polyethylene Electrets 108
Charge Stability of Annealed Electrets 113
Studies on Polarization Process of PVDF Electret 118
Structure in Electrically Polarized Polyacrylonitrile 123
Plate Electrets and Their Use in Condenser Type Headphones 128
A Review on Magnetoelectret 133
$Bi_{12}SiO_{20}$ Photoelectret in Optical Processing and X, γ Rays Imaging 138

III PIEZO- AND PYRO-ELECTRICITY

Polyvinylidene Fluoride - Viewed as a Polycrystalline Ferroelectric Material 143
Fine Structure of Drawn Polyvinylidene Fluoride and Its Poling Effects 148
On the Nature and Mechanism of Piezo- and Pyro-Electricity in Polyvinylidene Fluoride 153
Piezoelectricity, Pyroelectricity and Improper Ferroelectricity in PVF_2 158
Polarization and Depolarization Processes in Polyvinylidene Fluoride 165
Two Types of Pyroelectricity in a Copolymer of Vinylidene Fluoride and Tetrafluoroethylene 170
Use of Piezoelectric Polymers to Prevent Marine Fouling 175

IV CHARGE GENERATION AND TRANSPORT

Electron Transport and Breakdown in Polymers and Single Crystal of Long Chain Hydrocarbons 183
Electric Charges in Non-Polar Liquids 192
Pulsed Photoconduction in Some Polymers 203
Field-Controlled Photogeneration and Trapping of Carriers in Polyethylene-Terephthalate 208
On the Space Charge Injected from Electrode into Polymeric Materials at Low Temperature 213
High-Field Carrier Transport in Dielectric Liquids at Pre-breakdown and Breakdown Regions 218
Electron Mobility and Electron-Ion Recombination Rate Constants in Liquid and Solid Neopentane 223
Electron Transport Mechanisms in Dielectrics: Transition Caused by Density Changes 228
Transport of Space Charge in Naphthalene Crystals 234
Influence of Localized Electronic States on the Electrical Conduction in Iron Trifluoride Thin Films 239
Dielectric Studies on Ionic Charge Carrier in Amorphous Polymers 244

V THERMALLY STIMULATED CURRENTS

Some Aspect of Thermally Stimulated Polarization 251
Internal Polarization in Semi-Insulating GaAs 256
Thermally Stimulated Depolarisation Currents in $CaWO_4$ Crystals 261
Space Charge in $Pb(Zr, Ti)O_3$ Ceramics 266
TSC Studies of Charge Storage and Transport in Organic Crystals. A Critical Approach 271
Thermal-Depolarization-Current Study of Space Charge Polarization in Polymer Composites 277
Charge Carrier Behavior in Multilayer Dielectrics with Liquid Crystal 282
Model TSC Calculations for Amorphous Materials 287
Electric Charge in Thermal Electret of Carnauba Wax and Rosin Studied by Thermally Stimulated Current 292
Dielectric Relaxation Measurements in Vc: VAc Copolymer Thin Films 297
Thermally Stimulated Current of Polyethylene Terephthalate 302
Thermally Stimulated Current in Polyethylene 307
Thermally Stimulated Currents in Polypropyrene 312
Thermally Stimulated Current in Elongated Polyethylene by Gamma-Radiation 317
Carrier Trapping in Polymers 322
Electrical Conduction and Space Charge Accumulation in Polyethylene Terephthalate 327
Surface Voltage Decay Mechanism in Corona Charged PET Film 332
Thermally Stimulated Surface Potential of High Density Polyethylene Film Charged by Corona Irradiation 337
Surface Charge Decay and Thermally Stimulated Current in Polymers with Different Molecular Structures 342
Thermally Stimulated Current of Cellulose Electrets 347
Thermo-induced Electric Current from Paper Cellulose Sandwiched in between Metal Electrodes 352

VI ELECTRICAL CONDUCTION AND BREAKDOWN

A Classical Model for Electronic Conductivity of Polymers 359
A.C. Impedance Measurements on CdS:Cu Binder Layers 368
Space Charge Effects in Rutile Ceramics 372
Space Charge as an Instrument of Investigation of Electric Phenomena in Dielectrics 377
Electrical Conduction in Polymers 382
Electrical Conduction in Drawn Polymer 387
Non-Destructive Charge Release from High Volume Resistivity Polymers 392
Electrical Conduction and Free Radicals in Plasma Polymerized Styrene 397
Relation between Dielectric Breakdown and Morphological Structure in Polyethylene 402
On Nonlinear Dielectric Relaxation II 407
The Storage of Charge Caused by Conductivity Gradient in Electrically Stressed Semi-Insulating Materials, 412
Voltage-Induced Charge Transfer and Its Effects on Tree Initiation in Polyethylene 417

The Streamer Breakdown of Dielectrics in Divergent Field 422
Ionic Conduction Current Related to Neutralization at Insulator-Metal Interface 427
Interfacial Phenomena in Polymers and Their Application for Determining Carrier Species 432
Some Considerations on Electric Conduction and Charge Distribution in XLPE Cable Insulation 437
Auther Index 443

PREFACE

This book consists of the papers presented at the International Workshop on Electric Charges in Dielectrics held at Kyoto Kaikan, Kyoto, on October 9 to 12, 1978.

The Workshop aimed to promote the scientific and technological understanding of charge generation, storage and transport in dielectrics. Various problems were discussed in the meeting: such as on electrets, electrostatics, charge generation and transport, thermally stimulated current, piezo- and pyroelectricity, electrical conduction and breakdown, and so on.

The participants of the Workshop, numbering one hundred and seventy-two, came from thirteen nations. At the technical sessions which ran in a smooth and efficient manner at one hall on October 9 and at two halls on October 10 to 12, nine invited papers and eighty-six contributed papers were delivered. A technical exhibition was also held in connection with the Workshop. The participants enjoyed personal contacts and warm hospitality at the reception on the evening of October 9 at Kyoto Royal Hotel.

Planning of the Workshop was made by the joint effort of the Organizing Committee and the Executive Committee; the former included eleven foreign leading scientists and functioned as an international advisory board. Arrangements of the Workshop were carried out by the Executive Committee organized in the host country. The Workshop was financially supported by the Japan Society for the Promotion of Science and forty-five sponsoring companies in Japan, to which we owe more thanks than can be expressed individually.

Some pioneering investigations were developed in Japan in the field of electric charges in dielectrics; it was in 1919 that Prof. M. Eguchi prepared electrets of a mixture of wax and resin. Since then an increasing effort has been expended in Japan to meet scientific and technological requirements for the development in this field. Hence it seems natural that we had long cherished a desire to hold this kind of international meeting in this country. Foreign participants from all parts of the world are believed to deserve our gratitude for having made the Workshop so successful.

Y. Wada

I
ELECTROSTATICS

THE EFFECT OF CORONA GENERATED EXCITED MOLECULES ON SURFACE POTENTIAL DECAY
IN POLYETHYLENE *

M.M. Perlman, K.J. Kao and S. Bamji

Groupe de Recherches sur les Semiconducteurs et les
Diélectriques, Département de physique,
Collège militaire royal, Saint-Jean, Québec, Canada.

I. INTRODUCTION

In recent papers, Baum, Lewis and Toomer[1] reported that both light and excited neutral gas molecules, present in a negative corona discharge were effective in injecting negative charge from deep surface into more shallow bulk states in polyethylene. This led to a more rapid decay of surface potential at higher initial surface fields, resulting in the crossover phenomenon. The authors were led to these conclusions by observing differences in decay for short and long charging times, and between central and peripheral regions of their samples, in addition to differences observed when a current of air was passed over the face of the film to remove the excited gas molecules.

In this work, a special corona charger[2] was constructed to allow the samples to be charged in the absence or presence of corona light. This charger, together with blowing air currents, enabled us to separate the effects of photons from those of excited molecules. It was found that while light in a negative point or wire corona has no effect on negative charge decay (contrary to Baum et al[1]), excited molecules do, and are alone responsible for faster initial decays, and the crossover phenomenon. Neither has any effect on positive charge, in agreement with Baum et al[1]. Agreement with the latter authors was also obtained on decay differences observed for short and long-time charging, but our results attribute these differences to excited molecules only. If an external source of intense light is used on precharged samples in the range of energies where polyethylene absorbs, these photons also inject surface charge, changing the decay rate. The order of magnitude of the energies needed by the excited molecules in the corona, or photons in the external light source, to inject charge from surface states is $\gtrsim 6$ e.V.

II. EXPERIMENT

Dupont, 25μ, low density polyethylene films were painted with aquadag on one face, held firmly against a metal backplate by a partial vacuum.

Corona charger A (Fig.1) was used for the short and long time charging experiments of Figs. 2 and 3, and corona charger B (Fig.1) without light shield S and rod R for the experiments of Fig. 4. A and B were thin wire corona chargers, C (Fig.1) was a point charger.

In the experiments of Figs. 5 and 6, whose object was to determine the **separate** effects on decay of corona light, and corona generated excited molecules, samples were first precharged with charger A. A rotating fixture[3] swept the sample past this charger in a fraction of a second, allowing minimal time for corona light and excited molecules to act, and producing a uniform charge **distribution** over the sample area. The samples were allowed to decay for a time, and then held under corona charger B (Fig.1) for 100 seconds with or without light shield S and rod R, and with or without blowing air. During the latter experiments, a grounded grid was added to slot T in charger B which prevented any ions from passing through, but allowed either light or excited molecules, or both, to strike the sample. In the experiment of Fig. 7, whose object was to determine the effect on charge decay of an intense external light source, the samples were precharged with charger A, allowed to decay for a time, and then held for 100 secs., 25cm. in front of a 200W Xe-Hg Oriel lamp. After precharging and before and after conditioning, the samples were quickly rotated in front of a Monroe Electronics mini-probe and electrostatic voltmeter and recorder, to measure surface potential decay as a function of time.

All measurements were performed in laboratory air of relative humidity 55-65%, at a temperature of $21°C$.

III. RESULTS

Neither light nor corona generated excited molecules were observed to have any effect on positively charged polyethylene. Neither was any crossover observed for positive charge, agreeing with Baum et al[1], but contrary to the results of Ieda et al[4]. All figures in this paper are for negatively charged samples.

Figures 2 and 3 illustrate the differences between short time (a fraction of a second) and long time (30 sec.) charging, in the presence of light and excited molecules, using charger A. The short time charging curves of Fig. 2 show slower initial decay than those of Fig. 3, and the absence of any crossover effect, clearly evident in Fig. 3.

Figure 4 shows the surface potential decay differences obtained for different charging times, with the samples all charged to an initial potential of 1200 ± 50 volts, using charger B without shield and rod. For the shortest charging time (∼ 0.5 sec.), excited molecules have little time to act on the charge in surface traps, and the decay is relatively slow. For a 30 sec.

charging time, some charge has been injected into more mobile bulk states and the charge decay is more rapid, as previously found in Fig. 3. The longest charging times (200-600 sec.) are of the order of magnitude of the transit time of charge through the sample. There is then a large portion of the charge in the bulk, reducing the internal fields, with consequently slower decay.

Essentially no qualitative decay differences were found whether we shielded the corona light in charger B or not, showing that corona photons had no effect. This was proved in a more direct manner in the experiments that follow, where it is shown that it is excited molecules in the corona, generated by electron-gas collisions, that inject surface trapped charge.

In Fig. 5, two samples 1 and 2, both precharged to an initial potential of 1650 volts, (in a fraction of a second) using charger A, were allowed to decay for 125 sec., and followed essentially the same curve. Both samples were then held in front of corona charger B (without shield S and rod R but containing a grounded grid in slot T, which prevented any ions from passing through) for 100 sec. Sample 1 was under the influence of corona light and excited molecules, and one notes a large decrease in surface potential after the exposure. A strong air current was blown tangentially across sample 2's face during exposure to remove excited molecules in the corona wind, leaving the sample exposed to corona light only. The decay continues as if the exposure had not taken place, showing that corona photons have no effect on the charge. This may be because of insufficient intensity or energy. The same results were obtained when point charger C was substituted for wire charger B.

In Fig. 6, the experiment of Fig. 5 is repeated <u>including</u> the light shield and rod in charger B. Excited gas molecules alone pass through the grid, and are seen to cause large decreases in surface potential, whether the samples are exposed to them at an early (after 125 sec.) or late (after 350 sec.) stage in the decay. Negatively charged samples exposed to positive corona generated excited molecules during decay also showed decreases in surface potential, but the effect was about one-third as large as that due to negative corona generated molecules.

Measurements were taken on some three dozen samples in the experiments of Figures 2-6 inclusive.

Figure 7 shows the effect of a 200W Xe-Hg Oriel lamp on negative charge decay. The lamp had no effect on positive charge. Sample 1 was charged in a short time, and the decay followed for 600 seconds. Sample 2 was charged to the same initial potential, allowed to decay for 125 sec., and was then exposed to the lamp for 100 sec. The decrease in surface potential over and above normal decay in 100 sec. was about one-half that produced by the corona generated excited molecules. (Illumination at an earlier stage in the decay produces larger effects.) Sample 3 was charged to the same initial potential, allowed to decay for 125 sec., and was then exposed to the lamp containing a 220-410 mµ band-pass filter for 100 sec. The decrease in surface potential after illumination with the filter was about 20% of the decrease observed without the filter. When a 290-390 mµ filter was used, no decrease in surface potential was observed other than that due to normal decay in 100 sec. The absorption spectrum of polyethylene, and the transmission spectrum of the 220-410 mµ filter were measured, and are shown in Fig. 8. The output of the Oriel lamp rises from 200 to 230 mµ (to 0.2 µW/cm²Å at 50cm. distance), and may, for our purposes, be considered constant between 230 and 290 mµ.

From the above results, photons of wavelength greater than 290 mµ have no effect on charge decay, those between 220 and 290 mµ have a 20% effect (when the sample is exposed for 100 sec. after it has decayed for 125 sec.) and those between 200 and 220 mµ ($\sim$ 6 e.V.) are responsible for 80% of the decrease in surface potential.

As seen above, the excited molecules produced twice the decrease in surface potential caused by the Oriel lamp, even though the latter was hundreds of times more intense. This fact suggests that some of the excited molecules have wavelengths less than the lower limit of the lamp (200 mµ), i.e. energies $>$ 6e.V. This is confirmed by the fact that the molecules applied to precharged polyethylene at a late stage in the decay still produced relatively large decreases in surface potential (Fig.6), while the lamp did not. The latter would take place if the molecules and lamp were effective injectors of charge from deeper and shallower surface traps, respectively. More work is needed here.

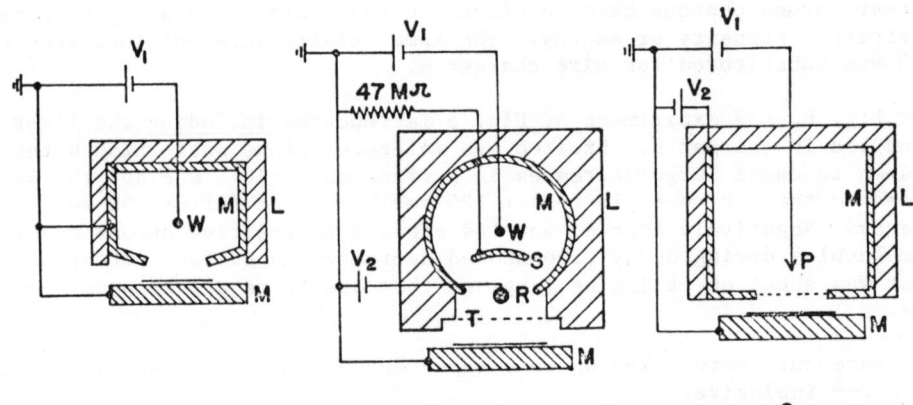

Fig. 1- Corona Chargers (roughly to scale) L-plexiglass, M-aluminum, W-thin tungsten wire, P-needle point, S-light shield, R-rod, T-grid.

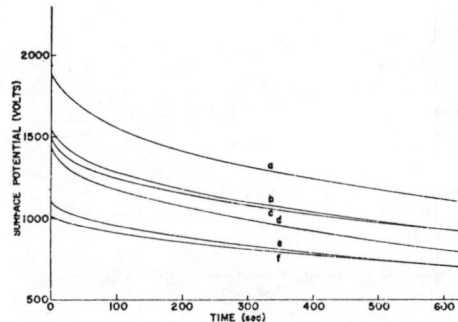

Fig. 2- Surface Potential Decay for short time (a fraction of a second) charging, Charger A. Different initial potentials obtained by changing $V_1(A)$ of Fig.1 → a-9.8KV, b-9.5KV, c-9.0KV, d-8.75KV, e-8.5KV, f-8.0KV.

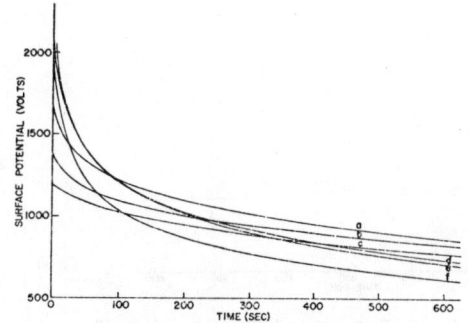

Fig. 3- Surface Potential Decay for long time (30 sec.) charging, Charger A. Different initial potentials obtained by changing $V_1(A)$ of Fig. 1 → a-7.5KV, b-7.0KV, c-6.5KV, d-7.8KV, e-8.0KV, f-8.2KV.

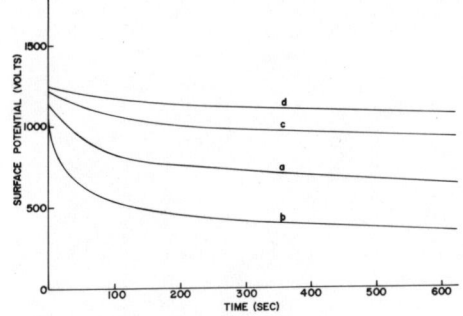

Fig.4-Surface Potential Decay for variable charging times. Charger B without shield and rod. a-0.5 sec., b-30 sec., c-200 sec., d-600 sec.

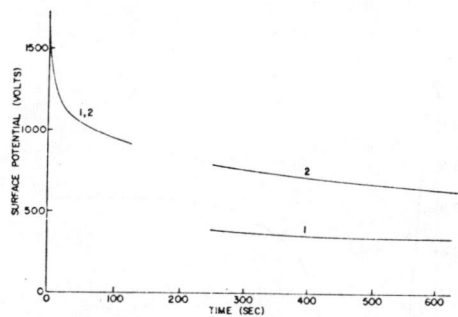

Fig. 5- Surface Potential Decay. Sample 1-exposed to corona light and excited molecules (no ions), after 125 sec. of decay, for 100 sec., using Charger B without light shield S, but with grounded grid T. Sample 2 exposed like sample 1, but with a tangentially blown air stream, so that corona light only is effective.

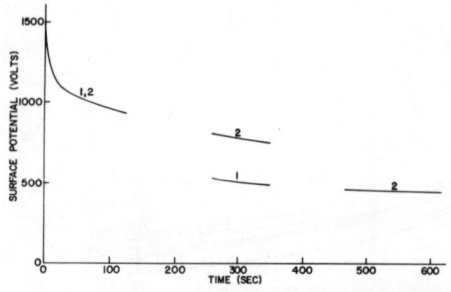

Fig.6-Surface Potential Decay. Sample 1-exposed to excited molecules (no light or ions) after 125 sec. of decay, for 100 sec., using Charger B with light shield S and grounded grid T. Sample 2-initially exposed like sample 1, but with a tangentially blown air stream, later (after 350 sec. of decay) exposed to excited molecules.

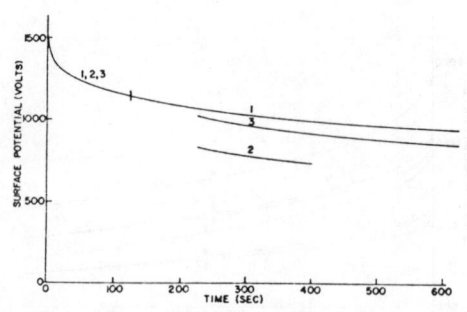

Fig. 7- Surface Potential Decay on exposure to 200W Xe-Hg Oriel lamp. Sample 1 - Ordinary decay followed for 600 sec. Sample 2 - Ordinary decay for 125 sec. exposed to lamp for 100 sec. Sample 3 - Ordinary decay for 125 sec., exposed to lamp with 220-410mµ band-pass filter for 100 sec.

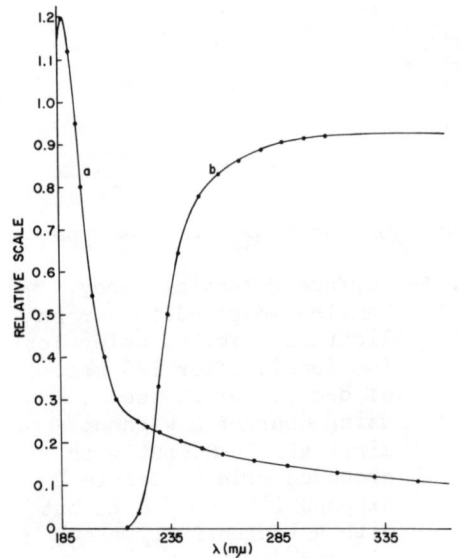

Fig.8- a- Absorption Spectrum of Polyethylene.
b- Transmission Spectrum of 220-410-mµ band-pass filter.

IV. DISCUSSION

While the results with samples cut from different sheets were not always perfectly reproducible quantitatively, differences such as the slow initial decays in Fig. 2, and the fast decays and crossover in Fig. 3, were easily discernible. Better reproducibility was obtained with samples from the same sheet.

The major ionic species in a negative corona at atmospheric pressure is the CO_3^- ion[5]. On landing on the polymer, the ion may be neutralized by giving up an electron to the conduction band near the surface[6]. The electron is subsequently trapped in a surface state, and needs some activation energy in order to move into the bulk. As stated by Baum et al[1], this energy can be supplied thermally, is lessened by the field of the deposited charge, and can also be supplied by excited molecules. The latter may take place through the emission of a photon of sufficient energy by the excited molecule, with subsequent photoinjection of the trapped charge.

The proportion of surface and bulk charge in the sample at the end of the charging period, together with the initial value of the propelling self-internal field, determines the subsequent decay[1]. As charge moves into the bulk after charging, the field at the surface and within the moving charge is reduced, and the decay slows down.

Since corona generated excited molecules are effective injectors of negative charge from surface traps into the bulk during the charging period, they play a major role in determining the type of decay in polyethylene.

References

(1) E.A. Baum, T.J. Lewis, and R. Toomer, J. Phys. D. (GB): Appl. Phys. 10, 487, (1977), ibid 10, 2525, (1977).

(2) L.F. Frank and G.W. Alley II, U.S. Patent 3,660,656, May 2, (1972).

(3) M.M. Perlman, T.J. Sonnonstine, and J.A. St. Pierre, J. Appl. Phys. 47, 5016 (1976).

(4) M. Ieda, G. Sawa, and U. Shinohara, Japan J. Appl. Phys. 6, 793 (1967), Elect. Eng. Japan 88, 67 (1968).

(5) M.M. Shahin, Ann. Report, Conf. on Elect. Ins. and Dielectric Phenomena, 85, (1968), Appl. Optics, Supplement 3 on Electrophotography 106, (1969).

(6) M.M. Perlman and T.J. Sonnonstine, Proceedings of International Symposium on Electrets and Dielectrics, Ed. Acad. Brazil de Ciencias, Rio de Janeiro, Brazil 337, (1977).

* Work supported by the Defence Research Board, and National Research Council of Canada.

CHARGE ABSORPTION AND CHARGE TRANSPORT IN POLYMERS WETTED BY AN INSULATING LIQUID AND SUBJECTED TO AN ELECTRIC FIELD

N.J. FELICI

C.N.R.S., Laboratoire d'Electrostatique

C.N.R.S., 166 X, 38042 Grenoble Cedex, France

Introduction

The problems raised by the heterogeneous structure of a dielectric were first considered by Maxwell and his treatment is still widely known as Maxwell's and Wagner's. It is based on the seemingly natural assumption that each homogeneous component can be univocally characterized by a set of data. For instance, in an impregnated capacitor, the sheets of paper are thought of as having permittivity ε_1 and conductivity γ_1 while the liquid has ε_2 and γ_2. Thus, polarisation and conduction would be given by the set ε_1 ε_2 γ_1 γ_2 making allowance for the continuity conditions for field and current at interfaces.

These assumptions are usually <u>wrong</u> for very important reasons we shall discuss presently. Before, it is useful to recall the physical basis of the conductivity concept.

The media in which this concept is thoroughly valid (e.g. metals and usual electrolytic solutions) are characterized by an <u>equilibrium</u> between charge carriers and their parent substance which cannot be appreciably shifted by an electric field. Thus, the concentration of charge carriers remains constant whatever the field applied. For the same reason, the <u>nature</u> of these charge carriers cannot be altered. A metal will always contain electrons, exclusively of any other type of carriers ; similarly, a sodium chloride solution will exhibit Na^+ and Cl^- ions and no others. When two different media, like a metal and an electrolyte, are in contact, the electric current is passed from one substance to the other by mere electron exchange at the interface and the basic nature of carriers in both media is thus preserved. It never occurs, for instance, that ions from an electrolyte would penetrate into a metal, no more that free electrons from an electrode could pervade an electrolytic solution. Thus, the nature and concentration of charge carriers in every medium being maintained irrespectively of the field applied, the conductivity of each component of the system remains a well defined quantity.

If we now turn to poor conductors like pure organic **liquids and insulating** polymers, we have a vastly different picture.

There is still an equilibrium between charge carriers and parent compounds in the absence of a field, but this equilibrium is entirely upset when a moderate voltage (by engineering standards) is applied to the sample. This is related to the slow kinetics of carrier generation. If ions are suddenly removed from an aqueous electrolyte, equilibrium is restored in a matter of nanoseconds ; in insulating liquids, it may take seconds, minutes, or even more.

No less surprising phenomena are observed at <u>interfaces</u>. Electron exchange, which readily occurs between highly conductive <u>media</u>, is very difficult or plainly impossible in the case of "insulators". Thus, carriers reaching an interface tend to proceed farther, while retaining their identity. Sometimes, they are prevented to do so by specific interactions, and pile up at the interface. This is the "Garton effect" we shall discuss in the next section. But, in many other cases, carriers from one phase penetrate into the other, sometimes pervading it entirely. To-day, we know many examples of that, in both directions, when a solid and a liquid are in contact.

We may conclude that ascribing a definite "conductivity" to the components of an insulating system is devoid of any meaning, as soon as an electric field is applied. Maxwell's problem cannot be solved without a thorough investigation of the physical mechanisms controlling :

- The generation and destruction of carriers within a given phase, with a field present.

- The interaction of carriers from one phase with the other(s).

We shall restrict ourselves to the second point, when organic liquids and insulating polymers are considered. In this case, the initial concentration of carriers is, as a rule, much larger in the liquid than in the solid. Thus, we may further limit our subject to the interaction of the ions from the liquid with the solid.

<u>A Case Story : Garton's effect</u>

After the advent of oil-impregnated cables and capacitors, electrical engineers soon discovered they did not behave according to Maxwell's scheme. In particular, tan δ showed a steep decrease when voltage was raised, instead of remaining constant. Mr. C.G. Garton gave the following explanation :
- The liquid phase is entirely out of equilibrium. Very few ions (if any) are created by dissociation while the field is on. Thus, the liquid is cleared of ions and has no conductivity during most of each half-wave.
- The ions swept by the field are merely stopped by the paper walls. They are ready to return into the liquid as soon as the field reverses.

Garton's theory was very well received by electrical engineers, if only because of its graphic simplicity. Similar ideas were expressed by Held and Wenzel in 1960 (1). The dependence of tan δ on voltage, ionic mobility (as a function of temperature) and frequency can be deduced by simple mathematics, and agreement is mostly fair to good. In particular, when temperature rises, ionic mobility increases as well, and tan δ must decrease (all other components of dielectric loss remaining constant). This prediction is confirmed by experiment, while the "conductivity" concept would lead to the opposite conclusion.

A straightforward check of the theory is allowed by oscilloscopic measurements of the instantaneous current flowing through the impregnated system under A.C. voltage. Strong peaks are observed at each field reversal. Thus, there is no reason to doubt the existence of ions trapped in the liquid layers and impinging on the paper walls at each half-wave, without any adherence or penetration. Similar phenomena are observed in aroclors and also nitrobenzene ($\varepsilon/\varepsilon_0 = 36$) and propylene carbonate ($\varepsilon/\varepsilon_0 = 70$).

Much criticism, however, may be leveled at the Garton-Held-Wenzel theory. Some electric facts are difficult to reconcile with it. For instance, the ionic extra-current should last no longer than the time required by the ions to cross the liquid layers by their mobility. Agreement with experimental results is good under A.C. voltage, but not at all with D.C. In this case, one would expect a step voltage of large magnitude to cause an extremely short ionic current and nothing else. This is never observed. The current lasts almost indefinitely, decreasing steadily, but its integral is entirely available at the discharge of the capacitor ; in other words, the ions penetrate into the paper and forge ahead slowly, to be stopped sooner of later, in any case before they reach an electrode. One may conclude that paper, as seen by the ions, is obviously no "wall", but rather a maze of channels ending in blind alleys, a picture which fits better the known structure of paper.

More serious objections come from other quarters. In gases, any ion approaching an insulating surface is attracted and remains trapped by image forces, since the gas has the lower permittivity. Thus, no ions should remain free in oil, whose permittivity ($\varepsilon/\varepsilon_0 \sim 2$) is much less than that of cellulose ($\varepsilon/\varepsilon_0 \sim 6$).

On the other hand, to be impregnated, capacitor paper must be a thoroughly porous material , allowing a liquid to pervade it completely. Why cannot ions get through the thinner channels, as they are no bigger than the molecules of the impregnant ? As a matter of fact, the blocking effect of paper depends on the nature of the liquid. In water and many other solvents, ions readily cross a paper "wall". To sum up we come to the conclusion that the blocking effect may be due to a solvent controlled interaction between ions and paper, and not to an imaginary "solidness" of the latter.

Recent work (2) (3) shows how intricate the picture may be at the molecular level :

Ions do not remain free when paper is wetted with a dilute electrolytic solution in an insulating liquid. Most of them adhere strongly. An adsorption equilibrium establishes itself, depending on concentration, as given by adsorption isotherms (Fig. 1)

- The affinity of an ion for paper depends on many factors : charge, sign, solvation state, chemical structure. Thus, both ions of a given electrolyte are not equally adsorbed. The paper surface takes a net electric charge, while the counter-ions form a diffuse layer in the neighbouring liquid, as shown by measurement of the electrophoretic mobility of small paper fragments or bundles of cellulose fibers immersed in insulating liquid with a variable amount or electrolyte added (Fig. 2). The above mentioned mobility reflects the surface charge of the paper bits. At very low concentrations, the charge always becomes negative. This is presumably due to the $-CO_2H$ groups present in cellulose.

- As a consequence, a paper layer will act as an <u>impassable barrier</u> for ions having the like sign, provided the pore diameter is smaller than the Gouy-Debye length, a condition easily satisfied at very low ionic concentrations as encountered in insulating systems. Paper thus behaves as a semi-permeable membrane.

- Garton's effect can be reproduced by inserting a weak cationic (carboxylic) ion-exchange membrane between two electrodes dipped in aroclor.

- Thus, Garton's effect is not at all related to the "solid" character of paper, but to the interaction of ions with the bound electric charge. When a D.C. voltage is applied to a paper capacitor, the current tends slowly to zero as ions reach locations where the pore channel is sufficiently narrowed for the exclusion effect to be fully felt. The ultimate resistivity may be 10^{16} Ω cm and more, 10 000 times larger than would be expected from the concentration of ions and the dimensions of pores.

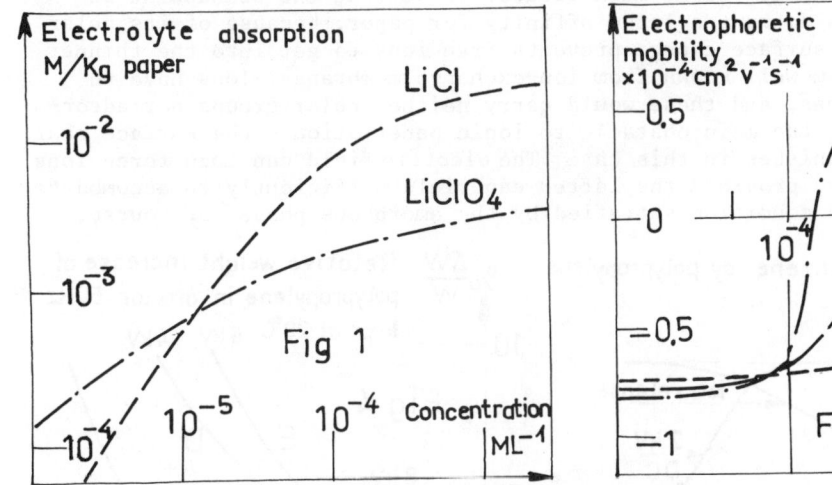
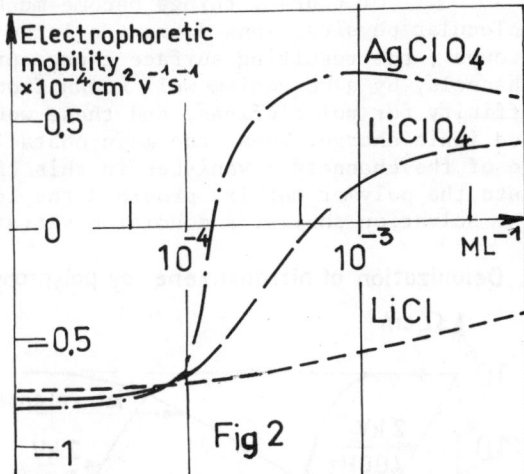

The Counter-Proof : Field Induced Absorption in non-polar polymers

As Garton's effect was ascribed to the "solidness" of paper, it was the more expected to occur with other materials whose structure is obviously compacter. Methodical experiments (4) (5) showed that many polymers, when wetted by an insulating liquid, would behave in a quite unexpected way :

- With A.C. applied, the conductivity of the liquid is a decreasing function of voltage, under steady-state conditions.

- There is no Garton effect present, however, save for a few seconds after voltage has been switched on. Oscillographic records do not show the peaks of current occuring after each field reversal and typical of the Garton effect. The liquid is permanently cleared of ions.

- The ions that disappear from the bulk of the liquid are not destroyed. They move (up to 99,9 %) into the solid and pervade its whole volume (Fig. 3)

- When voltage is switched off, ions return slowly (hours, days) but quantitatively into the liquid phase (Fig. 3)

- Ions take their solvation shells into the solid. Thus, the latter must accomodate a considerable amount of liquid. Weight increase (Fig. 4) and swelling are easily detected. Sometimes, the liquid would ooze out of the solid at the non-wetted surface.

From the standpoint of a "mechanical" theory, like Garton's, we have a perfectly paradoxical situation. Ions do not enter (or at least are blocked by) a notoriously porous material (paper) while they readily penetrate and pervade a much compacter polymer. Nevertheless, this is not due to some unexpected "affinity", since they are rejected and return into the liquid when the field is off.

Of course, things become much clearer if we take the standpoint of molecular physics. Ions do have a large affinity for paper, because of its polar groups ; the resulting surface charge prevents free ions to get into the thinner channels, by a mechanism well known from ion-exchange membranes. Ions have no affinity for polyolefines, and these would carry neither polar groups nor adsorbed ionic charge. Thus, the main obstacle to ionic penetration - the surface charge of the channels - vanishes in this case. The electric field can then force ions into the polymer matrix, provided the latter can yield sufficiently to accomodate the solvation shells, a condition satisfied by the amorphous phase. Of course,

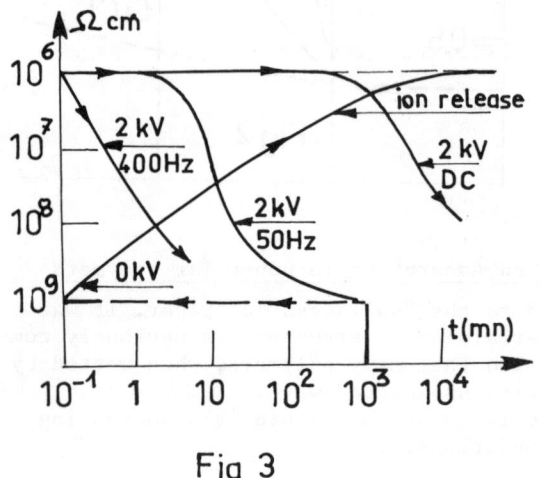

Fig 3

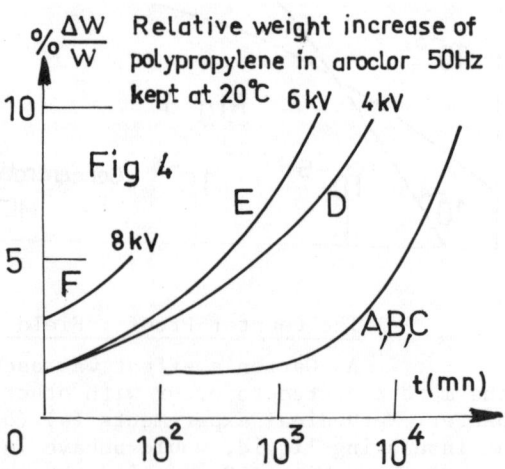

A: no V, no ions; B: V, no ions; C: ions, no V
D,E,F = V+ions (F ends in breakdown)

this process is much easier when the liquid itself has some affinity for the polymer, as we shall see presently.

Actually, the preceding results are valid for a variety of polymers (polypropylene, polyethylene, teflon) and for all "insulating" liquids (aroclors, nitrobenzene, propylene carbonate, water). The absorption phenomenon, however, largely depends on the specific interactions of polymer and liquid :

- If the affinity of the polymer for the liquid (as measured by equilibrium swelling without voltage applied) is <u>large</u>, the increase of the liquid content above swelling equilibrium caused by the field is not important. Absorption, however, is much faster with field on. The loss angle of the solid increases, as far as ions are available. It can reach 0,10 in an almost loss-free material like polypropylene. A good example of the preceding is polypropylene + aroclor, also nitrobenzene.

A part from increasing tan δ, ion absorption causes a significant fall of the elasticity modulus (50 %) and of the electric breakdown strength (50 to 60 %). Both are ascribable to enhanced swelling and plasticization.

- If the polymer/liquid affinity is <u>small</u> (e.g. polyethylene + water), the <u>relative</u> increase of liquid absorption is large (e.g. from 50 ppm without field to 2000 ppm with field). The phenomenon is as fast as in the preceding case (seconds, minutes). The loss angle also increases very much (for instance 0,0003 to 0,03). But the most spectacular effect is a disastrous drop in breakdown strength (up to 90 %). In this case, not only the ions, but the liquid itself is forcibly dragged into the solid. Thermodynamics shows this liquid will tend to congregate into small pockets subjected to strong electrostatic forces causing them to elongate and grow. This phenomenon may well explain the precipitous drop in breakdown strength and is probably related to water treeing.

All the preceding results are concerned with A.C. of low frequency (50 to 2000 Hz). Other experiments were carried out under D.C. voltage with polypropylene foils swollen by aroclor and sandwiched between various electrode arrangements (ion-exchange membranes, giving a unipolar ion flow ; electrolytic solutions in aroclor ; metallic electrodes).

The behaviour of the swollen foil cannot be described by a simple model and more experiments are needed before a well-founded interpretation can be proposed. Under steady-state conditions, between injector and collector, something reminiscent of a S.C.L.C. regime is observed ($I \alpha V^n$; $2 < n < 3$), ionic mobilities being of the order 10^{-10} to 10^{-8} cm^2/V.s.

With the same electrode arrangement, applying a voltage step does not give the classical Many and Rakavy curve (Fig. 5). The current decreases with time, though a blurred peak is detected (Fig. 5) which corresponds to the above-mentioned mobility. This shows there are "blind alleys" in polypropylene too - a hardly surprising fact - but many ions can nevertheless get through and give the ultimate, steady-state current. The corresponding "resistivity" may be 10^{11} Ωcm, in agreement with tanδ measurements, while paper can easily exceed 10^{16} Ωcm after a long exposure to D.C. voltage.

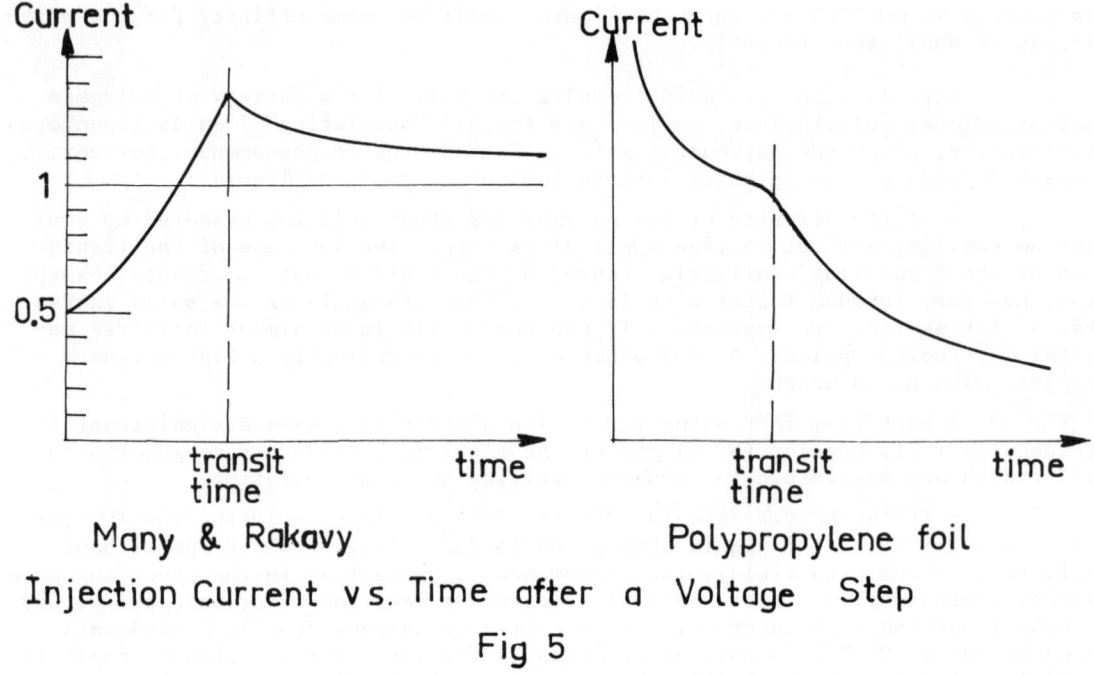

Injection Current vs. Time after a Voltage Step
Fig 5

To sum up, the striking contrast in behaviour between polar and non-polar polymers, when subjected to voltage and solvated ions, can be explained by the microscopic interactions of these with the polymer matrix, the bound charge playing a dominant role.

A Few Words on Ionizable Polymers

In this work, we take the standpoint of electrical engineering, and we are, therefore, only interested in _insulating_ materials. Nevertheless, it is necessary to mention the important ionic properties of ionizable polymers (polyelectrolytes, ion-exchange membranes) because they help us to discover why some materials or systems behave as good insulators and some others do not.

Paper owes its outstanding properties to the interaction of impregant and ions with its polar groups. These are hydroxyles (- OH), supplemented by a few - CO_2H due to oxidation at the end of polymeric chains.

What happens if other groups are substituted for these ? A large body of knowledge is now available, and we shall restrict ourselves to the case when the new groups are clearly ionizable, i.e. acidic or basic.

Naturally enough, such a substituted polymer behaves like an electrolytic conductor. For instance, if - SO_3H's are introduced, conductivity is due to the mobile H^+ ions, while the corresponding negative charge remains bound to the polymer matrix.

Needless to say, such a compound has strong exclusion properties. No negative ion can enter the matrix, while positive ones get through quite readily. In sheet form, the material will be called a semi-permeable ion-exchange membrane.

What if such a membrane, (stretched on a metallic electrode) is subjected to a D.C. field carrying its mobile ions towards a liquid ? It seems obvious that they are injected into the liquid, and so it actually happens. But another phenomenon of far-reaching significance takes place immediately : Fresh ions are <u>generated</u> by the polymer, sometimes indefinitely. The responsible mechanism is not yet fully understood, but it seems almost certain that residual water molecules are split by the electrostatic field of the $SO_3^- H^+$ and give H^+ and OH^- ions which are injected as such (H^+), or after undergoing electron exchange with the environment ($OH^- + M \rightarrow M^- + OH$, M being a reactive compound).

No heed has been paid so far by either physicists, chemists or electrical engineers to this remarkable fact. It explains, however, some of the problems encountered in engineering when dealing with a polar insulator like paper.

The Paper "Miracle". Are there other Candidates ?

Our interpretation of the wonderful insulating abilities of paper and of the puzzling facts related to ion penetration into polyolefines may help to foresee the behaviour of new or altered materials. For instance, it was already said that a <u>weak</u> ion-exchange membrane, carrying - CO_2H groups like those of paper can block ions and mimic Garton's effect. Conversely, we just mentioned that a polymer carrying stronger, ionizable groups (e.g. - SO_3H) can <u>generate</u> new ions and inject them into an otherwise insulating liquid. Actually, paper subjected to oxidizing reactants can lose its remarkable D.C. resistivity, presumably because of the formation of stronger acidic groups. Such a decay of the insulating properties of paper has been routinely observed with D.C. capacitors at higher temperatures, but was never given a plausible explanation.

Thus, to be a very good insulator, a polymer wetted by a liquid (always carrying ions) <u>must</u> have polar groups, but sufficiently <u>weak</u> ones, since stronger groups would trigger ion generation and injection. Some classical tricks of paper capacitor engineering (e.g. the importance of divalent metal traces as conduction inhibitors) may well be explained by the fact they tend to prevent residual injection from acidic groups grafted on cellulose.

Conclusion

The systems solid + liquid + ions, when subjected to an electric field, show a multifarious behaviour whose study is obviously relevant to important problems in engineering. Garton's pioneer work pointed out the two main facts Maxwell's treatment could not take into account : the field drives the system out of equilibrium and the ions from the liquid strongly interact with the solid. It has been necessary, however, to refine our understanding of this interaction by resorting to molecular physics instead of crude phenomenological schemes.

References

(1) W. HELD, K. WENZEL : Dielektrische Verluste durch Ionenleitung im geschichteten Dielektrikum, ETZ, February 1960, 4, p. 121.

(2) B. GOSSE, J.P. GOSSE, M. SAUVIAT : Garton's effect and its physicochemical interpretation, World Electrotechnical Congress, Moscow, June 1977.

(3) B. GOSSE, J.P. GOSSE, M. SAUVIAT : Physico-chemical study of Garton's effect. Sixth International Conference on Conduction and Breakdown in Dielectric Liquids, Rouen, July 1978.

(4) R. TOBAZEON, E. GARTNER : On the behavior of ions at insulator/liquid interfaces and its consequences for the losses in impregnated insulation.
1974, COnference on Electrical Insulation and Dielectric Phenomena, Annual Report, p. 404

(5) R. TOBAZEON, E. GARTNER : Dielectric behavior of solid polymeric materials in contact with slightly or highly conducting liquids. World Electrotechnical Congress Moscow, June 1977, Section 3A, Paper 26.

NEW METHOD FOR CHARGE DENSITY MEASUREMENTS IN DIELECTRIC LIQUIDS
——— CONTINUOUS SAMPLING METHOD ———

Satoru Kamoto

Denryoku-group, Central Research Lab., Mitsubishi Electric Corp.

80 Nakano, Minamishimizu, Amagasaki, Hyogo, Japan

Introduction

     We have needed some new measuring apparatus for examining various electrostatic hazards caused by flowing dielectric liquids; petrochemical tank explosion, insulation problems in electric installations, etc. The present paper describes a new measuring apparatus invented by the author for measuring charge density distribution in liquids. The measurement can be done in high accuracy, in high spatial resolution and in short response time.

     In order to measure charge density distributions in dielectric liquids, we have widely used or rather tried various probes. One of them is a probe composed of two grounded electrodes and one highly insulated center electrode, as shown in Fig.1. The composition is very simple, but the probe method generally has several problems:
1) The error due to the flow of the liquid.
       The center floating electrode causes static electrification on its surface when it is in a flowing dielectric liquid (1).
2) The error due to the external high potential nearby.
       The potential of the center electrode is affected by the current coming in through the opening between the two grounded electrodes.
3) Long response time.
       The response is delayed due to the high output resistance of the probe and the capacitance of the cable leading the potential of the center electrode to the field meter.
4) Limited spatial resolution.
       Since the measuring error due to the electrification increases as the distance between the floating electrode and the grounded electrodes decreases, the spatial resolution is principally quite limited as far as such a method as this probe is used.

Those are the reason why we considered a quite different type of measuring apparatus.

Basic Principle

This new apparatus is composed of a sampling tube, a metal cavity and a drain tube as shown in Fig.2. The sampling tube and the cavity are guarded from the external disturbances by shield electrodes (armature pipes in Fig.2). Small amount of charged liquid is continuously taken through the sampling tube into the cavity,

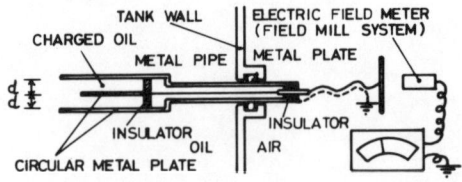

Fig.1. Probe

where the charge in the liquid relaxes to the wall of the cavity and the neutralized liquid is taken out to the drain. We measure the current leaving from the sampling tube as well as from the cavity.

Fig.2 also shows a gaussian surface which is set outside of the cavity and the sampling tube and inside the armature pipes. Since the armature pipes are grounded and the sampling tube and the cavity have an almost zero potential, there is not any electric field on the gaussian surface except at the inlet of the sampling tube. Let the average electric field there to be E. Generally we could say that the total current coming into the gaussian surface is constantly zero, if we include the capacitive current or the displacement current. So therefore, the current going out from the cavity and the sampling tube $i(t)$ is equal to

$$i(t) = \rho(t) q(t) - \rho'(t) q(t) + \kappa E(t) s + \varepsilon E(t) s \tag{1}$$

The first term is the convective current coming into the gaussian surface through the inlet, where $\rho(t)$ is the charge density in the oil which is sampled at a rate of $q(t)$. The second term is also the convective current going out of the gaussian surface through the outlet of the cavity with the liquid having a residual charge density of $\rho'(t)$. The third term is the coductive current coming into the gaussian surface through the inlet having an area of s, where κ is the conductivity of the liquid. The forth term is the displacement current coming into the gaussian surface through the inlet, where ε is the dielectric constant of the liquid. The second and the third terms bring errors if they are not small enough, while the forth term becomes a noise. If the second, the third and the forth terms are small enough, we have the charge density $\rho(t)$ in the sampled liquid as

$$\rho(t) = i(t)/q(t) \tag{2}$$

From the principle of charge relaxation of the liquid, the residual charge density ρ' is given by

$$\rho' \simeq (\rho + \rho'') e^{-T/\tau} \tag{3}$$

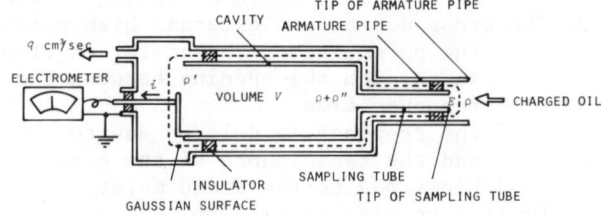

Fig.2. Principle of the new apparatus

where ρ'' is the charge density generated in the course of running through the sampling tube, T is the time interval for the liquid to flow from the inlet to the outlet, and τ is the relaxation time of the liquid which is equal to ε/κ.

The electric field in the third and fourth terms can be easily made small by making the armature pipe a little bit longer than the sampling tube. But it should not be too much longer than the sampling tube, since if it is, it generates additional new charges in the liquid before the liquid goes into the sampling tube.

We might have worried that the measurement is affected by the electrification of the liquid in the course of running through the sampling tube, but the equation (1) shows clearly that it is not. And also the equation (1) indicates that the output current $i(t)$ changes instantly as soon as the charge density $\rho(t)$ at the tip of the sampling tube is changed.

Design

Fig.3 shows the more details of the new measuring apparatus. The sampling tube is made of a stainless steel pipe of 2mm in inner diameter and about 2m long. Its surface is coated with an insulating film so that the sampling tube is insulated from the ground. After that, the sampling tube is inserted into a grounded armature pipe to be guarded from the external disturbances. Since the insulating film is flexible, the sampling part, after assembling into the armature pipe, can be bent to a curvature of 16cm in diameter.

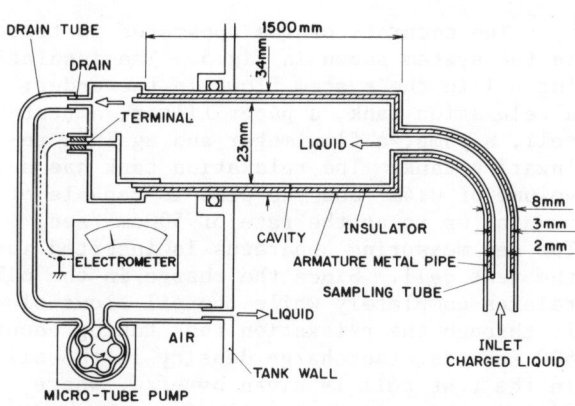

Fig.3. Schematic diagram of construction.

The cavity part having a volume of 700cm³ is constructed by an identical method to that of the sampling part. The effective volume of the cavity is assumed to be around 350 cm³, because some of the oil is stagnant in the cavity and in a laminor flow the distribution of the flow speed is nearly parabolic.

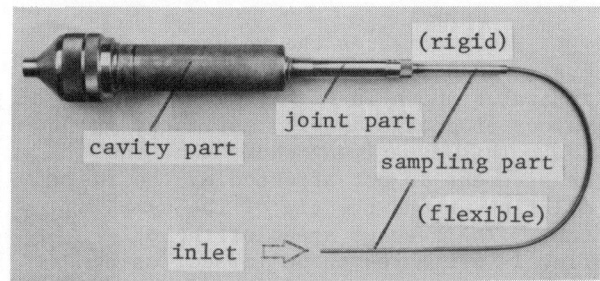

Fig.4. Photograph of the new apparatus.

In the case of insulating oils the dielectric constant ε is about $2.2 \cdot \varepsilon_0$ and the conductivity κ ranges from 10^{-15} to 10^{-13} S/cm. This means that the relaxation time τ of the oil is in the range from 2 seconds to 200 seconds. Then $e^{-T/\tau}$ ranges from almost zero to about 0.03. The charge density ρ'' generated in the sampling tube is very roughly estimated to be less than around 15pC/cm³, from the laminor flow approximation of pipe flow electrification. Therefore, we have from equation (3)

$$|\rho'| < 0.03 |\rho| + 0.5 \, pC/cm^3 \quad , \tag{4}$$

which shows that the charge in the oil relaxes enough while it is in the cavity.

Two types of stainless steel apparatus are developed. The fig.3 shows the schematic diagram of one type which has a long and slender cavity part, while the Fig.4 shows the picture of the other type which has a short and bulky cavity part. Since the sampling part can be as long as 2m and can be bent, we can measure the charge density at almost any place we want even in such a complicated system as an electric installation.

Performance

The accuracy of the apparatus is tested in the system shown in Fig.5. The insulating oil in the system flows in the order; a relaxation tank, a paper filter, a test cell, a pump, a flow meter and again a relaxation tank. The relaxation tank has a volume of $0.4m^3$ and the pump is capable of running up to at the rate of $500cm^3/sec$. The new measuring apparatus is inserted into the test cell. Since the charge in the oil relaxes completely while the oil flows slowly through the relaxation tank taking about 600 seconds, the charge density in the oil in the test cell is given by $-I/Q$, where $I(A)$ is the current leaving from the paper filter and $Q(cm^3/sec)$ is the flow rate of the circulating oil.

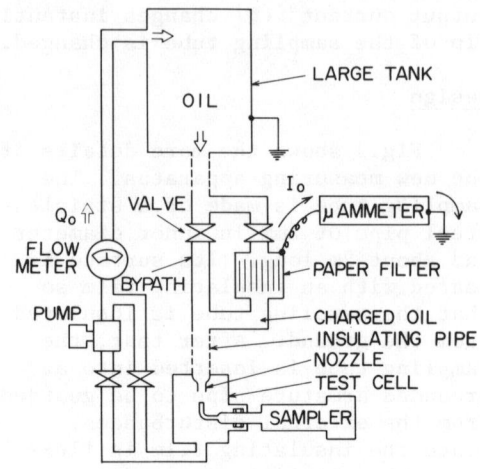

Fig.5. Calibration system.

Fig.6 compares the values of charge density ρ measured by the new apparatus shown in Fig.3, with the values of charge density estimated by I/Q. This experiment shows that the measurement is not affected by the fast oil flow around the tip of the sampling tube. The average speed of the oil flow is estimated to be 1m/sec, using the value of diameter 2.5cm of the nozzle from which the oil comes into the test cell. Fig.6 also shows the result of the accuracy test of the probe shown in Fig.1 tested under the same condition. The new apparatus is indicated in solid line and the probe is in the dotted line. The initial offset of the measurement is less than $0.8pC/cm^3$ for the new apparatus, while it is around $20pC/cm^3$ for the probe.

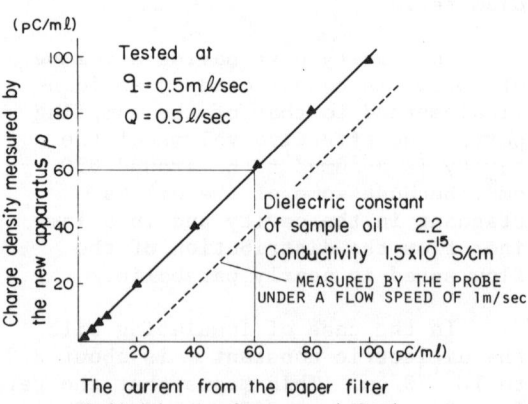

Fig.6. Result of calibration test.

A marked improvement can be seen in the present apparatus.

The charge relaxation in the cavity is examined by changing the sampling rate, with the charge density ρ to be 329 pC/cm and 3.8pC/cm. Fig.7 compares the measured current with the equation

$$i/\rho \simeq q\{1 - (1+\rho''/\rho)e^{-V/q\tau}\} \quad (5)$$

derived from the equations (1) and (3), where the electric field E is neglected, V is the effective volume (350cm^3) of the cavity. The value of ρ'' is estimated to be around 5.5pC/cm^3 by reversing the running direction of the sampling pump after keeping the pump off for more than 20 minutes.

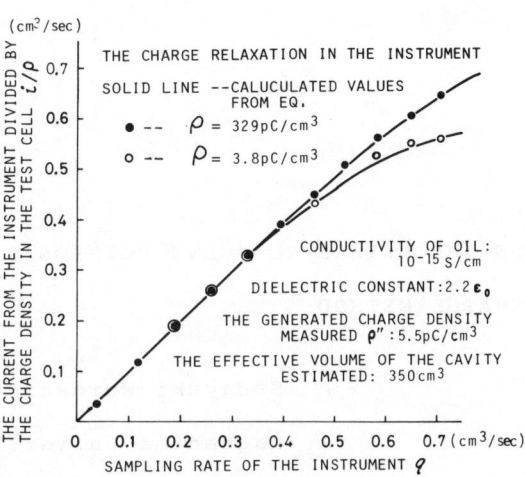

Fig.7. Charge relaxation in the cavity.

The response time is measured by decreasing stepwise the charge density in the oil in the test cell. This stepwise change can be made by suddenly opening the by-path valve at the filter part. The response time measured is less than 1 second.

The spatial resolution is also tested by scanning a region where a charge density is stepwise greater than the surroundings. This spepwise change is produced downstream after the nozzle from which the oil comes into the test cell. The output current from the apparatus shows the whole change in the traverse scanning of 1cm at the periphery of the jet flow.

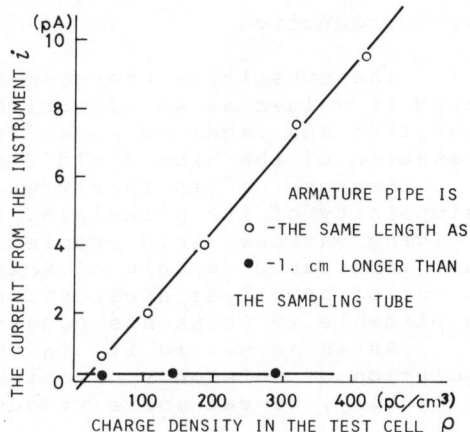

Fig.8. Error due to the high potential near by.

The high potential around the tip of the sampling part could be an error factor. High potential of around 20kV can be created in the oil in the test cell by the charge in it, where the diameter of the test cell is about 20cm. Two types of sampling part are served for the test, being set at the center of the test cell. One has an armature pipe of 1cm longer at the tip than the sampling tube, while the other has that of the same length as the sampling tube. The current from the apparatus is measured with the sampling stopped, since if it is, the equation (1) shows that the current i is equal to the current $\kappa E s$. Fig.8 shows that the normal apparatus having a longer armature pipe is free from the error, while the other has an error proportional to the charge density in the oil.

Reference
[1] A.Klinkenberg and J.L.van der Minne "Electrostatics in the petroleum industry", Elsevier Publishing Co. p.123(1958).

ON THE METHOD TO SOLVE POISSON'S EQUATION BY THE SUBSTITUTE CHARGE METHOD

Sadayuki Murashima

Kagoshima University

Kagoshima, Japan

1. Introduction

The substitute charge method presented by H. Steinbigler in 1969 is valued as an effective and simple method to solve Laplace's equation and regarded as an indispensable tool, especially in the research of the high field discharge phenomenon.

Because of its fairly wide applicability, high accuracy and simplicity of its principle, I believe, it will come to be used in solving various field problems in electrostatics. It is, however, not considered capable of solving Poisson's equation. On the view point of practical application, it is very desirable to make it applicable to Poisson's equation .

As an answer to it, in this paper, it is shown if a particular solution of Poisson's equation under consideration is obtained analytically, we can solve it just as we solve ordinary Laplace's equation.

Some researchers tried to express the particular solution, i.e. the effect of the space charges by means of the substitute charges and failed in obtaining satisfactory results . The test function used in the substiture charge method is essentially Laplace's field but not Poisson's field. This means that we must express the non-homogeneous term of Poisson's equation by other methods. Hence, in this paper, we express a particular solution analytically and the substitute charges are restricted to express the general solution.

2. Principle

Let us consider the Dirichlet problem of Poisson's type in two dimensions shown in Fig.1. Poisson's equation for this problem is:

$$\nabla^2 \psi(x,y) = -\rho(x,y)/\varepsilon, \quad (x,y) \in D \quad (1)$$

$$\psi(x,y)_\Gamma = f(s), \quad (x,y) \in \Gamma \quad (2)$$

where ρ is the charge density and ε is the dielectric constant. The letter D signifies the interior of the area under consideration and Γ signifies its boundary. We here assume that a particular solution $s(x,y)$ of Poisson's equation is obtained analytically such as:

$$\nabla^2 s(x,y) = -\rho(x,y)/\varepsilon, \quad (x,y) \in D. \quad (3)$$

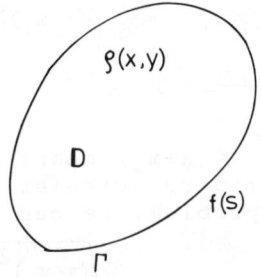

Fig.1 Dirichlet problem of Poisson's type

Then we express the solution $\psi(x,y)$ as a superposition of the particular solution and the general solution $\phi(x,y)$, i.e.,

$$\psi(x,y) = s(x,y) + \phi(x,y). \quad (4)$$

Substituting eq.(4) into eq.(1), we get

$$\nabla^2 \phi(x,y) = 0, \quad (x,y) \in D \quad (5)$$

$$\phi(x,y)_\Gamma = f(s) - s(x,y)_\Gamma, \quad (x,y) \in \Gamma. \quad (6)$$

This means that if a particular solution of the Poisson's equation is given, the Poisson's equation can be transformed into a Laplace's equation with regard to general solution. This Laplace's equation can be easily solved by means of the ordinary substitute charge method.

It is not so difficult to derive a particular solution because that it need not satisfy the boundary condition. Usually many different particular solutions are obtained. The accuracy of numerical analysis depends on which particular solution we use for superposition.

3. Examples

Let us demonstrate the technique proposed in this paper by studying two problems of different type which are both governed by Poisson's equation.

3-1. Current flow problem in two dimensions[4]

In this problem, the electric current I flows in a very thin eliptic semi-conductor plate of conductivity σ. The positions of the current source and sink on the plate are shown (x_+, y_+) and (x_-, y_-), respectively. The boundary condition is that the current does not flow across the boundary. The governing equations of this problem are as follows:

$$\nabla^2 \psi(x,y) = \frac{I}{\sigma}\{\delta(x-x_+)\delta(y-y_+) - \delta(x-x_-)\delta(y-y_-)\}, \quad (x,y) \in D \quad (7)$$

$$\frac{\partial \psi(x,y)}{\partial n}\bigg|_\Gamma = 0, \quad (x,y) \in \Gamma, \quad (8)$$

where $\delta(x-x_+)$ means the Dirac delta function and n is the unit vector perpendicular to the boundary. As a particular solution for this problem, we can easily obtain the following function:

$$s(x,y) = -\frac{I}{2\pi\sigma}\log\sqrt{\frac{(x-x_+)^2+(y-y_+)^2}{(x-x_-)^2+(y-y_-)^2}}. \quad (9)$$

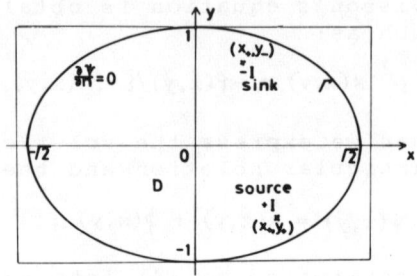

Substituting $\psi(x,y) = s(x,y) + \phi(x,y)$ into eq.(7) and (8), we get

$$\nabla^2 \phi(x,y) = 0, \quad (x,y) \in D \quad (10)$$

$$\frac{\partial \phi(x,y)}{\partial n}\bigg|_\Gamma = -\frac{\partial s(x,y)}{\partial n}\bigg|_\Gamma, \quad (x,y) \in \Gamma. \quad (11)$$

This problem of Neumann type is slightly difficult than that of Dirichlet type because the computation of the normal

Fig.2 Current flow problem in two dimensions

component of the gradient of potential function on the boundary is necessary. But this technique is included in the area covered by the conventional substitute charge method.

To use the particular solution eq.(9) does not give accurate results as the source or the sink approaches to the boundary. For such a case, the particular solution given below is preferable:

$$s(x,y) = -\frac{I}{2\pi\sigma}\left\{\log\sqrt{\frac{(x-x_+)^2+(y-y_+)^2}{(x-x_-)^2+(y-y_-)^2}} + \log\sqrt{\frac{(x-x'_+)^2+(y-y'_+)^2}{(x-x'_-)^2+(y-y'_-)^2}}\right\}, \quad (12)$$

where (x'_+,y'_+) indicates the image point of (x_+,y_+), assuming that the boundary is a circle with radius equal to the radius of curvature near the source and (x'_-,y'_-) has similar means as above. As the source and sink run far from the boundary, the second term of eq.(12) diminishes. In Fig.3, the potential distribution is shown. The maximum error on the boundary is 10^{-2} and the number of the substitute charges is 20.

The shapes of the modified boundary condition on one part of the boundary are computed from eqs.(9) and (12) and are depicted in Fig.4 (a) and (b), respectively. The two peaks of the curved line in Fig.4 (a) are the effects of the source and sink near the boundary. It is not easy to simulate accurately such a changefull curve as this. This difficulty can be removed by using the particular solution eq.(12) including the supposed images of the source and sink. It is shown in Fig.4 (b) that the shape of the modified boundary condition is perfectly smoothed. Some readers may think that the supposed images are two of the many substitute charges. They are not, however, in this case, because their magnitudes are previously known.

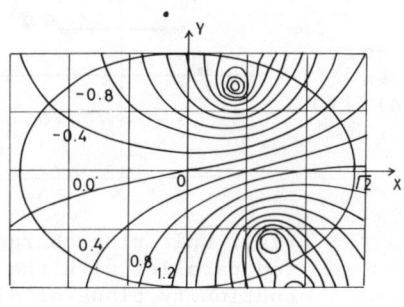

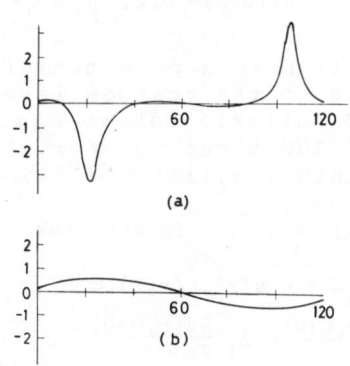

Fig.3 Potential distribution of the current flow problem

Fig.4 Shapes of the modified boundary condition

2-3. Axi-symmetric tank partly filled with charged liquid[5]

If a metal tank is partly filled with charged liquid, such as, hydrocarbons, the electrostatic potential can become very high and solve the following equations: (see Fig.5)

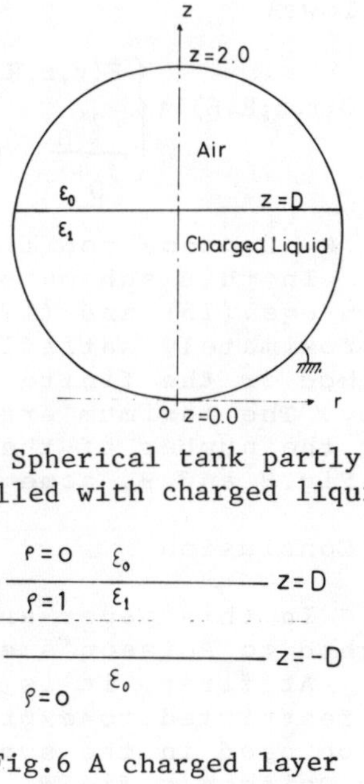

$$\nabla^2 \psi(r,z) = 0 \quad , \quad z \geq D \quad (13)$$

$$\nabla^2 \psi(r,z) = -\rho/\varepsilon \quad , \quad z < D \quad (14)$$

$$\psi(r,z)_\Gamma = 0 \quad , \quad (15)$$

$$\psi(r,D+0) = \psi(r,D-0) \quad , \quad (16)$$

$$\varepsilon_0 \frac{\partial \psi(r,D+0)}{\partial z} = \varepsilon_1 \frac{\partial \psi(r,D-0)}{\partial z} \quad , \quad (17)$$

where ε_0 and ε_1 are the dielectric constant in the air and the liquid, respectively. The ρ is the charge density of the charged liquid and the D is the depth of the liquid from the bottom.

As a particular solution, we use the following function ($\varepsilon_r = \varepsilon_1/\varepsilon_0$)

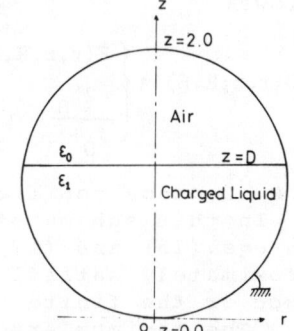

Fig.5 Spherical tank partly filled with charged liquid

Fig.6 A charged layer

$$s(z) = \frac{\rho}{\varepsilon_0} \begin{cases} D\{(D-z) - D/2\varepsilon_r\}, & z > D \\ -z^2/2\varepsilon_r, & -D \leq z \leq D \\ D\{(D+z) - D/2\varepsilon_r\}, & z < -D \end{cases} \quad (18)$$

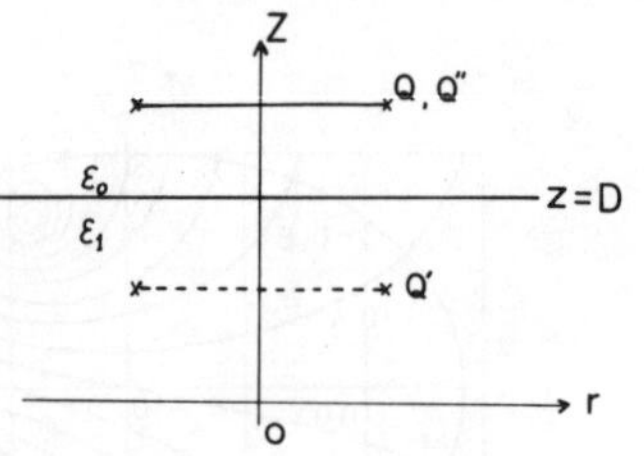

This function is a potential function corresponding to the charged layer shown in Fig.6 and satisfies above eqs.(13),(14)(16) and (17). Substituting $\psi(r,z) = s(z) + \phi(r,z)$ into eqs.(13) $\sim$ (17), we get

$$\nabla^2 \phi(r,z) = 0, \quad \text{in the tank} \quad (19)$$

$$\phi(r,z)_\Gamma = -s(z)_\Gamma, \quad (r,z) \in \Gamma \quad (20)$$

$$\varepsilon_0 \frac{\partial \phi(r, D+0)}{\partial z} = \varepsilon_1 \frac{\partial \phi(r, D-0)}{\partial z}, \quad (21)$$

Fig.7 Unit ring charge located parallel with the discontinuity plane of dielectrics

This derived Laplace's equation is slightly difficult to solve because that the dielectric constant has discontinuity at $z = D$. For such a case, we must locate several substitute charges at the both side of the discontinuity plane.

We can, however, remove the necessity of locating additional substitute charges by modifying the potential function $F(r,z;R,E)$ of the unit ring charge of radius R and placed at $z = E > D$ as follows:

$$G(r,z;R,E) = \begin{cases} F(r,z;R,E) + \dfrac{\varepsilon_0 - \varepsilon_1}{\varepsilon_0 + \varepsilon_1} F(r,z;R,2D-E), & z \geq D \\ \dfrac{2\varepsilon_0}{\varepsilon_0 + \varepsilon_1} F(r,z;R,E) & , z < D \end{cases} \quad (22)$$

For $E < D$, some modifications are necessary.

In this scheme of the substitute charges, the continuity condition eqs.(16) and (17) are exactly satisfied and only eq.(15) is approximately satisfied. This method is superior to the difference method or the finite element method in computing time and in accuracy. The maximum error at the boundary of the tank is 3×10^{-3} and 10^{-4} for the number of the substitute charges 13 and 21, respectively. In Fig.8 and 9, some potential distributions in the tank are shown.

4. Conclusion

In this paper an application technique of a substitute charge method to Poisson's equation is given.

At first, it is pointed out that the substitute charges should be restricted to express the general solution because the test function used in the substitute charge method is Laplace's field and not Poisson's field. Then superposing the general solution and the particular solution derived analytically, some of Poisson's problem are solved. The results are satisfactry in computing time and accuracy.

5. References

1) H. Steinbigler: Dissertation, T.H. Munchen, 1969
2) S. Masuda: Proc. IEJ Vol.1 (1977), p18
3) CIGRE 33-15: Electra No.23, p53(1972-6/7)
4) L.B. Valdes: Proc. IRE Vol.14(1954), p420
5) H. Kramer and G. Schon: Proc. -th World Petroleum Congress 1975, Tokyo, Vol.6(1976), p147

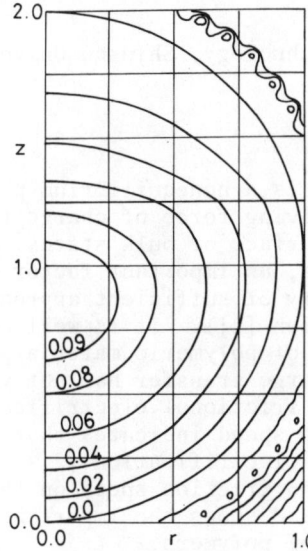

Fig.8 Potential distributions in the tank where the discontinuity plane is at z=1.25 and ρ/ε_0 =1, ε_r = 1.5.

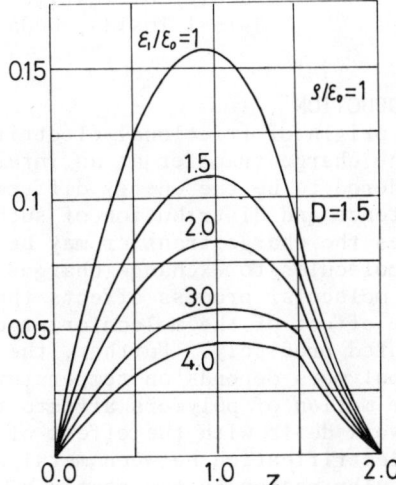

Fig.9 Potential distributions on the axial line of the tank where D=1.5, ε_r=1, 1.5, 2, 3 and 4, and ρ/ε_0=1.

TEMPERATURE AND FRICTION SPEED DEPENDENCE OF FRICTIONAL ELECTRIFICATION

BETWEEN METAL AND POLYMERS. CONTRIBUTION OF MOLECULAR MOTION

OF POLYMERS TO FRICTIONAL ELECTRIFICATION

Keiji Ohara

Faculty of Textile Science and Technology, Shinshu University
3-15-1 Tokida, Ueda 386 Japan

I. INTRODUCTION

The origin of frictional electrification which is a nonequilibrium phenomenon [1] is the charge transfer at an interface. The driving force of charge transfer is considered to be the energy difference between surface or bulk states. Besides the existence and distribution of such energy levels, an important factor which determines the charge transfer may be the probability of sufficient approach of surface molecules to exchange charges (2 nm separation [1]). It is well known that the molecular process affects the charge decay of polymeric materials. However, the effect of the molecular process on the charge transfer has not yet been investigated precisely. Further, the fact that the frictional electrification between polymers depends on temperature and friction speed indicates that the molecular motion of polymers affects the frictional electrification [2 - 4]. The present work deals with the effect of temperature and friction speed on the frictional electrification between metal and polymers, and shows the importance of the molecular motion in the static electrification of polymers.

II. EXPERIMENTAL

The apparatus used was shown schematically in Fig. 1. A polymer film was fixed on the grounded aluminum turn table (100 mm diameter). The lever on which an electrode (a chromium-deposited spherical steel tip, 5 mm diameter) was installed was balanced. By hanging a weight at the end of the lever through the pulley, the electrode comes into contact with the film at the position of 40 mm from the center of the turn table. As the turn table moves in relation to the electrode, charge is continuously transfered from the metal to the polymer, and vice versa. The transfered charge was accumulated in the condenser equipped in the electrometer and produced voltage (V) was recorded. Average values of V after 10 mm rubbings obtained in twenty runs of measurements at each condition were plotted. The sign of charge described in figures was that indicated by the electrometer. The friction speed (υ) at the contact position was changed from 10^{-3} to 10^2 mm/s. The apparatus was set in a temperature and humidity regulated chamber. Temperature (T) range was 10 to 100°C. The relative humidity of the ambient surroundings was kept at 30%.

Charge decay was investigated by stopping the turn table after obtaining an arbitrary value of V (V_0) and by measuring V_t as a function of time (t). The DC conductivity of the film was measured by an ordinary method.

Commercial polystyrene (PS) and nylon 6 (Ny) films of 0.03 mm thickness were used as samples without any heat treatment. Normal load at the contact point was 245 mN for PS and 61 mN for Ny. Before starting experiments, the surface of film was wiped with ethanol-dipped absorbent gauze for the purpose of eliminating the initial surface charge.

III. EXPERIMENTAL RESULTS

The T dependence of volume conductivity (σ_v) and surface one (σ_s) was shown in Fig. 2 and the charge decay in Fig. 3. For PS, σ did not change greatly and the charge decay was very small in the T range examined. For Ny, σ began to increase abruptly at about its glass transition temperature (42°C). The value of σ at 90°C was larger than that at 20°C by a factor of 10^4. However, V did not decrease rapidly within 60 s at each temperature. Interference fringe observed by multiple-beam interferometry that shows the condition of contact between the electrode and the PS film was shown in Fig. 4. The true contact was realized in a small area within the innermost ring. The deformation of the innermost ring which indicates the change in the condition of contact [5] could not be observed in the T and υ ranges examined and the diameter of the innermost ring was kept at 0.03 mm. Then it was concluded that there is no change in the condition of contact during the friction process in the T and υ ranges used in the present work.

PS/METAL SYSTEM

In the T dependence of V at various υ (Fig. 5), one peak or two were found. According to the well-known consideration that the observed charge is the sum of the generating charge and decaying one, it seems that the decrement of charge after a maximum has been taken is caused by the increase in σ at a high temperature. However, as seen in Fig. 2, σ of PS is very small (below 10^{-16} ohm^{-1}·cm^{-1}, or ohm^{-1} for $\sigma_{v,s}$) and temperature coefficient of σ is small in the examined temperature range. Further, if we assume that the contact is realized at most within the circle of 0.03 mm diameter, the time during which a point on the film is contacting with the electrode may be estimated to be 2.5 s for υ = 0.012 mm/s. In this time interval, the charge decay is negligible (Fig. 3). From this rough estimation and considering that the electrode is always contacting with a new surface of the film and that there are two peaks in the present temperature range, it was concluded that the effect of charge decay on the T dependence of V is small even in the high temperature region.

It is reported that in the T dependence of loss modulus (G'') of PS, G'' takes two maxima: one between 95 to 150°C and the other between 40 and 77°C depending on the frequency of mechanical stimulation [6]. The former peak is α dispersion which is related to the start of microbrownian motion of polymer segments at the glass transition temperature and the latter is β dispersion based on the local oscillation mode of the polymer main chain.

Comparing the present data with the viscoelastic behavior of PS, it is possible to attribute the peaks of V in the higher temperature region (at 50, 60 and 90°C for υ = 0.012, 0.1 and 1.07 mm/s, respectively) to the α dispersion and those in the lower temperature region (at 30 and 40°C for υ = 1.07 and 7.3 mm/s, respectively) to the β dispersion.

The υ dependence of V at various T for PS was shown in Fig. 6. It was found that V took maxima at two friction speeds (υ_{max}) and the position of both peaks shifted to the higher friction speed regions with increasing temperature. When log υ_{max} was plotted against 1/T, a linear relationship was obtained between them for two peaks (Fig. 7). On the other hand, it is well known that in the frequency (f) dependence of G'' of PS, there are two peaks (α and β dispersion peaks) in some frequency region [6]. Plotting log f_{max} against 1/T, where f_{max} is the frequency at which G'' becomes maximum, a linear relationship is satisfied between them. Comparing the log υ_{max} - 1/T plot with the log f_{max} - 1/T plot, it was concluded that the peak in the low υ region corresponds to the α dispersion and that in the higher υ region to the β dispersion. Assuming that the friction speed, which is proportional to the number of close approach in a unit time between two surfaces to exchange charges, corresponds to the frequency of the external mechanical stimulation in the ordinary viscoelastic measurements, then an Arrhenius type of equation (1) can be obtained by analogy with the famous equation, f_{max} =
$$\upsilon_{max} = A \exp (E/RT) \qquad (1)$$
A' exp (H/RT), where H and E are the apparent activation energy in the viscoelastic and frictional electrification processes, respectively.

From the result of Fig. 7 and eqn. (1), the apparent activation energy for the frictional electrification process was calculated to be 135 kJ/mol for the α process and 45 kJ/mol for the β process. The latter value is equal to E value (42 to 80 kJ/mol [6]) for the β process of PS in the mechanical behavior. Hence, the peaks in the higher υ region is related to the local oscillation of polymer main chain. The former activation energy is smaller than that obtained by the viscoelastic method (377 kJ/mol [6]) and the temperature at which the peak appears is lower than the glass transition temperature of PS (99°C). However, it may be reasonable to attribute the peaks to the α process or the start of microbrownian motion because the peaks were found in the lower friction speed region than that for β peaks.

Ny/METAL SYSTEM

Relationship between V and T was shown in Fig. 8 at various υ. Peaks of V were found at 45, 50, 60 and 70°C for υ = 0.02, 0.1, 1.07 and 7.3 mm/s, respectively. Increment of σ of Ny became large above its glass transition temperature. But approximate contacting duration obtained in the same way as PS was 1.5 s for υ = 0.02 mm/s. From this value and the decay curve in Fig. 3, it was concluded that there is no effect of charge decay on the T dependence of V.

It was reported that in the viscoelastic property of Ny measured at low frequency, there is a dynamic mechanical loss maximum at 77°C which is attributed to the α relaxation process of Ny [7]. The α transition is believed to originate with the rupture of interchain hydrogen bonding and is ascribed to the motions of long chain segments in the amorphous region of Ny [7]. By comparing the T dependence of V with that of the viscoelastic property, it is reasonable to consider that a maximum of V is caused by the molecular motion of main chain segments in the amorphous region of Ny. This identification is supported by the fact that the peak position shifted to the higher temperature region with increasing friction speed.

The υ dependence of V at various T was shown in Fig. 9. There was a peak of V in each curve. The peak position shifted to the higher υ region with increasing temperature. The log υ_{max} - 1/T plot was shown in Fig. 10, from which the apparent activation energy was estimated to be 88 kJ/mol. This value is smaller than the value obtained by the viscoelastic measurement (260 kJ/mol [8]). However, the peak can be attributed to the α process of Ny by comparing the temperature region where the peak appears in Fig. 10 with that in log f_{max} - 1/T plot of Ny [8]. Therefore, it is possible to say that the effect of molecular motion in the surface layer on the frictional electrification is prominent.

IV. DISCUSSION AND CONCLUSION

The peaks of charge were observed in the T and υ dependence of frictional electrification of PS and Ny. These phenomena can be explained as follows: the friction speed determines the number of close contact and separation between molecules on the two surfaces in a unit time. In this case, the "close contact" means the sufficient approach of molecules to exchange charges. When this condition is attained, molecular segments which appeared in one surface of polymeric material contact with molecules or atoms in the other surface and the charge transfer occurs. When the frequency of appearance of polymer segments to the surface by thermal motion, which is determined by temperature, coincides with the mechanical frequency (the number in unit time) of approach and separation of two surfaces, which is determined by the friction speed, the probability of molecular contact becomes largest and then the number of transfered charge becomes largest. If the temperature deviates from that at which two frequencies coincide at a constant friction speed or if the friction speed deviates from υ_{max} at a constant temperature, the probability of close contact becomes smaller. Then the transfered charge decreases.

Though it is impossible to discuss quantitatively the true frequency of thermal motion of polymer segments and the mechanical frequency of approach and separation between molecules on the two surfaces from the present data, it is sure that the probability of close contact of two surfaces plays an important role in the frictional electrification. The contribution of segmental motion will be large when the polymer molecule is an acceptor (as in the case of PS [9]) or a donor of charges. When other substances in the bulk concern the electrification, the migration of such substances to the surface becomes easier with the start of molecular motion and contributes to the observed charge values.

REFERENCES
1. F. R. Ruckdeschel and L. P. Hunter, J. Appl. Phys., <u>46</u> (1975) 4416, <u>48</u> (1978) 4898
2. K. Ohara, J. Appl. Polym. Sci., <u>21</u> (1977) 1409
3. K. Ohara, J. Electrostatics, <u>2</u> (1976/1977) 277
4. K. Ohara, J. Electrostatics, <u>4</u> (1978) 233
5. K. Ohara, WEAR, <u>39</u> (1976) 251, <u>50</u> (1978) 333
6. O. Yano and Y. Wada, J. Polym. Sci., A2, <u>9</u> (1971) 669
7. S. Kapur, C. E. Rogers and E. Baer, J. Polym. Sci.: Polym. Phys., <u>10</u> (1972) 2297
8. Y. Wada, J. Phys. Soc. Jpn., <u>16</u> (1961) 1226
9. C. B. Duke and T.J. Fabish, J. Appl. Phys., <u>49</u> (1978) 315, <u>48</u> (1977) 4256

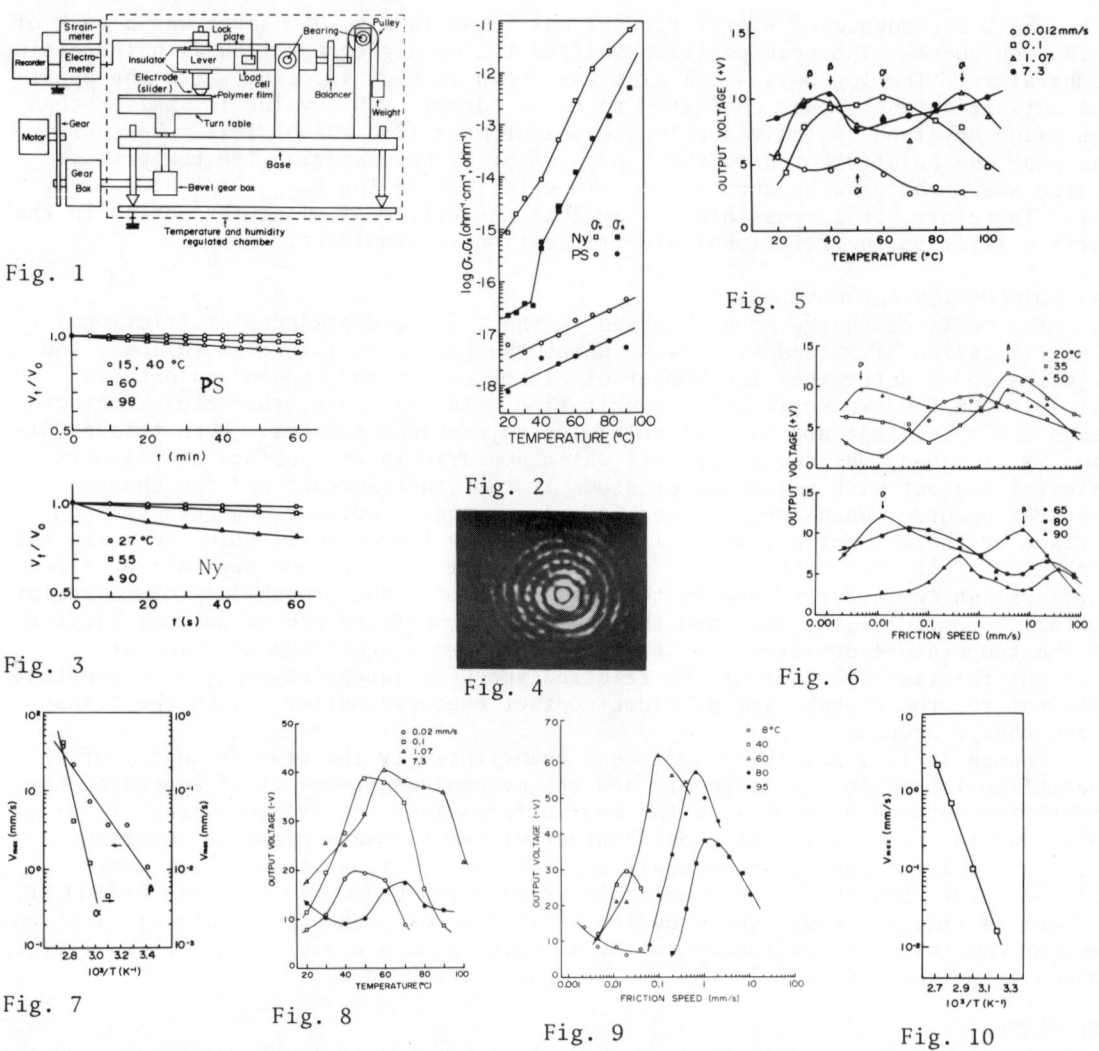

Fig. 1 Apparatus used in the present experiments.
Fig. 2 Temperature dependence of volume and surface conductivity for PS and Ny.
Fig. 3 Charge decay curves at various temperature.
Fig. 4 Interference fringe at an interface between PS film and electrode.
Fig. 5 Temperature dependence of charge at various friction speeds for PS.
Fig. 6 Friction speed dependence of charge at various temperatures for PS.
Fig. 7 log v_{max} - 1/T plot for PS.
Fig. 8 Temperature dependence of charge at various friction speeds for Ny.
Fig. 9 Friction speed dependence of charge at various temperatures for Ny.
Fig. 10 log v_{max} - 1/T plot for PS.

CORONA CHARGING OF A SPHERICAL PARTICLE

HAVING AN EXTREMELY HIGH RESISTIVITY

Senichi Masuda and Masao Washizu

Department of Electrical Engineering, Tokyo University

7-3-1 Hongo, Bunkyo-ku, Tokyo, Japan 113

1. Introduction
   Charge quantity imparted to a spherical insulating particle in an electric field by ion bombardment has hitherto been estimated by Pauthenier's equation [1]. In the derivation of this equation, the electrical field around the sphere is calculated by the superposition of:
   i) an external field distorted by the existance of the perfectly insulating particle, and
   ii) a self-induced field resulting from the charge imparted to the particle which is assumed to distribute uniformly on its surface.

These two assumptions, however, contradict each other in that no conductance is assumed in i), while a complete conductance is a prerequisite condition for ii).
   The only excuse for these contradictory assumptions may be the anticipation that the particle will rotate during charging process owing to the turbulence of gas flow.
   In this paper, a calculation is made rigorously without assuming uniform distribution of the imparted charge. Time increase and saturation values of the particle charge are derived.

2. Formulation of charging process
   In formulating the charging process, the following assumptions are made.
   i) A spherical particle with radius a and a specific dielectric constant ε_s

is located in a uniform external field E_0 where uniform positive ion space charge density ρ exists.

ii) The particle does not rotate.

iii) The particle is larger than $2\mu m$ in diameter, so that the charging is affected only by ion bombardment due to electric field (ion-field bombardment), and charging due to thermal diffusion can be neglected.

iv) Resistivity of particle is so high that movement of charge imparted to the particle can be neglected during the charging process.

Since the charing by positive ions is possible only in the region where the outer field at the particle surface is directed inward, assumptions ii) iv) require the continuity equations to hold at the surface regions speficied as follows:

$$-\frac{\partial \sigma(\theta,t)}{\partial t} = \rho\kappa E_r(\theta) \quad \text{for } \theta \text{ where } E_r(\theta) \leq 0 \quad (1)$$
$$= 0 \quad \text{for } \theta \text{ where } E_r(\theta) > 0 \quad (2)$$

where
- θ = angle (see Fig.1)
- $\sigma(\theta,t)$ = surface charge density,
- ρ = space charge density,
- κ = mobility of ions, and
- $E_r(\theta)$ = radial component of outer field at the surface

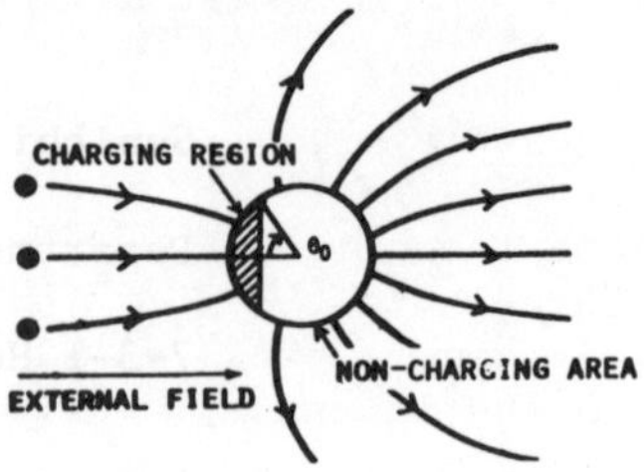

Fig.1 Charging region

or introducing a binary coefficient

$$F(\theta) = 0 \text{ for } \theta \text{ where } E_r(\theta) > 0$$
$$= 1 \text{ for } \theta \text{ where } E_r(\theta) \leq 0 \quad (3)$$

$$\frac{\partial \sigma(\theta,t)}{\partial t} = -\rho\kappa F(\theta) E_r(\theta) \quad (4)$$

Equation (4) is the basic equation of charging process.

3. Normalization

The following normalization is made to obtain the solution in generalized form where subscript n denotes normalized quantities:

i) distance r ; $r = a \cdot r_n$
ii) time t ; $t = (4\pi\varepsilon_0/\rho\kappa) \cdot t_n$
iii) charge Q ; $Q = Q_{\infty p} \cdot Q_n$

where
$$Q_{\infty p} = 4\pi\varepsilon_0 a^2 pE = \text{Pauthenier's saturation charge}$$
$$p = \frac{3\varepsilon_s}{\varepsilon_s + 2}$$

Using i)~iii), field E_r and charge density σ can be normalized as

iv) $E_r = (pE_0) \cdot E_{rn}$
v) $\sigma = (Q_{\infty p}/a^2) \cdot \sigma_n$

The reason why $4\pi\varepsilon_0/\rho\kappa$ and $Q_{\infty p}/a^2$ are used, instead of $4\varepsilon_0/\rho\kappa$ and $Q_{\infty p}/4\pi a^2$ in the normalization of time and surface charge density is that they provide simpler and clearer forms in the normalized equations to appear in this paper.

Substitution of i)~v) into equation (4) yields the normalized charging equation

$$\frac{\partial \sigma_n(\theta,t_n)}{\partial t_n} = -F(\theta) E_{rn}(\theta)$$
$$= -F(\theta)\{E_{ext}(\theta) + E_s(\sigma_n,\theta)\} \quad (5)$$

where $E_{ext}(\theta)$ = radial component of field due to external field (normalized), and

$E_s(\sigma_n,\theta)$ = radial component of field due to surface charge (normalized).
The calculations are carried out, expanding eq.(5) in Legendre polynomials as

$$\sigma_n(\theta,t_n) = \sum_{m=0}^{\infty} A_m(t_n)P_m(\cos\theta) \qquad (6)$$

$$E_s(\sigma_n,\theta) = \sum_{m=0}^{\infty} B_m(t_n)P_m(\cos\theta) \qquad (7)$$

$$E_{ext}(\theta) = -P_1(\cos\theta) \qquad (8)$$

and splitting eq.(5) into polynomial equations of equal order.

$$\frac{d}{dt} A(t_n) = -K(t_n)\{C+B(t_n)\} = -K(t_n)\{C+D\cdot A(t_n)\} \qquad (9)$$

where K = a matrix implying $F(\theta)$ in eq.(3)
       D = a matrix implying the relation between σ_n and E_s

4. Results

Time increase in particle charge imparted in a D.C. external field is calculated for different values of ε_s and shown in Fig.2 in normalized form. This figure indicates that for the case of particles with small ε_s saturation charge Q_∞ is considerably smaller than the value predicted by Pauthenier's equation. With the increase in ε_s the value of Q_∞ rises, finally to coincide with the Pauthenier's value at $\varepsilon_s \to \infty$.

The physical interpretation for this effect is exhibited in Fig.3. When ε_s is large, the outer field due to the particle's charge becomes almost uniform over the surface (Fig.3(A)), and fits approximately the assumption of Pauthenier's equation. On the other hand, when ε_s is small (Fig.3(B)), there occurs a local concentration of charge on the upstream side of the particle which concentrates the field lines and hampers further charging.

The time increase in particle charge is also calculated for the case of an applied rectangular A.C. field. The results obtained for normalized frequency $f_n=f/(\rho\kappa/4\pi\varepsilon_0)=10$ are indicated in Fig.4. The wavings appearing on the curves are resulted by periodical alternation of the rectangular A.C. field applied. Saturation charge for small ε_s's is approximately doubled compared with the case of uni-directional charging, using a D.C. field.

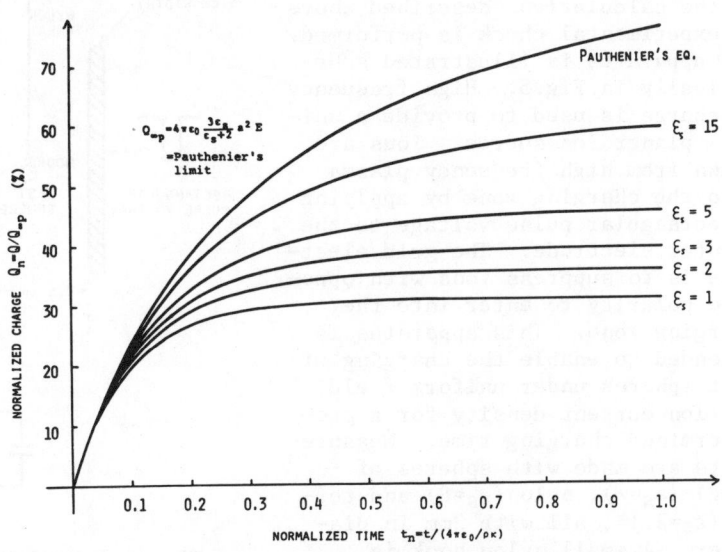

Fig.2 Time increase in particle charge in a D.C. external field

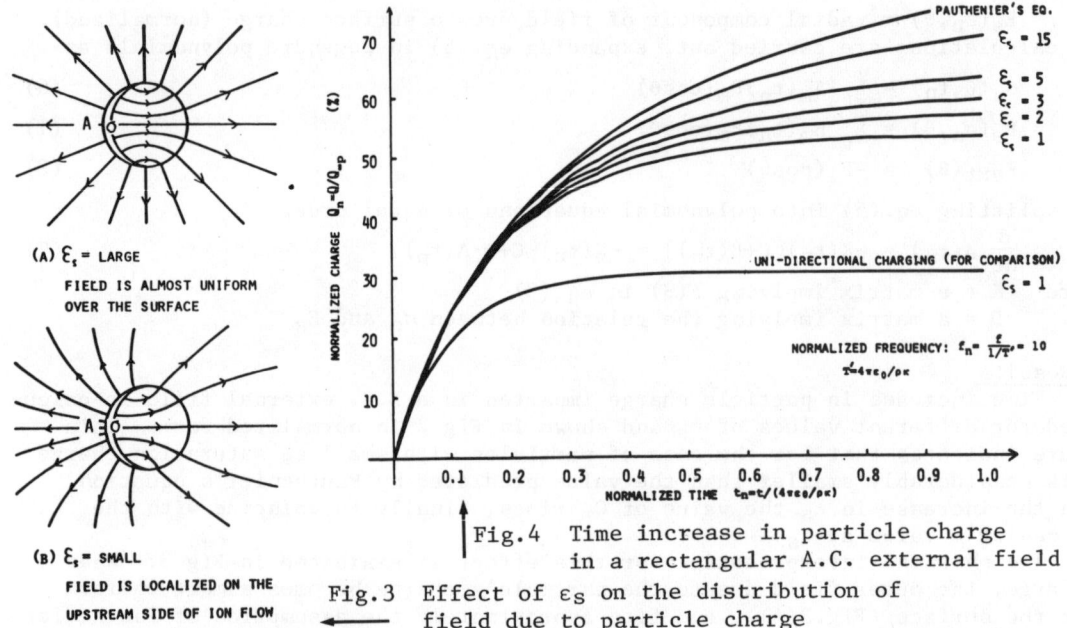

Fig.3 Effect of ε_s on the distribution of field due to particle charge

Fig.4 Time increase in particle charge in a rectangular A.C. external field

5. Experimental verification

In order to verify the results of the calculations described above an experimental check is performed. The apparatus is illustrated schematically in Fig.5. High frequency discharge is used to provide a uniform planer ion source. Ions are drawn from high frequency plasma into the charging zone by applying a rectangular pulse voltage to the counter electrode. The grid electrode is to suppress ions with opposite polarity to enter into the charging zone. This apparatus is intended to enable the charging of test spheres under uniform field and ion current density for a pre-determined charging time. Measurements are made with spheres of steel ($\varepsilon_s=\infty$), nylon ($\varepsilon_s=8$) and teflon ($\varepsilon_s=2.1$), all with 3mm in diameter. A small nylon hook is attsched to each sphere with which it is hung by a nylon thread of 100μm diameter. The thread is

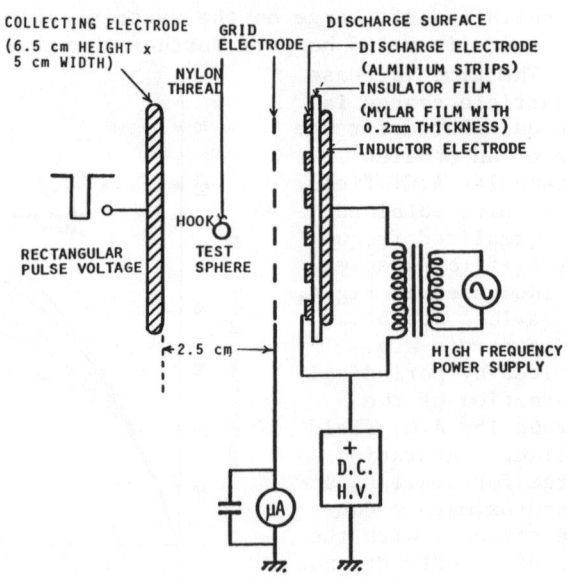

Fig.5 Experimental apparatus for particle charging

disconnected from the hook after charging, and the charge quantity measured with a Faraday Cage. Pulse width is varied so that the increase in charge with time can be measured. Here, the time of ion flight necessary to reach the test sphere from the ion source is negligibly small compared with the pulse width used.

Fairly large deviations are observed in the measured quantity of charge, in excess of ten percent at maximum. The mean values of 4~12 trials are plotted in normalized form in Fig.6. Solid lines indicate the calculated values. The calculated and the average of measured values show good agreement in spite of a large scatter in respective measured values. This scatter is found to be caused by irregularity in ion current distribution in space and time.

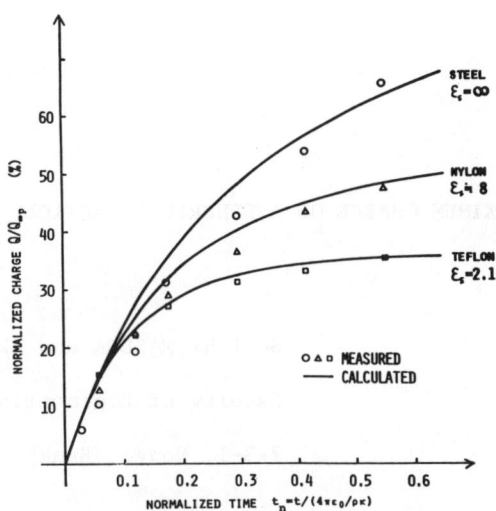

Fig.6 Comparison between average of measured values and calculated values in uni-directional charging

6. Conclusions

The charging process due to ion field bombardment is rigorously calculated for a spherical particle having a very high resistivity. The calculated results show that

i) if the external field is uni-directional, the saturation charge becomes considerably smaller than that predicted by Pauthenier's equation, especially when ε_s is small. For instance only 35% of Pauthenier's limit can be imparted to a particle with $\varepsilon_s = 2$.

ii) If the external field is bi-directional, saturation charge can largely be increased. 53% of Pauthenier's limit can be attained for a particle with $\varepsilon_s = 2$ in this case.

Comparison between the calculated values and average of measured values of particle charge is made with uni-directional charging, and a good agreement is obtained, providing support for our calculation.

7. Acknowledgement

The authors would like to thank Mr.K.Akutsu, and Dr.A.Mizuno for their helps and discussions.

References

[1] M.M.Pauthenier and M.Moreau-Hanot, La Charge des Particules Sphériques Dans un Cham Ionisé, Journal de Physique et le Radium 3, 590 (1932)
[2] S.Masuda and M.Washizu, Ionic Charging of a Very High Resistivity Spherical Particle, submitted to the Journal of Electrostatics (Elsevier Pub. Co., Amsterdam)
[3] S.Masuda, M.Washizu, A.Mizuno and K.Akutsu, Boxer Charger - A Novel Charging Device for Particles, to be presented at 1978 Annual Meeting of IEEE/IAS.

MAXIMUM CHARGE OF A SPHERICAL PARTICLE IMPARTED BY PULSE CHARGING

Senichi MASUDA and Akira MIZUNO

Faculty of Engineering, University of Tokyo

7-3-1, Hongo, Bunkyo-ku, Tokyo, Japan 113

1. Introduction
    Since pulse charging of particles by mono-polar ions proved to be a useful means in electrostatic processes treating high resistivity powders[1], its saturation charge has become a problem of great concern. The measurement by the authors showed that the saturation charge in pulse charging was not only smaller than that in a dc field with equal magnitude, but also it was affected by the quantities such as peak voltage, duration time and duty factor of the pulse voltage used. The authors considered that these peculiarities in pulse charging should be resulted by the occurence of an ion depletion zone which would be formed in the pulseless period due to repulsion of ions by the charge aquired by the particle. Based on this assumption, the authors calculated the saturation charge of a spherical particle imparted by mono-polar ion bombardment in pulse charging of two different modes, one with a repetitive pulse voltage of square wave form and another with the addition of a dc voltage to this pulse voltage. The former mode should be refered to as the "pure-pulse mode", whereas the latter mode as the "pulse-to-dc mode". The authors also carried out experimental examinations of the calculated results.

2. Formulation of charging process
    In formulation of charging process, the following assumptions are made:
  i) A spherical particle with radius a is located in a uniform pulse field of either mode described where uniform mono-polar ionic space charge exists, as illustrated in Fig.1.
 ii) The particle is larger than 2 μm in diameter, so that the charging is affected only by ion bombardment due to electric field (ion field bombardment)[2], and charging due to thermal diffusion can be neglected.
iii) Distortion of field due to ionic space charge can be neglected.

The wave form of field in pulse-to-dc mode in the charging space between the plane ionic source and plane electrode in Fig.1 is illustrated in Fig.2, where

t_1 = pulse daration time, t_2 = pulseless period, E_p = pulse height field, E_{dc} = dc field added, and E_o = peak field. In case of pure-pulse mode, $E_{dc} = 0$ and $E_p = E_o$. The duty factor, M, and ratio of dc field to peak field, E, should be defined as

$$M = t_1/(t_1 + t_2) \quad (1)$$

$$E = E_{dc}/E_o = E_{dc}/(E_p + E_{dc}) \quad (2)$$

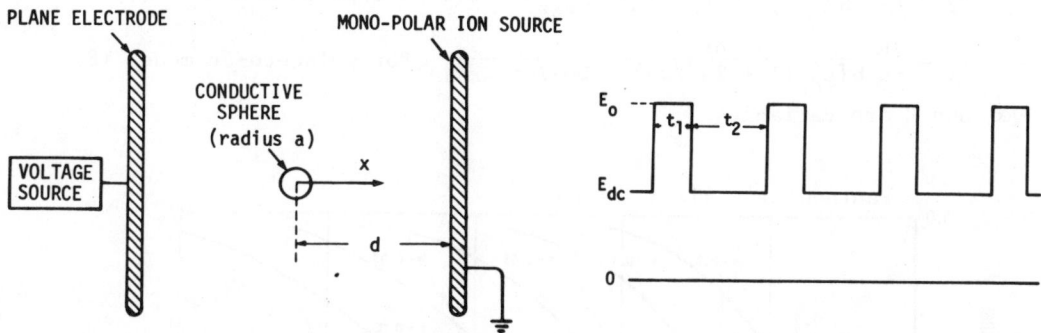

Fig.1 Calculation Model of Saturation Charge Imparted by Pulse Charging

Fig.2 Applied External Field

In pulse charging of either mode, charging is interrupted after pulse duration time t_1 is elapsed so that the charged particle is kept at the highest potential in the charging space during the pulseless period t_2. Hence, the ions are to be repelled from the particle to form an ion depletion zone around it. In the next charging cycle, charging does not begin until the perifery of this depletion zone arrives at the surface of the particle. With the increase in particle charge the range of this depletion zone will also increase, finally to reach such a level that its perifery can just arrive at the particle surface at the end of charging period t_1. This provides the saturation charge, so that its quantity could not reach the level of Pauthenier's limit[3] to be attained in a dc field with intensity E_o. In calculation of saturation charge, we only have to consider the movement of the perifery of depletion zone on x-axis in Fig.1, since the field intensity becomes maximum along this axis. The field intensity along x-axis is different for charging period t_1 and pulseless period t_2 as the following, under an assumption that surface charge is equally distributed at an instant (conductive sphere)[4]

i) charging period t_1: $\quad E_1(x) = -E_o(1 + \frac{2a^3}{x^3}) + \frac{Q_\infty}{4\pi\varepsilon_o x^2}$ (3)

ii) pulseless period t_2: $\quad E_2(x) = \frac{Q_\infty}{4\pi\varepsilon_o x^2}$ for pure-pulse mode (4)

$\quad E_2(x) = -E_{dc}(1 + \frac{2a^3}{x^3}) + \frac{Q_\infty}{4\pi\varepsilon_o x^2}$
for pulse-to-dc mode (5)

where Q_∞ = saturation charge of particle, b = ion mobility, ε_o = dielectric constant of vacuum space, and x = distance from the center of particle. When the saturation charge is attained, the maximum range of depletion zone R has to be traversed along x-axis by the perifery of the depletion zone in both charging period t_1 and pulseless period t_2, so that we get the following simultaneous equations for charging:

$$t_1 = \int_R^a \frac{dx}{bE_1(x)} = \int_R^a \frac{dx}{b[E_o(1 + 2a^3/x^3) - Q\infty/4\pi\varepsilon_o x^2]} \quad (6)$$

$$t_2 = \int_a^R \frac{dx}{bE_2(x)} = \int_a^R \frac{dx}{bQ\infty/4\pi\varepsilon_o x^2} \quad \text{for pure-pulse mode} \quad (7)$$

$$t_2 = \int_a^R \frac{dx}{b[E_{dc}(1 + 2a^3/x^3) - Q\infty/4\pi\varepsilon_o x^2]} \quad \text{for pulse-to-dc mode} \quad (8)$$

where $Q\infty$ and R are variables.

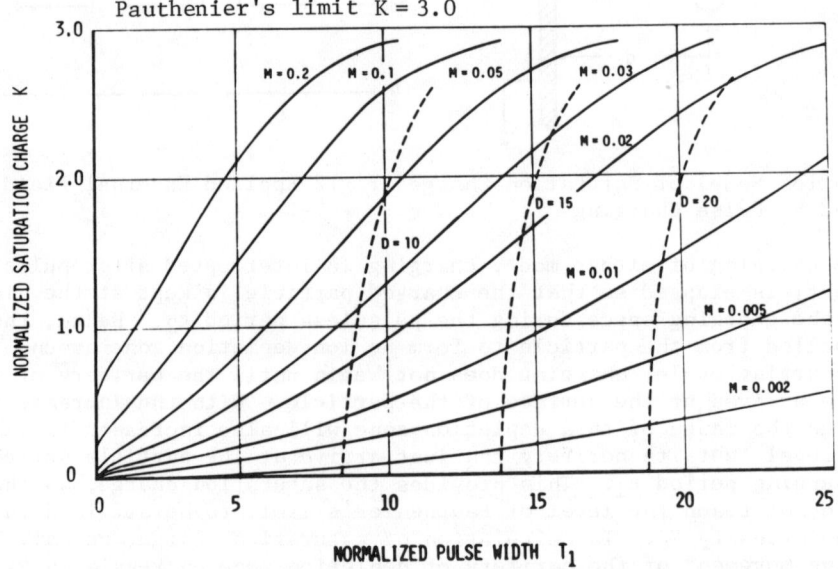

Fig.3 Normarized Theoretical Saturation Charge for Pure-Pulse Mode

Following normalizations are made to obtain the generalized solutions:

Saturation charge	$K = (Q\infty/4\pi\varepsilon_o a^2)/E_o$		(9)
Time	$T_1 = bE_o t_1/a$	pulse duration time	(10)
	$T_2 = bE_o t_2/a$	pulseless period	(11)
Distance	$x = r/a$		(12)
	$X_o = R/a$	position of the depletion zone	(13)

Using the duty factor M, following normalized equations can be derived from eqs. (6), (7) or eqs. (6), (8).

$$F(X) = 3K(1-M)/M * \int_1^X x^3/(x^3-Kx+2) \, dx - X^3 + 1 = 0 \quad \text{pure-pulse mode} \quad (14)$$

$$F(X) = (1-M)/M * \int_1^X x^3/(x^3-Kx+2) \, dx - \int_1^X x^3/(Ex^3-Kx+2E) \, dx = 0$$
$$\text{pulse-to-dc mode} \quad (15)$$

X_o can be obtained by solving these equations numerically for various values of the parameter M. T_1 is represented as

$$T_1 = \int_1^{X_o} x^3/(x^3 - Kx + 2) \, dx \quad (16)$$

Thus the relation between K and T_1 can be obtained. The solution for pure-pulse mode is indicated in Fig.3 in normalized form.

A particular case is considered here that ions can reach to the sphere from the ion source during the pulse duration time T_1. In this case, position of the depletion zone, X_o, is considered to be equal to the normalized distance D (=d/a) from the center of the sphere to the ion source and X_o should be calculated from eq. (7). Where d is the distance from the center of the sphere to the ion source. Hence, in this case, the saturation charge increases abruptly along the curve, on which

$$X_o = D \quad (17)$$

is satisfied. This curve is also plotted in Fig.3 by dotted line for different values of D.

Fig.4 shows the results obtained for the pulse-to-dc mode. In this case, the value of saturation charge is affected only by the parameter E within the practical range of E and M.

An experimental examination was made using a plane ion source with a grid electrode. The measured values were distributed within 10 % error of the theoretical values.

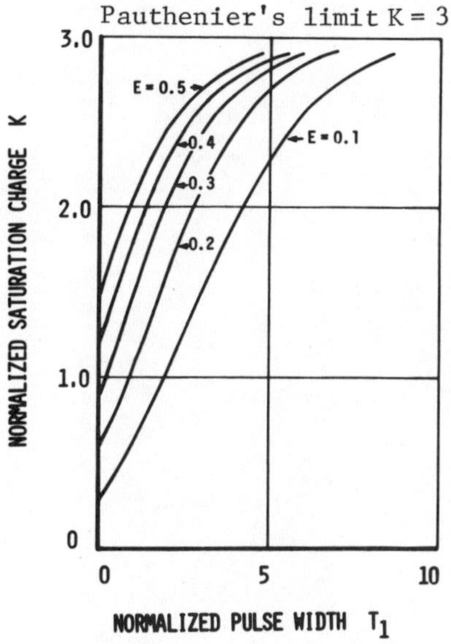

Fig.4 Normalized Theoretical Saturation Charge for Pulse-to-dc Mode

3. Conclusion

Theoretical value of saturation charge is calculated for two different mode of pulse charging system. Following conclusions are obtained.
i) Depletion zone of ion is formed in the vicinity of charged particle during pulseless period. Saturation occurs at the moment that ions existing outside the depletion zone cannot reach to the particle within pulse duration period.
ii) For the pure-pulse mode, saturation charge is affected by maximum field E_o, pulse duration period T_1, and duty factor M.

iii) For the pulse-to-dc mode, saturation charge is determined by E_o, T_1 and E (ratio of dc field to maximum field), and is not affected by duty factor M within practical range of E and M. In this case, range of the ion depletion zone is reduced, and a large increase in the saturation charge is resulted.

References

1. S.Masuda, I.Doi, M.Aoyama and A.Shibuya, Bias-Controlled Pulse Charging System for Electrostatic Precipitator, Staub-Reinhaltung der Luft, Bd. 36, Nr. 1, p.19, Januar, 1976
2. H.J. White, Industrial Electrostatic Precipitation, Addison-Wesley Pub. Co. 1963, p.137
3. M. Pauthenier and M. Morequ-Hanot, Rev. Gen. Electr. XIV(18) 1932, p. 583
4. S. Masuda and M. Washizu, Ionic Charging of A Spherical Particle having Extremely High Resistivity, submitted to the Journal of Electrostatics, IElsevier Pub. co., Amsterdam)

CORONA-CHARGING PROPERTIES OF POLYMER POWDERS

Manabu TAKEUCHI and Hideo NAGASAKA

Department of Electrical Engineering, Faculty of Engineering, Ibaraki University, Nakanarusawa-cho, Hitachi, Ibaraki, Japan

1. Introduction

Recently many studies have been made on the static electrification of polymer powders [1 — 4] and polymer films [5 — 7]. Most of the studies of the decay of the surface potential in charged polymer films have been conducted to investigate the electrical-transport processes in the polymers [8 — 12].

In the case of corona-charging of polymer powders, the behavior of the corona charges is very complicated, and a clear explanation has not been obtained on the mechanisms of the corona-charging and charge decaying. It is well known that the charge decay is influenced by the ambient humidity [13]. Because the powder sample has a large specific surface area, it can be assumed that the corona charging properties of polymer powders are influenced strongly by the ambient humidity.

The present paper describes a study of the effects of moisture absorption and adsorption on the rate of charge decay in polymer powder layers. The results of thermally stimulated surface charge decaying (TSCD) are also reported.

2. Experimental

The sample polymer powders used in this experiment and their particle sizes are tabulated in Table I. First, moisture sorption on the sample powders were studied. After sample powders were dried completely in an evacuated oven, they were then exposed to ambient air. The amount of moisture incorporated into the sample powders was weighed with balance at each humidity.

The decay of charge from the

Table I. Sample powders and their particle sizes.

Polymer powder	Particle size
polyester	3 — 50 μm
Nylon 12	5 — 50
Teflon	10 — 40
epoxy resin	10 — 200
polyethylene	50 — 200
polyvinyl chloride(PVC)	~100
polyvinyl alcohol(PVA)	—

sample powders was measured as follows: the sample powder was packed in a grounded aluminum cup, which was 50 mm in diameter and 1.5 mm in depth. The sample layer was brought under an electrode for corona discharging which was connected to the negative pole of a dc source of 5 kV, and the layer was charged negatively. After application of the negative dc voltage for 5 to 20 s, the layer was moved under a detecting probe within a definite time of 0.5 s. The detecting probe was connected to an electrometer and recorder. All measurements were made in an enclosed chamber at 20 °C, at various humidities.

3. Results and Discussions

Figure 1 shows changes in the moisture content incorporated into the polymer powders as a function of time after being left exposed to ambient air. The polyester powder took up a large amount of moisture, and there seemed to be no saturation.

Next, the surface potential decay was measured for various choices of the parameter of the moisture content. In this report all the results of the surface potential decay are plotted as a function of the square root of time. The decay of the surface potential was extremely slow for the Teflon powder layer. Figures 2 to 4 show the surface potential decay for the polyester, PVC and PVA powder layers, respectively. In these results, except for initial decay, the decay of the surface potential agreed well with the following equation presented by Inoue et al. [14]:

$$V = V_o \exp(-\alpha \sqrt{t}) \qquad (1)$$

where V and V_O are the surface potentials at $t = t$ and $t = 0$ respectively, t is the time after a cessation of corona charging and α is a constant. These results suggest that corona charges deposited on the powder layers are dissipated mainly by the diffusion through layers. If we take α as a decay rate, α can be determined from the slope of the straight line of V vs. $\sqrt{t}$. The initial surface potential V_O did not vary with the amount of the moisture content. On the other hand, the decay rate increased with an increase in the moisture content. Furthermore, the decay rate was found to increase exponentially with the moisture content as shown in Fig. 5. We obtain the following equation for the decay rate α:

$$\alpha = \alpha_o \exp(\beta m) \qquad (2)$$

where α_O is the decay rate at a completely dried state, m is the moisture content and β is a constant. This equation is analogous to the following equation presented by Shashoua [15]:

$$\tau = \tau_o \exp\{-b(RH)\} \qquad (3)$$

where τ is a halflife of charge decay, b is a constant, and (RH) denotes relative humidity. It is said that b in equation (3) is a constant which depends on only the chemical composition of the polymers. In our experiment, however, the straight lines in Fig. 5 bend at critical moisture content. It seems likely that the mechanism of moisture sorption by the polymer powders changes when the

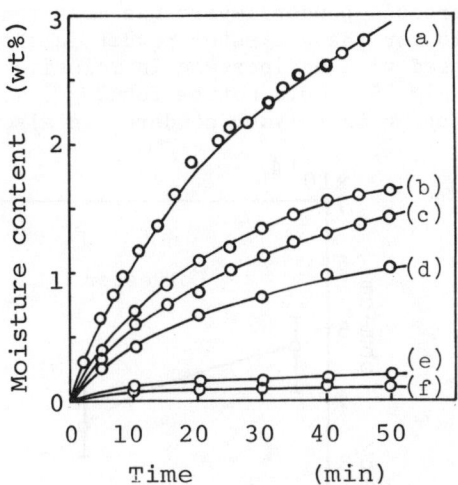

Fig.1. Moisture content change with time after being left exposed to ambient air.
(a) polyester, (b) Nylon 12, (c) Teflon, (d) epoxy, (e) polyethylene, (f) PVC.

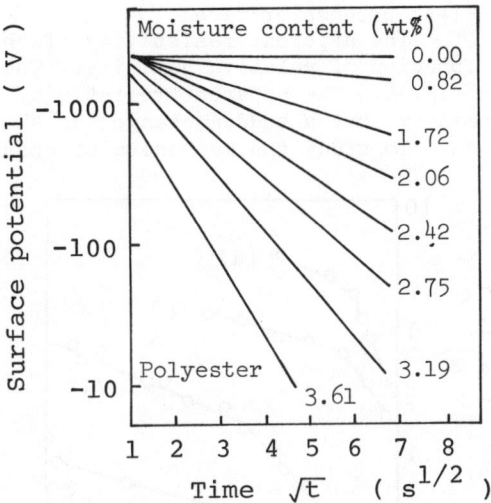

Fig.2. Decay of surface potential for the polyester powder layer.

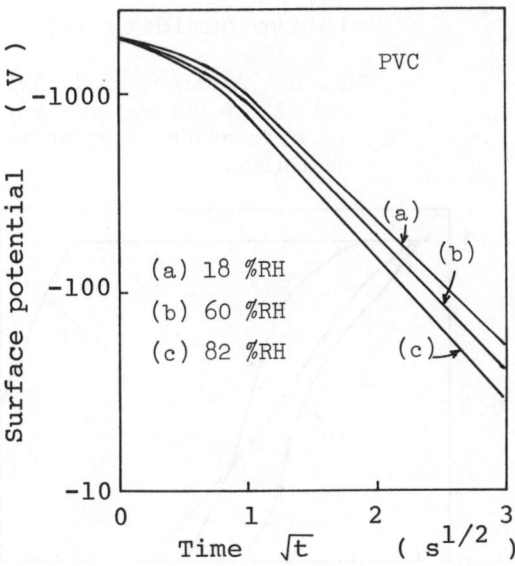

Fig.3. Decay of surface potential for the PVC powder layer.

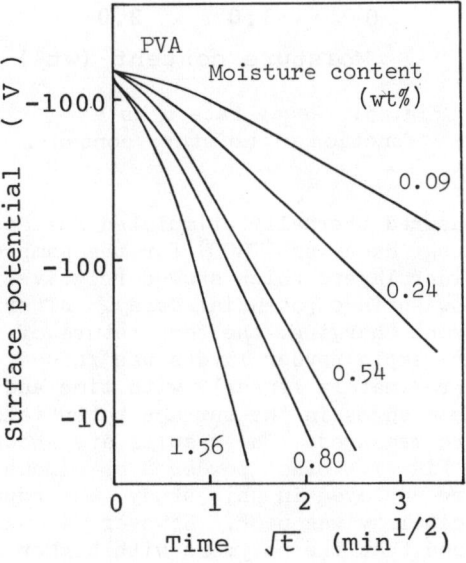

Fig.4. Decay of surface potential for the PVA powder layer.

content becomes critical.

The apparent resistivity of the compressed powder layers was measured as a function of relative humidity. The result for the polyester powder is shown in Fig. 6. The apparent resistivity decreased with an increase in relative humidity, but a sudden change, as shown in Fig. 5, could not be found.

To study the mechanism of charge trapping in polymer powders, we also

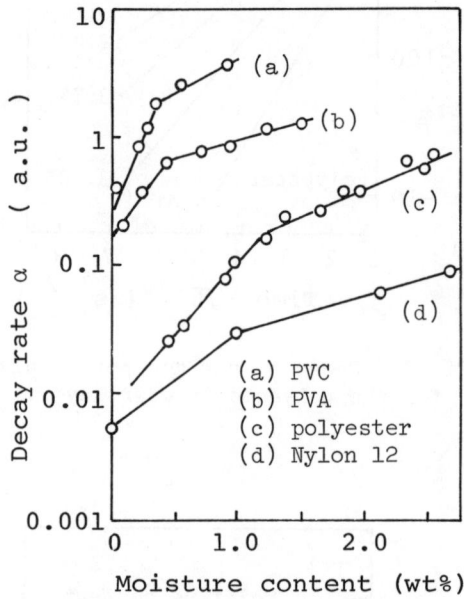

Fig. 5. Decay rate α as a function of moisture content.

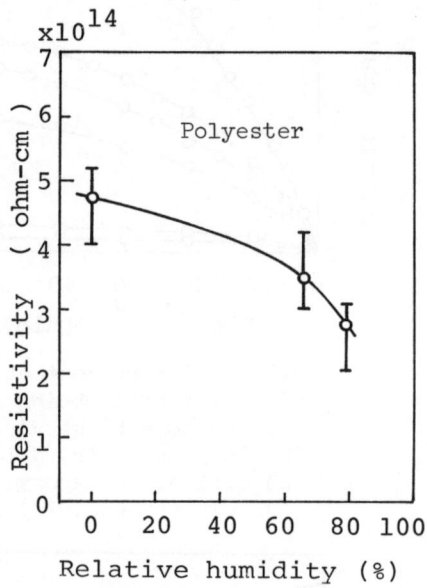

Fig. 6. Apparent resistivity of the polyester layer as a function of relative humidity.

measured thermally stimulated surface charge decaying (TSCD) for the sample powder layers which showed relatively slow surface potential decay. After corona charging, the temperature of the sample powder layers was raised approximately linearly with time and the changes in the surface potentials were measured. The results are shown in Fig. 7. Since powdered specimens were employed in this study, the reproducibility was poor. However, it was found that the polymers with higher melting points tended to show the beginning of the surface potential decay at higher temperatures.

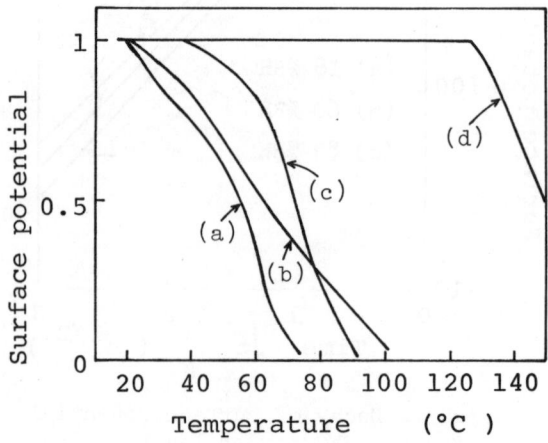

4. Conclusions

The surface potential decay in corona-charged polymer powder layers were studied in relation to the moisture sorption. Except for initial decay, the decay of the surface potential was found to agree with the equation, $V = V_0 \exp(-\alpha\sqrt{t})$. The decay rate α was found to increase exponentially with moisture content. The measurements of the thermally stimulated charge decaying were carried out, and it was found that the polymers with higher melting points tended to show the beginning of the surface potential decay at higher temperatures.

Fig. 7. Thermally stimulated surface charge decay. (a) Nylon 12, (b) polyethylene, (c) epoxy, (d) Teflon.

Acknowledgements

The authors are grateful to Mr. M. Sugai for his help in the experiments.

References

1) L.L.Spiller, J. Paint Technol., **44** (1972) 98.
2) M.Sugai, M.Takeuchi and H.Nagasaka, Japan. J. Appl. Phys., **15** (1976) 1563.
3) G.D.Cheever, J. Appl. Polymer. Sci., **19** (1975) 147.
4) L.B.Schein and J.Cranch, J. Appl. Phys., **46** (1975) 5140.
5) R.H.Partridge, J. Polymer Sci., Part A **3** (1965) 2817.
6) D.K.Davis and P.J.Lock, J. Electrochem. Soc., **120** (1973) 266.
7) A.Wahlin and G.Bäckström, J. Appl. Phys., **45** (1974) 2058.
8) M.Ieda, G.Sawa and U.Shinohara, Japan. J. Appl. Phys., **6** (1967) 793.
9) H.J.Wintle, J. Appl. Phys., **41** (1970) 4004.
10) H.J.Wintle, J. Appl. Phys., **43** (1972) 2927.
11) T.J.Sonnonstine and M.M.Perlman, J. Appl. Phys., **46** (1975) 3975.
12) E.A.Baum, T.J.Lewis, et al., J. Phys. D : Appl. Phys., **10** (1977) 487.
13) *For example*, R.G.C.Arridge, Brit. J. Appl. Phys., **11** (1960) 202.
14) E.Inoue, H.Kokado, et al., Denshishashin (Electrophotography) **1** (1957) 27 [in Japanese].
15) V.E.Shashoua, J. Polymer Sci., Part A **1** (1963) 169.

SURFACE POTENTIAL DECAY IN CdS POWDER LAYERS

Manabu TAKEUCHI

Department of Electrical Engineering, Faculty of Engineering, Ibaraki University
Nakanarusawa-cho, Hitachi, Ibaraki, Japan

1. Introduction

Electrophotography today has become a well-developed process of information-copying technology [1 — 3]. After charging the surface of the photoconductive layer, which is called an electrophotographic plate or photoreceptor, by means of a corona discharge and exposing to an optical image which dissipate the charges in the light-struck areas, one produces an invisible electrostatic image. The image can be developed by dusting with an electrically-charged toner. Most electrophotographic plates consist of an amorphous selenium layer on a metallic substrate or a zinc-oxide pigmented binder layer on paper. Corona charging and photoinduced discharging characteristics of these plates have been widely investigated by several workers [4 — 10].

Recently a CdS binder-type plate has been developed [11 — 15]. When a photoconductive CdS powder is applied to an electrophotographic plate, its corona charging ability plays an important role in a photographic performance.

To study the corona charging behavior of a CdS electrophotographic plate, the dark decay of the surface potential in CdS powder layers was studied. The effects of moisture adsorption and heat treatments on the dark decay of the surface potential are reported in this paper.

2. Experimental

The sample CdS powders were precipitated from aqueous solutions of cadmium sulfate through the **following reactions:**

$$CdSO_4 + H_2S \longrightarrow CdS + H_2SO_4 \qquad (1)$$

where the initial sulfate acidity was chosen to be 0 N, 2 N and 8 N. The precipitates were washed thoroughly with distilled water and dried at room temperature. All the CdS powders were found to have mixed structures of the cubic and hexagonal phases. The mean particle sizes of the CdS powders prepared from 0 N, 2 N and 8 N solutions were about 2.0, 6.7 and 20 μm respectively and these specimens are noted as CdS-ON, CdS-2N and CdS-8N, respectively, in this paper.

First, moisture adsorption on the CdS powders were studied. After sample powders were dried completely in an evacuated oven, they were left exposed to ambient air. The amount of moisture adsorbed on the sample powders was weighed with balance at each humidity.

The dark decay of the surface potential in corona-charged CdS powder layers was measured as follows: the sample CdS powder was packed in a grounded aluminum cup, which was 50 mm in diameter and 1.5 mm in depth. The sample layer was brought under an electrode for corona discharging, and they were charged negatively in the darkness. In principle the corona discharging voltage was -5 kV and the charging time was 15 s. The surface potential decay after a cessation of corona charging was measured with a vibrating reed surface potential meter. All measurements were made in the darkness at 20 °C, at various ambient humidities after equilibrium was obtained.

3. Results and Discussions

Figure 1 shows the change in moisture contents adsorbed on the CdS powders with time after being left exposed to ambient air. A relatively large amount of moisture was adsorbed on the CdS-ON powder. This is due to the large specific surface area of the CdS-ON. The moisture adsorption was relatively low for the CdS-2N and CdS-8N specimens.

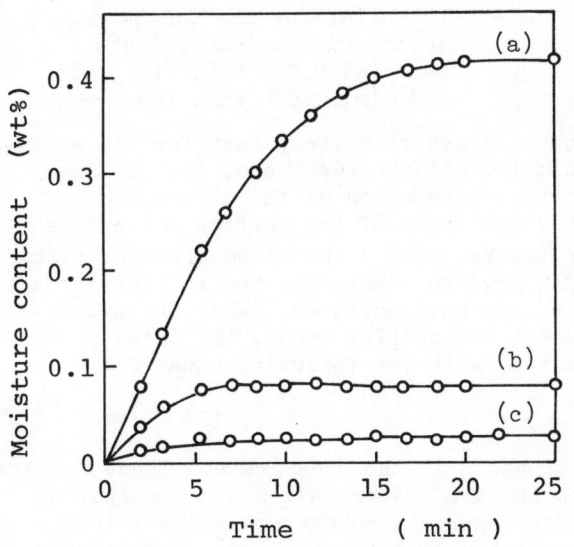

Fig.1. Moisture content change with time after being left exposed to ambient air. (a) CdS-ON, (b) CdS-2N, (c) CdS-8N.

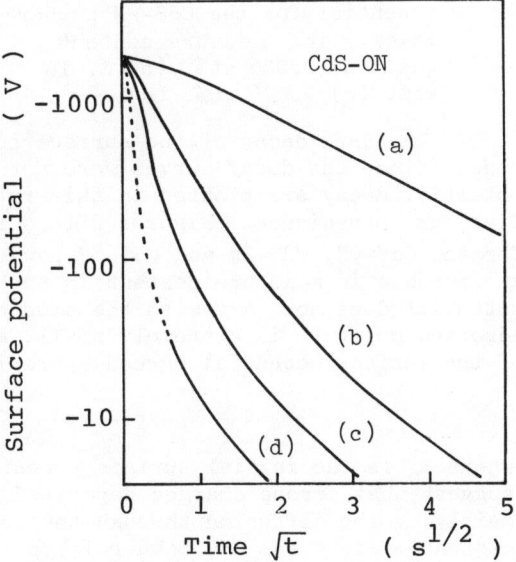

Fig.2. Decay of the surface potential for the CdS-ON powder layer. The moisture content was: (a) 0.000 wt%, (b) 0.092 wt%, (c) 0.151 wt%, (d) 0.232 wt%.

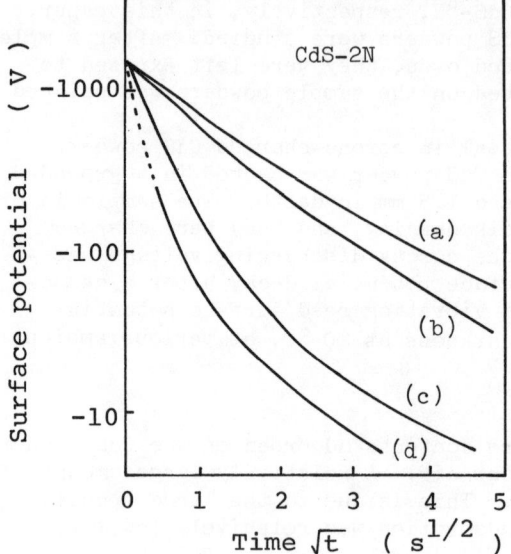
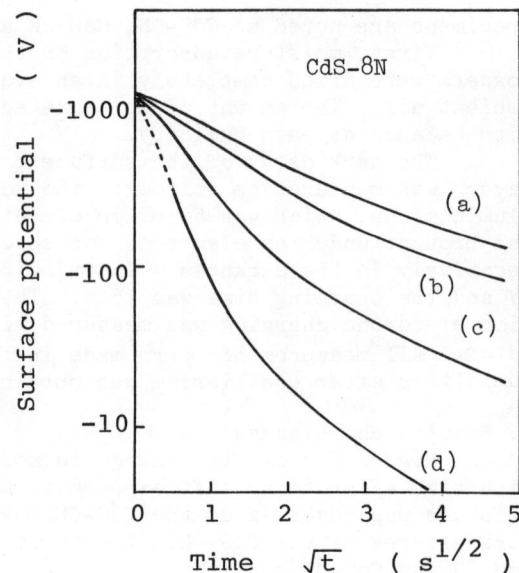

Fig.3. Decay of the surface potential for the CdS-2N powder layer. The moisture content was: (a) 0.006 wt%, (b) 0.016 wt%, (c) 0.096 wt%, (d) — .

Fig.4. Decay of the surface potential for the CdS-8N powder layer. The moisture content was: (a) 0.001 wt%, (b) 0.015 wt%, (c) 0.020 wt%, (d) — .

The dark decay of the surface potential was relatively fast for all specimens. Since the decay curves were not simple, all the results of the surface potential decay are plotted in this report as a function of the square root of time for convenience. Figures 2 to 4 show the decay of the surface potentials for the CdS-0N, CdS-2N and CdS-8N powder layers. The decay becomes faster with an increase in moisture content in the CdS powders. However, the initial surface potential does not vary with the amount of adsorbed moisture. When the amount of adsorbed moisture is extremely small, except for initial decay, the decay curves of the surface potential agreed approximately with the following equation:

$$V = V_o \exp(-\alpha \sqrt{t}) \qquad (2)$$

where V_o is the initial surface potential and α is the decay rate. These results suggest that corona charges deposited on the CdS powder layers are dissipated mainly by the diffusion through the powder layers at an extremely low moisture content. Since the CdS powder layer is relatively thick (1.5 mm), the carrier transportation caused by the electric field cannot be considered to be dominant. At medium and high moisture contents, however, the decay curves do not agree with the equation (2). One of the reasons for the departure from the straight lines in these results may be the carrier transportation caused by the electric field through the continuous (bulk) layers of adsorbed water. Carrier trapping in CdS can be considered as another reason for the departure. The charge carriers captured by shallow traps during their transportation can be released at room temperature and continue their movement again. The decay of surface potential has consequently been delayed. The carriers captured by deep traps will remain in the centers for a long time and bring about residual surface potentials.

To change the surface conditions of the CdS particles, the heat treatments of the CdS-2N powders have been carried out at 400 °C in an atmosphere of argon or air. The moisture adsorption was reduced drastically for the heat-treated CdS powders. Figure 5 shows the decay curves of the surface potential as a function of the square root of time for the heat-treated CdS-2N powders. The surface potential decay slowed down. Since the mean particle size of the CdS powder was not affected by the heat treatments, the change in the moisture adsorption may be due to the change in the surface conditions of the CdS particles. It was found, as shown in Fig. 5, that the CdS powder heat-treated in argon was more insensitive to the ambient humidity than the specimen heat-treated in air. It is reasonable to assume from these results that the density of the adsorption site for water can be reduced by heat treatments and that the heat treatment in argon is more effective in the reduction in the adsorption sites than the heat treatment in air.

The influence of corona charging time on the surface potential decay was also studied. Figure 6 shows the surface potential decay in CdS-2N powder layers for various corona charging times. The initial surface potential was found to be saturated by the corona charging within 5 s and was little affected by the charging time above 5 s. However, the surface potential decay slowed down with an increase in corona charging time.

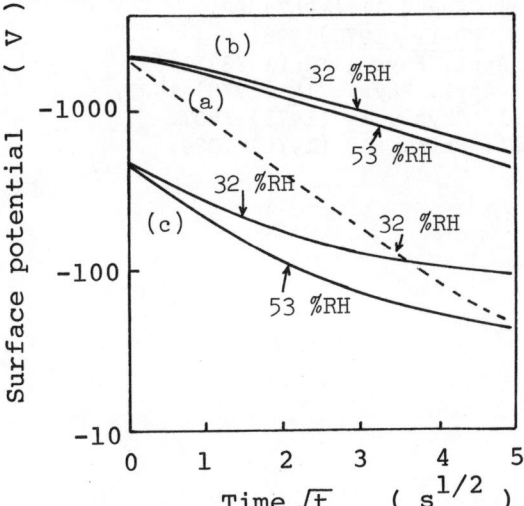

Fig.5. Decay of the surface potential. (a) original CdS, (b) CdS heat-treated in argon at 400 °C for 30 min, (c) CdS heat-treated in air at 400 °C for 30 min.

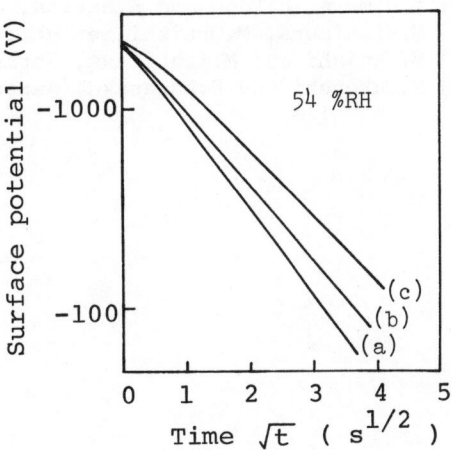

Fig.6. Decay of the surface potential for the CdS-2N powder layer. The corona-charging time was: (a) 10 s, (b) 60 s, (c) 240 s.

4. Conclusions

The surface potential decay in corona charged CdS powder layers was measured in darkness. The adsorption of water accelerates the surface potential decay but does not have effects upon the initial surface potentials. The surface potential decay was found to slow down in the heat-treated CdS powder layers. It was assumed that the heat treatment in argon is more effective in the reduction in the adsorption sites for water than the heat treatment in air.

Acknowledgements

The author is grateful to Mr. K. Matsuzaki for his help in the experiments. The author also wishes to express his appreciation to Professor S. Hayakawa of Tokyo Institute of Technology for his helpful discussion and to Professor H. Nagasaka of Ibaraki University for his continuing guidance and encouragement.

References
1) P.M.Cassiers, IEEE Trans. Electron Devices, ED-19 (1972) 524.
2) J.Viscakas, IEEE Trans. Electron Devices, ED-19 (1972) 531.
3) H.Kokado, IEEE Trans. Electron Devices, ED-19 (1972) 540.
4) K.Hauffe, J. Electrophotogr. Sci., 10 (1962) 321.
5) V.M.Fridkin and Yu.N.Barulin, Soviet Phys.-Doklady, 7 (1963) 629.
6) H.Tung Li and P.J.Regensburger, J. Appl. Phys., 34 (1963) 1730.
7) P.J.Regensburger, J. Appl. Phys., 35 (1964) 1863.
8) H.Kiess, J. Appl. Phys., 40 (1969) 4054.
9) H.Seki, IEEE Trans. Electron Devices, ED-19 (1972) 421.
10) A.Kondo and T.Kitamura, IEEE Trans. Electron Devices, ED-19 (1972) 462.
11) D.W.Chapman and F.J.Stryker, Photogr. Sci. Eng., 11 (1967) 22.
12) M.J.Mitsui, IEEE Trans. Electron Devices, ED-19 (1972) 396.
13) Y.Mimura, H.Endo and K.Narita, Japan J. Appl. Phys., 12 (1973) 1165.
14) M.Nishimura, M.Ohnishi, et al., Japan J. Appl. Phys., 14 (1975) 1401.
15) M.Ohnishi and M.Yoshizawa, Japan. J. Appl. Phys., 14 (1975) 1407.
16) M.Takeuchi and H.Nagasaka, Japan. J. Appl. Phys., 13 (1974) 1029.

ELECTRIFICATION OF LIQUIDS IN LAMINAR FLOW THROUGH A SMALL DIAMETER METAL PIPE

I.Chonan, A.Ohashi, M.Ueda

Nagoya university

Furo-cho, Chikusa-ku, Nagoya, Japan

Introduction

The process of the charge generation during the flow of liquids in pipes has been a subject of considerable interest for a long time. The streaming current depends upon the characters of the liquids such as the electrical conductivity, permittivity etc.. To study this point, the various liquids such as ethylalcohol, water, acetone, silicone, n-heptane, benzene etc. were used for the charge generation. And surfactant was used to change the conductivity of the liquids.

It has been generally said that the charge generation occurs when the diffuse double layer which is built up at the surface of the pipe is swept away by the flow. The thickness of the double layer is called "Debye length". It is an interesting problem that what kind of phenomena will be observed when the dimension of the pipe is equivalent to or smaller than the Debye length. So, the pipe of small diameter 0.2mm was used for the measurement.

Measuring system

The measuring system is shown in Fig.1.. The vessel made of glass or copper is supported by teflon from the earth. The stainless steel pipe is the one of the 0.2mm diameter and 5cm long. The flow velocity is controlled by the air pressure from 0 to 0.5atm.. The streaming current from the pipe to the earth (I) and the charge of the liquid flowing out of the pipe (Q) are mea-

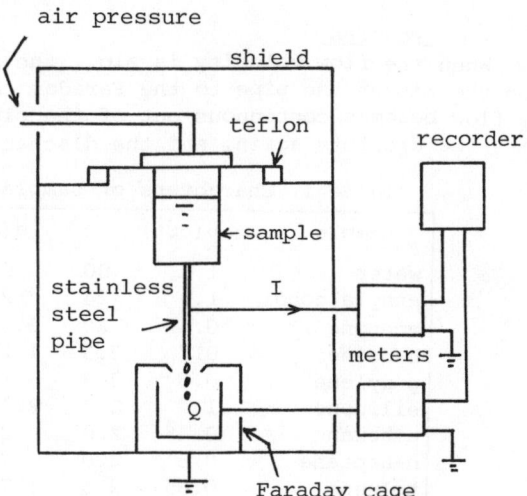

Fig.1. apparatus

sured by the electrometers. The lowest detectable level of the current is limited to about 10^{-14}A because of the noise.

The sample liquids usually have the initial charges when they are poured into the vessel. And it is necessary to keep liquid still to discharge them before the measurement. At least half an hour is needed for the relatively high conductivity liquids and more time lapse for the lower ones. Then I is equal to $-dQ/dt$. All measurements are carried out at the room temperature.

Characters of samples

The samples used for the measurement are listed in Table 1. with their characters. The notations are listed at the end of this paper. The conductivity was measured using the standard cylindrical electrodes under D.C. stress (30V/cm). It is generally said that the conduction current decreases with time lapse because the initial impurity ions in the liquids are swept out to the electrodes by the application of D.C. stress. So the value of conductivity of the liquids seems to vary with the time lapse. To calculate the most correct value of conductivity the backward exterpolation of conduction current to the zero sec. must be used. But generally the curve of current decay is nearly hyperbolic, so the exterpolation mentioned above is very difficult. We did not use the initial value but the steady state value.

The maximum value of the Reynolds number is 1500 of acetone. So the flow must be always the laminar one.

The Debye length (λ) is calculated from the expression

$$\lambda = \sqrt{\frac{\varepsilon kT}{2Ne^2}} = \sqrt{\frac{\varepsilon kT\mu}{\sigma e}} \qquad \sigma = Ne\mu$$

supposing roughly that T=300K, $\mu=10^{-9}$m^2/V·s. The maximum value of λ in Table 1. is 0.18mm of toluene. It is smaller than the diameter of the pipe 0.2mm. So it needs the thinner pipe to investigate the diameter dependence of the charge generation associated with the Debye length.

Results

When the flow velocity is slow, the sample liquid falls down as droplets from the tip of the pipe to the Faraday cage. As the velocity becomes higher, the flow becomes continuous out of the pipe. But at the halfway of fall, it becomes the droplets again, and the discharge from the liquids in the Faraday cage

Table 1. Characters of samples and results

sample	ν (cSt)	ε	σ (S/m)	λ (μm)	Re	n
water	1	80	5×10^{-5}	0.019	300	3.6
ethylalcohol	1.4	24	2×10^{-7}	0.17	390	3.6
acetone	0.4	20	6.5×10^{-8}	0.26	1500	3.6
benzene	0.7	2.3	1.2×10^{-12}	21	740	0.70
p-xylene	0.6	2.3	1×10^{-12}	23	830	0.72
silicone	1	2.3	2.4×10^{-13}	44	920	0.17
n-hexane	0.45	2.0	2×10^{-14}	150	1200	-
n-heptane	0.6	2.0	2×10^{-14}	150	1000	-
toluene	0.65	2.4	1.7×10^{-14}	180	770	-

to the pipe does not occur.

Typical data are shown in Fig.2., where the streaming current is plotted against the volumetric flow rate for various liquids. The current is negative except ethylalcohol and water. On ethylalcohol, the current is negative in case of the continuous flow, but is positive in case of the droplets. On water, it is always positive.

It can be seen from the data that the every curve falls into two groups. Acetone, water, ethylalcohol belong to the first group, and the other liquids belong to the second one. For the first group, the streaming current is proportional to about 3.6 powers of flow velocity when the flow is continuous. It is independent of the kind of the liquid, though they have different characters each other. These liquids have D.C. conductivity greater than 10^{-8} S/m and permittivity greater than 20. As the current increases rapidly with increasing the volumetric flow rate, the charge generation can become very large.

On the other hand, for the second group the streaming current is proportional to n powers of flow velocity, where n depends on the kind of the liquid sample and is usually smaller than 1. These liquids are high resistivity nonpolar or weak polar liquids and have D.C. conductivity smaller than 10^{-12} S/m. For n-heptane, toluene, n-hexane, the streaming current is so smaller than 10^{-14} A that it can not be detected by the electrometer. Generally the charge generation is very small when the conductivity is extremely low. These results are summa-

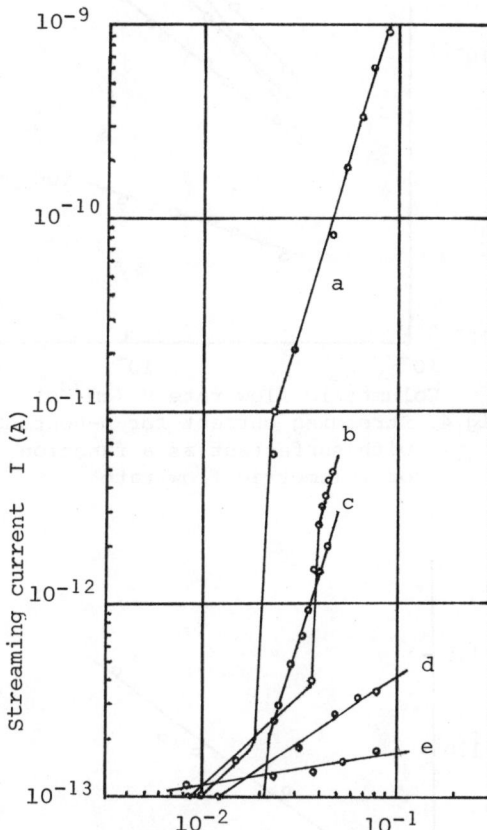

Fig. 2 Streaming current for various liquids as a function of volumetric flow rate

a: acetone b: water
c: ethyl alcohol d: benzene
e: silicone

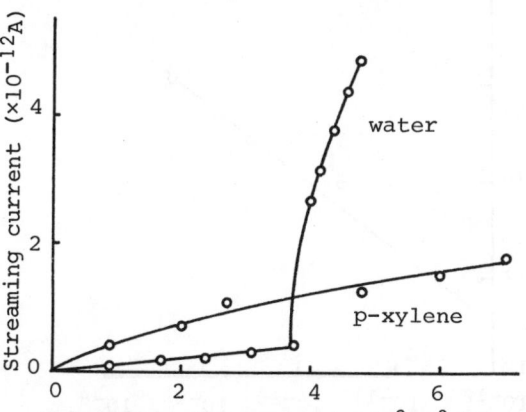

Fig. 3. Linear plot of streaming current

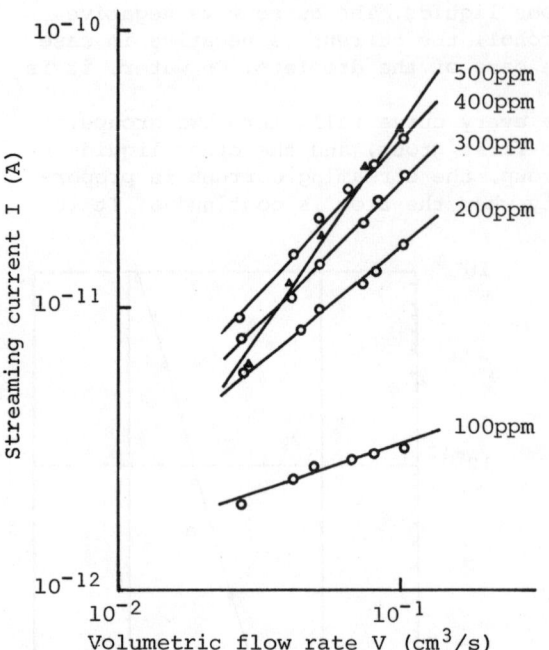

Fig.4. Streaming current for n-heptane with surfactant as a function of volumetric flow rate

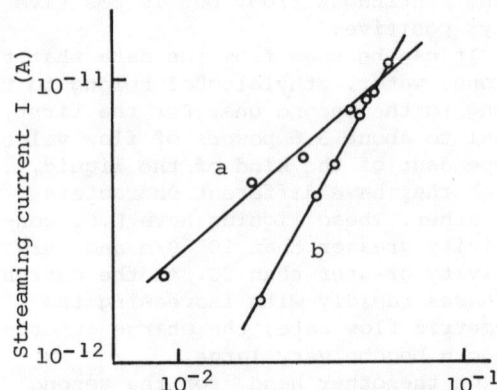

Fig.6. Streaming current for silicone with surfactant as a function of volumetric flow rate

a: $\sigma = 10^{-11}$ (S/m). $I \propto V^{0.86}$
b: $\sigma = 4 \times 10^{-8}$ (S/m). $I \propto V^{2.2}$

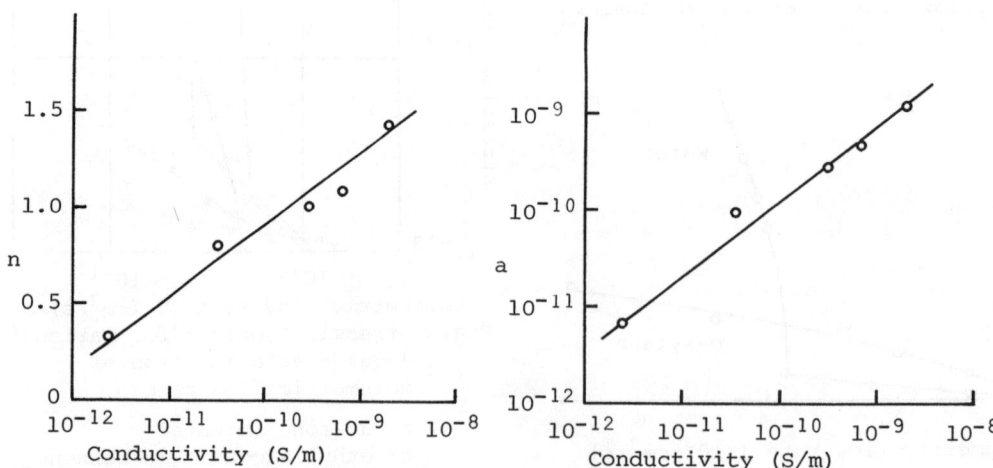

Fig.5. Conductivity dependence of n and a for n-heptane with surfactant

rized in Table 1..

Another difference between two groups is noticed at their velocity dependence of the current. The streaming current is replotted against the volumetric flow rate on linear scale in Fig.3.. It can be seen from the curve of water that the streaming current is proportional to the volumetric flow rate at the low velocity, and it increases abruptly at the transition point from the droplets to the continuous flow. This is the common feature to the first group.

On the contrary, there is no abrupt increase of the current at the transition point from the droplets to the continuous flow for p-xylene same as the other liquids of the second group.

This phenomenon seems to be associated with their conductivity. That is, the liquids classified into the first group have the high conductivity, so the charge of the droplets generated in the pipe will relax at the tip of the pipe, and the value of the streaming current will be smaller than that of the continuous flow.

The streaming current is strongly affected by the conductivity of the liquid. The conductivity of n-heptane was changed by addition of the surfactant (Sorbon S-80) from 100p.p.m. to 500p.p.m.. The conductivity of n-heptane increases with the amount of the surfactant. The conductivity of pure n-heptane is smaller by about five orders than that of n-heptane with 500p.p.m. surfactant. The results are shown in Fig.4., where the streaming current is plotted against the volumetric flow rate on log-log scale. As mentioned above, the streaming current of n-heptane without surfactant is much smaller. The curves follow the expression

$$I = a \cdot (V/V_0)^n$$

where V is volumetric flow rate, n is exponent, a is a constant, $V_0 = 1 cm^3/s$.

It can be seen from the data that both n and a increase with increasing conductivity. The conductivity dependence of n and a is shown in Fig.5.. The velocity dependence of the streaming current becomes similar to the one of the first group when the conductivity becomes high. The permittivity of the samples was unchanged by the surfactant.

The similar data for silicone are shown in Fig.6.. The current for the sample without surfactant is much smaller. And the slope of the curve increases with increasing conductivity like the data of n-heptane.

It has been generally said that the streaming current is proportional to about 2 powers of flow velocity. But in our case, the exponent can be much larger or smaller and is variable by the characters of the samples. The theoretical approach is under way at present.

Notations
ν: viscosity
ε: permittivity
σ: conductivity
λ: Debye length
Re: Reynolds number
n: slope of curve
N: ion concentration
μ: mobility
k: Boltzmann constant
T: temperature

V: volumetric flow rate
e: charge of electron
I: streaming current
Q: charge of liquid

References
1) A.Klinkenberg and J.L.van der Minne: Electrostatics in the Petroleum Industry, Ch.5 & Ch.6, Elsevier, Amsterdam (1958).
2) N.Gibson: Static Electrification, Conf. Ser.11, p.71, Inst. Physical Soc., London (1971).

ELECTROKINETIC PHENOMENA IN ELECTRICAL INSULATING OIL/IMPREGNATED CELLULOSIC

INSULATING PAPER SYSTEMS

Sachio Yasufuku, Takeshi Tanii and Yoshiyuki Inoue

Toshiba Corporation

2-1, Ukishima-cho, Kawasaki-ku, Kawasaki, Japan

INTRODUCTION

A new method was already developed by the authors for measuring the zeta potential which exists on the surface of cellulosic pressboard impregnated with electrical insulating oil, while the oil streams through the capillary hole in the pressboard (1). The metal salts of di-(2-ethyl-hexyl) sulphosuccinate were dissolved in a purified insulating oil, and the zeta potential and the ac conductivity were investigated. As a result, the dependence of the zeta potential on the type and concentration of the salt was clarified by them. It was concluded that electrokinetic phenomena in insulating oil are thought to be identical with those in aqueous solutions.

Moreover, a modified Klinkenberg's apparatus was also proposed for measuring the electrostatic charging tendency of cellulosic paper/electrical insulating oil system under a triboelectric condition (1). And it was suggested by them that the dependence of the electrostatic charging tendency on the concentration of the salt appears to show a trend similar to that of the zeta potential on its concentration.

In this paper the authors deal with the measurement of the streaming potential emerging on the surface of impregnated cellulosic paper, when the oil flows across its surface in the above apparatus, and its dependence on the water content of the paper.

EXPERIMENTAL

The electrical insulating oil used in this study is a naphthenic base transformer oil which satisfies JIS C 2320 (1). The electrical insulating paper used is a Kraft paper with a thickness of 0.15 mm which satisfies JIS C 2304.

The modified Klinkenberg's apparatus was slightly changed for the purpose of measuring the streaming potential emerging on the surface of the paper, when the oil flows across its surface. Figure 1 shows a schematic diagram of this set-up, which is placed in such a constant environment condition as RH 42% at 23°C, referring to ASTM D 2679 - Test for Electrostatic Charging Tendency. In this Figure an alphabetical order describes the detail of this apparatus as follows: A, an upper reservoir: B, an adjustable rod for the purpose of flowing oil or stopping oil flow: C, a metal capillary: D, oil-impregnated cellulosic insulating papers: E, an earthed copper plate: F, an electrode ring made of brass, which is supported above D by a combination of press-board pieces with a Teflon ring: G, a Faraday cage: H, insulators made of Teflon: I, an outer cage: J, a Laboratory Recorder Type 3046 of Yokogawa Electric Works, Japan and V, Electrometers Type TR-8651 of Takeda Riken Industry, Japan, which have an input impedance of 10^{14} Ω and are capable of measuring dc voltages from 1mV to 100V, respectively.

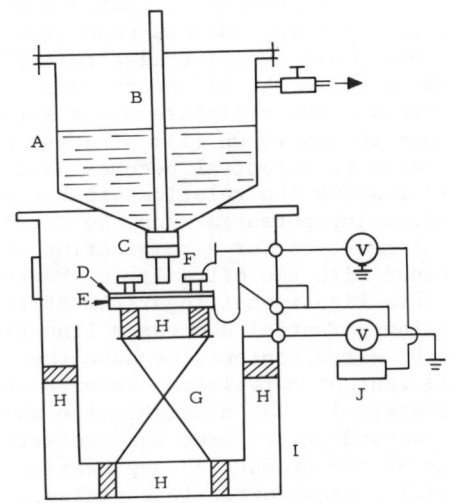

Fig. 1: Schematic Diagram of Apparatus

The evacuated oil was provided and its water content was measured by the Karl Fischer method in accordance with ASTM D 1533 - Standard Test Method for Water in Insulating Liquids (Karl Fischer Method). The papers were dried in an oven and subsequently vacuum impregnated with the evacuated oil. Their water contents measured by the Karl Fischer method were conditioned in the range of 0.1 to 9.0%.

The oil specimen was flowed vertically onto the center of the paper in a constant flow rate through the capillary. Since the separation of electrostatic charge happens at the interface of the oil and the paper, the charging tendency of the oil, Q (pC/min.) was calculated on the grounds that Q=CV, where C, Capacitance of the overall apparatus, Farad and V, the reading of the Electrometer in Volt, measured for the oil flowed in a Faraday cage. At the same time, apparent streaming potential per one liter of the oil, which flowed across the surface of the paper was directly measured by another Electrometer. To obtain real streaming potential instead of the apparent one, known dc voltage supplied from a dry cell was applied to the electrode with the same configuration as the afore-mentioned without the paper and the surface potential was directly measured by this electrometer. Then a correction factor was determined. Thus, the apparent streaming potential was corrected by this factor and eventually the real one was obtained.

In addition, after the above experiment was finished, the dc resistivity of the paper was measured at 22°C by use of a High Resistance Meter Type 4329A made by Yokogawa-Hewlett Packard, Japan, with the electrode EY-02559 made by Nisshin Electric, Japan, specified in JIS C 2111. Moreover, its water content was also measured.

RESULTS AND DISCUSSION

In the measurement of streaming potential that the authors conducted, the paper bears negative electrostatic charge, whereas the oil bears positive one, because the cellulose has a strong tendency to adsorb OH^- (1). Therefore, the streaming potential becomes negative. Figure 2 shows the relationship between the streaming potential and the water content measured for a combination of the paper with the oil. It is observed from this Figure that logarithm of the streaming potential decreases linearly with the water content, because the coefficient of correlation becomes approximately -1. The straight line was drawn according to least squares method. As far as the production engineering of oil-filled transformer is concerned, it is an important result that the smaller the water content becomes, the higher the streaming potential, because it is stated that the cellulosic materials in oil-filled transformer should be dried as much as possible. Figure 3 shows the relationship between the charging tendency of the oil and the water content of the paper. It is observed from this Figure that the charging tendency appears independent of the water content of the paper, when the latter covers a range of 0.2 to 9.0%.

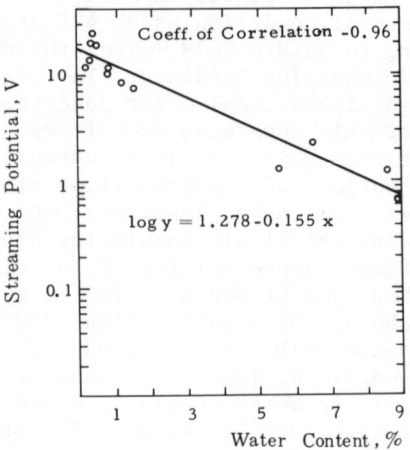

Fig. 2: Streaming Potential vs. Water Content

There are a lot of researches dealing with the relationship between the dc electric conduction and the water content of cellulosic material so far. It was stated that the water in cellulose commands easiness of ion transport in it, when the ion moves in it by the aid of an electric field. According to Walker (2), in a low water content range logarithm of the dc resistivity of cellulose is inversely proportional to its water content. The authors also confirmed that this relationship is available for their case.

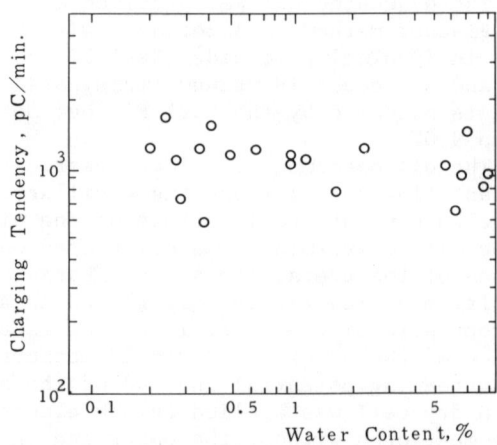

Fig. 3: Charging Tendency vs. Water Content

Thus, it follows from the above relationship that logarithm of the streaming potential could be assumed to correlate linearly with that of the dc resistivity. Fugure 4 shows the relationship between the streaming potential and the dc resistivity of the paper. It is observed from this Figure that logarithm of the streaming potential increases linearly with that of the dc resistivity, because the coefficient of correlation becomes approximately 1. The straight line was drawn according to least squares method. This Figure reveals that the afore-mentioned assumption of the authors seems reasonable. Thus, it became obvious that dryness of the paper strongly governs occurrence of the streaming potential on the surface of the paper under the triboelectric condition, suggesting an important role of the dc electric conduction happened in the paper.

Since it is difficult to deal with the above phenomena as such theoretically, the authors try to envisage these on the analogy of their results on the electrokinetic phenomena in electrical insulating oil/impregnated cellulosic pressboard systems (1). The Helmholtz-Smoluchowski equation under laminar flow condition is described as follows (1):

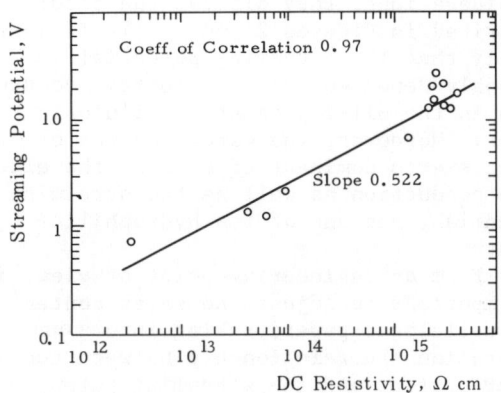

Fig. 4: Streaming Potential vs. DC Resistivity

$$\zeta = \frac{4\pi \eta Hks}{P\varepsilon} = 8.476 \times 10^8 \frac{\eta H}{\varepsilon P} ks \qquad (1)$$

where ζ, zeta potential (mV): η, coefficient of viscosity of the oil (poise): ε, dielectric constant of the oil: H, observed electromotive force produced by streaming the oil through a capillary existing in the pressboard (mV); P, hydrostatic pressure for streaming the oil through the capillary and ks, specific electric conductivity of the capillary with the oil in it ($\mho$ cm^{-1}).

To eliminate experimental uncertainties in the determination of the hydrostatic pressure, they rewrote Eq.(1) as follows, using Hagen-Poiseuille equation (1).

$$\zeta = 3.55 \times 10^{13} \frac{R^4}{8\ell} \frac{1}{\varepsilon} \frac{H}{V} ks \qquad (2)$$

where V, flow rate through a capillary (cm^3/sec); R, radius of a capillary (cm) and ℓ, length of a capillary (cm).

Employing an improved Gortner's apparatus (1), they measured the dependence of the electromotive force on the water content of the pressboard, on condition that the water content of the oil was kept at 10 ppm. Figure 5 shows such a relationship. It is observed from this Figure that logarithm of the electromotive force linearly decreases with the increase in the water content, indicating a strong resemblance to Fig. 2. The zeta potential calculated by Eq.(2) decreased with the increase in the water content. While the specific electric conductivity increased with the increase in the water content, the electromotive force rapidly decreased as the afore-mentioned. In the result, it appears to follow from Eq.(2) that the zeta potential is inclined to decrease gradually with the increase in the water content. It is worth noting that the zeta potential also depends on the water content of the oil. Furthermore, it should be emphasized that surface conductance at solid/liquid interface plays an important role in the electrokinetic phenomena stemming from streaming liquid through a capillary (1), (3).

Based upon the authors' fundamental considerations, they discuss the results described in Figures 2 and 4. It is most likely that the streaming potential is strongly dependent on the electric conduction in the oil-impregnated cellulosic paper. Moreover, the water content of the paper exerts dominant effects on the electric conduction as well as the streaming potential, because of its hydrophilic nature.

From an engineering point of view, it is important to adjust the water content of the cellulosic paper, taking into consideration the relationship between the water content and the streaming potential.

Finally, it was experimentally approved that the water content of the cellulosic paper is not influential to the changing tendency as such, apart from the effect of the water content of the oil.

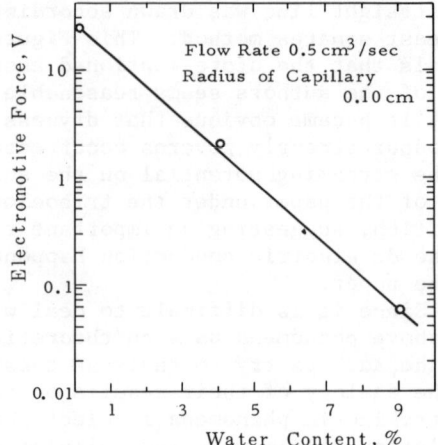

Fig. 5: Electromotive Force vs. Water Content

CONCLUSION

In electrical insulating oil/impregnated cellulosic insulating paper system, a new method for measuring the streaming potential was developed by the authors and its dependence on the water content of the paper was demonstrated. The relationship between the streaming potential and the dc electric conductivity was clarified. In addition, the above results were successfully explained by their fundamental considerations based upon the Helmholtz-Smoluchowski equation for the electrokinetic phenomena stemming from streaming the oil through the capillary.

ACKNOWLEDGMENT

The authors wish to thank Professor A. Kitahara of Tokyo University of Science, Tokyo, Japan for his helpful discussion and suggestions.

REFERENCES

(1) S. Yasufuku, J. Ise, Y. Inoue and Y. Ishioka:
    IEEE Trans. Elec. Insul., EI-12, 5, p.3 70, 1977
(2) A.C. Walker; J. Textile Inst. 24, T123, 1933.
(3) J. Th. G. Overbeek and P.W.O. Wijga: Rec. Trav. Chim., 65, p.556, 1946.

MEASUREMENT OF SPACE CHARGE QUANTITY IN DIELECTRIC FILM BY MAXWELL STRESS METHOD

Tatsuo Takada, Yotsuo Toriyama and Takao Sakai

Department of Electrical Engineering,
Musashi Institute of Technology

Tamazutsumi, Setagaya-ku, Tokyo, Japan

1   Introduction

Basic electrical properties of dielectric material are generally represented by quantities of (1) dielectric permittivity and dielectric loss, (2) electric resistivity and (3) electric breakdown strength. These quantities are commonly used for designing the insulation of electric power machines, transmission line cables and other electric apparatus. The resistivity of polymeric insulator is generally greater than 10^{14} Ω cm and is non-linear depending on the intensity of applied electric field. Thus, it is difficult to define the resistivity as a single value. The electric carriers are generated in a complicated mechanism and drifted in the polymer by the external applied field and also by the field produced by the space charge of trapped carriers.

When a high d.c electric field is applied to an insulation layer of an electric apparatus, the space charge accumulates in the layer, and in consequence the insulation could fail. For this reason, it is important to obtain the space charge quantity stored in the dielectric material rather than its resistivity. We believe that the charge accumulation property is as important as the above described three basic electric properties.

However, an equipment to measure directly the quantity of space charge distributed in dielectric film is not available in market. When the space charge is stored in a dielectric film, the induced charge on both electrodes which sandwich the film is opposite polarity, and is equal to the total quantity of space charge. Thus, the dielectric film with electrodes is electrically neutral. Hence, it is difficult in practice to measure the quantity of space charge in a polymer film by an electrometer of Farady cap method or other techniques available up to now.

This paper reports a new technique of measuring the space charge quantity in a dielectric film. When the space charge is present in a dielectric film, the electrodes are stressed mechanically by Maxwell stress which can be detected with piezo-electric devices attached to the electrodes. There is a linear relation between the Maxwell stress and the field intensity E at the interface between the electrode and the film. The space charge quantity can be determined from the difference between E(+) and E(-) at the positive and negative electrode, respectively.

2 Principle of Measurement

As shown in Fig.1, a voltage V_o was applied to a polyethylene-telephthalate film. It was assumed that charge carriers are either injected into the film an electrode by Schottky effect, or generated in the bulk of the film by Poole-Frenkel effect. Then, the space charge accumulates in the film by the trapping mechanism of these charge carriers.

Let $\rho(x)$ [c/m^3] be the space charge density, a function of the distance x from the cathode in the direction of the thickness of the dielectric film. From the solution of Poisson equation, the electric field intensities at the positive and negative electrodes, E(+) [V/m] and E(-) [V/m], respectively, are given by

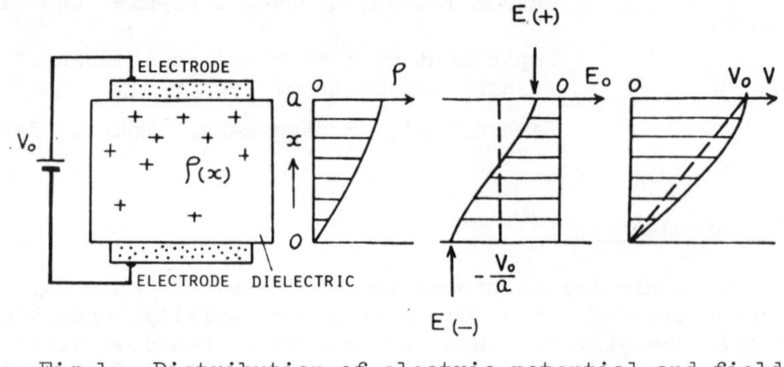

Fig.1 Distribution of electric potential and field intensity by space charge in dielectric film.

$$E(+) = -\frac{V_o}{a} + \frac{1}{\varepsilon}\int_0^a \frac{x}{a}\rho(x)\,dx \qquad (1)$$

$$E(-) = -\frac{V_o}{a} - \frac{1}{\varepsilon}\int_0^a \frac{a-x}{a}\rho(x)\,dx \qquad (2)$$

where a [m] and ε [F/m] are the thickness and the permittivity of the dielectric film, respectively. The difference between E(+) and E(-) equals the total quantity of the space charge divided by ε, as shown in the following equation.

$$E(+) - E(-) = \frac{1}{\varepsilon}\int_0^a \rho(x)\,dx \qquad (3)$$

If each field intensity at the surface of both electrodes can be measured experimentally, the total quantity of the space charge accumulated in the dielectric film can be obtained. The measuring technique of the electric field intensity employing the Maxwell stress method is described below.

As shown in Fig.2, two components of field are superposed to the dielectic film. One is the static field $E(V_o, \rho)$ and the other is the alternating field $e_o \sin\omega t$. Here, E is a function of both external applied voltage V_o and distributed space charge density ρ, as shown in the equation (1) and (2), and e is equal to v/a where v is the alternating applied voltage of the frequency of 8MHz.

Maxwell stress shown in the equation (4) acts on the electrode surface which is in contact with the dielectric film.

$$P = \frac{1}{2}\varepsilon[E(V_o,\rho) + e_o\sin\omega t]^2 = \frac{1}{2}\varepsilon E^2 + \varepsilon E e_o\sin\omega t + \frac{1}{2}\varepsilon e_o^2\sin^2\omega t \qquad (4)$$

The first term represents the static Maxwell stress, the second term is the alternating stress with the fundamental frequency f_o= 8MHz, and the third the alternating stress with the second harmonic frequency $f=2f_o$. As the second term is proportional to the product of the static field E and the alternating field $e_o\sin\omega t$, the signal of the second term can be easily amplified by a high-frequency-amplifier. To do this, a quartz disk with the resonance frequency of 8MHz is pasted onto the other side of each disk electrode. The voltage U_o induced in the quartz is given by

$$U_o = K \varepsilon E\, e_o\sin\omega t \qquad (5)$$

where K is the transfer constant from the mechanical stress to the electrical out-put of the quartz. Hence, the intensity of the electric field at the surface of electrode can be obtained from the induced voltage of the quartz.

3 Experimental Apparatus

A block diagram of the experimental apparatus is shown in Fig.2. The both external d.c. and alternating voltage are applied to the dielectric film, so that the alternating Maxwell stress acts on the surface of electrode. The external static voltage V_o is applied to the electrode through the series resistor, R = 1MΩ, and the alternating voltage v is also applied to the electrodes through the coupling capacitor, C = 0.001µF. The values of these elements of R and C were chosen so that they satisfy the following relations, $Rs \gg R$, $R \gg 1/\omega C$, and $1/\omega Cs \gg 1/\omega C$, where Rs and Cs are the resistance and the capacitance of the dielectric film, respectively. The alternating Maxwell stress of the second term of equation (4) acts on the electrode surface, and the stress of elastic wave of 8MHz is transmitted through the bulk of the disk electrode to the quartz from the electrode surface facing the film. Both electrodes are made of aluminium in the shape of disk 10 mm thick and 30 mm in diameter. The frequency of the alternating field was adjusted such that it was exactly equal to the resonance frequency of the quartz, 8 MHz The induced voltage in the quartz is amplified by a linear integrated circuit, rectified and filtered by a diode and capacitor circuit.

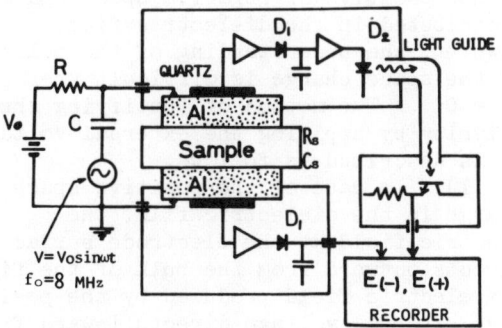

Fig.2 Block diagram of experimental apparatus.

The out-put voltage of the quartz at the high voltage electrode was tranduced to the light signal by a photo-diode and was guided to the recorder at the earth potential by a light guide. The out-put voltages were recorded for various applied voltages.

4   Results and Discussion

The specimens used are polyethylene-telephthalate (PET) film of 50 μm thick. Both side of the PET film was evaporated with aluminium as electrode of 30 mm in diameter. At the temperatures shown in Fig.4, the PET film is biased with the voltage, Vb = 4000 V, for 10 minutes before the measurement. Then, the space charge builds up in the specimen by the charge injection from the electrodes or by the charge generation in the specimen bulk.
Fig.3 illustrates the experimental results. The abscissa represents the voltage V_o from which the polarity of the space charge can be justified, and the ordinate is the field intensity E and the induced charge density $\sigma = \varepsilon E$ on the electrode as well. In Fig.4 indicated are the various experimental conditions and corresponding space charge distribution obtained from the results of Fig.3.

Even if $V_o = 0$, the induced voltage of the quartz disk can be measured because the Maxwell stress acting on the electrode surface is produced by the either positive or negative space charge distributed in the dielectric film. However, the understanding of the polarity of the space charge is difficult when $V_o = 0$. The method of justifying the polarity by applying the external voltage V_o is described as follows.

(1) In case of the positive space charge in the dielectric film, the electric field at the electrode surface directs outward from the bulk of the film. The electric field produced by the positive external voltage directs inward from the electrode. The voltage induced by both field is decreased until they are in balance. As the result, the null condition of the induced voltage is obtained.

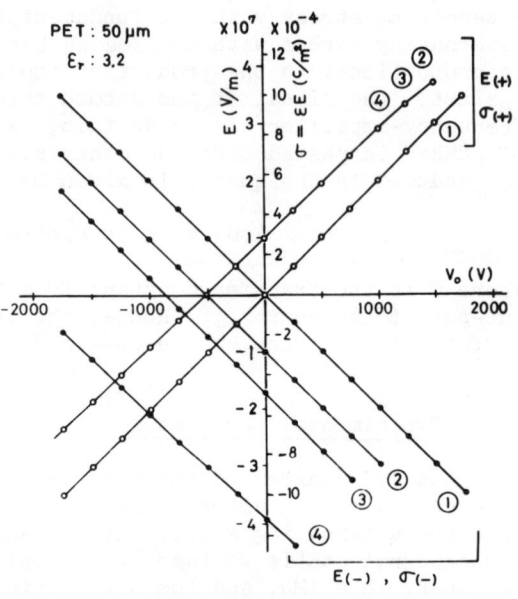

Fig.3   Dependency of electric field E and induced-charge density σ on external applied voltage V_o.

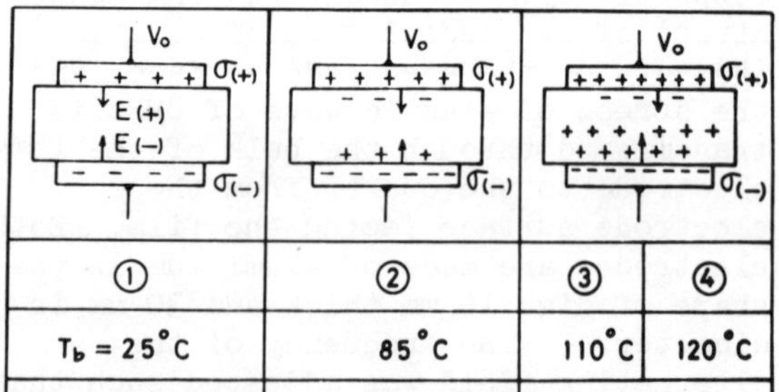

Fig.4   Experimental condition for results shown in Fig.3, and distribution of space charge in PET film.

If the external field is further increased continually after passing the null condition, the induced voltage again increases.

On the other hand, if the negative external voltage is applied to the electrode, the induced voltage starts at a certain value at $V_o=0$ and increases linearly, and the null condition can never be obtained. Hence, it can be justified that the polarity of the space charge is positive in case that the above described characteristics are obtained.

(2) In case of the negative space charge, the electric field at the electrode surface directs inward. By adjusting the negative external voltage, the null condition of induced voltage is obtained in the same manner as described earlier.

(3) If there is no space charge in the bulk of the film, no induced voltage is detected. By increasing the positive or negative external voltage, the induced voltage increases. The null condition in this case is obtained only at $V_o = 0$.

In Fig.3 shown are the induced electric field E and the induced charge density σ versus the applied voltage V_o.

The curve ① for the specimen prepared at 25°C exhibits a straight line passing through the origin. Consequently, there is no space charge in the film as shown in the case of ① in Fig.4. The curves ②, ③ and ④ which are obtained at the temperatures higher than 85°C are also linear but being away from the origin. From the curve ② which exhibits $E(+) = -E(-) = 1.0\times10^7$ [V/m] and $\sigma(+) = -\sigma(-) = 2.8\times10^{-4}$ [C/m^2] at $V_o = 0$, it is considered that two kinds of the space charge with equal quantity but opposite polarity exsist in the film near each electrode. However, the net space charge in this cace is zero. The model of the charge distribution for this condition is illustrated in Fig.4 - ②. For the curve ④ in Fig.3 which shows $E(+) = 1.0\times10^7$ [V/m] $\sigma(+) = 2.8\times10^{-4}$ [C/m^2] and $E(-) = -3.9\times10^7$ [V/m], $\sigma(-) = -11.2\times10^{-4}$[C/m^2] at $V_o=0$, it is also thought that the space charges with opposite polarity and with different quantities are present in the film. The model of the charge distribution for this case is shown in Fig.3 - ④. For specimen prepared at a higher temperature than 110°C, the quantity of positive charge is more than that of negative charge. Hence the net space charge of positive polarity of 2.8×10^{-7} [C] is obtained from equation (3).

It can be seen from Fig.3 that the negative space charge quantity saturates and reaches a certain value. The reason for this may be that the negative electronic carriers which were released from trap centers are neutralized with the induced positive charges at the electrode. On the other hand, the positive space charge consists of immovable parent-ions which were created by the electron release, and thus accumulates gradually at a higher temperature.

It is thought that these charge carriers stored in PET film are produced by Poole-Frenkel effect from the bulk of the PET film rather than by Schottky effect, and that the difference in quantity between positive and negative space charges is attributed to the asymmetric-ionic-polarization in the dielectric film.

SURFACE POTENTIAL OF SiO$_2$ DURING ION IMPLANTATION

Moto'o Nakano and Ryoiku Togei

IC Division, Fujitsu Limited

1015 Kamikodanaka, Nakahara-ku, Kawasaki, Japan

Introduction

An ion implantation is one of the most useful technique for producing semiconductor devices. But there are some problems associated with high ion current implantation (1). One of the most vital of them is the dielectric breakdown due to built-up charges (2) on an insulator film which is used as a mask for a selective ion implantation. This work reports on the characteristics of the charge transport and the suface potential of SiO$_2$ during ion implantation in order to obtain a better understanding of the break down phenomena.

Experiment

The measurement technique is illustrated in Fig. 1. The oxide used were grown on 4Ω-cm $\langle 100 \rangle$ silicon to thickness of 0.65μ and 1.0μ. In order to determine an implanted region of 2L in width, a surrounding electrode made of polycrystalline silicon was formed on the SiO$_2$, and was held at the ground potential during the ion implantation. A stripe shaped electrode for measuring potential and current was placed at the center of the implanted region. The boron ions were scanned perpendicular to the center electrode mostly at 30Hz in frequency in this work. There were small differences of results as varying the frequency from one to 900Hz. The current which corresponds to the flow along the SiO$_2$ surface (I_s) and that to the substrate through the SiO$_2$ (I_{sub}) were measured simultaneously.

Results and discussion

(a) Transport paths of ion charge

The SiO$_2$ is a typical insulator having an energy gap of about 9eV that do not give rise to any carriers at room temperature. Therefore the implanted charges will remain there without being neutralized and thus increase the surface potential of SiO$_2$ more and more. And it may finaly cause the dielectric break down. However, ions implanted into SiO$_2$ can have sufficient energy to be able to generate electron-hole pairs in the conduction band and the valence band respectively. Since these generated carriers exist at least during the ion implantation, the implanted charges into the SiO_2 could be transported by these carriers and this would decrease the potential of the SiO_2 than expected in the above case. Under such situation, charges would flow out by way of two paths, the one directly through to the silicon substrate and another along the surface of the SiO_2.

Figure 2 represents the ratio of I_s and I_{sub} as a function of the ion energy and the width of the implanted SiO_2 region at 0.65μ thick SiO_2 above 1×10^{15} ions cm^{-2}. When the energy is so low as the penetration depth of ion is shallow compared with the SiO_2 thickness, it is noted that ion charges flow out along the SiO_2 surface. But it is realized that I_{sub} increases rapidly as the depth of ion becomes near the SiO_2 thickness. The average depth of the ion is also shown in Fig. 2. These results suggest the presence of a conductive layer which is limited in the ion penetrated region on SiO_2 under scanned ion beams.

Figure 3 reveals the ratio of I_s and I_{sub} as a function of ion doses in the case of thick SiO_2 relative to the ion depth. It is noted that I_{sub} decreases abruptly with increasing ion doses, and I_s dominates the current flow. These results suggest that damages created by ion collisions have influenced the charge transport mechanism of SiO_2 during ion implantation. So the most of measurements were done under the condition that I_s was dominant at doses above 1×10^{15} ions cm^{-2}.

(b) V-I characteristics of surface conductive layer

Figure 4 shows the typical V-I characteristics using the measurement apparatus shown in Fig. 1. The current I_{sc} flow out from the center electrode was observed, which decreased linearly with increasing supplied voltage. This characteristic can be understood as follows. We will assume a constant carrier density on the SiO_2 surface at any time and position suggested by Fig. 2. Across sectional view of the sample and the potential distribution on SiO_2 are shown in Fig. 5. The charges implanted into SiO_2 flow to the center electrode or to the surrounding one as a surface current. In a steady state, $I(\chi)$ which is the current at a distance χ from the center electrode corresponds to the charges implanted into the region from χ_p to χ where χ_p is a position to give a peak value of the surface potential. This is expressed as

$$I(\chi) = \int_{\chi_p}^{\chi} K l \, d\chi' = K l (\chi - \chi_p) \qquad (1)$$

where K and l are the beam intensity and the beam diameter respectively. The gradient of the surface potential, $dV/d\chi$, at χ is required to let the $I(\chi)$ pass through the point, so we get a equation,

$$I(\chi) = en\mu(-dV/d\chi)S \qquad (2)$$

where e is the electron charge, n is the carrier density, μ is the mobility and S is the cross sectional area of the implanted region. Since the center electrode is supplied with V_a and the surrounding electrode is held at ground potential we get

$$V(0) = V_a \qquad (3)$$

and $\qquad V(L) = 0 \qquad (4)$

Combining Eqns (1), (2), (3) and (4), Eqn. (2) can be solved for $\mathcal{X}_p$,

$$\mathcal{X}_p = \frac{L}{2} - \frac{en\mu s}{K l L} \cdot V_a \qquad (5).$$

Because the current flow to the center electrode corresponds to charges implanted into the region, $-\mathcal{X}_p < \mathcal{X} < \mathcal{X}_p$, I_{sc} is give by

$$I_{sc} = K l L - \frac{2en\mu s}{L} V_a \qquad (6).$$

Equation (6) represents that I_{sc} must decrease as the bias voltage increases and is well in accordance with the experimental characteristics as seen in Fig. 4. Table 1 reveals values of the experimental I_{sc} when $V_a = 0$ in comparison with the theoretical values given by Eqn. (6). The differences of them are very small. These results are understood as a confirmation for the presence of the surface conductive layer on SiO_2 during ion implantation.

(c) Maximum surface potential of SiO_2

From Eqn. 6 we can get the maximum surface potential V_s of the SiO_2, spanned over a distance $2L$. The V_s is defined as the voltage of the center electrode under a condition that I_{sc} reduces to I_o which corresponds to a current directly injected into the center electrode having a certain width, 2mm in this experiment. The V_s is shown in Fig. 6 as a function of L, which has a relation

$$V_s \propto L^{1.8} \qquad (7)$$

On the other hand when $I_{sc} = 0$, Eqn. (6) leads to the relation,

$$V_s = \frac{Kl}{2en\mu s} L^2 \qquad (8)$$

Equation (8) represents that the surface potential increases proportionaly to the squre of L and is in a close agreement with the experimental relation given by Eqn. (7).

(d) Surface potential as a function of beam intensity

It can be resonably assumed that the carrier density n is propotional to the beam intensity K, and then the ratio K/n in Eqn. (8) should be a constant of K. It can be thus predicted that the surface potential is independent of the beam current. The experimental data are shown in Fig. 7. It is found that the surface potential keeps a constant value at the beam intensity above 2×10^{-7} A/cm^2, but below that value, the surface potential reduces slowly as decreasing the beam intensity.

(e) Charge transport model

In the foregoing arguments we have explained the experimental results assuming the presence of the counductive layer on the SiO_2 surface under scanned ion beams. Now we will discuss about a model of the charge transport mechanism. We assumed that the implanted ion creates electron-hole pairs. As it simultaneously makes the potential of SiO_2 positive, the electric field in the direction to the substrate arises. This field sweeps out holes from the surface to the substrate and attracts electrons to the surface as well that can neutralize the implanted positive charges. This flow of holes to the substrate should be measured as the substrate current I_{sub}, but it was observed only at the first step of the ion implantation and then decreased rapidly as shown in Fig. 3. It is understood that the generated holes are trapped immediatly at the deep traps created by atomic collisions before they can reach the undamaged SiO_2 region. In order to explain the surface current as noted in Figs 2 and 3, the trapped holes must be contributed for the charge transport. The hole transport in the SiO_2 has been reported by H. E. Boesch, Jr, et al.[3] and R. C. Hughes[4] who explain the experimental results on the basis of continuous-time random walk model (CTRW). This model treats the conduction in an insulator via phonon-assisted tunneling between localized states in the optical band gap of the insulator. According to the CTRW model, the surface current is considered to be a charge transport by the holes tunneling between states which are created by the ion collisions, and their energy levels are deeper than that of the undamaged SiO_2. Then holes trapped at these deep level states can transport to another same level states but hardly transport to the shallow level states in the undamaged SiO_2. This damaged layer is supposed to be the conductive layer during there are excess carriers in that layer and this assumption may be applicable to the argument of the V-I characteristics.

Conclusion

The surface current and potential of the thick SiO_2 during ion implantation were measured. These characteristics can be well discribed by assuming the presence of conductive layer on SiO_2. The conductive layer is considered as the damaged layer created by the ion collisions and has excess holes generated by the ion beam.

References

(1) T. Ikeda, Proc. of 6th Conference on Solid-State Devices, Tokyo 1974, p.311
    P. H. Rose et al, J. Vac. Sci. Technol., vol. 13, No. 5, 1030 (1976)
(2) Y. Wada and K. Sato, Japan J. Appl. Phys., vol. 15, 2289 (1976)
(3) H. E. Boesh, Jr., F. B. McLean, J. M. McGarrity, G. A. Ausman, Jr., IEEE Trans. Nucl. Sci, vol. NS-22, 2163 (1975)
(4) R. C. Hughes, Phys. Rev., vol. 15, No. 4, 2012 (1977)

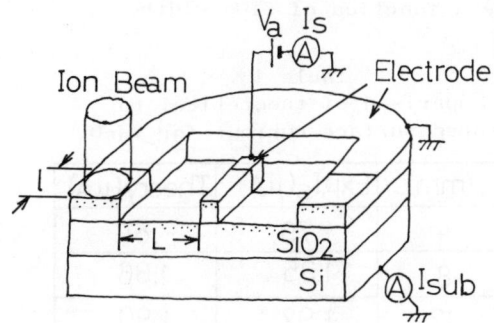

Fig.1 System for measuring surface potential and current.

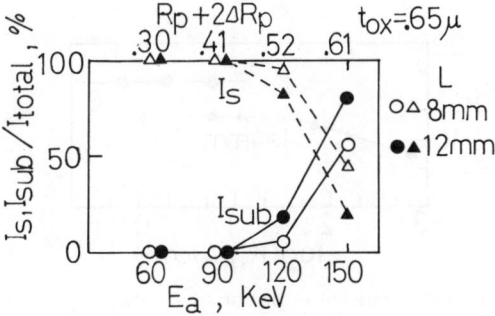

Fig.2 Ratio of I_s and I_{sub} as a function of ion energy.

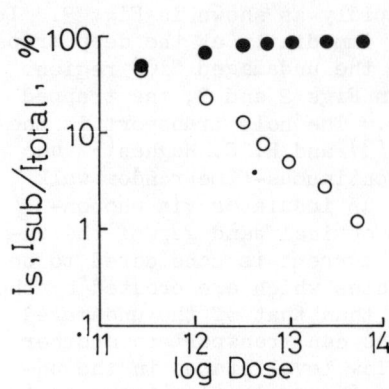

Fig.3 Ratio of I_s and I_{sub} as a function of ion doses.

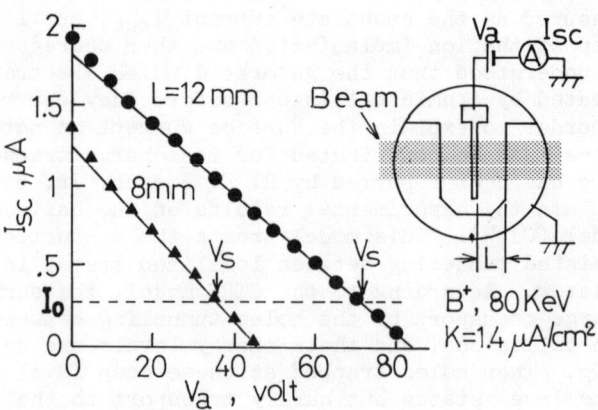

Fig.4 Typical surface current vs. applied voltage characteristics.

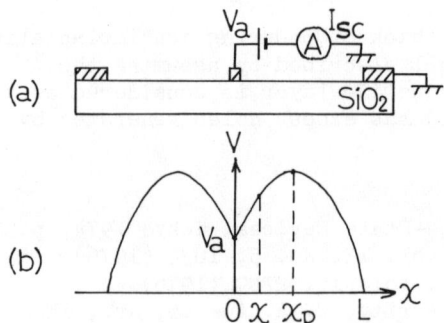

Fig.5 (a)Cross sectional view of sample (b)Potential distribution on SiO_2.

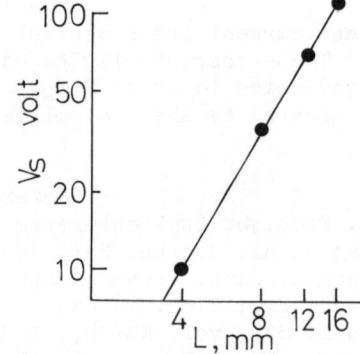

Fig.6 Surface potential as a function of SiO_2 width.

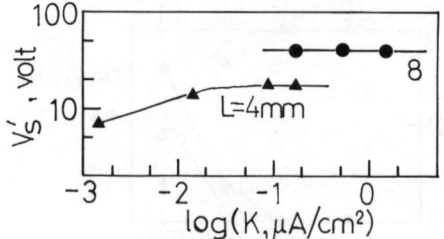

Fig.7 Surface potential vs. beam intensity.

Table 1
Comparison of theoretical to measured surface current for $V_a=0$.

L(mm)	Exp.I_s(μA)	Theory(μA)
4	.79	.81
8	1.35	1.36
12	1.92	1.90

TRIBOELECTRICAL PROPERTIES OF TONER AND CARRIER

Toshiaki Narusawa, Seiji Okada and Hirofumi Okuyama

FUJITSU LABORATORIES LTD., Kawasaki, Japan

Introduction
Tribo-charge has been discussed in reference to the development process for electrophotography, and is probably the most important factor in making high quality prints[1]. The amount of toner deposited is closely related to the charge-to-mass ratio of the toner (tp), which is the most practical parameter for describing the toner tribo-charge. The tp depends on toner particle size and chemical composition[2], and particle size distribution[3]. In this report, the relation between tp and geometrical factors such as specific surface areas of toner and carrier, and toner density on the carrier surface are discussed.

Experiment
Three sorts of positive type toner were prepared with the same chemical composition (epoxy resin, channel carbon and nigrosine dye). These toners have different values of specific surface area (St) which is defined by Eq.1, where λ is the average sphericity of toner particle. λ is considered to be 10.4 by assuming that the toner particles are composed of several polyhedrons { regular tetrahedron ($6\sqrt{6}$), regular hexahedron ($6\sqrt{3}$) and regular octahedron ($6\sqrt{3}$) } and sphere (6). R and f(R) are the toner particle size and distribution function in the form of numerical fractions. The values of St of three samples were determined to be 6.6×10^3, 5.5×10^3 and 3.9×10^3 cm^2/g and in deposited toner after development to be 6.4×10^3, 4.6×10^3 and 2.3×10^3 cm^2/g, respectively. Specific gravity (ρ) was 1.14.

$$St = \frac{\lambda \int R^2 f(R) dR}{\rho \int R^3 f(R) dR} \qquad (1)$$

For the carrier, two sorts of iron beads coated with a 1μm layer of styrene butadiene copolymer were used. These beads have a plate-like shape, and their specific surface areas were 1.21×10^2 and 6.3×10 cm^2/g, as determined by scanning electron micrographs.

Fig.1 showes scanning electron micrographs of the typical toner and carrier adopted in this study.

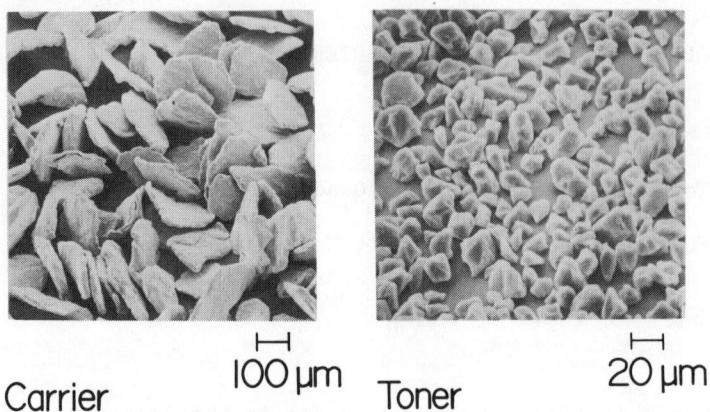

Fig.1 Scanning electron micrographs of samples

To study the dependence of toner tribo-charge on geometrical factors of the toner, the relation between tp and St has been examined. The tp was measured by using a capacitor and a blowing-off process[4].

Results and discussion

It has been shown that the tp was proportional to St for deposited toner as shown in Fig.2. This linear relation can be described as

$$tp = \sigma \cdot St \qquad (2)$$

where the proportional coefficient σ can be regarded as toner surface charge density, assuming that the toner charge exists only on its surface.

The relation between σ and toner amount adhering on a unit carrier surface area (Wt/ScWc) was also examined. Here, Wt and Wc are weight amounts of toner and carrier, respectively. In this examination, the Wt/Wc dependence of tp at a constant Sc (Fig.3) and the Sc dependence of tp at a constant Wt/Wc (Fig.4) were measured. Both results are in good agreement, as shown in Fig.5. Therefore, the σ can be given by

$$\sigma = K \left(\frac{Wt}{Sc \cdot Wc} \right)^N \qquad (3)$$

where K and N are constants.

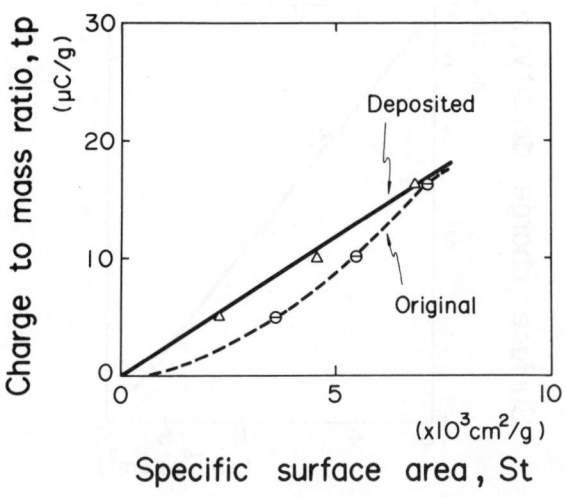
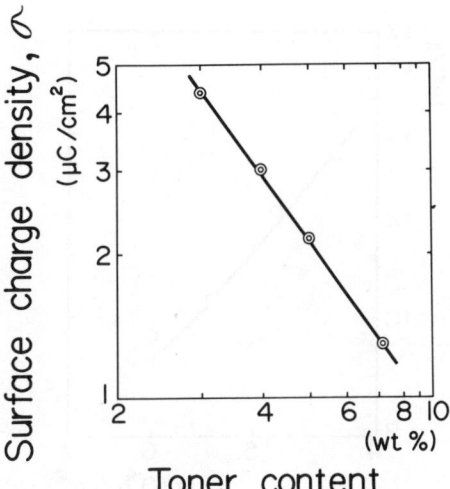

Fig.2 Charge-to-mass ratio as a function of the specific surface areas of original and deposited toners.

Fig.3 Surface charge density determined by Eq.2 on several developers of 3 to 7 wt% of toner content, at constant Sc of 1.04×10^2 cm^2/g.

Substituting Eqs.1 and 3 into Eq.2, we derive the Eq.4.

$$tp = \sigma \cdot St = K \left(\frac{Wt}{Sc \cdot Wc}\right)^N \times \frac{\lambda \int R^2 f(R) dR}{\rho \int R^3 f(R) dR} \quad (4)$$

Equation 4 showes that the tp can be estimated by the product of toner specific surface area calculated from toner particle size distribution and the N-th power of toner density on the carrier surface.

Conclusion

It is clear that the toner and carrier can be characterized by their specific surface areas. Also, the charge-to-mass ratio of toner can be controlled as a product of surface charge density and specific surface area. The surface charge density is calculated as a function of the toner density on the carrier surface. The specific surface area is calculated from the distribution function of toner particle diameter.

This evaluation method will contribute to adjustment of toner tribo-charge for various latent image conditions.

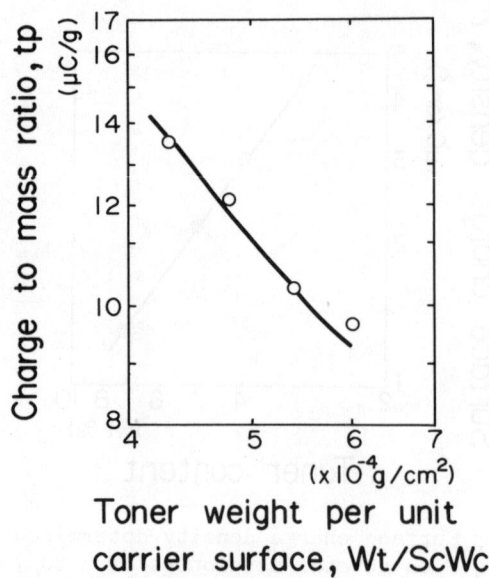
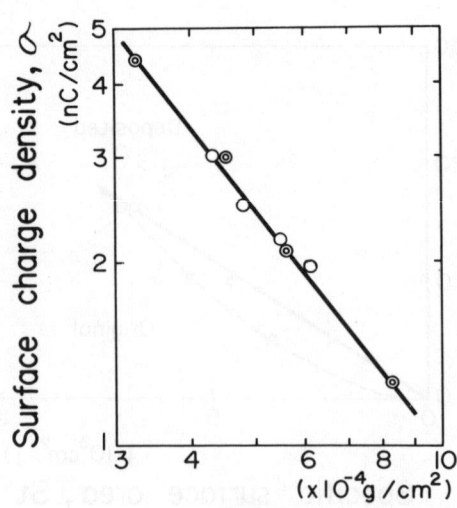

Fig.4 Charge-to-mass ratio as a function of toner weight per unit carrier surface area.

Fig.5 Toner surface charge density as a function of toner amount adhering on a unit carrier surface area.
⊚, measured by varying Wt/Wc.
o, measured by varying Sc.

Acknowledgement
    The authors wish to thank Dr.Z.Henmi and Mr.K.Murakawa for their encouragement and suggestions.

References
1) G.Harpavat; IEEE-IAS Conference Record, 128 (1975).
2) C.R.Raschke, The Conference Record of Second International Conference on Electrophotography, 104 (1975).
3) H.Okuyama and T.Narusawa; IEEE-IAS Conference Record, 443 (1977).
4) J.Nakajima, K.Aikawa, M.Horie and H.Takahashi; The Transactions of the IECE of Japan, J59-C3, 175 (1976).

A NEW IDEA FOR IMPROVED ELECTRICAL PRECIPITATOR (EP) DESIGN

K. Hashimoto, T. Adachi, M. Kawasaki* K. Ohtsuka**

Yamaguchi Univ., *Nishinippon Tech. Univ., **Hitachi Research Laboratory
Japan

1. INTRODUCTION

The electrical precipitation process is one of main corona-aided works, and has been employed widely in various fields. The purpose of these installations may be divided into two main parts as general classification. The first class example will be found as for improving of the plant efficiency, and these works belong to the category of profit production. The second case example may be found in public nuisance prevention. Notwithstanding marjor part of these works are profitless, it should be done in conformity with "Pollutant Pay Principle".

In most cases of these applications, it became more important problem to design the EP more compact. Otherwise it may become disproportional in comparison with the main installations. Because the main installations are becomingly scaled down with rapid strides. For example, in steam boiler design they adopt higher steam temperature and pressure or outdoor style construction and do it compact. While the condition of the flue gas remains unchanged and so bulky.
As for the effective ways to design EP more compact following items are under study. (1)Superinpose of electrical and mechanical forces in the precipitation field. (2)Construction of the passive electrode for suppressing the reentrainment of precipitated dust under improved gas velocity. (3) Improvement of the precipitation field intensity with the same voltage. (4) Improving the field intensity with an automatic control of the applied voltage. (5) Electrode rapping techniques for maintaining the corona discharge stable and vigorous.

2. MAIN FORCES AFFECT ON THE PARTICLE MOTION IN AN EP.

In Fig-1, the main forces under respective condition are as follows.
Range - ① $E < E_c$: Applied Voltage E is less than the corona starting voltage E_c. The all of the affairs arise in this range are of electrostatics. (a) Gradient force $F_G = K_G \varepsilon D_p^2 (dG^2/dx)$ grows rapidly as E increased. Thereby the particles

suspending in the field begin to settle to
the active electrode surface by F_G. (b) In
other hand, balling of dusts due to the F_G
on the particle surface grows too, and co-
agulation of dusts proceeds. Thereby, falling motion $V_g=(\pi D_p^2 g)/(18\mu)$ of coagulated
dust ball due to the gravitational force F_g
will accelerate. Because, the effect of increase of apparent size D_p is more superior
to that of decrease of apparent density γ.

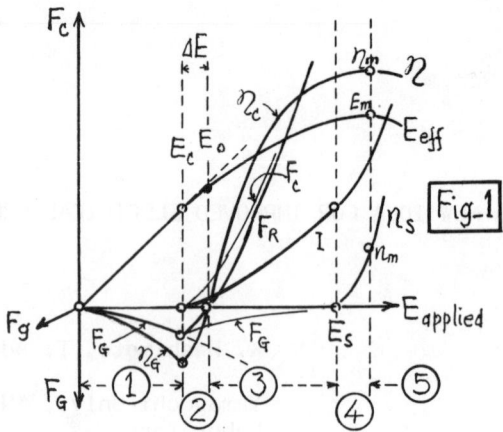

Range- ② $E=E_c$; As E reaches to Ec, corona
discharge comes in appear, and then the discharge current I begins to flush with an
epoch. (a) While F_G begins to decay because
of increasing of the space charge generated
by corona discharge, which makes the field
gradient uniform. Namely, the effect of
the field intensity dG^2/dx falls rapidly.
(c) Then, in the other hand, the Coulomb force Fc begins to grow rapidly.
  Range- ②/③: $E=E_o$; When E reaches Eo, that is, slightly over Ec, F_G and Fc
become to have the equal magnitude and opposite sense. Therefore, as the result,
the electrical force becomes none. That is, $\vec{F}_G + \vec{F}_c = 0$.
  Range- ③ $F_c \gg F_G, F_g$: Fc grows far strong compared with F_G or $F_{\hat{g}}$. Consequently, Fc
becomes the main force, and dust particles begin to settle to the passive e-
lectrode. Hereby, the normal electrical precipitation performance proceed effectively. (b) While, as for the coagulated ball of dusts F_g is far strong than
the Fc. Therefore, considerable amount of coagulated ball of dust fall out of the
gas stream before they arrive at the passive electrode(5).
  Range- ③/④ $E=E_s$: As E rises up to the sparking voltage Es, frequent flashing
come to appear bet. the both electrodes. However, as a sparc flashes, the dusts
deposited on the passive electrode jump into the gas stream in a moment and form
a heavy dust cloud. Therewith the spark is extinguished immediately, not extending to an arc discharge. Therefore, even under such condition, notwithstanding the
effective voltage Eeff continues to increase as E increase. While the inclination
of the I-Eeff performance curve increase with an epoch. Thereafter, the relation
$I=A(E_{eff})^\beta$ does not hold.
  Range- ④ $E>E_s$: Notwithstanding the flashing grows more violent as E increase,
$E_{eff}=[(\int^T e^2 dt)/T]^{\frac{1}{2}}$ continues to increase and the collection efficiency
improves too.
  Range- ④/⑤ Hereby, the spark counts Ns exceeds the optimum value Nm, and
then Eeff begins to decrease. Consequently, η begins to fall too. That is
said, Em or η_m does correspond to the optimum condition.
  Range- ⑤ Normal operation of EP becomes difficult because of heavy flashes.

3. PROCESS OF IMPROVING THE COLLECTION EFFICIENCY

3.1 Collection Efficiency

As it is well known, Dr. Deutsch and others had gave an equation[1][2] for η of
an EP as "exponential law" under such assumption that the suspending fine parti-

cles always distribute uniformly in the precipitation field, notwithstanding dust settling to the passive electrode proceeds. That is, $\eta = 1 - \exp(-K_1 V_d t)$, wr., $V_d = (\varepsilon D_p^2 G^2)/(12\pi\mu)$ is the migration velocity of a dust particle, and t is the treatment duration in the electrical field. Substituting V_d with the corresponding field intensity G, we will get the equation $\eta = 1 - \exp(-K_2 G^2 t)$. K_1 or K_2 is said a precipitation constant wich depend upon the properties and states of the aerosols to be treated, and also on the applied voltage. As for a specified aerosols, K_1 or K_2 may be considered as a constant practically under sufficient field intensity and ample amount of charge density.

This formula is essentially useful with good agreement for estimation of the η of an EP that is servicing under different working condition. Where D_p in the equation should refers to that of coagulate state, not of the elementary(5).

As for such electrical field that E_{eff} is high enough and sufficient amount of charge being supplied rough approximation $G_s = E_{eff}/S$ is possible. Where S is the spacing bet. the active and passive electrodes. Then the equation may be reduced as,

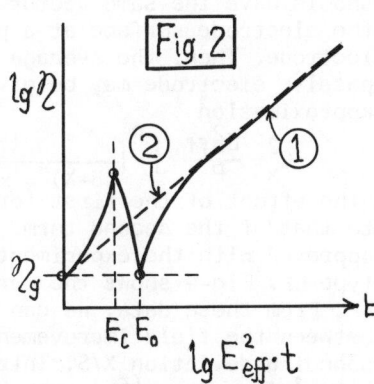

Fig.2

$1 - \eta = \exp(-K_3 E_{eff}^2 t)$. Hence, these are lineally porportional amounts on a semi-log graph. Consequently, under such condition that G or E_{eff} being held at a sufficient intensity constantly, while t or the gas loading varies, the performance curve of the EP may be shown as curve ② in Fig-2. While in the case of t is held constant and E varies, the performance curve may be shown as curve ①.

3.2 Means of improving the collection efficiency

Now, we can think of that there are two effective ways for improving η. The one is to take the value of t large, and the other is to keep G or E_{eff} higher and stable. The former will not be desirable. Because, that will not economical and will result disproportionally bulky set as has been mentioned previously. Moreover, it may be expected from above equation that, increasing G by 5% will give actually the same result with increasing t by more than 10%.

Now, the later may be yielded by two different processes. The one is to keep the E_{eff} about the optimum value Em controlling it automatically. In this case Em may be detected easily on a corona-voltmeter. This process has been fairly well improved. The other one is to develop some new idea about field construction and improve the field intensity with the same E_{eff}. That is the way we fixed our eyes upon and are to describe hereof.

3.3 Consideration of the electrical field construction and the field intensity therewith

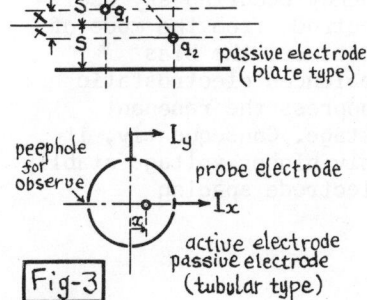

Fig-3

Now, let us consider the electrical field constructed with parallel plates and wires or a tube and a wire arranged as shown in Fig-3. The case of no deviation, that is X=0 corresponds to the ordinary design type emploied hetherto.

In Fig-3, it may be considered, as the lines of force from the charges q_1 and q_2 on the active electrodes should have the same vector sense perpendicular to the electrode surface at a point P on the passive electrode. Then, the average field intensity G_x on the passive electrode may be given as follows with some approximation

$$G_x^2 = \frac{E_{eff}^2}{P} \int^P \left\{ \frac{1}{(S-X)^2 + x^2} + \frac{1}{(S+X)^2 + (p-x)^2} \right\} dx$$

The effect of the first term on G_x is more superior to that of the second term. This fact has been approved with the experiment(3) carried on the plate type EP. Fig-4 shows the performance of a tube type EP. From these data, we can derive the relation between the field improvement ratio G_x/G_{sx} and corresponding deviation X/S, introducing a thought of $I \propto G^2$ after Perry(6), as shown in Fig-5. Here, we are to discriminate G_{xs} (X=0) against G_s, because they are not agree always.

Moreover, as for the electrode arrangement used up to this time, that corresponds to the case of X=0 in Fig-3 or Fig-4, the wire electrode is held in the nutral position with balanced tension due to the electrostatic force. Therefore, the rise and fall of unstable corona dischrge, that say, change of accessory ionic wind will readily excite the resonant oscillation of the wire electrode, and then periodic flashover will continue so far as the applied voltage exist on. Arising of such happening may be readily expected from the Lissajous' figure or the mode of Ix and Iy on their oscillogram in Fig-3. Consequently, under such condition, keeping of stable supply of sufficient voltage will become difficult soon after EP start.

While in the case of deviated field construction, that is X>0, for example X/S = 38%, Ix increase about fourfold and we would not feel an expectation of tendency occuring any oscillation of the wire electrode from the mode of change of Ix and Iy. Because, the bias tension due to the unbalanced electrostatic force contribute to suppress the resonant oscillation in early stage. Consequently, it makes possible to supply higher voltage stablly notwithstanding the electrode spacing reduced.

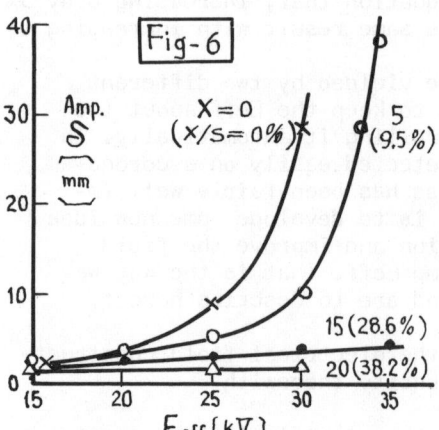

4. Experimental approval of effectiveness of the newly developed design

The experiment has been carried, keeping the gas loading (t) constant, Eeff vary.

As the result, we have gotten a lot of η-value for respective electrode deviation as shown in Fig-7. Rearranging these data we would have Fig-8.

Through the observation about these data we will see two remarkable facts. The first, the existence of such a condition that the electrical force results none at some applied voltage E_o, refering to Fig-7. The second, deviated field construction is very effective for improving the collection efficiency under the same Eeff, refering to Fig-8.

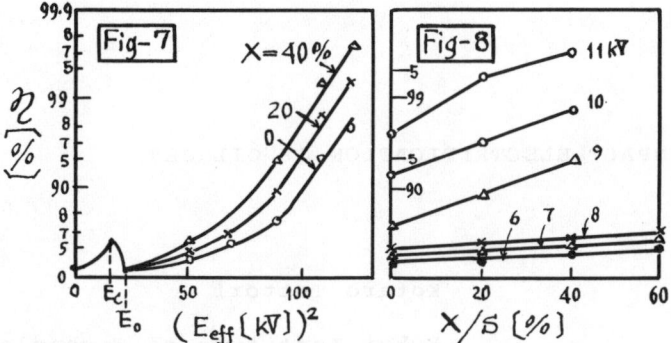

5. Summary

The affairs described hitherto may be summarized as follows.
(1) In general, the characteristics of the electrical discharge through gas dielectrics may be expressed as $I=A(Eeff)^\beta$. Where β is a const. depending upon the mode of the electrical discharge. As the reversal, the discharge mode may be presumed from the value of the constant in the performance characteristics.
(2) The electrical precipitation will proceed with F_G up to E_c, and then, passing such condition that no electrical precipitation happening, break into the normal performance with F_c. And then, passing the optimum condition the performance character falls away with heavy sparking.
(3) The collection efficiency of an EP servicing with normal performance may be improved by increasing the porduct of t and G^2. The way to increase G is far sperior to increase t as design technique.
(4) There are two ways for improving the field intensity G of an EP. The one is to keep Eeff as high as possible with automatic control. The other is to produce more high field intensity under the same Eeff, applying the newly developed idea for field construction technique.
(5) The effectiveness of the newly developed idea has been approved with a series of experiments employing some model EP.

Reference

(1) S: Fukuda: JIEE, 1 (1930) (2) W. Deutsch : Ann. d. Phys., 335 (1922)
(3) T. Adachi, M. Kawasaki, Others: JNIT, 4 (1974) 111
(4) K. Hashimoto, T. Adachi: Japan-Patd., 49-64071(1949)
(5) Ditto: Electrostatics and its Ind. Applications, Tokyo El.Mach.Univ.Pub.(1973)72
(6) J.H. Perry: Chem.Engr's HB, 3rd Ed., McGrowHill Pub.(1950)1040

IMPACT ELECTRIFICATION OF OIL JET

Kotaro Tottori

Fukui Institute of Technology

3-618, Gakuen, Fukui, Japan

1. Introduction

In January 1975, a 10,000 ℓ tank lorry for naphtha was burnt after the elapse of 60 seconds from the start of oil-discharging near the Japanese National Railways Fukui Station as shown in Fig.1. The fire might have been ignited through the driver's mistake by the static electrification of splashing oil flowing out from a 3 inch diameter pipe at the rear of the tank-lorry.

To ascertain the cause of this electrification, the following experimental measurements were performed at the request of the Identification Division of Police Head Quarters, Fukui Prefecture.

Fig. 1 The scene of the accident to the tank lorry and the ensuing fire showing the initial fire-fighting.

2. Experimental details

2.1. The first measurement

The first photographs shown in Figs.2 and 3 show the aspect of a water colliding state during the elapse of 60 seconds at the rear of the same type of the tank-lorry. Fig.2 shows the state of the water colliding during the first 30 seconds after the opening of the base valve of the lorry, and Fig.3 shows the state in the following 30 seconds after operating the feeding pump located at the receiving site.

ELECTROSTATICS

Fig. 2 The state of the water colliding for the first 30 sec. after opening the base valve.

Fig. 3 The state of the water colliding during the 30 sec. after the operation of the feeding pump.

To ascertain this electrification of an oil jet, a pattern based on a well known electric double layer could be assumed as shown in Fig.4.

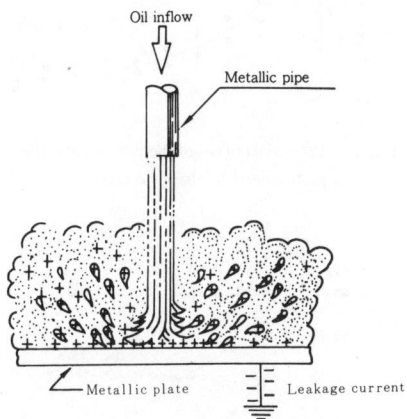

Fig. 4 Impact electrification pattern of oil jet

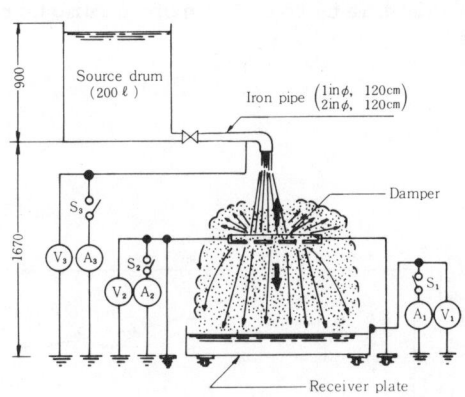

Fig. 5 Schema of the electrical measurement in the experimental set.

The experimental set for this assumption was made to reproduce the condition as close as possible. Fig.5 shows the schema of the electrical measurement in this set. For the present scale of this experiment, light oil was used instead of naphtha for safety and two pipes of 1 inch and 2 inch diameters were used instead of 3 inch diameter pipe.

Further, the static capacities of the damper and the flat receiver plate were measured as 100 and 185 pF respectively.

In Fig.6, charging rates of the damper and the receiver plate were indicated as potential curves measured simultaneously, as static energy curves which could be calculated.

Thus, the potentials of the damper and the receiver plate at

60 seconds were -1kV and +0.4kV respectively. Then, the impact electrification pattern of the oil jet could be clearly indicated.

Fig.7 shows the potential and the earth current curves measured simultaneously in the damper and the outflowing pipe. The damper was isolated from earth and was charged up to 3kV after 60 seconds of oil collision, while the earth current from the damper flowed continuously during oil collision and varied from 10mμA to 8 mμA. It seems that the oil jet generates a constant current.

Fig.8 shows the comparison of the damper potentials for outflow from 1 inch and 2 inch diameter pipes. The damper potential during the first 60 seconds for 2 inch diameter pipe was about twice as high as that for 1 inch diameter pipe.

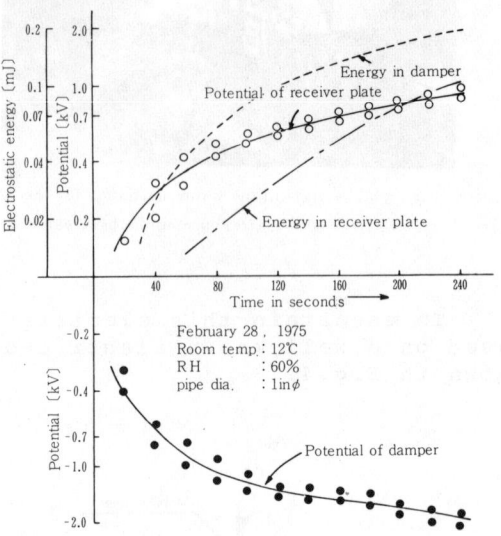

Fig. 6 Potential curves of the damper and the receiver plate (average charging rate)

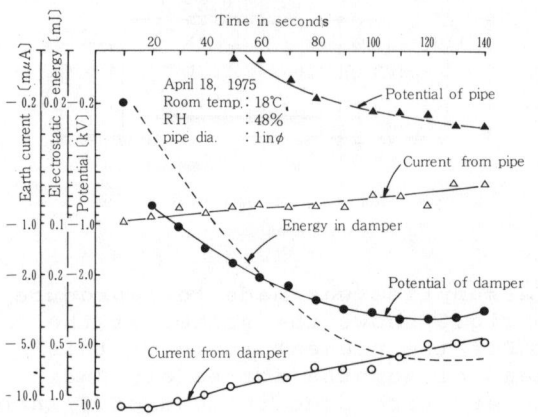

Fig. 7 Potential and earth current curves of the damper and the outflowing pipe (max. charging rate)

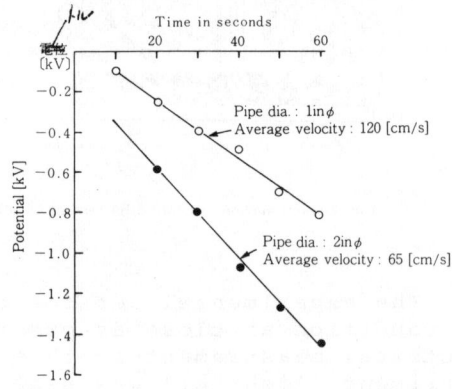

Fig. 8 Comparison of the damper potentials between 1 in and 2 in diameter pipes for oil flowing out.

2.2. The second measurement

Furthermore, charging rates should be confirmed for various oils refined by the same refinery. Fig.9 shows the comparison test-

2.2. The second measurement

Furthermore charging rates should be confirmed for various other oils refined by the same refinery. Fig.9 shows the comparison testing set used.

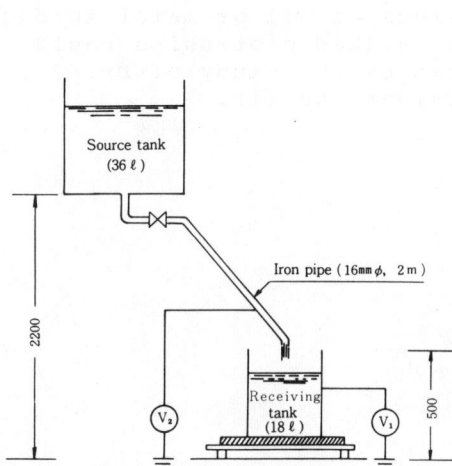

Fig. 9   Comparison teasting set for charging rate.

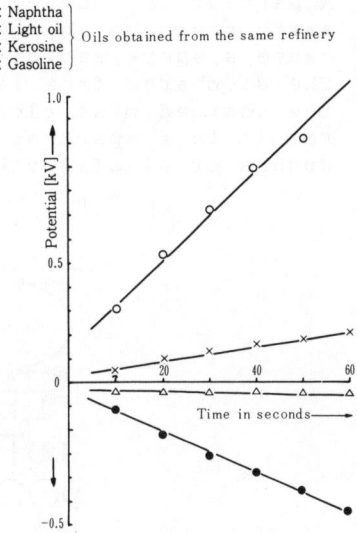

Fig. 10   Charging rates for different kinds of oil.

Fig.10 shows the charging rates for various oils measured by this testing set. The charging rate during the first 60 seconds for naphtha was about five times higher than that for light oil.

2.3. The third measurement

Moreover, leakage resistances and static capacities were measured at each point and part at the rear of the same type tank-lorries, presently driven, as shown in Fig.11.

The maximum values of the leakage resistance was $3 \times 10^{10} \Omega$, and that of static capacity was 142 pF respectively.

3. Conclusion

From the extrapolation of the above results, the potential of the damper for naphtha from a pipe of 2 inch diameter in 60 seconds would be as follows:
 [1] 20 kV for an insulated damper, and
 [2] 3 kV for a high insulation resistance and a high static capacity.

[3] The static energy in the latter case would be 0.63 mJ.

Now, the tank-lorry in the accident mentioned has a 3 inch diameter pipe as the naphtha feed port, and the circumference of the damper was filled with inflammable gas vaporized from naphtha.

From such conditions, the ignition source in the tank-lorry could be concluded to be as follows:

[1] A part of the damper having a high leakage resistance to earth could be charged up to a high voltage which would cause a spark and/or

[2] The discharge from isolated slugs of oil or metal through the charged mist clouds to an earthed protrusion could result in a spark as determined by the study of Dr. J.F. Hughes et al of Southampton University (1).

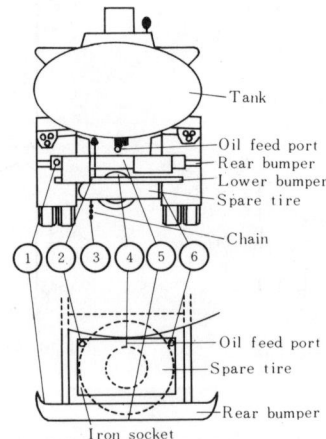

Fig. 11 Measuring points and parts at the rear of the tank lorry.

Reference

1) J.F. Hughes, A.W. Bright, B. Makin and J.F. Parker, "A Study of electrical discharges in a charged water aerosol" J. Phys, D: Appl. Phys., Vol. 6, 1973.

BEHAVIOUR OF FLUIDS UNDER ELECTRIC FIELD

— PROMOTION OF COMBUSTION, VAPORIZATION, HEAT TRANSFER —

Yukichi Asakawa

ex-Prof. Nihon University, Asakawa Laboratory

5-6-9 Kasugacho, Nerima-ku, Tokyo

1. Increase of Combustion Velocity

The author tried to put an orifice of the Bunsen burner under a dc source as shown in Fig.1; flowing out of town gas was observed to increase[1]. Measurements with a Chattock micromanometer indicated an increase of "negative water head" of maximum value of 0.60-0.70mm under an electric field of dc 70-80kV. This increase of gas flow was similar to an increase of flowing down of oils from a narrow nozzle, 1.0mm in diameter when placed in an electric field in the same way to Fig.1[2]; namely, an excited flowing down stream of diesel oil is widened cyclically corresponding to the cycle of ac source as shown in (b) of Fig.2, while spontaneous flowing down stream goes down in a straight line as shown in (a) in the same figure. Yet, the length of flowing out flame diminished as shown in Fig.3 as the intensity of electric field became larger[3], especially in the reducing flame portion. The shortening of the flame may be regarded as an indication of "time reduction" in the phenomenon of combustion: then, this is nothing but an indication of promotion of combustion velocity. The diminish of the reducing flame means that we can expect more complete combustion.

Then combustion experiments about liquid fuels were carried out; fuels were held in an open container, one of the terminals of the secondary

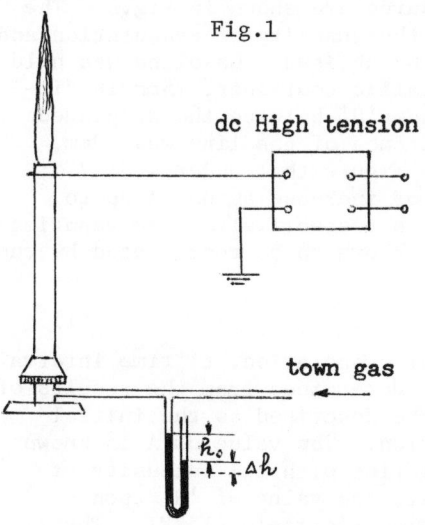
Fig.1

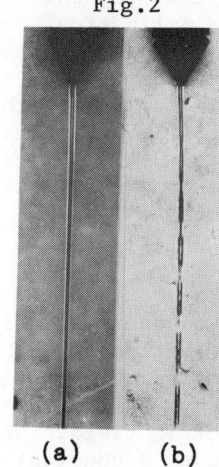
Fig.2
(a)   (b)

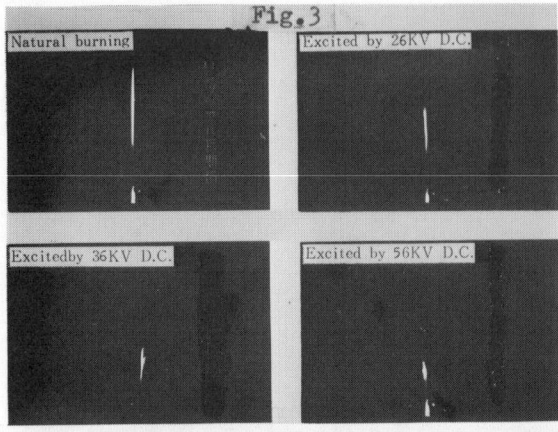

Fig.3

Fig.4

coil of high tension was connected to the electrode which was suspended above the surface of gasoline and the other electrode was grounded at the container's bottom. It was observed that the burning gasoline flame was simultaneously lengthened and widened when placed under an electric field as shown in Fig.4[4]. Under 7-10kV, a time required for burning 6cc gasoline was 11 seconds instead of 62 seconds, the time at zero field, as shown in Fig.5. Similar results about other fuels are also shown in the same figure.

2. Increase of Vaporization Velocity

Owing to the widely accepted idea that combustion contains successive vaporization, the evaporation of liquid fuels and liquid media was experimented[5]. The author here dealt with gasoline and distilled water.

2.1 Gasoline   Quantity of vaporization at specific time intervals was measured by a chemical balance; so-called vaporization-time curves under various intensities of ac source are shown in Fig.6. The ordinate represents the quantity of evaporation and the time is denoted by absissa. Gasoline was held in a cylindrical metallic container, 75mm in diameter and the distance 'H' between the suspended electrode and the surface of gasoline was 30mm. It is known from the figure that under 4.5; 7.5; 10.5; 15kV the rate of increase amounted up to 4.0; 8.0; 11; 12 times respectively. The vaporization-time curves are known to be represented by the formula[6];

$$Q = A\, t^n \qquad (1)$$

where, Q: quantity of evaporation, t: time interval, A: constant which is determined from the results of observation and can be described as the initial velocity of evaporation. The value of A is known to have a linear relation with the intensity of electric field. Next, the value of 'H' upon evaporation rate was determined under a constant electric field (15kV). The

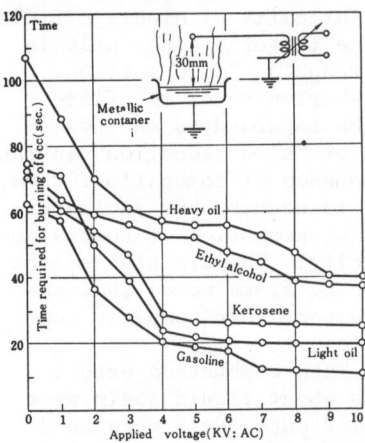

Fig.5
Relation between burning time and applied voltage of various liquid fuels.

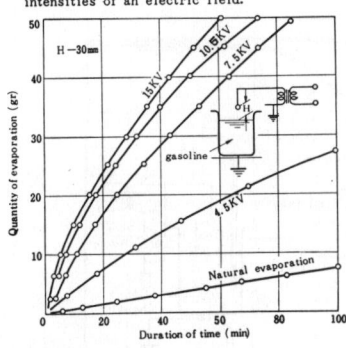

Fig.6
Vaporization curves of gasoline under various intensities of an electric field.

results thereof showed that 'H' has a direct bearing upon the rate of evaporation; as the distance became larger, the rate of evaporation smaller, as anticipated referring to the distribution map of intensity of electric field. In case 'H' was 105mm, the rate of evaporation, however, was still as large as two times

2.2 Distilled water below boiling point   Vaporization-time curves obtained at various temperatures substantiate the above equation (1). It is also ascertained that increase under 15kV excitation really reaches 50 times the normal rate when 'H' being 10mm, but it descends logarithmically as the temperature rises and at 90°C the rate is 1.8 times as shown in Fig.7. Not only distilled water, but sea water behaved just in the same manner. This may open up a new promising application for freshening sea water. About other liquids and camphor (sublimation) too, similar results as above were obtained.

2.3 Comment upon promotion of vaporization   The author tried to plot the value of $\log_e A$ (A: constant in Equation (1)) against the reciprocal of the absolute temperature (1/T) about water. The result is shown in Fig.8. Here, it must be mentioned that there exists a linear relation between $\log_e A$ and 1/T both in the ordinary and in the excited (15kV) vaporizations; the ordinary vaporization is indicated by the line M and the excited vaporization by the line N. These linear relations can be written to take the formula[7]

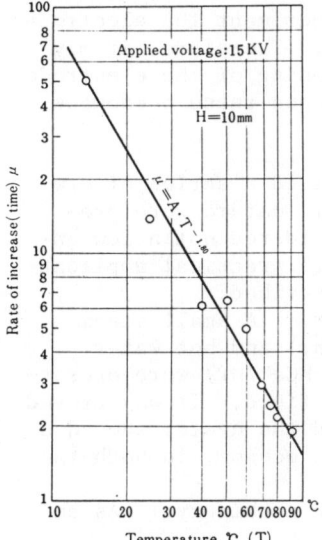

Fig.7
Relation between the rate of increase of vaporization and temperature about water.

$$A = A_o \exp[(-m)\,1/T] \tag{2}$$

If the inclination (m) be taken as $m = \Delta E/R$ (ΔE: activation energy, R: gas constant), it can be seen that the equation just obtained is the same with the Arrhenius formula. Then, there ixists a similarity between the behaviour of water vaporization and that of chemical reaction. It is to be noted however that the value of inclination of the line N (in Fig.8) is less than that of the line M. According to the theory of reaction velocity, the decrease of the value of inclination $\Delta E/R$ corresponds to the effect of catalyser in a chemical reaction[8].

3. Increase or Decrease of Heat Transfer

3.1 The author came across[8] the following observation as shown in Fig.9. We know that the electro-excited (15kV) heating curve (b) rises decisively more than the ordinary heating curve (a): this is nothing but a proof of promotion of heat transfer about water and encourages us to come to such an interpretation that the wall of the metallic container besides the air just situated between the wall of the container and heater when placed under electric field may have transferred much more heat energy per unit time. Then, experiments upon metal and air were

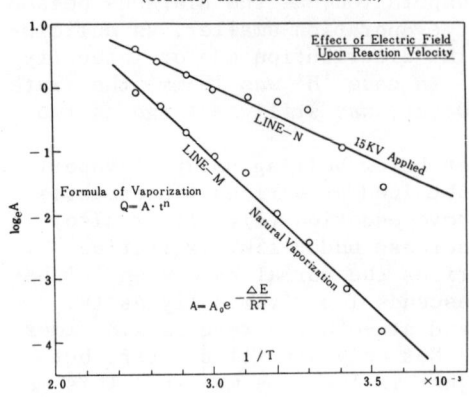

Fig.8 Behaviours of vaporization (ordinary and excited condition)

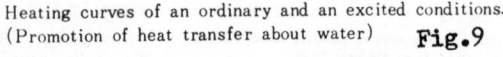
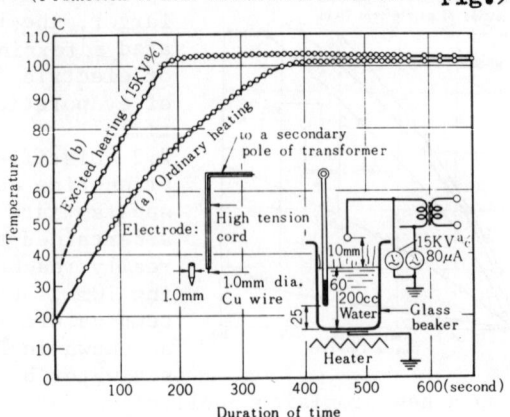

Fig.9 Heating curves of an ordinary and an excited conditions. (Promotion of heat transfer about water)

carried out respectively and we got affirmative results therefrom; typical ones of them are introduced in Fig.10 and Fig.11 which give us a clear idea about promotion of heat transfer about iron pipe and air respectively[10]. Lastly added to be noted, Fig.12 shows two continuous heating and cooling curves about an iron piece which was placed under ordinary and excited states respectively. We can see that in the heating zone the excited heating curve (b) rises more rapidly than the ordinary heating curve (a), but in the cooling zone the excited cooling curve descends more slowly than the ordinary cooling curve; namely, in the heating zone promotion of heat transfer occurs under electric field while in the cooling zone retardation of heat transfer really occurs; in fact after 200 minutes the iron piece attained 155°C instead of 78°C as the effect of excitation. Here, it must be noted that in this experiment the electrode was held inside the body. Therefore, it may be concluded that positioning of the electrode has an important effect upon these phenomena.

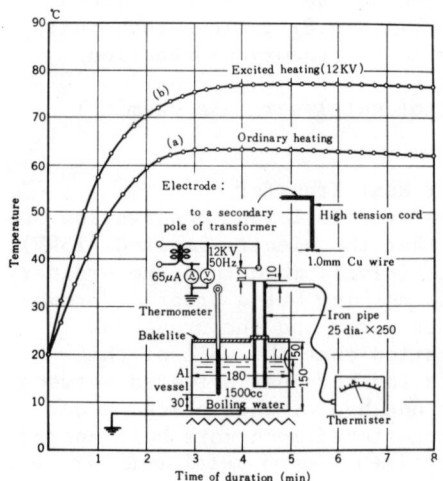

Fig.10 Heating curves about an iron pipe. (Promotion of heat transfer about an iron pipe)

4. Conclusion

Under an electric field a series of new phenomena of thermodynamic nature were presented; there may be considered not a few applications therefrom. Here practical applications will be briefly described.

4.1 Application to boiler

A small steam boiler (capacity: 35 Kg/hr) and hot water heater (capacity: 20,000 KCal/hr) were operated under a 15kV electric field. It was proved that the steam boiler and the heater showed a gain of about 17% of fuel economy in both the cases.

4.2 Application to drying machinery

As an

Fig.11

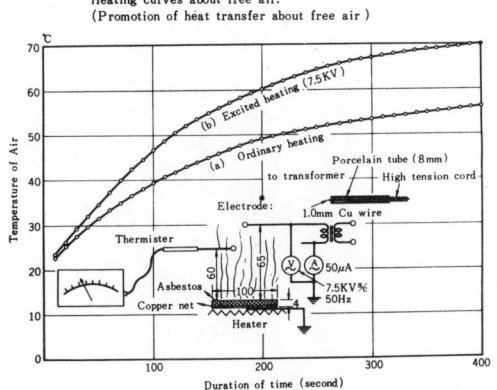

Heating curves about free air.
(Promotion of heat transfer about free air)

Fig.12 Heating and cooling curves of an iron piece.

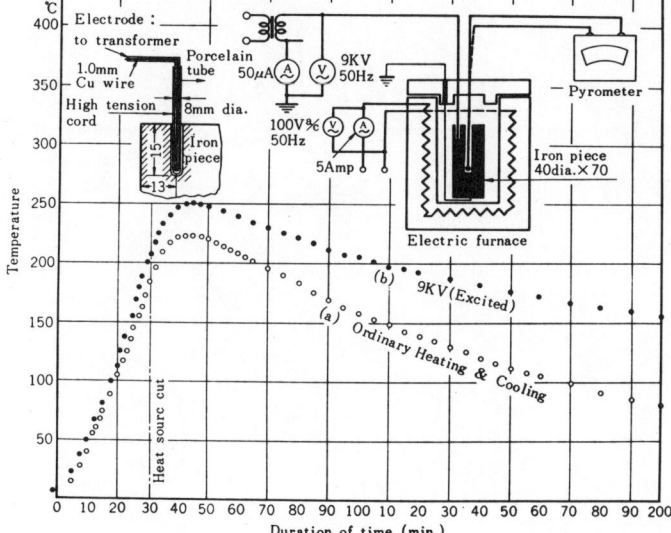

example, a rice-cake drying machine (80Kg/one charge) was operated under 15kV electric field. The shortening of time was 25%.

References
(1) Y. Asakawa: new study of improvement of flowing out of a fluid (2nd report). Japan Society of Mechanical Engineers (JSME). (39)No.225 p.10, 1936 (in Japanese)
(2) Y. Asakawa: First report of (1). "Kogyo Zashi" No.866 p.422-424, 1934 (in Japanese)
(3) The same to Ref.1
(4) Y. Asakawa et al: JSME 1964.4 General meeting. Advanced Print No.110 (in Japanese)
(5) Y. Asakawa: JSME 1965.5 General meeting. Advanced Print No.135 (in Japanese)
(6) The same to Ref.5
(7) Y. Asakawa et al: JSME 1966 General meeting. Advanced Print No.150 (in Japanese)
(8) Y. Asakawa: JSME 1967 Semi-International Symposium 1967.9.4-8, Tokyo, Promotion of Vaporization by Means of Application of Electric Field. Papers p.95-102
(9) Y. Asakawa et al: 7th Japan Symposium of Heat Transfer 1970.5 I-12.2 p.257-260 (in Japanese)
(10) Y. Asakawa: Nature Vol. 261, No.5557 p.220-221 1976.3.20

II
ELECTRETS

II
ELECTRETS

BIOELECTRETS AND BOUND WATER IN BIOLOGICAL MOLECULES

Sergio Mascarenhas, S. Quezado and S. Celaschi

Institute of Physics and Chemistry, CP 369, S. Carlos, SP
13560

I - INTRODUCTION

Electrical properties of biological molecules may be fundamental for their role in life processes. On the other hand the interaction of bound water with biological molecules is also essential to characterize their conformational and electrical properties. Thus, the study of electrical properties of the fundamental molecules of life becomes essentially connected with the bound water problem. Among the many electrical properties that are important, dielectric constants, conductivity, piezoelectricity and the electret state have been studied. Fukada and coll. have pioneered the investigations of piezoelectric properties of biopolymers (1) and also studied their properties in different degrees of hydration, thereby calling attention to the importance of this parameter for the understanding of their piezoelectric behaviour. The electret state was found to be a general property of most biopolymers (2) and the term BIOELECTRET was proposed to indicate the biological and biophysical importance of the concept (3). Bioelectrets have been proposed to play an important role in membranes, enzyme action and other biological systems (4). The role of bound water was found to be fundamental for the characterization of the bioelectret properties. This is due to several causes: bound-water interacts with the sources for polarization storage in the macromolecule. These interactions may be connected with the polyelectrolyte behavior of the macromolecule and/or conformational changes induced by hydration. Bound water dipoles contribute strongly to the polarization as has been shown to be the case for all biopolymers so far studied such as proteins, polysacharides and polynucleotides. It can also be shown that non-linear effects in transport and polarization storage are strongly dependent on hydration. The non-linear bioelectret may simulate ferroelectric hysteresis curves. On the other hand dielectric and piezoelectric measurements which are usually made with "harmonic" types of instruments, pre-supposing linearity should be carefully checked, since non-linearity specially at very low frequencies may set in at very low fields depending on hydration.

In this paper we discuss some of these aspects as related to our results with lysozyme(5) and DNA and RNA(6).

II - THERMAL STIMULATED PRESSURE(TSP) AND HYDRATION CONTROLLED ELECTRET INVESTIGATIONS

In view of what was discussed above we have introduced(5) TSP as an auxiliary technique in electret investigations along with the usual thermal stimulated depolarization(TSD) and isothermal thermal decay(ITD). For the TSP the rate of water desorption is followed as a function of temperature and the experimental details are described elsewhere(5). This is a very simple technique which may be very usefull for investigations not only with biological molecules but with any material capable of hydration and desorption. Our results for lysozyme indicate that:
a) water dipoles, at low hydration levels (up to 40 mg/water per gr of protein) are bound in two different sites which can be characterized by their relaxation times and activation energies.
b) bound water dipoles are responsible for most of the polarization storage of the bioelectret
c) for the first time a new electrical effect was found due to the desorption of previously oriented water dipoles.

It was possible to investigate the latter effect by the simulataneous application of TSP and TSD.
d) By the concurrent use of gravimetry and thermal gravimetric analysis (TGA) it was possible to conclude that TSP and TSD are usefull techniques for hydration studies even at very low hydrations levels. Some of the results for lysozyme are shown in figs. 1 and 2. Similar results have now been obtained in our Laboratory for other biopolymers. In the following we discuss DNA and RNA(6).

III - DNA and RNA are BIOELECTRETS and NOT FERROELECTRICS

DNA was reported to be a ferroelectric(7). Fukada and coll. (8) also studied the piezoelectric properties of oriented DNA films. Polonsky and coll. briefly reported a persistent polarization in DNA but turned their attention to the apparent ferroelectricity of the material. This was inferred by the use of Sawyer-Tower circuit techniques. RNA was similarly reported to be a ferroelectric by Stanford and Lorey(10). Previously Brot and coll.(9) had shown that DNA non-linear electrical transport was responsible for the apparent ferroelectricity hysteresis cycles. We have confirmed their results for DNA and extended them for RNA. By applying the previously discussed techniques such as TSD, TSP, IPD at controlled hydrations we have shown that
a) DNA and RNA in the solid hydrated state are strong bioelectrets (polarization storage as high as 10^{-8} coul/cm^2 for fields as small as 1V/cm at a few percent hydration levels).
b) The polarization storage depends drastically on hydration varying almost exponentially for hydrations above 10 percent per weight. Space-charge effects are very important and we have a non-linear electret. c) the non-linearity of the polarization storage and of the current transport can be followed as a function of hydration and frequency by the use of a controlled hydration chamber in a very convenient way. This is very difficult to achieve for normal TSD measurements in view of the desorption phenomena indicated by our TSP measurements. In TSD there is a continuous change of the hydration level of the

sample. The IPD measurements were also associated with hysteresis studies
using the Sawyer Tower technique. For both DNA and RNA we have found hysteresis
behaviour strongly dependent on hydration. The measuring frequency was also very
important. Non-linearity sets in at very low fields for very low frequencies
such as 1Hz and below. Fig. 3 ilustrates the Sawyer Tower circuit and fig. 4
typical hysteresis curves for RNA at different frequencies. At very low
frequencies a XY recorder was used in place of the osciloscope.

Our simultaneous TSP measurements on both DNA and RNA permitted the
interpretation of the apparent Curie Temperature reported as a dehydration effect
peaking at the 50-60 C range. We have also confirmed the original general
piezoelectric behaviour found by Fukada and coll.(1) while the temperature
dependence of the piezoelectricity will have to take into account the hydration
dependence as observed by our TSP measurements. Our conclusions besides
confirming Brot el all.(9) results for DNA and extending them to RNA invalidate
the sugestions often found in the literature that DNA and RNA(10) ferroelectricity may be important for information storage mechanisms(11). However in view
of the strong bioelectret state found for both DNA and RNA as herein reported,
the speculation may be raised that information storage via the bioelectret
effect may be important in memory mechanisms and that hydration properties and
non-linearity of the bioelectret may provide interesting properties for such
speculated mechanisms.

CONCLUSIONS AND DISCUSSIONS

Hydration studies are proven to be fundamental for investigating the
behavior of bioelectrets. This is certainly also the case for electrets in
general when they are hydrofilic. The use of thermal stimulated pressure (TSP)
is found to be very usefull for such investigations. Also classical Isothermal
Polarization Decay techniques(IPD) are very convenient to use with controlled
hydration conditions and lead to information on non-linearity of polarization
storage and of the electrical transport properties as well. Neglecting
hydration changes during TSD measurements or piezoelectric, pyroelectric and
thermal differential analysis or calorimetric studies of hydrated samples may
lead to serious errors and artifacts. In the case of hydrated samples
non-linearity should always be investigated and may set in at very low fields
specially at very low frequencies (below 1 Hz). The non-linear properties can
bevery conveniently studied by the use of a Sawyer-Tower system again using
controlled hydration chambers.

For most materials, the non-linear electret behaviour is the rule, and this
field of investigations will certainly grow in importance. The state of the
water of hydration for may biopolymers can be investigated by these techniques.
In the case of strongly bound-water with long relaxation times (of the order of
1 second for ex) electret studies are very convenient since these relaxations
cannot be observed by NMR or "normal" dielectric measurements. As to the state
of the bound-water(up to 10 percent hydration) it may be speculated as that of a
frozen gas attached to hydrofilic sites or as amorphous ice for higher
hydrations. It is interesting to remark that natural electret behavior was found
for amorphous ice (12).

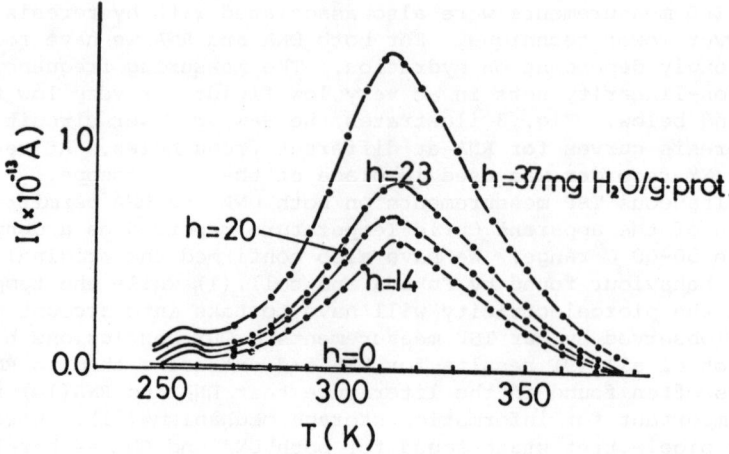

Fig. 1 - TSD curves for lysozyme for several hydrations (5kV/cm)
Observe absence of effect for dry sample. Small peak
at low temperatures is due to ice formation on electrodes.
Circles are calculated points for two hydration sites.

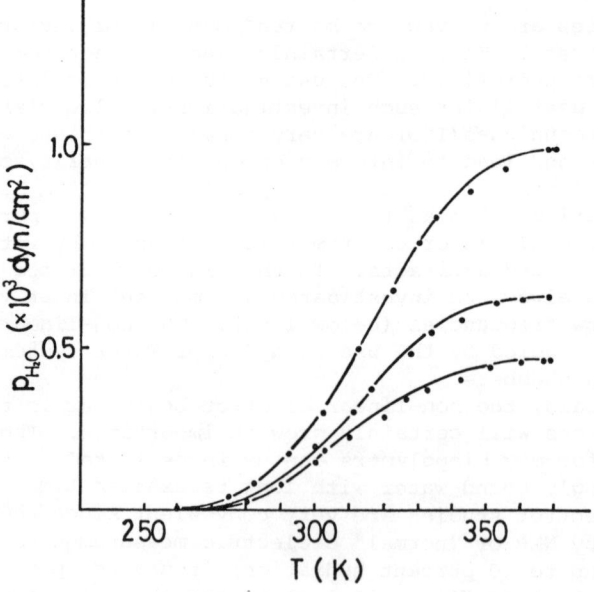

Fig. 2 - TSP curves for lysozyme for several hydrations.
Circles are points calculated from TSD curves.

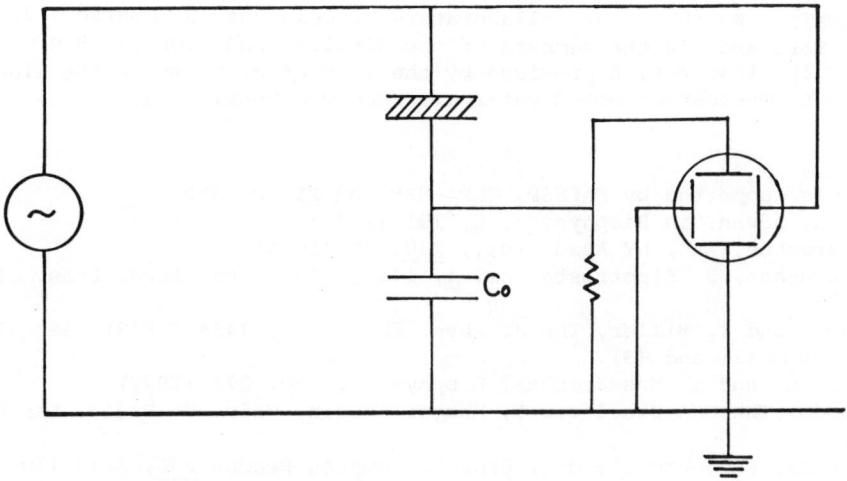

Fig. 3 - Sawyer Tower circuit to observe hysteresis curves.

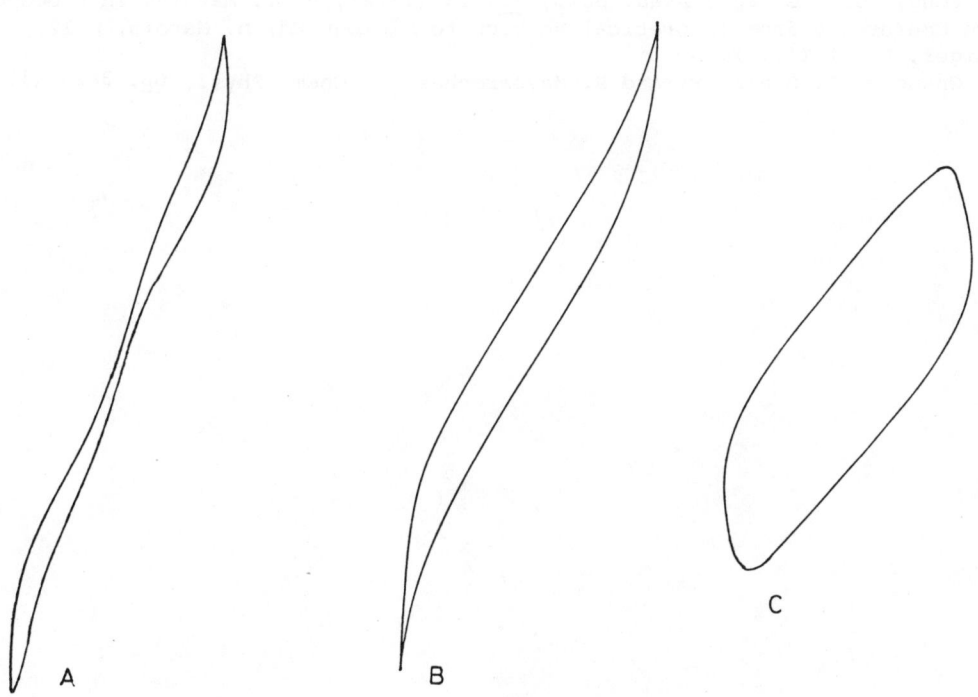

Fig. 4 - Typical hysteresis curves for RNA: A: 5 Hz, B: 1 Hz, C: 3 Hz showing strong non-linear behavior.

ACKNOWLEDMENTS - We thank our collaborators S. Celaschi, S. Quezado, J. N. Onuchic, S. Sanches Vera and all the members of the Electret and Biophysics groups at S. Carlos. The inspiration provided by the work of B. Gross on the electret and of the late L. Onsager on bound-water are deeply acknowledged.

REFERENCES

This work was supported by FAPESP, CNP_q-NSF and FINEP - BID
1. E. Fukada, Advan. in Biophysics, 6, 121 (1974).
2. S. Mascarenhas, Ann. NY Acad. Sci., 238, 36 (1974).
3. S. Mascarenhas, J. Electrostatics, 1, 141 (1975); Ann. Acad. Bras. Ci., 47, 223 (1975).
4. C. Linder, and I. Miller, The J. Phys. Chem., 76, 3434 (1972); see also references in (2) and (3).
5. S. Celaschi, and S. Mascarenhas, Biophys. J., 20, 273 (1977).
6. S. Quezado, The DNA Bioelectret, Master Thesis, Univ. S. Paulo, Sao Carlos, (1978).
7. J. Polonsky, P. Douzou, and C. Sadron, Comptes Rendus 250, 3414 (1960); also P. Douzou, J. Franca, J. Polonsky, C. Sadron, Comptes Rendus, 251, 976 (1961).
8. See reference (1).
9. C. Brot et all., J. Chem. Phys., 43, B603 (1965).
10. A. Stanford, and A. Lorey, Nature, 219, 1250 (1968).
11. P. Fong, Bull. Georgia Acad. Sci., 30, 13 (1972), B. T. Mathias in Proceed. 3rd Conference from Theoretical Physics to Biology Ed. M. Marois, p.12, Karger, Basel (1973).
12. L. Onsager, D. Staebler, and S. Mascarenhas, J. Chem. Phys., 68, 3823 (1978)

CHARGE DISTRIBUTION AND TRANSPORT IN POLYMERS

G. M. Sessler and H. von Seggern

Technische Hochschule Darmstadt

Darmstadt, W-Germany

1. Methods for Measuring Charge Distribution

Several methods have been suggested to measure the distribution of charges in thin polymer films in the direction perpendicular to the plane of the films.[1-4] Two of these methods appeared particularly promising to yield resolution in the μm-range: the thermal-pulse technique[1] and the virtual-electrode method.[3]

The former technique depends on an evaluation of the voltage generated by the nonuniform thermal expansion of a one-sided metalized charged dielectric upon short illumination on the metalized side. While it is not possible to use such experiments for the calculation of unique charge distributions,[1,5,6] a small number of spatial Fourier coefficients can be obtained, depending on measuring accuracy and location of the charge.[6]

The virtual-electrode method[3] is based on the generation of a conductive region within a two-sided metalized and short-circuited electret by means of an electron beam. If the front of this region, which forms a virtual electrode, is slowly moved through the sample by increasing the beam energy, the charge distribution may be found from a measurement of the variation of the induction charge on the rear electrode. In some materials such as Teflon PTFE and FEP, the room-temperature Schubweg of holes in the fields generated by trapped-charge layers is of the order of 10^{-3} cm, giving a corresponding resolution limit. Better resolution can be obtained in materials with smaller Schubwegs.

2. Hole Transit in Teflon

The long Schubweg of the holes in Teflon gives rise to transit effects in thin films of this material under sufficiently large applied fields.[7] The setup for an appropriate experiment is shown in the inset of Fig. 1. By irradiating the positively biased surface of the sample with a short electron pulse of small penetration depth, a plasma of secondaries is generated in the irradiated region, from which holes are extracted. A hole packet now propagates through the sample causing currents such as shown in Fig. 1. The location of the shoulder represents the transit time of the carriers. This yields a mobility of 2×10^{-9} cm^2/Vs, independent of applied field for the fields under consideration (80 to 240 kV/cm). This mobility is in good agreement with a previously determined value,[8] but considerably smaller than values determined from short pulse transit experiments.[9] The long tails of the current traces after the transit time indicate a dispersion of hopping times, as discussed by Scher and Montroll.[10] Plotting of the data on normalized log-log scales, shown in Fig. 2, yields a master curve consisting of two straight line sections. Thus, universality is observed with respect to the applied field. The slopes of the straight lines add up to -1.7, differing somewhat from the expected value of -2 but located within the range found for other materials.[11] The ready transit of holes through Teflon shows that the observed[12] storage of positive charge in this material must be due to surface traps.

3. Electron Transport in Teflon

As opposed to positive charges, electrons in Teflon are deeply trapped in volume states, resulting in very small trap-modulated mobilities. A model of the transport of negative charge in a one-sided metalized dielectric, considering deep trapping with a finite capture time, has recently been discussed.[13] The initial conditions of this model, when rigorously applied, lead to the transit of an infinitely thin sheet of free carriers followed by a tail of detrapped carriers. A more general model,[14] replacing instantaneous injection by gradual injection according to an exponential process, is outlined in Fig. 3. The equations have been solved numerically on a digital computer. Of the four parameters, τ_0 and t_0 can be determined independently (t_0 indirectly) while τ_Q and τ_T are selected to fit experimentally determined voltage decay curves. The evolution of the charge distribution, following from this model, is shown in Fig. 4. The free-charge density, having a time-independent value at $x = 0$, shows a steep front (somewhat "rounded" due to the numerical process) which moves with constant velocity. The trapped—charge density at $x = 0$ increases initially proportional with time since the release from traps is very slow.

4. Surface and Bulk Traps for Electrons in Teflon

The distribution of surface and bulk traps in Teflon was evaluated with thermally-stimulated current (TSC) techniques.[14] Samples 25 µm thick having one metalized surface were charged on their nonmetalized sides with a corona method or with electron beams of energies between 3 and 10 keV. The samples were then subjected to an open circuit TSC process, some after annealing. The measured currents are shown in Figs. 5 and 6. For the corona-charged sample not annealed, peaks at 155 and 200 C appear, as seen from Fig. 5. The 155 C peak shifts to 170 C (shoulder) for medium annealing and disappears for longer annealing. Since the corona process generates only surface charges which move into the material upon annealing, the 155 and 170 C peaks must be due to surface traps and near-surface traps, respectively. These findings are substantiated by the TSC results for the electron-beam charged samples shown in Fig. 6. Charging with low energy (3 keV) fills essentially surface traps, which show again a peak at 155 C. Higher energies (5 and 7 keV) fill near-surface traps discharging at 170 C while charging at 10 keV fills only volume traps. From the penetration[15] depth of the electrons one can estimate the range in which the traps are found. The results are listed in Table 1.

Table 1: Distribution of electron traps in 25 μm Teflon FEP-A.

Peak temperature	Location relative to charged surface	Kind of trap
155 C	0 to 0.5 μm	surface trap
170 C	0.5 to 1.8 μm	near-surface trap
200 C	1.8 to 25 μm	bulk trap

Acknowledgments

The authors are grateful to Drs. B. Gross and J. E. West for many stimulating discussions and for their permission to use Figs. 1 and 2 of this paper prior to joint publication. They also wish to thank the Deutsche Forschungsgemeinschaft for support of this work.

References

(1) R. E. Collins, Rev. Sci. Instruments, 48, 83 (1977)
(2) P. Laurenceau, G. Dreyfus, J. Lewiner, Phys. Rev. Lett. 38, 46 (1977)
(3) G. M. Sessler, J. E. West, D. A. Berkley and G. Morgenstern, Phys. Rev. Letters, 38, 368 (1977)
(4) B. Andress, P. Fischer and P. Röhl, Progr. Coll. and Polym. Sci., 62, 141 (1977)
(5) A. S. DeReggi, C. M. Gutman, F. I. Mopsik, G. T. Davis and M. G. Broadhurst, Phys. Rev. Letters, 40, 413 (1978)
(6) H. von Seggern, Appl. Phys. Lett. 33, 134 (1978).
(7) B. Gross, G. Sessler, H. v. Seggern and J. West, to be published.
(8) B. Gross, G. Sessler, J. West, J. Appl. Phys. 47, 968 (1976)
(9) K. Hayashi, Y. Yoshino, and Y. Inuishi, Jap. J. Appl. Phys. 14, 39 (1975)
(10) H. Scher and E. W. Montroll, Phys. Rev. B12, 2455 (1975)
(11) G. Pfister and H. Scher, Phys. Rev. B15, 2062 (1977)
(12) G. M. Sessler and J. E. West, J. Appl. Phys. 43, 922 (1972)
(13) P. W. Chudleigh, J. Appl. Phys. 48, 4591 (1977)
(14) H. von Seggern, to be published, J. Appl. Phys. (1979)
(15) B. Gross, G. M. Sessler, and J. E. West, J. Appl. Phys. 48, 4303 (1977)

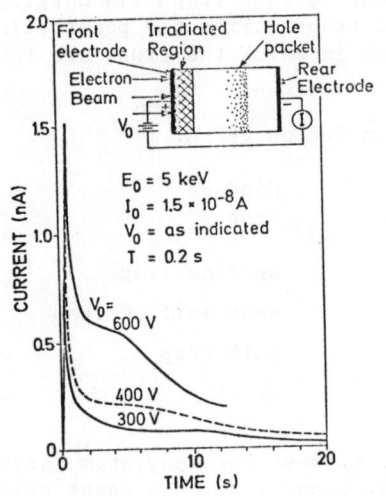

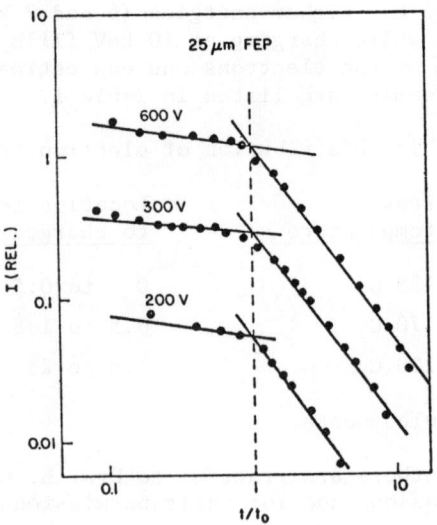

Fig. 1: Hole currents from 25 μm Teflon FEP after irradiation with a 0.25 s, $1.5 \cdot 10^{-8}$ A burst of 5 keV electrons. Irradiated area 20 cm^2.
Inset: Experimental setup.

Fig. 2: Log-log plot of hole currents from 25 μm Teflon FEP after electron-beam irradiation.

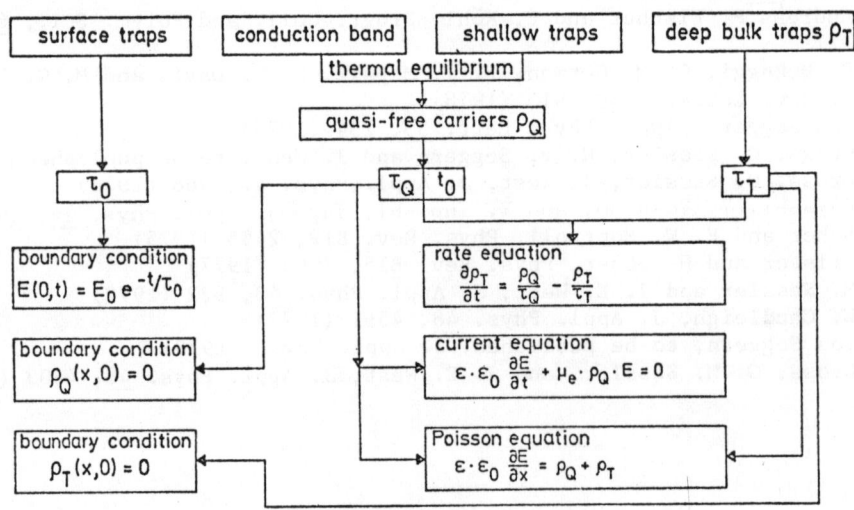

Fig. 3: Model for charge transport in Teflon.

Fig. 4: Evolution of free and trapped charge densities in Teflon. Metal electrode at x/s = 1. Parameter values: $\tau_0/t_0 = 0.1$, $\tau_Q/t_0 = 0.3$, and $\tau_T/t_0 = 4.0$.

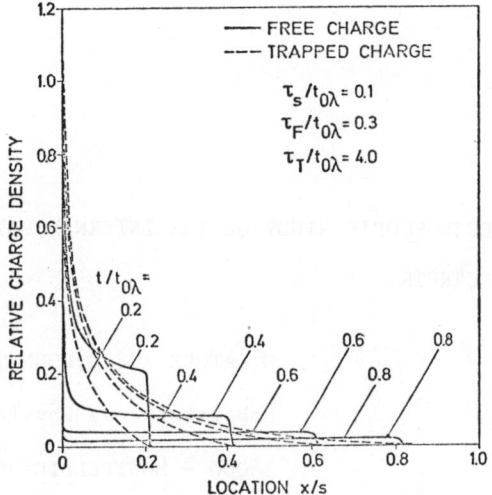

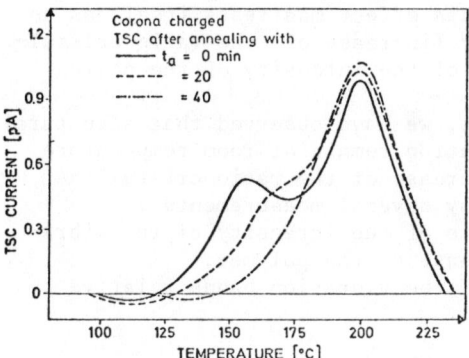

Fig. 5: Thermally-stimulated currents in open circuit from corona-charged Teflon after annealing (annealing times indicated).

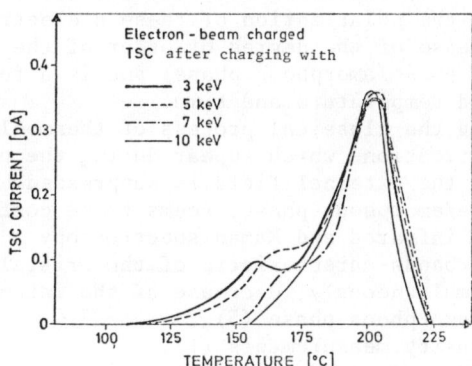

Fig. 6: Thermally-stimulated currents in open circuit from electron-beam charged Teflon (beam energies indicated).

SPECTROSCOPIC STUDY OF THE INTERNAL POTENTIALS TOPOGRAPHY FOR POLYETHYLENE ELECTRETS.

M.LATOUR and G.DONNET

Laboratoire de Physique Moléculaire et Cristalline, U.S.T.L.

34060 - MONTPELLIER Cedex - France.

INTRODUCTION :
Using Raman and infrared spectroscopy we have observed in several polymers (1-2-3-4) including Polyethylene (2) that structural modifications occur during the polarization of these dielectrics. This effect manifests itself as an increase of the degree of order of the polymer (increase of the ratio cristalline phase/amorphous phase) and is a function of the intensity of the poling field temperature and time.
Using the classical process of thermoelectrets, we have observed that structural modifications which appear during the polarization remain at room temperature when the external field is suppressed. The increase of the ratio cristalline phase/amorphous phase, seems to be confirmed by several measurements :
- in infrared and Raman spectroscopy : increase of the intensity of the vibration bands caracteristic of the cristalline phase of the polymer,
- simultaneously, decrease of the intensity of the vibration bands relative to the amorphous phase (5),
- density measurements (1),
- X-Ray measurements (6-3).
The change in polymer spectra produced by poling can be calibrated in terms of field intensity, time and temperature. This data can thence be used to deduce the history of the electrical stress for samples of the polymer used in high voltage insulation.

EXPERIMENTAL :
The polymer used in these experiments is a low density polyethylene doped by 500 ppm of NaCl. Samples are provided by ATO CHIMIE (8). Sheets of 25 μ and 4mm are used respectively for the infrared and Raman spectroscopy. Cylindrical samples of coaxial cable insulated with Polyethylene 12mm thick also are used. The Raman measurements are carried out using the classical right angle scattering method in polarized light. We record spectra in the frequency range : 2500 to 3000, 1000 to 1500 cm^{-1}. The Raman spectrometer used is a Coderg PH 1 with a

Helium – Neon Laser (P = 90 mW). The infrared measurements are carried out using a 577 Perkin Elmer spectrometer, in polarized light, in the frequency range : 1800 to 600 cm^{-1}. The attribution of the vibration bands is classical and has been detailed in previous papers (5-9) for the Raman and for the infrared spectra. A special cell (9) adaptable on the spectrometers permits the polarization of the samples in a control atmosphere, for temperatures varying from 80°K to 293°K and for potentials up to 100 kV. A micrometric translation system in the three fundamental directions permits the measurement of spectra in any point of the sample.

RESULTS AND DISCUSSION :
In previous papers (5-7) we have show that the spectral changes provide a non destructive method to establish the topography of field-induced internal structural modifications of a dielectric; from this the topography, the internal potentials can be derived with the aid of a prearanged standard. For electrets this method is based on the comparison of the spectra obtained for the electret and for a control sample wich has been subjected to the same thermal treatment but not to the polarizing field. By vertically shifting the sample it is possible to compare the spectra obtained in different points of the dielectric and to deduce the corresponding field distribution.

The same experimental procedure also allows one to record the evolution of the structure of a particular point of the dielectric during its polarization. For example we record the variations of the peak of a cristalline band observed by Raman or infrared spectroscopy during the process of polarization.

Fig.1 and Fig.2 show the typical spectral modifications observed on doped low density Polyethylene during its polarization. Fig.1 shows the Raman spectrum between 2500 and 3000 cm^{-1} for a sample polarized at 80°C, 100 kV/cm, during 45min.; the most important vibration band at 2882 cm^{-1} is attribuated to the CH asymetric stretching. Fig.2 shows the infrared spectrum recorded between 600 and 1800 cm^{-1} for two sheets of 25 μ thick cuted out at (a) 0.7mm., and (b) 0.5mm. from the internal electrode of a coaxial cable. The cable is polarized during 11 hours at 60°C under 100 kV/cm. The results given on Fig.3 show the internal field repartition deduced from the spectral analysis of a planar sample. The sample was polarized in one case using metallic electrodes (gold evaporated), and in the other case using "blocking contacts": a 3 μ thick teflon film was interposed between the metallic electrode and the sample. We can see that the internal repartition of the field is relatively uniform in the center of the sample when it is polarized between "blocking contacts", and presents important perturbations located near the electrodes in the case of metallic electrodes. We note that this perturbation affects more or less a region of 1 mm of the dielectric near the electrodes.

Fig.4 shows the topography of the internal field obtained for coaxial cables used for the transport of direct high voltages. On Fig.4 the cable has been polarized during 171 hours under 750 kV/cm at 60°C, at the end of the polarization a breakdown accured after an inversion of polarity of the high voltage in the cable. As in the case of a planar sample, we observe rather important changes of structure located near the electrodes, the structure of the central region of the dielectric being more uniform.

Fig.5 shows the evolution of the peak of the cristalline band 2882cm^{-1} observed by Raman spectroscopy during the process of polarization of the sample at 80°C under 100 kV/cm. The same graph shows the cycle of the thermal treatment and of the poling field. We observe a change in the degree of cristallinity of the polymer after the process of polarization, this effect is permanent and remain after heating cycles at a temperature of 80°C. We did not reheat above 80°C.

REFERENCES :
(1) M.LATOUR
J.of Electrostatics 242-248 (1976)
Translation in J.of the British Post Office Telecommunications N°3471 (p.1-11)(1977)
(2) M.LATOUR
Polymer,18,3 277-280 (1977)
(3) D.K.DAS GUPTA, T.NOON
Inst.Phys. Conf. Sem. 27, 122-129 (1975)
(4) M.AOZASA, K.KIMURA and K.YAHAGI
J.Phys. Soc. Japan 34, 568 (1973)
(5) M.LATOUR, G.DONNET
J.Phys. Lett. 37 - L145-148 (1976)
(6) G.T.DAVIS,J.E.Mc KINNEY,M.G. BROADHURST and S.C.ROTH J.Appl.Phys.to be published
(7) M.LATOUR, G.DONNET, Revue Générale de l'Electricité (to be published)
(8) in a contrat sponsored by E.D.F. and C.N.R.S. (1975 to 1978)

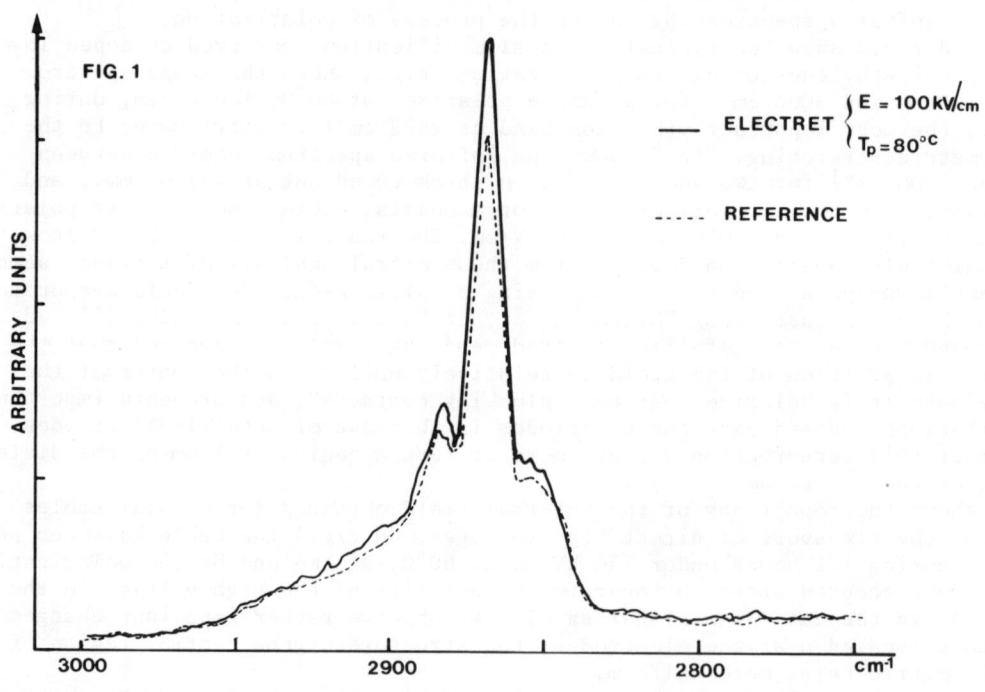

FIG. 1 — ELECTRET $E = 100\,kV/cm$, $T_p = 80°C$ ----- REFERENCE

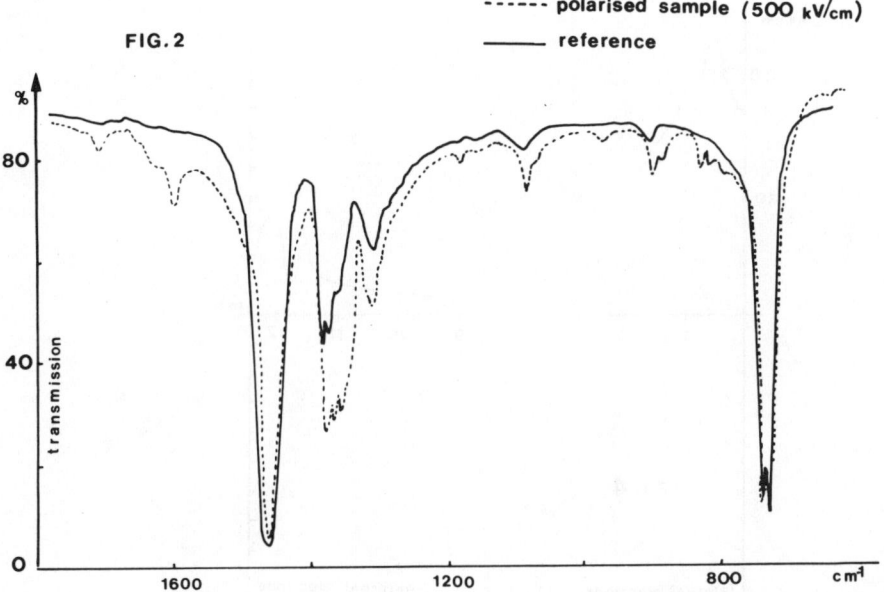

FIG. 2 ----- polarised sample (500 kV/cm)
——— reference

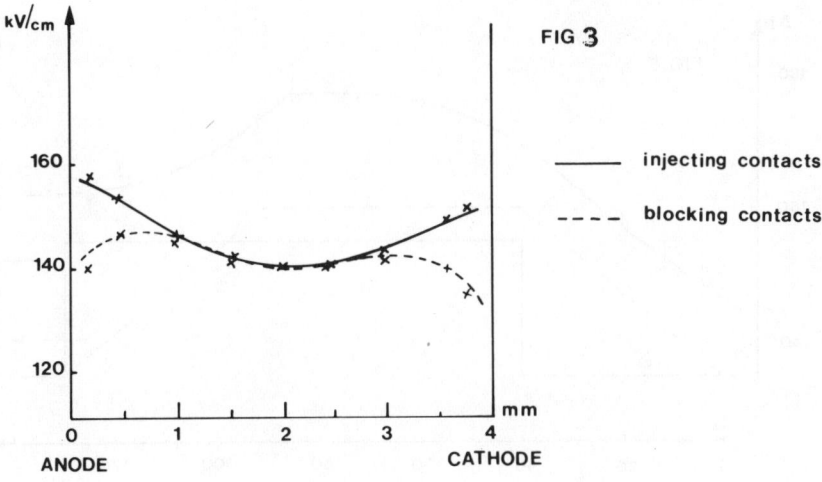

FIG 3 ——— injecting contacts
----- blocking contacts

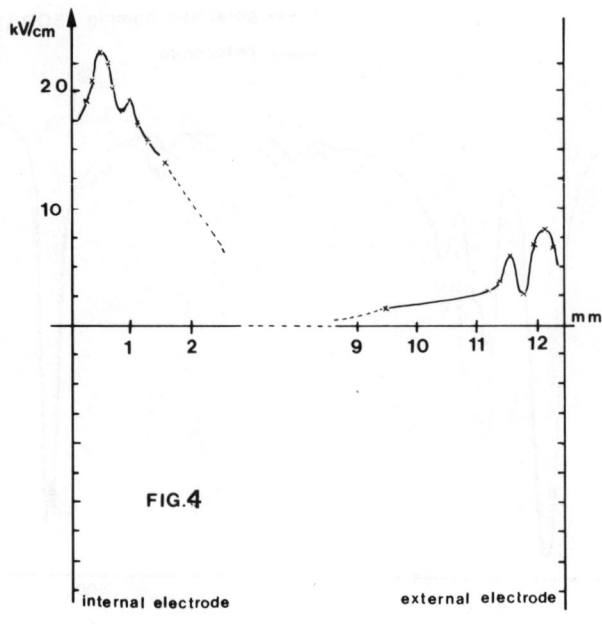

FIG.4

internal electrode   external electrode

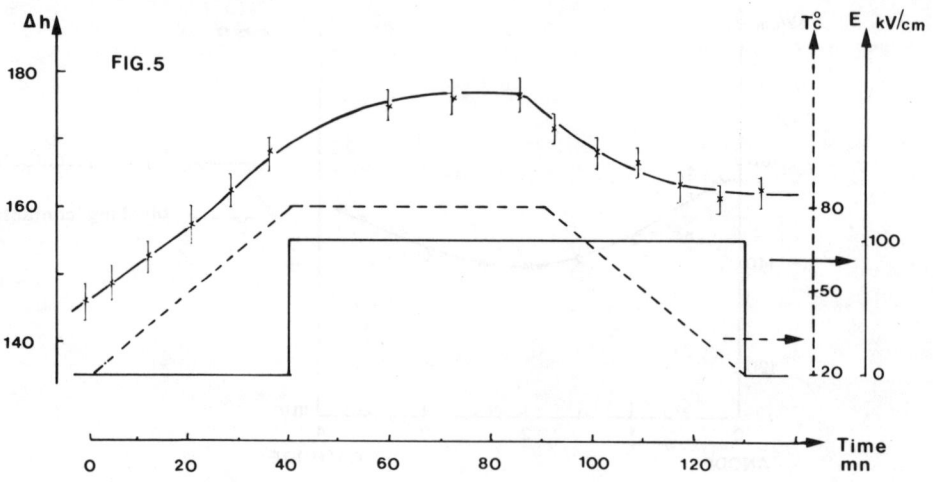

FIG.5

CHARGE STABILITY OF ANNEALED ELECTRETS

Yoichi Kodera and Tetsuo Toyoda*

Sony Corporation Research Center

174 Fujitsuka-cho, Hodogaya-ku, Yokohama, 240 Japan

1. Introduction

   The non-polar polymers, such as Teflon FEP, are widely used for electret materials, because of their excellent charge stability over the wide temperature and humidity range. These electrets are formed by thermal, corona, and electron bombardment methods.
   The sensitivity of an electret microphone depends on the charge density of the electret diaphragm. Although, a FEP film metallized on one side can be charged up to 100 nC/cm^2 by the corona method, spark breakdown between the charged surface and nearby metal may occur to produce the charge drop by its high electric field. Therefore, a moderate charge density of the order of 10 nC/cm^2 is suited to electret microphones. The corona method controlled by a screen electrode[1] and thermal methods give such an amount of charge. It has been known that the annealing of FEP electrets decreases the surface charge but improve the charge stability.[2,3,4] Corona charging and subsequent annealing seems to be the best way to make the stable electret of a moderate charge density.
   In this paper, the charge stability of annealed Teflon FEP electrets formed by corona discharge is discussed.

2. Experimental

   The Teflon FEP films used in this study were 12.7 μm in thickness with an evaporated metal electrode on one side. The metallized films were mounted on metal rings, 17 mm in diameter. The samples were charged by corona discharge with a screen electrode which controls the amount of surface charges. After charging, some of the samples were set into an oven, metallized side down, and annealed by heating them to a temperature of 120-150°C for periods of 10 min-2hr. Annealed and non annealed electrets were studied for their electret properties using thermally stimulated charge decay (TSCD) and thermally stimulated

*Present address: Audio Research Co., Yamanashi, Japan

currents (TSC). The current and charge were measured using a Keithley 610C electrometer. The measuring system is shown in Fig. 1. For TSC, silver paste electrode was painted on the charged surface of the sample. Another electrode, formed by vacuum evaporation was also tested. We could not find any difference in TSC's from different electrodes.

3. Results

The charge decay curves of corona charged electrets at different heating rates are shown in Fig. 2. The surface charges at high temperatures depend on the heating rate, and have no relation to the initial charge density.

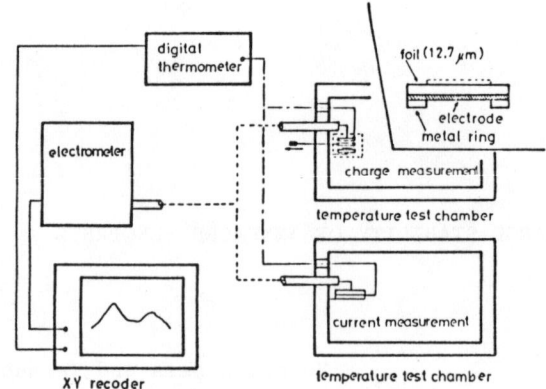

Fig. 1 Apparatus for measuring the thermally stimulated current and charge decay. The sample crosssection is shown in the inset.

The curve B' shows the charge decay of the electret with the initial charge density of 16.6×10^{-8} C/cm^2. This electret lost its charge at a higher rate than the lower charged electret shown in curve B, until the charge densities coincided each other. In the following experiments the heating rate was about 1°C/min.

The corona charged electret with initial charge density of about 16×10^{-8} C/cm^2 were heated up to 150°C and then rapidly cooled to a room temperature. At the 2nd heating run, the surface charge density remained constant up to the final temperature of the first heating run. The present result is the same as those reported by Turnhout[3] on the electrets charged by electron bombardment. We also measured the TSC of the sample subjected the same heat treatment as 1st heating run for TSCD. The result is shown in Fig.4. The low temperature peak about 60°C was still observed after this pre-heating.
TSC obtained had two current peaks about 2×10^{-12} A. The released current flowed in a direction which negative charges moved toward the charge injected surface. On the other hand, the current reversal was observed for a not annealed sample. The total current released was smaller than that of the annealed sample. These two electrets had the same initial charge densities of 8×10^{-8} C/cm^2, the TSC curves, however, were quite different from each other.

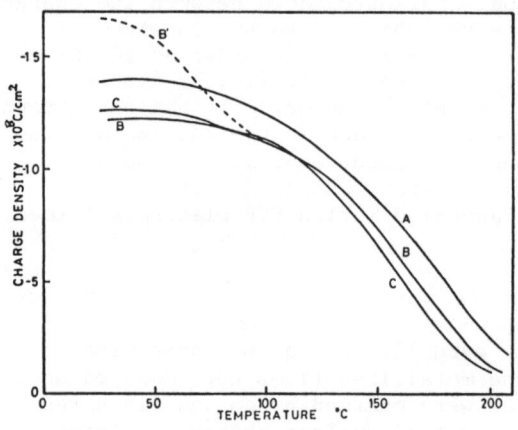

Fig. 2 The thermally stimulated charge decay curves of corona charged FEP electrets at different heating rates. A: 1.6°C/min., B,B': 0.8°C/min., C: 0.4°C/min.

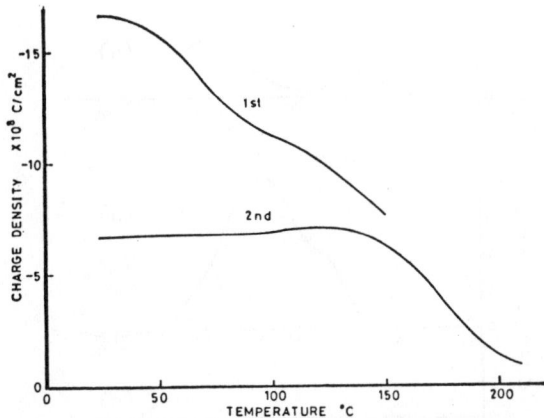

Fig. 3 Surface charge decay at the heating rate of 1°C/min. After heating up to 150°C (1st run), the sample was rapidly cooled to a room temperature and reheated (2nd run).

The similar experiments were carried out with samples annealed at a constant temperature. The TSC curves obtained from samples charged by corona discharge and annealed at 120°C for different periods of time are in Fig. 5 with the TSC curve of a sample not annealed. Change in charge due to annealing is listed in Table 1. The current reversal was observed for not annealed sample. An increse in current was also found for annealed electrets.

The results obtained from samples annealed at 150°C are shown in Fig.6. The maximum released charge was obtained from the sample annealed for 30 minutes. The TSCD curve from the electret annealed at 120°C for 2hr is compared with that of a not annealed sample in Fig.7.

Usually the higher charged electret loses its charge at a higher rate than lower charged electret. However, we observed that the charge remained constant for the annealed sample with higher charge density.

Fig. 8 shows the plots of temperatures T_0 vs. initial charge density. T_0 is a temperature at which retained charge decrease under the initial charge density level. The higher T_0 was obtained for the annealed samples.

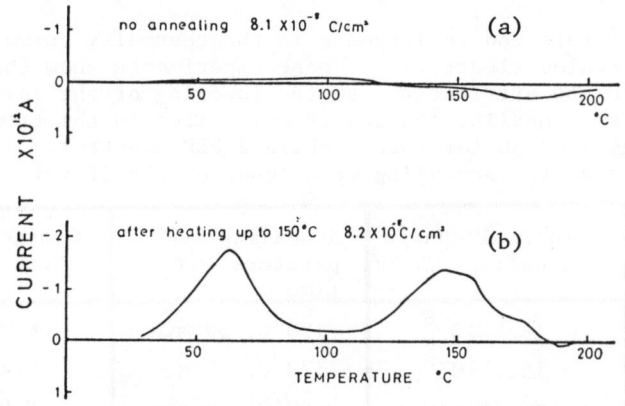

Fig. 4 TSC of corona charged electrets.
(a) not annealed sample.
(b) the sample pre-heated up to 130°C at the heating rate of 1°C/min.

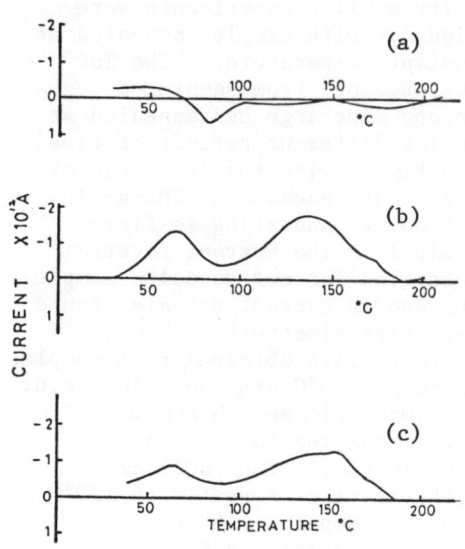

Fig. 5 TSC of FEP electrets.
(a) not annealed, charge density of 13.9X10^{-8} C/cm^2. (b) annealed at 120°C, for 30 min. (c) annealed at 120°C, for 2hr.

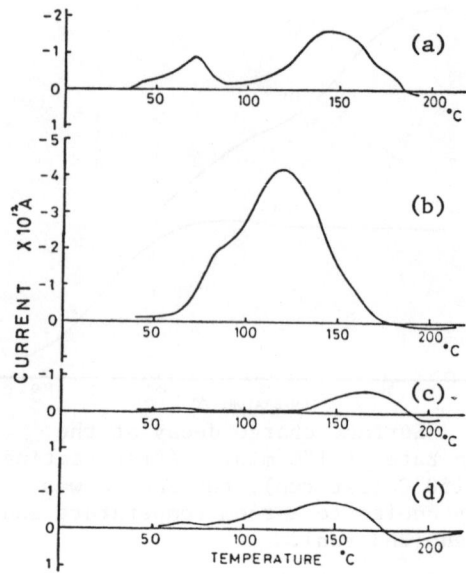

Fig. 6 TSC of annealed FEP electrets. The samples are annealed at 150°C, (a) for 10 min., (b) for 30 min., (c) for 1 hr, (d) for 2 hr.

4. Conclusion

A slower charge decay rate and an increase in the thermally stimulated current were found for annealed electrets. These experiments show that this charge stabilization cannot be attributed to the lowering of the initial charge density.   It seems that an annealing changes storage sites in the films and gives a more stable charge storage for corona charged FEP electret.

It can be concluded that the annealing of an charged FEP film is   useful

Table 1. Surface charge densities before and after annealing.	Initial charge density C/cm^2	Annealing temperature and time	Charge density after annealing C/cm^2
	16.8X10^{-8}	120°C, 30 min	8.9X10^{-8}
	15.7X10^{-8}	120°C, 2 hr	9.2X10^{-8}
	14.7X10^{-8}	150°C, 10 min	8.9X10^{-8}
	15.3X10^{-8}	150°C, 30 min	7.4X10^{-8}
	12.3X10^{-8}	150°C, 1 hr	5.1X10^{-8}
	14.3X10^{-8}	150°C, 2 hr	5.3X10^{-8}

for controlling the charge density and increasing the charge stability.

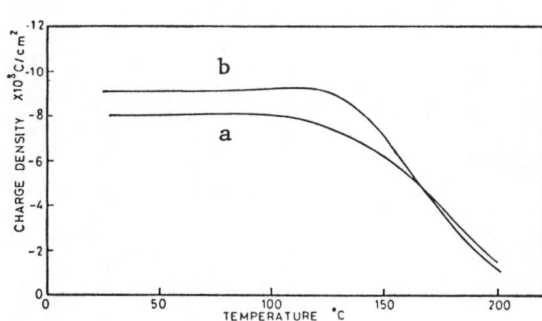

Fig. 7 Surface charge decay of corona charged electrets.
(a) not annealed, (b) annealed at 120°C for 2 hr.

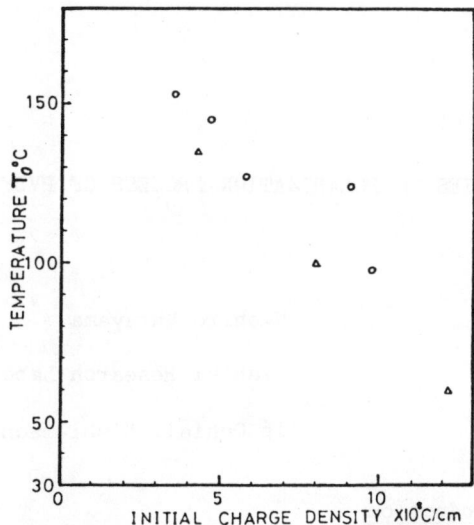

Fig. 8 Plots of the initial charge density vs. temperature at which the retained charge decreases under the initial charge value.
o annealed.
△ not annealed.

References

(1) M. Ieda, G. Sawa and U. Shinohara, J.I.E.E.J. 88, 1107 (1968).
(2) Y. Kodera and T. Toyoda, Japanese Patent No.873986, Field 30 April 1971.
(3) J. Van Turnhout, Electret, edited by M. M. Perlman (Electrochemical Society 1973), pp.230-251.
(4) B. Gross, G.M. Sessler and J. E. West, J. Appl. Phys. 46, 4674 (1975).

STUDIES ON POLARIZATION PROCESS OF PVDF ELECTRET

Naohiro Murayama

Nishiki Research Laboratories, Kureha Chem. Ind. Co. Ltd.

16 Ochiai, Nishi-machi, Iwaki-shi, Fukushima-ken, Japan

Introduction

Poly(vinylidene fluoride) (PVDF) has a large dielectric constant, due to the large dipole moment of CF_2, and is one of the most polar dielectric polymers. The polymer is characteristic of forming the electret which has large piezoelectric, pyroelectric, and non-linear optical effects. PVDF is the most interesting and unique material to investigate electric properties of polymers.

There are five polarization mechanisms proposed to explain the large piezoelectric effect of the electret of PVDF(especially predominantly containing ß-form crystal):

(1) orientation of the spontaneous polarization of ß-form crystal as a unit cell (ß-form crystal is ferroelectric);
(2) orientation of ß-form crystallite, including the appearance of ß -form crystals due to crystallization (ß-form crystal has a spontaneous polarization which is not ferroelectric);
(3) orientation of amorphous chains;
(4) macroscopic charge separation due to trapped charges;
(5) induced dipole due to trapped charges (microscopic charge separation).

The casual relation between the piezoelectricity and the polarization mechanism for PVDF electret has not been established yet, probably because induced dipole is difficult to be distinguished from permanent dipole.

The piezoelectricity in PVDF electret is known to be affected mainly by polarizing electric field and polarizing temperature.[1] There have been many contradicting data on the polarizing phenomena for PVDF electrets, although they are obtained under similar conditions concerning with field and temperature. In this paper, the polarizing process of PVDF electret was examined in details by means of the observation of the piezoelectricity. The investigation of the polarizing process can give different kind of information on the piezoelectricity for PVDF electret and throw light on the polarization responsible for the piezoelectricity. The investigation of the piezoelectricity in PVDF electret could surely give rise to a true understanding of charge storage phenomena for polymers, or furthermore dielectrics in general.

Experimental

PVDF is KF-Polymer(#1100) produced by Kureha Chemical Ind. Co. Ltd. Film A (30 microns) was produced by stretching a melt cast sheet at 110°C to a stretching ratio of 5.1, and Film B (30 microns) was prepared by stretching another melt cast sheet at 110°C to 4.5. Both films contain predominantly ß-form crystal. The infra-red absorgance ratio D_{530}/D_{510} is 0.21 and 0.26 for Film A and Film B respectively. The films were provided with evaporated aluminum electrodes. Film A was mainly used for experiments without referring so.

The pre-heat-treatemnt was carried out in a thermostat at various temperatures for various times. The metallized films with or without heat-treatment were put in a thermostat at a polarizing temperature (T_p) subjected to a high electric field (E_p) for a polarizing time (t_p), and cooled to room temperature quickly over less than 3 seconds, after which the field was removed. All heat-treatment and polarization procedures were conducted without any care to prevent deformation.

The piezoelectricity was measured as d_{431} with the frequency of the applied stress of 15 Hz and the static stress of 10^{43} g/cm.

Result

The polarization responsible for the piezoelectricity can be separated into fast and slow processes.[9(2,3)] In order to investigate the two processes, the original film (Film A) was heat-treated at various temperatures for 15 hr. and polarized at 130°C and 600kV/cm. The saturated values, d(1) and d(2), and the time conctants, $\tau(1)$ and $\tau(2)$, for the fast and slow processes, were measured ed.

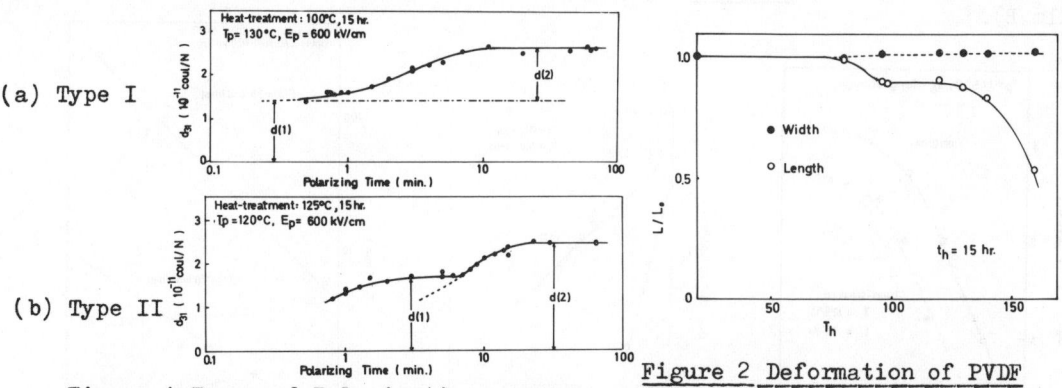

Figure 1 Types of Polarization process

Figure 2 Deformation of PVDF film during heat-treatment

There are two types of polarization phenomena: Type I appears when the heat-treatment temperature (T_h) is lower than T_p and d(2) will start after d(1) has been completed; on the other hand Type II appears when T_h is higher than T_p and the formation of d(2) is independent of d(1)(Figure 1). d(1) for Type I is too fast to be followed by our techniques, but that for Type II forms rather slowly. It is very interesting that for Type II it appears as if d(1) would be replaced by d(2). The original film can be polarized at the fastest rate among the films. It takes a fairly long time for the heat-treatment effect to be completed. The heat-treatment time (t_h) of less than 10 min. is not enough to make a transition from Type I to Type II at 130°C. When the original film is heat-treated, the film shows some shrinkage in both stretching and thickness directions.(Figure 2) The orientation of β-form crystals become slightly worse and the crystallinity becomes slightly larger. The Young's modulus of the film becomes lower by the heat-treatment at high temperatures, so the amorphous region comes to be more relaxed.

In order to investigate the case of Type II, the film A was heat-treated at 125°C for 15 hr. and polarized at various E_p and T_p. It is found that the effect of E_p on $\tau(1)$ is larger than that on $\tau(2)$. The dependence of d_{31} on E_p shows a threshold around 150kV/cm. On the other hand, the dependence of d_{31} on E_p for the original film electret does not show any that sort of threshold and d_{31}^p increases linearly. Although the piezoelectricity depends on E_p, sometimes linearly, sometimes having a threshold, it is found that the threshold appears when the film is heat-treated.(Figure 3)

The same film was polarized at various temperatures, and d(1) and d(2) with $\tau(1)$ and $\tau(2)$ were obtained.(Figure 4) The apparent activation energies, E(1) for $\tau(1)$ and E(2) for $\tau(2)$, were calculated from the temperature dependence data (Figure 5). The apparent activation energies for various heat-treated films are given in Table 1. E(1) is considerably affected by T_h, but E(2) doesnt change so much, especially below 145°C. Both E(1) and E(2) become smaller as T_h becomes higher.

The effect of ionic impurity on the polarization processes was examined. The ionic doping was carried out using the polymer powder and methanol containing KBr. The addition of KBr (20ppm) does not decreas the rate of polarization, but even increases it. The activation energies E(1) and E(2) for the doped film electret are listed in Table 2; they are lower than those for control electret (Film B).

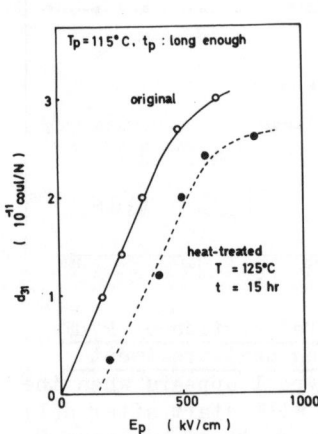
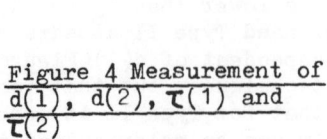
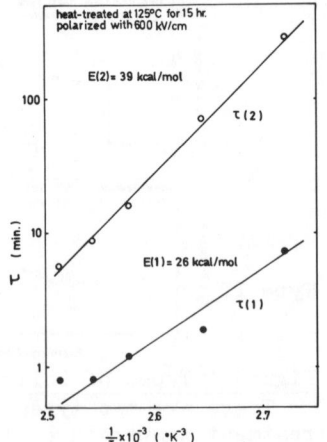

Figure 3 Dependence of d_{31} on electric field strength

Figure 4 Measurement of d(1), d(2), $\tau(1)$ and $\tau(2)$

Figure 5 Activation energy for fast and slow processes

Table 1 Apparent activation energy of polarizing processes for heat-treated PVDF film electret (600kV/cm)

Heat-treatment temperature(°C)	Heat-treatment time (hour)	E(1) (kcal/mol)	E(2) (kcal/mol)
125	15	26	39
130	14	19	39
133	15	19	39
145	15	14	34

Table 2 Apparent activation energy of polarizing processes for KBr-doped PVDF film electret (T_h=140°C, t_h=14hr., E_p=600kV/cm)

	E(1) (kcal/mol)	E(2) (kcal/mol)
control	15	26
KBr-doped (20ppm)	10	24

The films electroded with aluminum and gold were compared as electrets and no difference was observed.

The original film was polarized under fixed conditions at 130°C; d(2) and τ(2) are 2.9x10^{-11}coul/N and 0.5 min. respectively. These are nearly the same with the data, d(2)=2.6x10^{-11}coul/N and τ(2)=0.6min., obtained for unfixed conditions.

Discussion

The heat-treatment accompanies with the following characteristics: the polarization rates are slowed, but the activation energies are lowered; τ(1) is more dependent on E_p than τ(2); E(1) depends on T_h, although E(2) does not change. The slow process for Type II seems to be due to the crystalline nature.

The behaviour of the polarization process for Type II does not arise from the change in charge distribution in the thickness direction due to space charges because the addition of ionic impurity did not depress the rate and the piezoelectricity does not affected by the electrode metals. It can be concluded that the fast process may be related to the molecular motions of the amorphous or semicrystalline regions.

Both fast and slow processes for Type I are assisted by the drastic molecular motions during the polarization. The apparent deformation (shrinkage) is not necessary for Type I. The characteristics are quite different between Type I and Type II.

The heterocharge which appears after the apparent homocharge has disappeared is known to be anomalously large and stable. As such a heterocharge is quite unexpected, it is called as "anomalous heterocharge".[4] In order to explain, the characteristic phenomena of the anomalous heterocharge, it should consist of a hidden homocharge and a hidden heterocharge and be observed as a difference; $P_{homo}-P_{hetero}$. The anomalous heterocharge observed under an identical condition is closely related with the stable piezoelectricity which is different from the piezoelectricity in other polymer electrets such as PVC, PMMA, etc. It may be speculated that a rather unstable polarization is at first formed which is related to the orientation of dipoles in the amorphous or semicrystalline regions, and then the polarization is converted into the more stable polarization in combination with the homocharge. Recently, Fukada et al. reported that there are two types of piezoelectricity; one is depending on the orientation of dipoles and the other is independent of the orientation. The latter type of piezoelectricity is more stable than the former one when the electrets are subjected to high temperature.[5]

The amount of polarization in PVDF electrets are reported to 2.2×10^{-6},[6] 8×10^{-6},[7] and 5×10^{-6} coul/cm^2.[8] Recently we have measured the polarization by the depolarization technique for electrets polarized at low temperatures; the value is 3.2×10^{-6} coul/cm^2 for $d_{31} = 3.0 \times 10^{-11}$ coul/N.[9] Since the crystallinity of β-form film is 50% at most and the calculated spontaneous polarization of β-form crystal is 1.3×10^{-5} coul/cm^2, the spontaneous polarization according to (1), (2), or (3) should be less than 0.65×10^{-5} coul/cm^2. Either amorphous chains or β-form crystal should be oriented by about 50%. No such evidence has been observed yet. The mechansim (4) is precluded because the depolarization current of the polarization responsible for the piezoelectricity is positive, and the polarization due to the mechanism (4) should show the negative current. Therefore, it can be concluded that the polarization responsible for the piezoelectricity in PVDF electret originates from mechanism (5).

Literature

(1) N.Murayama, Microsymposium on Electrical Properties of Polymers, Tokyo (January, 1972)
(2) N.Murayama et al., J.Polym.Sci.(phys) 13 1033 (1975)
(3) M.Oshiki and E.Fukada, J.J.A.P. 15(1) 43 (1976)
(4) N.Murayama, J.Polym.Sci.(phys)13 929 (1975)
(5) E.Fukada et al., 175-ACS National Meeting, Anaheim, U.S.A. (March, 1978)
(6) M.Tamura et al., J.Appl. Phys. 45, 3768 (1974)
(7) P.Buchman, Ferroelectrics 5 39 (1973)
(8) N.Murayama and H.Hashizume, J.Polym.Sci. (phys), 14 989 (1976)
(9) N.Murayama, to be published

STRUCTURE IN ELECTRICALLY POLARIZED POLYACRYLONITRILE

S. I. Stupp and S. H. Carr
Department of Materials Science and Engineering
and Materials Research Center
Northwestern University
Evanston, Illinois 60201 U.S.A.

Introduction

Persistent electrical polarization in polyacrylonitrile (PAN) has been studied because of the potential of PAN to form highly active polymeric electrets. This prediction is based on the large dipole moment possessed by the nitrile side-groups. However, measured values of piezo- and pyroelectric constants (1) for PAN are smaller than those found for many other polymers, and so our work has been concerned with the reasons why this potentially high activity fails to be realized with this polymer.

Thermally stimulated discharge currents (TSDC) have been obtained from polyacrylonitrile polarized under a very wide variety of conditions (2-5). Three peaks, centered about $90°C$, $145°C$, and $190°C$, have been characterized by various physical techniques and have been assigned to: loss of a preferred orientation of dipoles, the onset of chain segmental mobility (and the attending onset of small ionic diffusion), and the onset of solvent diffusion, respectively. It is seen that the amount of dipolar orientation which is developed is relatively small and therefore contributes a very small (less than 1%) proportion to the total electrical polarization. Persistent polarization that is lost at higher temperatures is likely to be of only one sign, and therefore this material appears to be simply a monopole electret.

Theory

Although the upper bound to polarization arising from preferentially oriented nitrile side-groups in PAN is 14.8 $\mu C/cm^2$ (6), the largest value of the $90°C$ TSDC peak is 0.52 $\mu C/cm^2$. This suggests that the nitrile side-groups have only a slight preferred orientation in the direction of polarization. Use of optical birefringence (3) and infrared dichroism studies (5) have confirmed this low level of nitrile side-group orientation, although there appears to be excess orientation when compared with what should be predicted simply by equation #1:

$$P = \langle N\mu \cos \theta \rangle \qquad (1)$$

where N is number of dipoles per unit volume, μ is their permanent dipole moment, and $\langle\cos\theta\rangle$ accounts for the average inclination of side-groups with regard to the polarization direction. Interestingly, information on the orientation of dipoles still produces a three-fold overestimate of polarization when compared with what is measured experimentally. It is possible that this is due to local inhomogeneities (structure ?) thought to exist in PAN (7).

Predictions of persistent polarization due to preferred orientation of dipoles can be used as tests for the validity of measurements such as obtained in this study. Assuming a very simple case of nitrile groups immersed in a nonpolar medium that is incapable of influencing the impressed electrical field at the local scale, one can use the Langevin equation, as follows:

$$P_{Lan} = \frac{N\mu^2 E_p}{kT_p} \tag{2}$$

where E_p is the polarizing field strength, k is the Boltzmann constant, and T_p is the temperature at which polarization was developed. The result of this equation is $P_{Lan} = 0.05\ \mu C/cm^2$. The Onsager equation (8), which has also been used to predict polarization, is as follows:

$$P_{Ons} = \frac{N\mu^2 E_p}{kT_p} \cdot \frac{\varepsilon_o (2\varepsilon_o + 1)(n^2 + 2)^2}{9(2\varepsilon_o + n^2)^2} \tag{3}$$

where ε_o is the dielectric constant at 0 frequency and n is the index of refraction of the medium. Making no presuppositions about the orientational aspects of the nitrile side-groups, one predicts $P_{Ons} = 0.15\ \mu C/cm^2$. On the other hand, if one assumes that dipoles are constrained to orientational motions by rotations about axes perpendicular to the electrical field, this value can be increased to $P_{Ons} = 0.23\ \mu C/cm^2$. The Frölich equation,

$$P_{Frö} = \frac{N\mu^2 E_p}{kT_p} \cdot \frac{\varepsilon_o (n^2 + 2)^2}{9(2\varepsilon_o + n^2)} \cdot g \tag{4}$$

is capable of taking into account not only how the local environment of a dipole modifies the impressed electrical field but also how the local environment imposes a force field upon the dipole. The factor g appearing in equation 4 takes this consideration into account. In the case of numbers reported in the previous paragraph, g would need to have values in the range of 3 in order to reconcile experiments with such theories as represented by equation 4.

Experimental Results

Values of g in excess of unity strongly imply the existence of some specific characteristics of the environment locally surrounding dipoles. It has long been suspected that nitrile side-groups in PAN associate with each other, both intramolecularly and intermolecularly, and, judging from calculations presented above, such seems likely to be the case in the specimens studied here. High resolution infrared absorption spectroscopy has been applied to characterize in greater detail the state of nitrile groups in these specimens. The instrument used was a Perkin Elmer Model 180 Spectrophotometer, and expanded scale portions of spectra in the vicinity of 2239 cm^{-1} are shown in Figure 1. It is immediately apparent that a splitting of the main nitrile absorption peak is found in spectra obtained from electrically polarized samples. Such a splitting suggests either that two populations of nitriles are present in PAN or that all nitrile side-groups are sufficiently closely associated with others that coupling of their oscillatory motions can occur. In either case, one envisions that at least some fraction of the nitrile side-groups find themselves in a new invironment which is not the same as had existed prior to electrical polarization. Such a case is, obviously, consistent with the implication of g-factors greater than unity.

Figure 2 shows data on the contributions to polarization arising from dipoles having a preferred orientation. The ordinate is expressed as percent of saturation polarization which might exist as an upper bound at each of the temperatures in question. Data are obtained, once again, from infrared absorption spectroscopy. It is seen that decay of the preferred orientation becomes essentially complete at different temperatures for specimens prepared under different conditions. As-cast films, which have never been heated to elevated temperatures of the kind used during electrical polarization, lose their small amount of orientational anisotropy in the vicinity of 80°C. Samples polarized under modest-strength electrical fields require heating temperatures near 120°C in order to lose their dipolar polarization, while specimens polarized under stronger electric fields lose their orientational anisotropy rather slowly, with some orientation still present at temperatures where chemical decomposition begins to occur ($\geq$165°C). The different decay kinetics exhibited by specimens polarized under different field strengths suggests that the imposed external electrical field is capable of creating an electrically biased state in which dipolar side groups require higher temperatures before they can undergo relaxation motions.

Wide-angle X-ray scattering data were also obtained on PAN films polarized under these same electrical fields. The experiments were performed using a Philips Model XRG 5000 X-ray Diffractometer, using Cu K$_\alpha$ radiation. It is immediately apparent from the data in Figure 3 that the stronger field induces a densification of the average chain packing in the PAN material. Densification calculated on the basis of change in interatomic packing is on the order of 1.5%. It appears from the data that a single kind of environment for the PAN chains exists, as opposed to the possibility of two different populations. However, the rather broad peak-width still suggests a considerable distribution in the kinds of environments present. One, further, notices that the splitting of the X-ray scattering maximum is slightly different for the two specimens studied here. This might suggest a kind of texturing of the interchain packing which is induced by the electrical polarization. Finally, one observes that the high-angle peak above 30° becomes more easily distinguished in the high field case. This would be consistent with a better-developed interchain packing for the high-field polarized specimens.

Summary

Testing of theoretical predictions of polarization in polyacrylonitrile reveals that some local structure must be developed by the polarized state. Corroboration of this prediction appears to be obtained from infrared spectroscopy in the form of a fine structure in the nitrile absorbance maximum and a need to go to higher temperatures to cause relaxation of the nitrile orientations in the more strongly polarized specimens. X-ray diffractometry, likewise, suggests that a densification of the PAN is induced by the electrical polarized state of this material. It appears, therefore, that chains in electrically polarized PAN have a packing arrangement which cannot be achieved in specimens prepared by any previously studied thermo-mechanical method.

References
1. M. G. Broadhurst, et al., Polymer Preprints, Am. Chem. Soc. 14(2), 820 (1973).
2. S. I. Stupp and S. H. Carr, J. Appl. Phys. 46, 4120 (1975).
3. R. J. Comstock, S. I. Stupp, and S. H. Carr, J. Macromol. Sci.-Phys. 13, 101 (1977).
4. S. I. Stupp and S. H. Carr, J. Polym. Sci., Polym. Phys. Ed. 15, 485 (1977).
5. S. I. Stupp and S. H. Carr, J. Polym. Sci., Polym. Phys. Ed. 16, 13 (1978).
6. S. I. Stupp, Ph.D. Thesis, Northwestern University, Evanston, Illinois, August, 1977.
7. S. H. Carr, 1975 Conference on Electrical Insulation and Dielectric Phenomena, U.S. National Acad. Sci., Washington, DC, 1978, p. 45.
8. N. E. Hill, et al., Dielectric Properties and Molecular Behavior, Van Nostrand-Reinhold, London, 1969, Chapt. 1.

Acknowledgements
This work is supported by the Office of Naval Research. Use was made of facilities of the Northwestern Univ. Materials Research Center, which is supported under the NSF-MRL program, grant DMR76-80847.

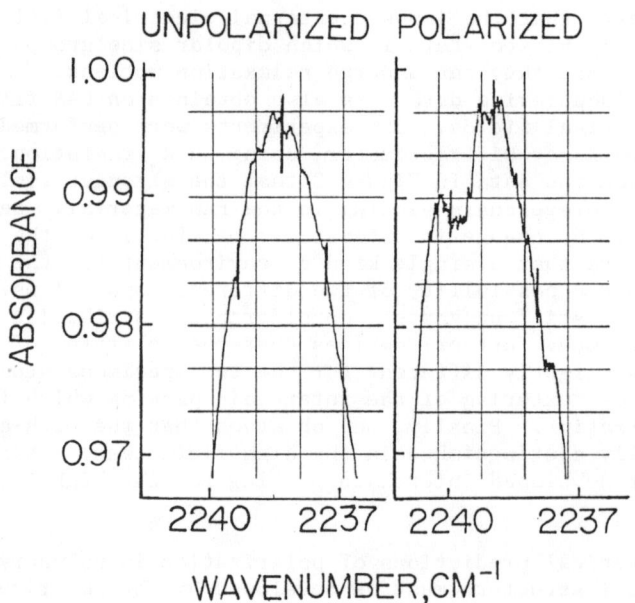

Figure 1. Infrared nitrile absorbance maximum, obtained from an unpolarized and an electrically polarized (5×10^5 V/cm) specimen.

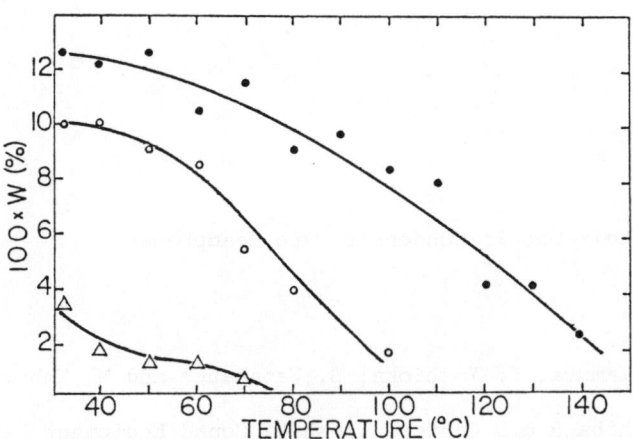

Figure 2. Temperature-dependence of nitrile orientation in films polarized at 145 C for 30 min under fields of 5 x 10^5 V/cm (●), 5 x 10^4 V/cm (○), and 0 V/cm (△). Orientations were obtained from infrared dichroism measurements.

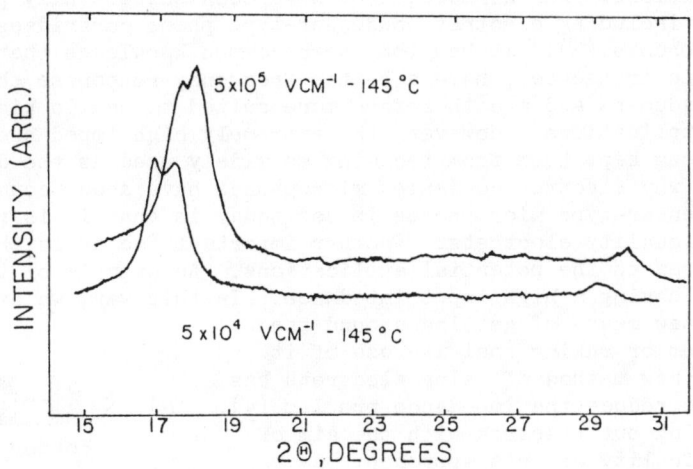

Figure 3. X-ray diffractometer scans from specimens polarized at 145 C for 30 min.

Plate Electrets And Their Use In Condenser Type Headphones

S. Tamura, H. Yoshioka, S. Watanabe* and M. Kobayashi*

Toshiba R & D Center, *Toshiba Sound Equipment Div.

1, Komukai Toshibacho, Saiwai-ku, Kawasaki City Kanagawa, 210
Japan

1. Introduction

Plate shaped electrets have already been developed having a negative surface potential and boasting over 100 years of electret voltage life under normal ambient temperature conditions. Furthermore, they have been successfully put to work in various products, including electret condenser-type phono cartridges (1) (2), microphones and headphones (3). It has long been common knowledge that electrostatic electroacoustic transducers have a better frequency responese than dynamic electroacoustic transducers and are therefore more suited to use in high fidelity audio reproduction applications. However, the extremely high impedance of electrostatics has long kept them from becoming as widely used as the dynamic type.

One of the reasons why electret condenser microphones have about cought up to the majority of condenser-type microphones in use today is that it is possible to mass produce high quality electrets. Another important factor is the related technology which increases the potential applications. An example of this is the development of FET's having a high input impedance. In this way, we are constantly looking for new means of getting around the high impedance problem or making positive use of it.

In recent years, a new method of using electrets has been devised so as to reduce the impedance problem (4). (a)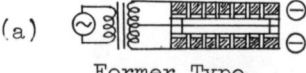

Former Type.

We have been carrying out research with the aim of increasing the practicality of this approach.

Specifically, this new approach is a design of the type shown in Fig. 1(a), as opposed to the conventional push-pull type shown in Fig. 1(b). (b)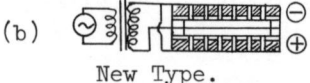

New Type.

As the inventors describe in the patent specification this configuration offers much better performance than the old design, when used in such electroacoustic transducers as headphones, loudspeakers, and the like. Fig.1 Cross sectional View

However, Fig. 1(a) and Fig. 1(b) are compared, notice of the electret condenser that, in the new design, the surface potential of the two headphones electrets must be of opposite polarity. Until now, noboday has been able to come up with a positive electret having the necessary performance characteristics. Because of this single fact, it has not been possible to incorporate this way transducer in any practical products.

Improving our product line, attempts have been made to develop a more reliable positive electret.

After much research, we have finally been successful in producing a positive electret of high surface voltage having high reliability on a per with our negative electret.

This report describes a method of manufacturing this new electret, its application in headphones, and hte results obtained so far.

2. General description of new type of electrostatic electroacoustic transducer:

Electrostatic electroacoustic transducers can be broadly divided into two categories. The first is the "single" category, in which the condenser configuration is formed by one fixed and one vibratable electrode. The second is the "push-pull"category, in which one vibratable electrode stands between two fixed electrodes.

Figure 2(a) shows an equivalent circuit for the conventional electrostatic electroacoustic transducer employing electrets shown in Fig.1(a). Fig.2(b) shows an equivalent circuit for the new electrostatic electroacoustic transducer that appears in Fig.1(b). The following letter symbols will be used in subsequent discussion.

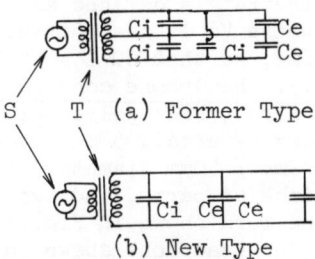

Fig.2. Equivalent circuit of electret condenser headphones

- C_e: Capacitance actuslly contributing to electroacoustic transduction.
- C_i: Capacitance not contributing to electroacoustic transduction. For example, the combined capacitance present in lead lines, distributed capacitance between the secondary windings of the transformer, and the electrostatic capacitance between the fixed electrodes and the nonmoving portions of the vibrating electrode opposite the fixed electrodes.
- S: Signal source   T: Step-up transformer

As can be seen from Figs. 1(b) and 2(b) the adoption of this new configuration finally makes it possible to greatly reduce the impedance of the electrostatic electroacoustic transducer. This also results in an improvement in the ratio between the capacitance effectively contributing to electroacoustic transduction (C_e), and the combined capacitance (C_i) that does not contribute to the process of electroacoustic transduction. In the case of headphones, this would mean an improvement in the ratio between the power converted to sound waves and the power dissipated as heat.

3. Experiments:

3.1. Manufacturing the electret

First, the basic electret material is obtained by affixing FEP Teflon film (0.075mm thick) to an aluminum plate (0.7mm thick). Then, this layered material is cut into a disk 62mm in circumference, and holes are punched to allow sound wave radiation (2mm ϕ X 120 holes). Next, the material is processed so that its surface becomes finely corrugated,

After ultrasonic washing and drying, corona discharge is used to create the electrets.

3.2. Judging electrets

Assuming that the electret is to be used in an electrostatic electroacoustic transducer, the most important qualities to consider are the strength of the surface potential, and the life of the surface potential. Surface potential was measured in the usual way, employing a vibrating reed electrometer: It was possible to judge how long the surface potential would last under normal temperature conditions by heating the electret in a metal box so as to speed up the loss of its charge.

Specifically, the life (τt) of the charge at each temperature "T" (°K) was measured and plotted on the graph. The straight lines drawn by $\log(\tau t)$ allow one to estimate electret life at normal temperatures (τrt). The life of the electret is defined as the time it takes for the surface potential to drop 3dB (where Vo is the initial potential and Vt is the potential after some amount of time, so that $20 \log Vt/Vo = -3$).

3.3. Headphone construction

First, PET (polyethylene terephthalate) is affixed to a metal ring (diameter: 60mm external, 50mm internal) to form the vibrating electrode. A PET ring (62mm / 50mm diameter) is used as a spacer and for insulation. With two more kinds of parts, and two negative plate electrets, the headphones is assembled in the form shown in Fig. 1(a).

The headphone shown in Fig. 1(b) uses all the same parts, but one of the plate electrets is positive. Furthermore, for the vibrating electrode, two approaches were tried. In one case, only one side of the PET film was coated with aluminum. In the other case, both sides were coated. The distance between the vibrating electrode and the electret surface is 0.35mm (consisting of the 0.3mm thick metal ring and the 0.05mm thick insulation spacer ring).

3.4. Judging headphone performance

Headphones built in the manner described above were tested under simulated human hearing conditions using simulated ears and standard test microphones (B&K 4134). The main items covered were output sound pressure versus input signal voltage, frequency response, and secondary (harmonic) distortion.

4. Test results and considerations :

Electret surface potential increases in proportion to the thickness of the electret material when the voltage supplied to the corona discharge electrode is maintained at a set level. If the unstable charge on the surface is removed by using electro-conductive rubber, the result is as shown in Fig. 3. This agrees with the results of leaving it for a day under 60°C and 95% relative humidity conditions. This is thought to be because the charge distribution within the electret film varies with thickness, but further investigation is under way.

Figure 4 shows the change in surface potential, with time lapse, for the 0.075mm thick FEP Teflon electret that has been employed in practical applications. Figure 5 shows the relationship between surface potential and elapsed time for the recently developed positive electret. Figure 8 shows the useful life of FEP electrets of different thicknesses.

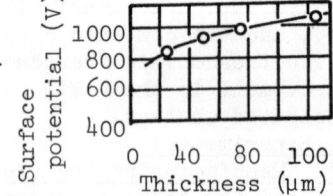

Fig.3. <u>Surface potential of FEP-electrets VS FEP films thickness</u>

The tests show that the thicker the FEP Teflon film, the longer the electret's life. Another result is that the negative electrets last longer than the positive ones of the same thickness. However, this in considered to be a characteristic of FEP Teflon, as has been suggested by other researchers (5) (6).

If the present development aim is an electret having an initial surface potential of +800V and a life of 50 years or more at normal temperature, then one can see from the test results that these requirements could be met by using 0.075mm thick FEP Teflon film.

There are certain problems that occur in the new design (Fig.2) using both positive and negative electrets that bear discussion. These are problems not present with the conventional design (Fig. 1(a)).

Since the vibrating electrode has to be able to conduct electricity, it is common to coat either one or both sides with metal. Furthermore, since the headphone must be able to faithfully reproduce a wide frequency range, the diaphragm (vibrating electrode) cannot be stretched more tightly than necessary.

Therefore, during large signal inputs, or if there is strong air pressure from one side, there is a danger of the diaphragm contacting the surface of one of the electrets. In a headphone, where such contact can occur, it is obviously necessary to consider the changes that may occur in performance as a result.

After much study of this phenomenon, the idea of corrugation was arrived at. This corrugation concept has been found to be very effective. A patent has been granted on this applicaiton (7). When using negative electrets it wasn't necessary to worry too much about the material used as the vibrating electrode.

However, as can be see, in Fig. 7, when a positive electret is employed, the loss of surface potential varies according to the material contacted. These results were arrived at using the apparatus shown in Fig. 7. The data for Fig. 7 were obtained by repeatedly contacting and separating the diaphragm and electret. While it may seem that the charge leaks away in this process, it is more correct to think that a new charge is created on the electret surface.

This is true because, if a negative charge has been generated on the film surface through repeated contact between the metal and FEP, it explains the lowered surface potential. Furthermore, this unstable charge can be removed by touching the surface with electro conductive rubber. While it is possible to use the same diaphragm material as that used in the cross section of the construction shown in Fig. 8, for headphones employing a positive electret (Fig. 2), it should be clear from the above discussion that it is much better

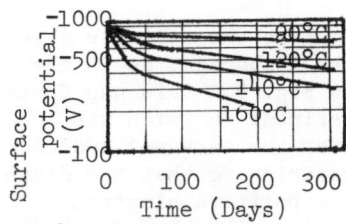

Fig.4. Decay of surface potential for **negative** charged electret

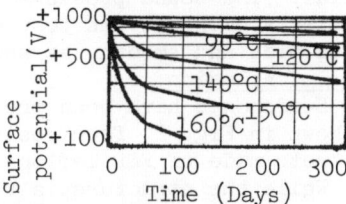

Fig.5. Decay of surface potential for positive charged electret.

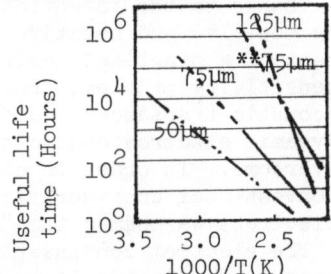

Fig.6. Life of FEP-Teflon electrets. (**negatively charged electret)

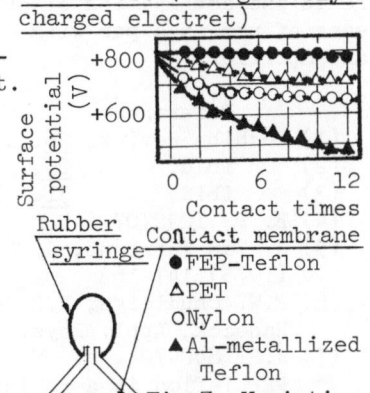

● FEP-Teflon
△ PET
○ Nylon
▲ Al-metallized Teflon

Fig.7. Variations of surface potential vs contact times.

to have the poritive electret facing the PET side of
the diaphragm, as in the cross sectional view shown
in Fig. 9. This was further established by an experiment in which the headphones were actually worn
over the ears and sudden pressure was applied to the
exterior so that the diaphragm membrane would contact
the positive electret.

Figure 10 shows test results for the headphones whose
structure is shown in Fig. 1(b), with electrets having
opposite polarity but the same absolute surface potential. The sound pressure produced was the result of
using a 400Hz 30Vrms test signal. Frequency response
was wide and smooth, as shown in Fig. 11.

6. Conclusion:

Headphones have been produced, whose structure is
shown in Fig. 2, from positive and negative fixed electrets made of FEP Teflon film.

While the structure is virtually the same as Toshiba's
other electret condenser headphones (Aurex HR-910, 810,
etc.), this new model has a voltage sensitivity
6dB higher thasn the previous models (101dB/3Vrms).

Thanks to the invention of the new configuration using positive and negative electrets, and the development of a practical positive electret, the voltage
sensitivity problem, from which electrostatic electroacoustic transducers suffered in comparison with
dynamic electroacoustic transducers, has finally been
overcome. In order to clearly distinguish it from
conventional configurations, this new way of using
electrets was named the "complimentary electret" design.

The electret condenser headphones described in this
report are already in practical use in Toshiba products
(Aurex HR-810II, HR-X1, HR-F1). Their greatest feature is the combination of outstanding audio reproduction performance, along with
ease of use on a per with dynamic headphones, which can be directly plugged into
an amplifier.

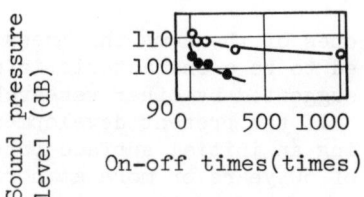

Fig.8. Electret-condenser headphone sensitivity as a function of times puting on off

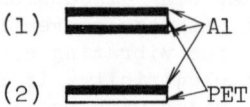

Fig.9. Construction of diaphragm membrane

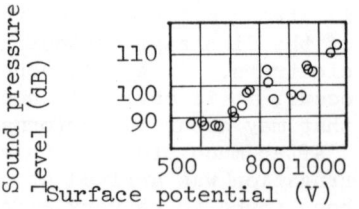

Fig.10. Sensitivity of electret condenser headphone. Input Signal: 400Hz 30V,.

Reference
(1) Toshiba review, <u>26</u>, 1308 (1971)
(2)   ibid,   <u>29</u>, 598 (1974)
(3)   ibid,   <u>31</u>, 172 (1976)
(4) B. P 1292707
(5) J. Van Turnhout, J. Electrostatics, <u>1</u>, (1975) 147-163
(6) P.W. Chudleigh, R.E. Collines, and G.D. Hanocck, Appl. Phys. Left., <u>23</u>. 211-212
(7) J.P. 840972
* FEP Teflon is a registered trademark of E.I. duPont de Nemours Co.,
* Aurex is a registered trademark of Toshiba Corp.,

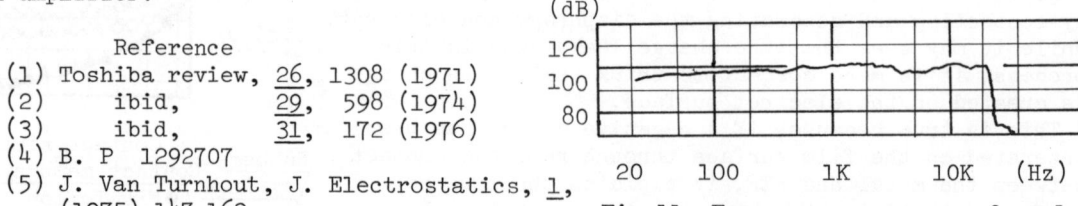

Fig.11. Frequency response of newly made electret condenser headphone Aurex HR-F1

A REVIEW ON MAGNETOELECTRET

C.S. Bhatnagar, M.L. Khare, M.S. Qureshi

M.A.C.T. BHOPAL 462007 (INDIA)

1. Introduction

Magnetoelectret(ME) is a dielectric body on which electric charge appears due to joint treatment of heat and magnetic field (1) and the change persists for long after the treatment (2). Various materials as waxes (3) plastics (4), organic and inorganic substances are studied for this effect. Samples in the form of thin disc with flat surfaces normal to the field, tubes with curved surfaces parallel to the field and cubes with two surfaces normal and four parallel to field have been studied.

2. Charge Characteristics

Charge on ME is of the order of 10^{-10} Coul/cm^2. It's gross features are phenomenologically described by the two-charge hypothesis (5) according to which it consists of intrinsic isocharge and intrinsic idiocharge. Intrinsic isocharge is a charge of the same sign on the opposite surfaces of the samples. Intrinsic idiocharge is polarization. Thus the surface charge density $\sigma = \sigma(\text{iso}) \pm \sigma(\text{idio})$. In disc samples prepared at lower forming temperature (T) below softening point the charge is mainly isocharge. Contribution of idiocharge increases on increasing (T) (Fig. 1a). Greater idiocharge is obtained on keeping opposite surfaces of samples insulated from each other during formation than on keeping them shorted. The isocharge does not seem to vary with forming magnetic field (H) but idiocharge varies with it (Fig. 2a). The variation pattern depends on the material and (T). In perspex samples for (T) above softening point, a reversal in the sense of polarization at some (H) is seen (Fig. 2a). This transition field varies linearly with (T). The role of (H), in addition to help promote the charge, is also to stablize it so that its persistence is prolonged. In tubular samples the charge is mostly isocharge even at higher (T). It's variation for outer and inner surface does not show mirror image relationship but the two curves have similar nature (Fig. 3). The charge has maximum value for some (H), which depends on the material. In cubical samples the charge on surfaces normal to the field has similar features as on disc samples, but the lateral surfaces mainly get isocharge. It seems that surfaces parallel to lines of force get only isocharge while those normal to them get both iso and idiocharge.

3. Charge Decay and Reversal

The charge decays with time and within a month it takes up a virtually constant value. In disc samples the decay is somewhat exponential. If the initial charge is idio, a reversal of sign on only one surface invariably takes place during decay and the final stable charge is iso, the sign of which is characteristic to the material. If initially it is the characteristics isocharge, no reversal takes place. Decay curves of tubular samples exhibit peaks and valleys before settling down to gradual fall. In sulphur and napthalene charge decay is much faster under illumination with mercury light. Instead of days it disappears in minutes. Semilog plot of photo decay (Fig. 4) gives two decay constants τ_1 and τ_2. The decay is governed by the relation

$$\sigma = \sigma_1 \exp(-t/\tau_1) + \sigma_2 \exp(-t/\tau_2)$$

4. Change in Magnetic Susceptibility

Materials undergo decrease in their diamagnetic susceptibility (overall increase) just after ME formation (6), due to polarization induced paramagnetism. In some cases the induced effect is so large that the material becomes paramagnetic. In many materials the susceptibility recovers on storage, but in carnauba wax the change increases with lapse of time, finally becoming constant after fifteen days (3). In some cases the material becomes more diamagnetic than normal after the disappearance of induced paramagnetism.

5. Change in Dielectric Constant

About 5% change in dielectric constant is observed after ME formation in some materials (7). Generally it decreases in audio frequency range. The decrease ($\Delta \epsilon$) is related to ME charge (Fig. 1b). Its variation with field closely follows the variation in charge irrespective of sense of polarization. Emperically $\Delta \epsilon = k_1 |\sigma_{iso}| + k_2 |\sigma_{idio}|$ is found to hold good, where k_1 and k_2 are constants of the material. There is partial recovery in dielectric constant on storage, and about 2% ($\Delta \epsilon$) is permanent.

6. Change in Refractive Index

A permanent change in refractive index upto $\pm 0.3\%$ is observed along field direction in transparent plastics (8) after ME formation. For lower (H) and (T) the refractive index increases. At moderate fields it decreases and then again increases for higher (H) and (T) (Fig. 5).

7. TSC-Study

Thermally stimulated current (TSC) of perspex ME has two peaks α and φ (Fig. 6) which correspond to two trap levels. Calculation of activation energy by three different methods give consistent values. Though TSC peak height vary irregularly with (H) but time constant at room temperature increases regularly with (H) confirming its charge stabilizing role.

8. Miscellaneous Effects

X-Ray diffraction patterns taken with X-ray beam perpendicular to (H) show preferred orientation in ME but photographs taken with beam parallel to field do not show such orientation. In ceresin wax a new type of orientation, D-type has been detected (9). However, preferred orientation is not uniquely related to ME charge (10). Composite ME -a two-layer sample made of two different materials - develops more charge and is more stable (11). A decrease in resistance is also found after ME formation. Most recently a change in the plane of polarization is noticed in optically anistropic material.

9. Discussion

Intrinsic isocharge is akin to frictional charge and may be due to contact between electrode and dielectric. In disc samples this charge transfer takes place along the magnetic lines of force and so it is not influenced by them. In tubular samples, however, the radial flow of charge carriers from metal to dielectric or vice versa, is perpendicular to the magnetic field so the carriers get deflected circumferencially. This increases trap probability near the surface giving more of isocharge. At certain (H) when carriers trajectory coincides with tube curvature, the trapping is maximum. This explains occurrence of isocharge maxima. Magnetic field can indirectly influence trapping by itself generating trap sites as visualized by Hasegawa (12).

The idiocharge may be due to space charge polarization and/or dipole orientation. At higher (T) space charge separation is possible due to greater ionic mobility and dipoles are also free to rotate. Thus greater idiocharge is expected at higher (T), as has been observed. Due to molecular diamagnetic anisotropy, the long chain molecules get oriented with their length perpendicular to magnetic field. If the field is in x direction, the molecules will be in yz plane. This would not necessarily produce polarization because electric dipoles may still have random orientation in yz plane. Hence preferred orientation should not be uniquely correlated with ME charge as has been reported (10). Idiocharge may be visualized due to orientation of side chains along field direction. Due to thermal excitations, atoms in side chains may lose electrons and may become paramagnetic. The side chain will therefore try to align in the field direction. It seems that at lower (T) only side chains get twisted and orient parallel to the field. At higher (T) main chains orient normal to the field causing greater alignment of side chains along field direction. In tubular samples idiocharge would not be observable on cylindrical surfaces. Idiocharge may also arise due to asymmetric temperature gradients at the surfaces during

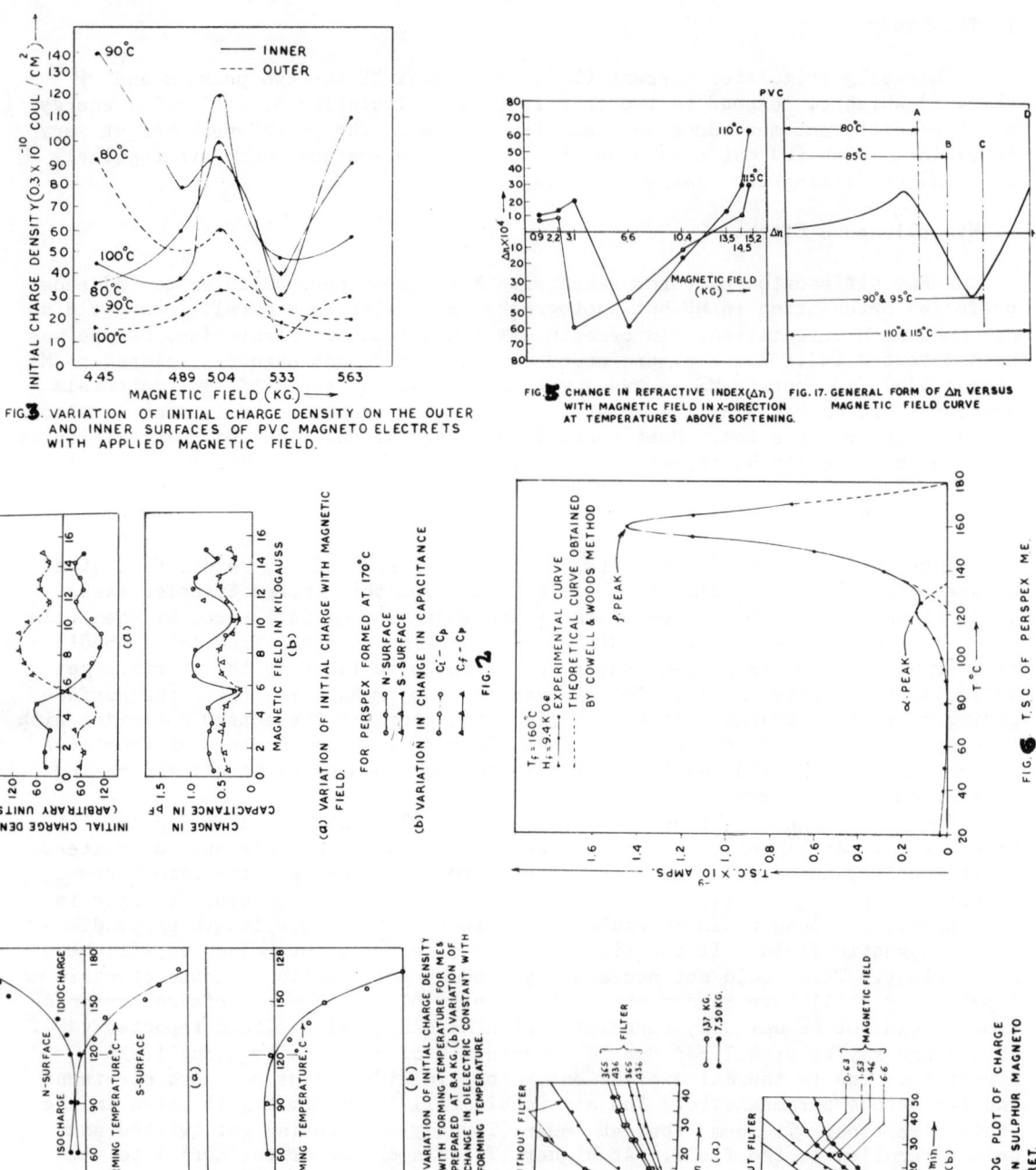

cooling.

References
1) C.S.Bhatnagar, Indian Jour. Pure Appl. Phys., 4, 355 (1966)
2) PKC Pillai, Kamlesh Jain, & VK Jain, J. Electrochemical Soc. (USA), 118, 1675 (1971)
3) SD Chatterjee, KD Roychaudhury, H.De & K.Banerjee, Naturweissenschaften, 56, 324 (1969)
4) MS Qureshi & CS Bhatnagar, Indian Jour. Pure Appl. Phys., 9, 361 (1971)
5) ML Khare & CS Bhatnagar, Indian Jour. Pure Appl. Phys., 8, 700 (1970)
6) M.Bandopadhyay & ML Khare, Indian Jour. Pure Appl. Phys., 13, 19 (1975)
7) JK Quamara, ML Khare & CS Bhatnagar, Indian Jour. Pure Appl. Phys., 15, 617 (1977)
8) JK Quamara, ML Khare & CS Bhatnagar, Indian Jour. Pure Appl. Phys., 15, 138 (1977)
9) DS Parashar & CS Bhatnagar, Indian Jour. Pure Appl. Phys., 13, 131 (1975)
10) NR Pan, Indian Jour. Pure Appl. Phys., 14, 800 (1976)
11) M Bandopadhyay & ML Khare, Indian Jour. Pure Appl. Phys., 14, 163 (1976)
12) H.Hasegawa & M.Nakamura, Jour. Phy. Soc. Japan, 26, 1362 (1969)

$Bi_{12}SiO_{20}$ PHOTOELECTRET IN OPTICAL PROCESSING AND X,γ RAYS IMAGING

F. MICHERON

Thomson-CSF BP.10 - 91401 ORSAY France

C. MAYEUX

COMPAGNIE GENERALE DE RADIOLOGIE -

3, rue d'Amiens - 93240 STAINS France

$Bi_{12}SiO_{20}$ (and isomorph $Bi_{12}GeO_{20}$) are known as electrooptic, piezoelectric and highly photoconductive crystals (1). Under non uniform illumination, photoelectrons self diffuse or drift under an applied field before being trapped. At room temperature, drift mobility is $u = 10^{-2}$ V^{-1} $cm^2 s^{-1}$ and lifetime = 1.7 10^{-5}s, therefore photoelectron displacements are 1.7 μm per KV applied.

The photoinduced space charge (photoelectret state) may be observed either by changes in the refractive index (photorefractive effect), or selective toner deposition (electrophotography). The photorefractive effect allows images and holograms recording as refractive index spatial modulations, and readout is performed without any development process, in real time and in situ. The photon-electron quantum efficiency is 0.7 at the blue line of the argon laser ; the photosensitivity is comparable with the high resolution (≥ 2000 lines pairs/mm) photographic plate sensitivity (100-300 μJ/cm^2) (2,3,4,5). Storage time is nearly the dielectric relaxation time, i.e. 30 hours in the dark ; uniform illumination causes erasure, and no fatigue effect is observed during write-erase cycles.

In the longitudinal electrooptic configuration (electric field parallel to the light beam), and between blocking electrodes, $Bi_{12}SiO_{20}$ is used as incoherent-coherent light converters for data input in optical processors (PROM, 6). In the transverse electrooptic configuration, (electric field perpendicular to the light beam), and using the holographic technique, $Bi_{12}SiO_{20}$ provides completely new possibilities in optical processing : real time holographic interferometry, image differentiation and substraction (7,8,9).

Since these crystals contain heavy Bismuth ions, they show large X and γ rays absorption. High photoconductivity efficiencies have been measured in these crystals : more than 3000 electrons are detected per 30 KeV photon, therefore these crystals may be used as sensitive X and γ rays imaging devices : in

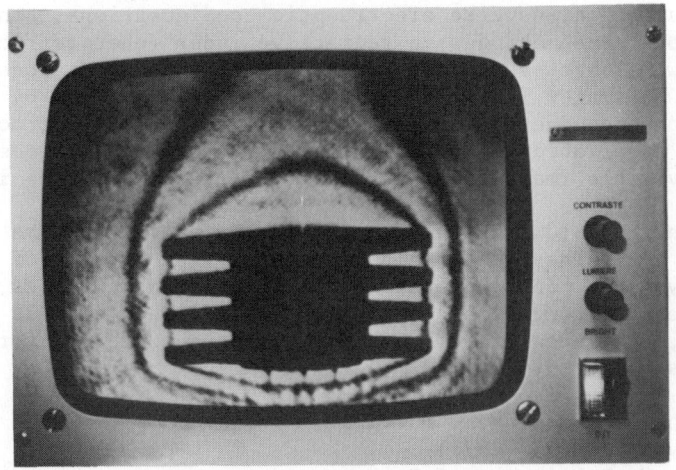

Real time holographic interferometry of a heating transistor
BSO crystal, λ = 514nm (Ref.7)

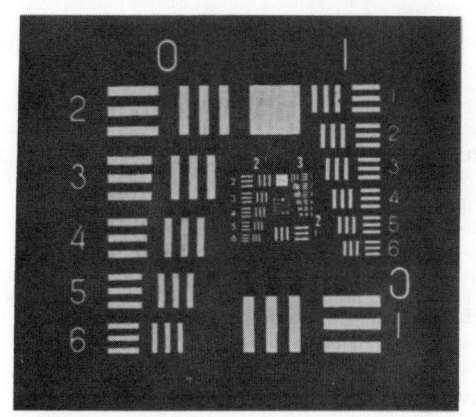 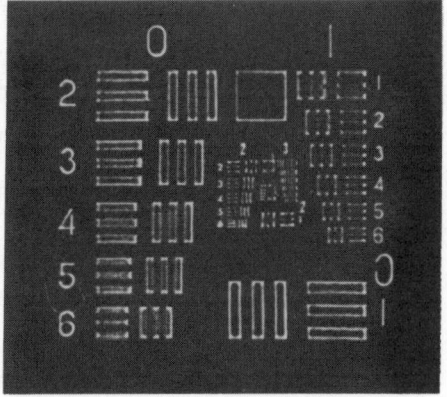

Image differentiation in BSO crystal using the image-hologram
technique at λ = 514nm.

both longitudinal and transverse electrooptic configurations, non uniform X, γ rays illuminations cause changes in refractive index (photorefractive effect at X and γ wavelengths), which are used in direct conversions of X, γ images into visible images (10). The same results are obtained by combining $Bi_{12}SiO_{20}$ (the photoconductor) and a liquid crystal (the electrooptic material 11). The memory associated with the photoelectret state makes possible observations of the visible image after a short time exposure to the ionizing radiation.

Polycristalline $Bi_{12}SiO_{20}$ has been prepared for the first time using hot pressing techniques (12). These materials show nearly the density of single crystals ($<$ 1% porosity), but can't be obtained as transparent ceramics, since the cristallites are birefringent and optically active. When used as Xerographic photoconductors, they show higher sensitivities than Selenium for high energy X rays (energies $\geqslant$ 80 keV).

(1) R.E. Aldrich, S.L. Hou, M.L. Harvill
    J. Appl. Phys. $\underline{42}$, 493 (1971)
(2) F. Micheron, M. Peltier and J.P. Huignard
    Opt. Comm. $\underline{42}$, 216 (1976)
(3) J.P. Huignard and F. Micheron
    Appl. Phys. Letters $\underline{29}$, 591 (1976)
(4) M. Peltier and F. Micheron
    J. Appl. Phys. $\underline{48}$, 3683 (1977)
(5) F. Micheron
    Ferroelectrics - $\underline{18}$, 153 (1978)
(6) R.A. Spragne and P. Nisenson
    Opt. Engin $\underline{17}$, 256 (1978)
(7) J.P. Huignard and J.P. Herriau
    Appl. Opt. $\underline{16}$, 1807 (1977)
(8) J.P. Huignard, J.P. Herriau and T. Valentin
    Appl. Opt. $\underline{16}$, 2796 (1977)
(9) J.P. Herriau, J.P. Huignard and P. Aubourg
    Appl. Opt. $\underline{17}$, 1851 (1978)
(10) C. Mayeux and F. Micheron
     French Pat 77 23 441
(11) C. Mayeux, F. Micheron, J.P. Huignard, M. Hareng
     French Pat 77 28 738
(12) A. Morell and A. Hermosin
     Ann. Meeting of the Amer. Ceram. Soc., Detroit, May 1978
     - To be published J. Amer. Ceram. Soc.

III
PIEZO- AND PYRO-ELECTRICITY

POLYVINYLIDENE FLUORIDE - VIEWED AS A POLYCRYSTALLINE FERROELECTRIC MATERIAL*

M. G. Broadhurst

National Bureau of Standards

Washington, D. C.

Piezoelectric and pyroelectric polyvinylidene fluoride (PVDF) is being used increasingly in transducer applications. It's advantages over ceramic material include light weight, flexibility, toughness, ease of fabrication and low permittivity. Detailed descriptions of the mechanisms whereby 1) a high electric field creates activity in PVDF, 2) an active specimen of PVDF responds electrically to changes in temperature and mechanical stress and 3) the activity decays slowly with time, have not yet been agreed upon. Such descriptions are needed to facilitate the optimization of PVDF transducers. This paper reports some results of recent work from our laboratory (1,2) which helps clarify the above mechanisms. We propose here that 1) a large electric field causes crystallized segments of PVDF to rotate individually about their molecular axes and repack in polar crystal lamellae with preferential alignment in the direction of the applied field, 2) these stable aligned crystals change their net polarization when temperature or stress are changed largely because of dimensional changes and 3) the aging of the activity is due to untwisting of the molecular links between crystalline segments where stress was concentrated during the original poling process.

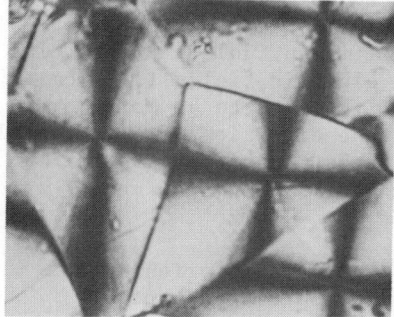

A change in polarization can occur through a change in moment or volume. Most proposed mechanisms for activity in PVDF focus on changes in moment. Here we emphasize effects of volume changes where oriented polar crystals are present. Fig. 1 is a photomicrograph of a melt-crystallized PVDF film taken between crossed polaroids, and showing spherulitic structures. A typical spherulite consists of parallel layers of alternating crystal and liquid material, each layer 10 to 20 nm thick. In the crystals the molecular segments are linear and roughly normal to the large crystal surfaces.

Figure 1. Photomicrograph of a rapidly cooled melt of commercial unoriented PVDF between crossed polaroids.

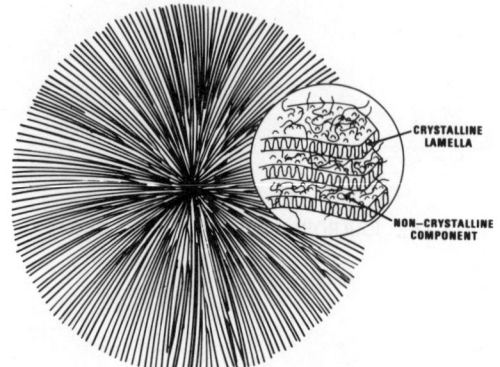

Figure 2. (left)

A schematic diagram of a spherulite and a detail of a section emphasizing the lamellar structure of the radiating branches.

These linear molecular segments can have either a tgtg' conformation (as in Form II) or an all-t conformation (Form I) as shown in Fig. 3. The dipole moment for all-t is about 2.1D (1D = 3.3×10^{-28}Ccm) normal to the chain axis and for tgtg' it is 1.2D normal to the chain axis and 1.0D along the chain axis.

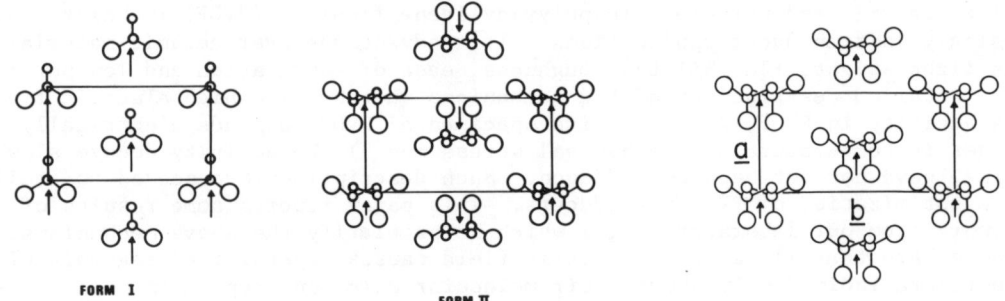

Figure 3. (Above) Projection of poly(vinylidene fluoride) chain onto ab plane of unit cell for polymorphic forms I and II and for polar form II. Larger circles represent fluorine, smaller circles represent carbon, and hydrogens have been omitted. Arrows indicate net dipole moment normal to chain axis.

The Form I unit cell is polar and the Form II unit cell is antipolar. Both forms can be made piezo- and pyroelectric by applying large electric fields (1 MV/cm). The molecules rotate about their long axes (individually) so as to be more favorably aligned with respect to the field, thus changing the direction of the crystal moment. At fields near 1 MV/cm antipolar Form II crystals transform to polar Form II with no change in conformation. After poling near 5 MV/cm the chain conformation has been changed to produce Form I.

Xray diffractometer scans obtained from 12μm thick biaxially oriented commercial films are shown in Fig. 4 and 5. Results are for specimens that are unpoled (solid line), and poled at 1.25 Mv/cm (dashed line) and 5 MV/cm (dotted line). Comparison of the solid and dashed reflections shows that 100 decreases

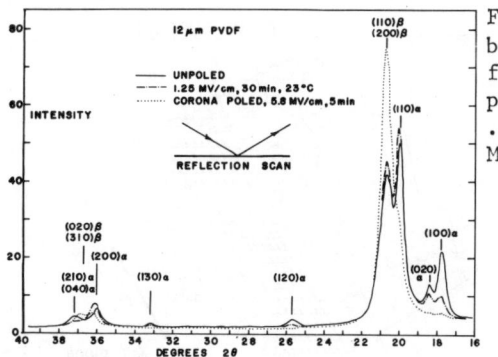
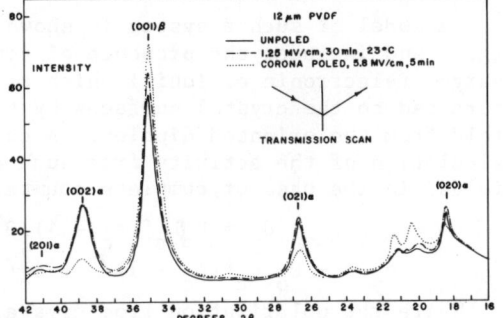

Figure 4 (left) X-ray diffractometer scan from biaxially oriented PVDF film in symmetrical reflection mode. ——Before poling;——·——After poling at 1.25 MV/cm for 30 minutes at 23°C; . . .after poling in corona discharge at 5.8 MV/cm for 5 minutes at 23°C.

Figure 5 (right) X-ray diffractometer scan from same films as in Fig. 3 but tilted 90° to examine in symmetrical transmission mode. Symbols same as in Fig. 3.

and 200 increases in intensity so that the results are not due simply to unit cell orientation. Structure factor calculations for the polar Form II relative to antipolar Form II predict decreased 100, 011, 120, 121, and 210 reflections, unchanged 020 and 040 reflections and an increase in the remaining reflections in accord with the data. The antipolar to polar transformation at 1.25 MV/cm gave no changes in the infrared spectrum, as shown in Fig. 6. The poled samples yielded hydrostatic piezoelectric and pyroelectric coefficients of d = 6.9 pC/N and p = 1.4 nC/cm^{2}K (poled at 1.25 MV/cm) and approximately 15 pC/N and 3.5 nC/cm^{2}K (poled at 5 MV/cm). Thus it appears plausible that the basis for the activity in both forms of PVDF is the presence of oriented dipolar crystals.

Figure 6. (right) Infrared spectra from single thickness of 12 μm thick biaxially oriented PVDF before and after poling. Same symbols apply as used in Figs. 4 and 5. Data after poling at 1.25 MV/cm are coincident with that from unpoled speciman over most of the range in frequency.

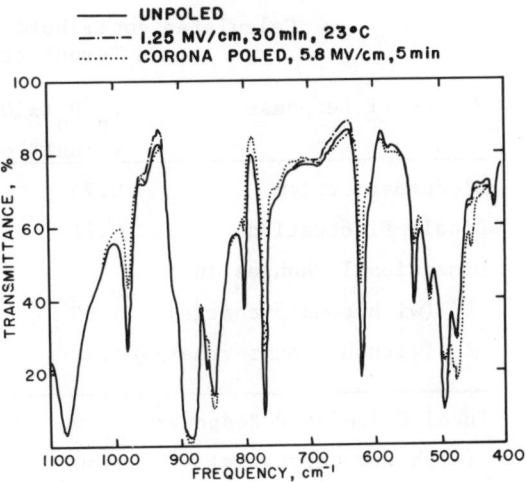

Figure 7. (right) A schematic diagram showing dipole alignment and countercharge on a lamellar section of polar crystal. In this paper we calculate the response from a preferentially aligned array of such objects in a semicrystalline sample.

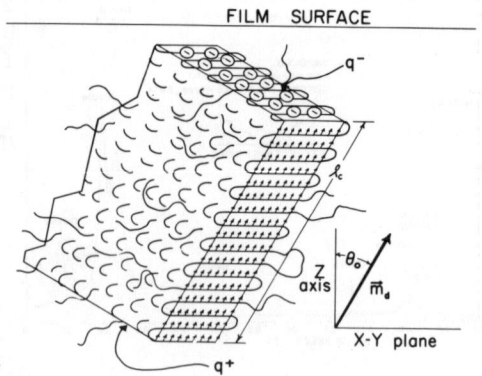

A model of such a system is shown in Fig. 7 and includes the presence of countercharges (electronic or ionic) which are attracted to the crystal surfaces by the field from the oriented dipoles. A careful calculation of the activity from such a model yields, in the case of complete countercharge

$$d_p = P_o \beta_c [(\epsilon_c - 1)/3 + \phi_o^2 \gamma/2 + \partial(\ln \ell_c)/\partial(\ln v_c)]$$

$$p_y = -P_o \alpha_c [(\epsilon_c - 1)/3 + \phi_o^2 (\gamma + (2T\alpha_c)^{-1})/2 + \partial(\ln \ell_c)/\partial(\ln v_c)]$$

Where the polarization from crystal dipoles is,

$$P_o = \Phi(\epsilon_c + 2) N \mu_o J_o (\Phi_o) <\cos \theta_o>/3V_c$$

For the case of no counter charge the crystal length ℓ_c is replaced by the sample thickness ℓ_s. The equations were evaluated for Form I crystals, using the following experimental values: Temperature, T = 300K, volume expansion coefficients for the crystal α_c = 1.7 x 10^{-4}K^{-1} and sample α_s = 4.2 x 10^{-4}K^{-1}, volume compressibilities for the crystal β_c = 1.1 x10^{-10}m^{2}N^{-1} and sample β_s = 2.39 x 10^{-10}m^{2}N^{-1} crystal relative permittivity, ϵ_c = 3, Grüneisen constant, γ = 5, dipole fluctuation amplitude ϕ_o = 16°, volume fraction of crystals, Φ = 0.5, crystal polarization for Form I, $N\mu_o/V_c$ = 12 x 10^{-6} Ccm^{-2}, d_p/P_o = 2 x 10^{-6}cm^{2}N^{-1} and p_y/P_o = 4 x 10^{-4}K^{-1}. These equations yield the results shown in Table I.

TABLE 1

Calculated contributions to Piezoelectricity and Pyroelectricity in PVDF

Source of Response	d_p/P_o (x10^{-6}cm^{2}N^{-1}) value	(% of exp. total)	p_y/P_o (x10^{-4} K^{-1}) value	(% of exp. total)
Electrostriction	0.74	(37)	1.14	(28)
Dipole Fluctuations	0.21	(10)	0.98	(24)
Dimensional Changes in				
ℓ_c (with countercharge)	0.55	(28)	0.85	(21)
ℓ_s (without countercharge)	1.2	(60)	2.1	(52)
Total Calculated Response				
(with countercharge)	1.50	(75)	2.97	(74)
(without countercharge)	2.15	(107)	4.22	(106)

This model predicts quite well the observed activity of Form I PVDF. Varying the amount of countercharge assumed present at the crystal surfaces has only a secondary effect on the results. It is seen that a close relationship between the piezoelectric and pyroelectric coefficients is predicted and their predicted ratio $d_p/P_y \approx .005 \text{Kcm}^2\text{N}^{-1}$ in good agreement with experiment (3,4). The model also predicts that all of the piezoelectricity and about 85% of the pyroelectricity is secondary (i.e. dependent on volume change). This prediction is in apparent contridiction to the results of Kepler and Anderson (5) who concluded that no more than 50% of the pyroelectricity in PVDF is secondary. We believe this discrepency is due to an assumption of isotropic mechanical properties (5). We do not see that an additional mechanism such as reversible temperature and pressure induced crystallization is necessary to account for the activity in PVDF.

* Work supported in part by the Office of Naval Research

1. G. T. Davis, J. E. McKinney, M. G. Broadhurst and S. Roth, J. Appl. Phys. 49, 4998, (1978).

2. M. G. Broadhurst, G. T. Davis, J. E. McKinney, and R. E. Collins, J. Appl. Phys. 49, 4992, (1978).

3. P. D. Southgate, Appl. Phys. Letters 28, 250 (1976).

4. J. E. McKinney and G. T. Davis, Organic Coatings and Plastics Chemistry, 38 271 (1978) (Preprints of ACS Meeting in Anaheim, CA, March 1978.)

5. R. G. Kepler and R. A. Anderson, J. Appl. Phys. 49 4490 (1978).

Fine Structure of Drawn Polyvinylidene Fluoride and Its Poling Effects

Nobuyuki Takahashi and Akira Odajima

Department of Applied Physics, Hokkaido University

Sapporo 060, Japan

I. INTRODUCTION

The form-I modification of polyvinyliden fluoride (PVDF), the crystal studied in this work, consists of two planar zig-zag chains per the orthorhombic (pseudo-hexagonal) unit cell. In this crystal structure, the dipole moments of the monomer units are all parallel to the b-axis. The film of form-I PVDF prepared by stretching exhibits a strong piezo- and pyroelectricity after the poling process.

A number of studies have been attempted to elucidate the orientation of dipoles in the crystalline phase occuring during the poling. Kepler et al.(1,2) observed changes enduced in the X-ray pole figures of the PVDF film by poling and also the changes in the aximuthal angle dependence of the intensity of the {110, 200} composite reflection. Similar results were obtained by IR-spectroscopy(3,4). Fukada et al.(4) gave the relation of the infrared dichroism and the piezoelectric constant d_{31}. But, conclusive evidence of the crystallite orientation has not been obtained experimentally.

One of the reason is that PVDF is a semi-crystalline polymer with low crystallinity and small crystal size, and its crystalline lattice is severely distorted by chemical sigularities (head to head and tail to tail) included in its molecular chains. Nevertheless, little has been investigated about defects in this crystal so far. We have made X-ray diffraction studies on the PVDF stretched film, in order to search the influence of the poling process on it from view point of defect. In this paper, studies on the crystallite orientation, the micro-texture and defect structures in the PVDF stretched film and their poling effects are reported.

II. EXPERIMENTS

The PVDF film with 30 μm thickness is the KF-polymer supplied by Kureha Chem. Ind. Co. Ltd. As-drawn films were stretched uniaxially to approximately five times their original length at 100°C under the constant width by the use of a rolling mill. Typical poling conditions used for the present experiment were appling of a DC electric field of 7.0×10^5 V/cm to the sample at 120°C for 1 hour. The sample was subsequently cooled to room temperature with the field applied. The lengths of the sample were kept constant during this poling process. For comparison annealed samples were also made by the same heat treatment as the poled. All of sample preparations were made at Nishiki Research Laboratories, Kureha Chemical Industry Co. Ltd.

The X-ray source was a Rigaku Denki RU-100PL rotating anode X-ray unit and was operated at 40KV and 80mA. The copper radiation was nickel-filtered. X-ray diffraction profiles were measured by using an X-ray diffractometer with a scintilation counter and a pulse-height analyser in terms of a step scanning method. The specimen is either a stack of films with the desired thickness for the diffractometry or a rectangular prism ($0.5 \times 0.5 \times 2$ mm) made of stacking films for the X-ray photography. The arrangement of the slit system was carefully made to avoid aberations in the intensity measurement of the X-ray diffraction for the determination of the orientation distribution.

(Table 1) Structure factors of (hkl) in form-I PVDF.

hkl	$m|F|^2$	$\dfrac{m|F|^2}{\Sigma m|F|^2}$
200	408	0.23
110	1372	0.77
310	728	0.64
020	403	0.36
400	87	0.55
220	71	0.45
600	62	0.25
330	190	0.75
620	56	0.51
040	54	0.49
201	0	0.00
111	1276	1.00
311	174	0.23
021	488	0.77
401	260	0.73
221	94	0.27

III. ANALYTICAL METHOD OF BIAXIAL ORIENTATION

The orientation of the normal vector to the (hkl) plane in a crystallite, $\vec{r}_i$ with respect to the sample-fixed coordinate system is conveniently specified by the polar angle χ_i and the aximuthal angle η_i. Herein, $\chi_i = 0$ denotes the draw direction and $\eta_i = 0$ the normal to this film surface. The orientation distribution of $\vec{r}_i$ is given by the normalized intensity:

$$q_i(\chi_i, \eta_i) = I_i(\chi_i, \eta_i) / \int_{-\pi}^{\pi} \int_{-1}^{1} I_i(\chi_i, \eta_i) \, d(\cos\chi_i) d\eta_i , \qquad (1)$$

where $I_i(\chi_i, \eta_i)$ means the integrated intensity around the Bragg angles. A general theory in the determination of preferred orientation of crystallities has been obtained by Roe(5) in terms of an inversion method based on spherical harmonics.

In this work we assume, for simplicity, that the c-axis of crystallites is parallel to the draw direction. Practically, it is adequate to deal with the intensity averaged along the polar angle χ_i as follows:

$$\bar{q}_i(\eta_i) = \int_{-1}^{1} I_i(\chi_i, \eta_i) \, d(\cos\chi_i) / \int_{-\pi}^{\pi} \int_{-1}^{1} I_i(\chi_i, \eta_i) \, d(\cos\chi_i) d\eta_i . \qquad (2)$$

We consider the vector $\vec{r}_i{}'$, the projection of $\vec{r}_i$ on to the plane normal to the draw direction. The present method results in the determination of the angular distribution $\bar{q}_i(\eta_i)$ of the vector $\vec{r}_i{}'$ about the fibre axis of the sample. However, its determination for form-I PVDF is not easy, since almost reflections are composed of several independent (hkl) planes, having so nearly the same Bragg angle that they cannot be resolved. The orientation distribution of the composed vector $\vec{\gamma}_i = \sum_j \vec{\gamma}_{ij}$ ($j = 1, 2 \ldots n$) is expressed by

$$\bar{q}_i(\eta_i) = \sum_{j,\pm} C_{ij} \times \bar{q}_{ij}^{\pm}(\eta_i) , \qquad (3)$$

where $C_{ij} = m_{ij}(F_{ij})^2 / \sum_j m_{ij}(F_{ij})^2$ (F_{ij} is the structure factor of the (hkl) plane corresponding to the vector, $\vec{\gamma}_{ij}$, and m_{ij} the multiplicity factor). Each vector $\vec{\gamma}_{ij}$ is oriented with the azimuthal angle, $\pm \Phi_{ij}$ from the b-axis in the crystal-fixed coordinate system. Therefore, the angle $\phi^{\pm} = \eta_{ij} \pm \Phi_{ij}$ specifies the azimuthal angle of b-axis in the crystallite, to which the vector belongs, with respect to the sample space. If $\bar{q}(\phi)$ denotes the pole distribution of the b-axis, the orientation distribution, $\bar{q}_{ij}^{\pm}(\eta_i)$ in the right side of eq.(3), is given by

$$\bar{q}_{ij}^{\pm}(\eta_i) = \bar{q}_b(\eta_i \pm \Phi_{ij}) \equiv \bar{q}_b(\phi^{\pm}) . \qquad (4)$$

Then, the biaxial orientation coefficient $\langle \cos 2\phi \rangle_{\bar{q}_b}$ of b-axis is obtained from

$$\langle \cos 2\phi \rangle_{\bar{q}_b} = \langle \cos 2\eta_i \rangle_{\bar{q}_i} / \sum_j C_{ij} \cos 2\Phi_{ij} . \qquad (5)$$

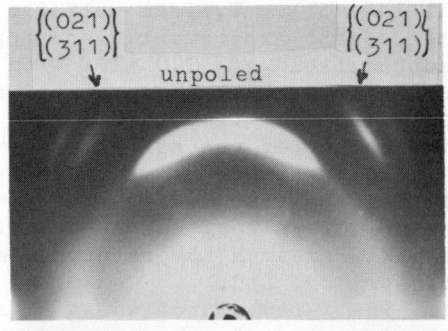
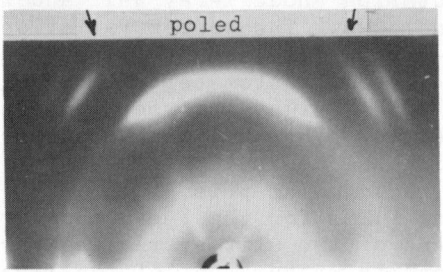

(Fig.1) X-ray photographs of PVDF poled and unpoled.

IV. BIAXIAL ORIENTATION OF FORM-I PVDF

The structure factors of form-I PVDF are shown in Table 1, which are calculated based on the Hasegawa et al.'s crystal structure(5). Since almost all the strong reflections on equator are composite reflections as shown in Table 1, they are not suitable for the determination of the biaxial orientation. Although the {021, 311} and {401, 221} reflections are the composite ones, the intensities of (021) and (401) in them are about 3 times as large as those of the other (311) and (221), respectively. The projected vector $\vec{\gamma}_i'$ from their normals is parallel either to the a-axis or to the b-axis and the coefficients $\sum_j C_{ij} \cos 2\Phi_{ij}$ in eq.(5) are comparatively large. Therefore, it is most convenient to work with these reflections.

The X-ray photographs of the unpoled and poled PVDF stretched film are shown in Fig.1. They were taken with the incident beam with the direction of $\chi = 84°$ and $\eta = 45°$ in order to observe the intensities of {021, 311} and {401, 221} in the arrangement of $\eta = 0°$ (the left side) and $\eta = 90°$ (the right side). The unpoled sample was annealed at 120°C for 1 hour with the same condition as the poled one mentioned before. Figure 1 shows that the b-axis of crystallites is preferentially oriented parallel to the film surface in the unpoled sample and perpendicular to it in the poled, since for the former the {021, 311} intensity on the right side ($\eta = 90°$) is stronger than the one on the left side ($\eta = 0°$) while the relation is the reverse for the latter. Thus it is clarified that the poling process affects the b-axis orientation even though the intensity changes are weak.

These findings are confirmed from the observed values of $\langle \cos 2\phi \rangle_{\bar{q}}$. To do this, the $\{311, 021\}$ and $\{401, 221\}$ intensities at both $\eta = 0°$ and $90°$ were measured by the X-ray diffractometry. If distribution of b-axis orientation is known, $\langle \cos 2\eta_i \rangle_{\bar{q}_i}$ can be evaluated by using the observed intensities at these two angles in terms of the least squares method. When the degree of orientation is low, the distribution function can be approximated by two terms in Fourier series:

$$\overline{q}_b(\phi) = a_o + a_1 \cos 2\phi . \qquad (6)$$

The biaxially orientation coefficients of b-axis, $\langle \cos 2\phi \rangle_{\bar{q}_b}$ are obtained as 0.15 and -0.25 for the poled and unpoled samples, respectively. The former shows the b-axis orientation, while the latter the a-axis orientation.

V. ORIENTATIONAL DISORDER AND DISCUSSTION

(Fig.2) Calculated and observed intensity ratios between the poled and unpoled samples.
(R: reflection, T: transmission)

Intensities of equatorial reflections for the poled and the unpoled samples with the orientation coefficients, 0.15 and -0.25, respectively were calculated at both the transmission (T) and the reflection (R) modes, based on the structural factors given in Table 1. Figure 2 shows a comparison of the calculated intensity ratios between the poled and unpoled samples with respect to the corresponding observed ratios. There are fairly good agreements between the observed and calculated ratios except for the R $\{400, 220\}$ and T $\{400, 220\}$ reflections.

Figure 3 indicates that the observed profile of R $\{400, 220\}$ for the unpoled sample is broader than that for the poled. On poling the height and area of the R $\{400, 220\}$ profile increase. Examinations of the observed as well as calculated intensity relations among equatorial reflections revealed that the observed intensity of the R $\{400, 220\}$ for the unpoled sample was abnormally weak(7). It is suspected that the abnormal intensity decreasing of the R $\{400, 220\}$ in the unpoled sample is associated with the first kind distortion, e.g. some type of disorderness in its lattice. It may be related to this disorder that the a-axis of crystallites in the film is preferentially oriented normal to the film surface as mentioned above by the stretching process with a rolling mill. We propose a disordered alignment such that a portion of long molecules in the lattice is oriented about its chain axis together with the c/2 translation, resulting in a chain twisting. Preliminary investigation shows that the chain molecule can move from the regular position to a meta-stable position with the chain dipole orientation from the regular direction to 120-180°, accompanying with a slight lateral shift of its chain axis. The disordered arrangement of a chain molecule with the orientation angle $\varphi = 150°$ in the lattice is illustrated in Fig.4. The calculation of intensities scattered from such a distorted structure revealed that a marked reduction for only $\{400, 220\}$ yields in the case of the 135° orientation(7).

As shown in Fig.5 the T 400, 220 profile for the poled film indicates a doublet structure. It was ascertained that the doublet is attributed to the lower angle shift of the (400) reflection in the crystallite, the b-axis of which is orientated to the electric field (7). Futher studies on the structural change induced by poling is now in progress.

(Fig.4) Orientational disorder model. ($\varphi = 150°$)

(Fig.3) $\{400, 220\}$ profiles of the poled and unpoled PVDF in the reflection arrangement.

(Fig.5) $\{400, 220\}$ profiles of the poled and unpoled PVDF in the transmission arrangement.

ACKNOWLEDGEMENTS

The authors wish to thank Dr. N. Murayama and Dr. K. Nakamura, Nishiki Research Laboratories, Kureha Chemical Industry Co. Ltd., for their kindly supplying of samples and cooperation throught this work. This work was supported in part by the Grant-in Aid for Scientific Research from the Ministry of Education.

REFERENCES

1) R. G. Kepler: Piezo. and Pyro. Symposium-Workshop, Gaithersburg, U.S.A. (1975).
2) R. G. Kepler, R. A. Anderson: J. Appl. Phys., 49 (1978) 1232.
3) M. Tamura, S. Hagiwara, S. Matsumoto and N. Ono: J. Appl. Phys., 48 (1977) 513.
4) E. Fukada, M. Date, T. Furukawa: 175-ACS National Meet., Anaheim, U.S.A. (1978).
5) W. Ruland, W. Wiegand: Polym. Symposium, 58 (1977) 43.
6) R. Hasegawa, M. Kobayashi, H. Tadokoro: Polym. J. 3 (1972) 600.
7) N. Takahashi, A. Odajima, K. Nakamura and N. Murayama: to be published.

ON THE NATURE AND MECHANISM OF PIEZO- AND PYRO-ELECTRICITY IN

POLYVINYLIDENE FLUORIDE

D. K. Das-Gupta and K. Doughty

School of Electronic Engineering Science, University College

of North Wales, Dean Street, Bangor, Gwynedd, LL57 1UT, G.B.

Polyvinylidene fluoride is a semi-crystalline polymer of the monomer unit ($CH_2=CF_2$), and it may have simultaneously at least two stable crystalline structures, i.e. a polar Form 1 (β-form) and a non-polar Form 2 (α-form).[1,2] The Form 1 has a nearly planar zig-zag chain conformation with two monomer units per unit cell (figure 1) and a space group $CmC2m(C_{2v}^{14})$. It has a net dipole moment of 2.1D ($\equiv 7.01\times10^{-30}$ C-m) per monomer unit perpendicular (i.e., b-axis) to the chain axis (i.e., c-axis) which leads to a spontaneous polarization of $0.13 C/m^2$ for an unpoled single crystal.[3] The Form 2 crystallite belongs to the space group $P2_1(C_{2h}^5)$ in which each molecule has a T-G-T-G' chain conformation. Although each molecule of Form 2 structure has a dipole moment normal to the chain axis, however, the adjacent chains pack in an anti-polar array which results in no net dipole moment along the b-axis (figure 2). Of all the polymers PVF_2 has attracted special attention as a potential transducing material because of the large piezo and pyro-electricity which may be induced in thin films (PVF_2) by suitable poling in a high electrical field. It has been suggested[3] that the piezo- and pyro-electricity in PVF_2 may arise either from the heterogeneity of trapped charges in the bulk or from dipolar orientations. Murayama et al[4] show that for poled films of PVF_2, originally containing a mixture of both Form 1 and Form 2 crystallites, the piezo-electric response increases with increasing Form 1 content. However, Ohigashi[5] has observed considerable piezo-electric responses from poled films of PVF_2, originally containing only Form 2 crystallites. Thus the origin of the piezo- and pyro-electricity in PVF_2 has yet to be established unambiguously.

In a previous report[6] it has been shown that after corona poling of PVF_2 films containing both Form 1 and Form 2 crystallites, at fields not exceeding $\sim 2\times10^8 Vm^{-1}$, a conversion of the non-polar Form 2 crystallites into an intermediate polar form occurs in which the T-G-T-G' conformation of the molecular chain is still maintained.[7] This may be accomplished by only a rotation of the alternate molecular chains about the c-axis through $180°$. The proposed intermediate polar form is shown in figure 3.[6,8] At fields in excess of $2\times10^8 Vm^{-1}$ a progressive conversion from the intermediate polar form (figure 3) to Form 1 structure occurs.

The present work is a continuation of the previous structural study[6] in which the changes in X-ray diffraction profiles of corona charged 25μm thick PVF_2 films (Kureha) have been measured up to a Bragg angle (2θ) of 40°. Furthermore, infrared spectral studies of corona charged PVF_2 films have also been made.

The variation of the intensities of the major diffraction peaks with surface potential (corona charging) is shown in figure 4. It may be observed (figure 4) that for a surface potential of ∼5KV the Form 2 (110) peak reaches a maximum whilst the Form 2 (100) becomes vanishingly small. Furthermore, figure 5 shows that the Form 2 (200) also reaches a maximum after corona charging at a surface potential of 5KV. The observed behaviour of Form 2 (110), (200) and (100) is in agreement with the structure of the proposed intermediate polar form (figure 3). However, the observed reduction in the peak heights (Form 2) of (020) (figure 4) and (040)/(210) (see figure 5) may not be explained by the proposed conversion of the non-polar Form 2 into an intermediate polar form (figure 3). By a comparison of the rate of changes in the Form 2 peak values of (100) and (020) due to corona charging at 5KV surface potential (figure 6), it may be observed that in addition to the conversion into an intermediate polar form (figure 3) a further process must also be occurring. The reduction of (020) and (040) peaks may be explained by a process of b-axis orientation by an interaction with the applied field. This has been confirmed by the X-ray transmission diffraction photographs which show the presence of these peaks. [8,9] The result of both these processes will lead to an internal polarization which may account for the significant increases in the piezo- and pyro-electric responses observed for such poling conditions.

It has been shown[10] that the intermediate structure is stable up to a temperature of 90°C and that less than only 10% of the piezo-electricity is lost by cycling the poled specimen up to this temperature. Furthermore, the piezo-electricity was observed to be proportional to the pyro-electricity up to this temperature[10] which suggests that a common origin may be responsible for both the phenomena. It may also be noted that the piezo-electricity due to a thermal expansion against its constraints may be responsible for a significant part of the observed pyro-electricity in this polymer.

Figure 5 shows that the intermediate polar form gets converted in Form 1 structure on poling the specimen with corona charging at a surface potential of 10KV. This corresponds to a change in the chain conformation from a T-G-T-G' to a zig-zag type. This is confirmed (figure 7) by the reduction of infrared absorption peaks of Form 2 structure which are in agreement with Southgate.[11] It may also be observed (figure 7) that the Form 1 absorption peak at $510 cm^{-1}$ increases after poling. This may only be due to a conversion from Form 2 to Form 1 structure because Naegele and Yoon[12] have shown that with conventional poling the absorption at $510 cm^{-1}$ decreases very slightly due to an orientation of CF_2 dipoles along the direction of the poling field.

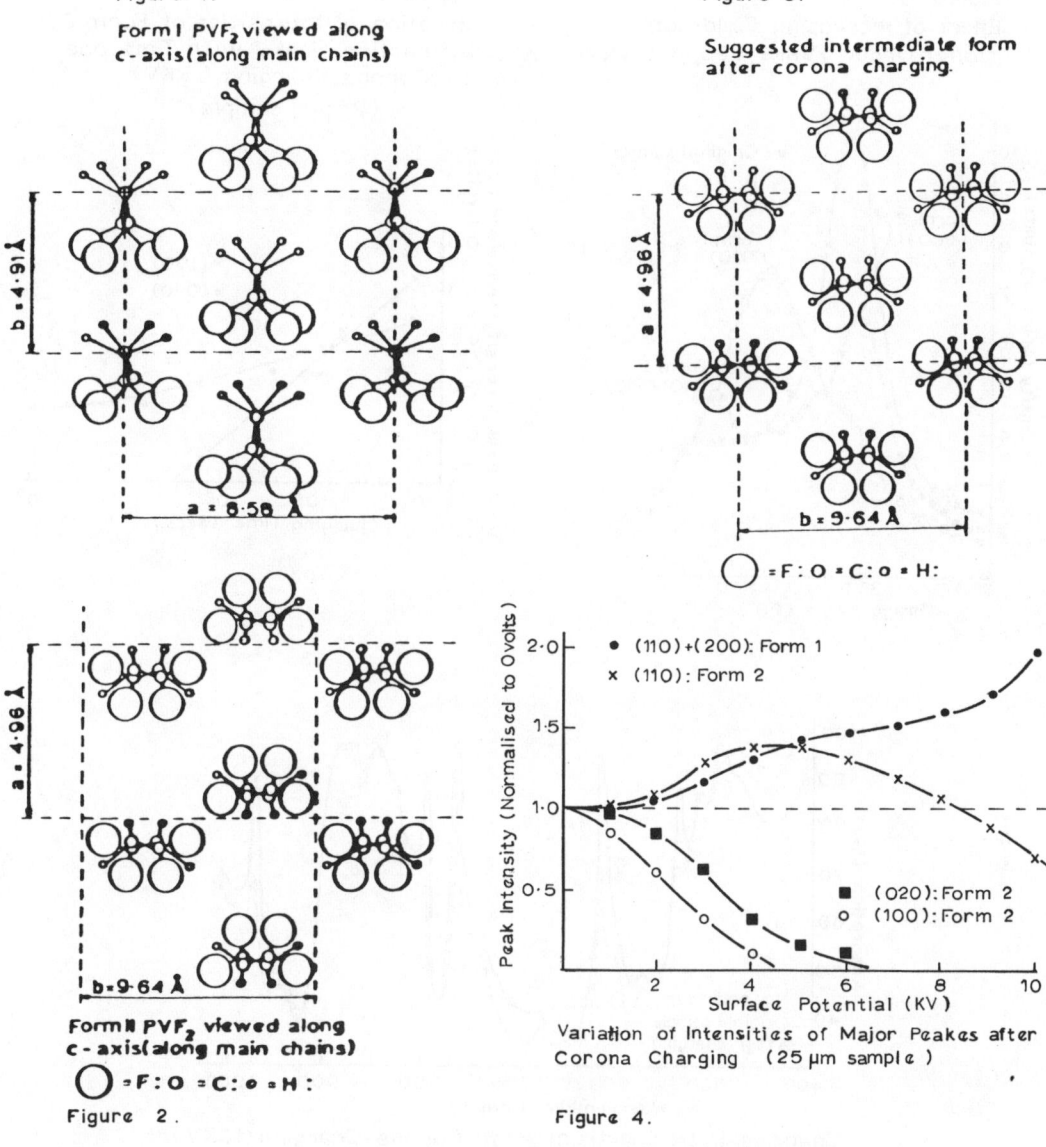

Figure 1. Form I PVF$_2$ viewed along c-axis (along main chains)

Figure 3. Suggested intermediate form after corona charging.

Form II PVF$_2$ viewed along c-axis (along main chains)
○ = F : O = C : • = H :
Figure 2.

Figure 4. Variation of Intensities of Major Peaks after Corona Charging (25 μm sample)

- • (110)+(200): Form 1
- × (110): Form 2
- ■ (020): Form 2
- ○ (100): Form 2

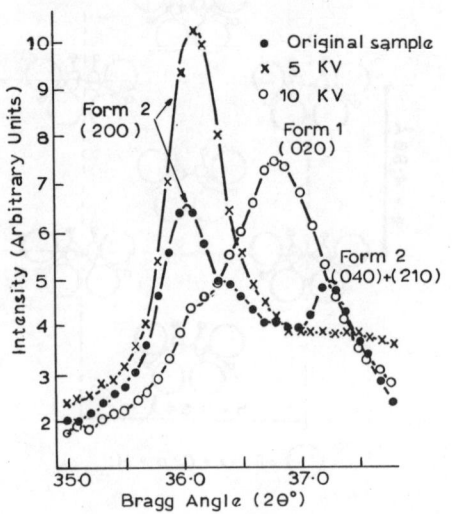

Figure 5.
Effect of Increasing Fields on Higher Order Peaks. (25μm sample)

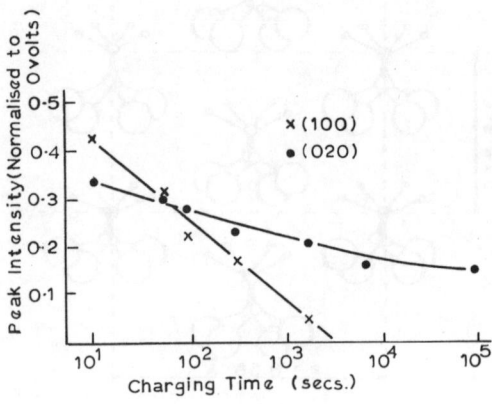

Figure 6.
Variation of Intensities of Form 2 Diffraction Peaks with Time due to Corona Charging (5KV).
(25μm sample)

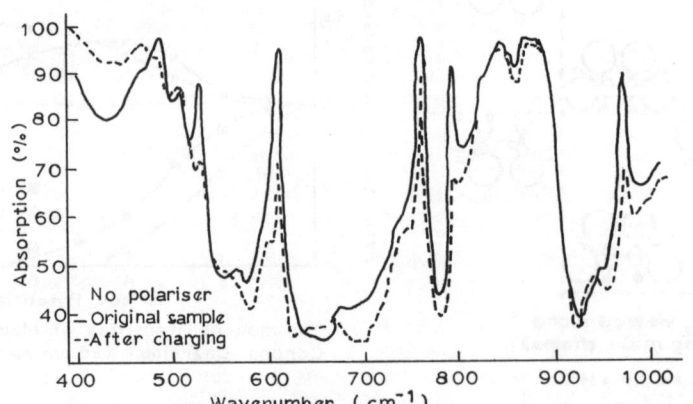

Changes in IR Spectrum after Corona Charging (10KV)
(25μm sample)

Figure 7.

These results demonstrate that the origin of piezo- and pyro-electricity in PVF_2 is dipolar orientation rather than the trapped space charge.

Acknowledgement

This work is supported by a research grant from the US Army. One of the authors (KD) is also grateful to the Science Research Council of Great Britain for a research studentship. The authors express their gratitude to Kureha Chemical Co. Ltd. of Japan for the generous supply of PVF_2 films.

References

1. J.B. Lando, H.G. Olf and A. Peterlin, J. Polym. Sci. A-1, 4, 941 (1966).
2. R. Hasegawa, Y. Takahashi, Y. Chatani and H. Tadokoro, Polym. J. 3(5), 600 (1972).
3. Y. Wada and R. Hayakawa, Jap. J. Appl. Phys., 15(11) 2041 (1976).
4. N. Murayama, T. Oikawa, T. Katto and N. Nakamura, J. Polym. Sci., Polym. Phys. Ed., 13, 1033 (1975).
5. H. Ohigashi, J. Appl. Phys., 47, 949 (1976).
6. D.K. Das-Gupta and K. Doughty, J. Appl. Phys., 49(8), 4601 (1978).
7. R.G. Kepler and R.A. Anderson, J. Appl. Phys., 49(3), 1232 (1978).
8. G.T. Davis, J.E. McKinney, M.G. Broadhurst and S.C. Roth, J. Appl. Phys., (to be published).
9. K. Doughty, Ph.D. Thesis, University of Wales (1978).
10. D.K. Das-Gupta and K. Doughty, J. Phys. D., (to be published).
11. P.D. Southgate, Appl. Phys. Lett., 28, 250 (1976).
12. D. Naegele and D.Y. Yoon, Appl. Phys. Lett., (to be published).

PIEZOELECTRICITY, PYROELECTRICITY AND IMPROPER FERROELECTRICITY IN PVF$_2$

F. MICHERON, G. BICHON, C. LEMONON, H. FACOETTI and M. ROYER

THOMSON-CSF

BP.10 - 91401 ORSAY - France

In this model, we assume that piezoelectric and pyroelectric properties of phase I PVF$_2$ depend only upon the dipolar remanent polarization P, the mechanical compliances s_{ij} and thermal expansion coefficients α_{ij}. The piezoelectric strain tensor components d_{ij}, the pyroelectric coefficient p, and permittivity χ_{ij} are first expressed in terms of P, S_{ij} and α_{ij}, using a thermodynamical model similar to the Devonshire model of ferroelectric BaTiO$_3$ (1) ; the origin of P is found as resulting of cooperative dipolar interactions in the plane perpendicular to the chain axis. The experimental values of d_{ij} and χ_{33} are approached by using averaging formulas which take into account the interlamellar amorphous phase.

Thermodynamical model

We consider the non polar orthorhombic class mmm as the prototype of phase I PVF$_2$, (with the free energy Go(T), which is slightly distorted by strains x_{ij} and polarization P_i. Using these parameters as independant variables, the free energy G of the distorted phase is :

$$G - G_0(T) = \frac{1}{2} \eta^x_{ij}(T) P_i P_j + \frac{1}{4} b_{ijkl} P_i P_j P_k P_l + \frac{1}{2} c^P_{ijkl}(T) x_{ij} x_{kl} + \frac{1}{2} \gamma_{ijkl} x_{ij} P_k P_l.$$

where the development G-Go is limited to fourth order in P_i (2). We assume that only inverse susceptibilities η_{ij} and stiffnesses C_{ijkl} are temperature dependant, while anharmonic polarization coefficients b_{ijkl}, and polarization-strain electrostriction components γ_{ijkl} are not.

In the orthorhombic symetry class, η_{ij} has 3 diagonal independant components, while fourth rank tensors have 6 diagonal independant plus 3 off diagonal (12, 13,23) components (3).

By assuming that equilibrium conditions are zero applied field ($E_i = -\partial G/\partial P_i = 0$) and stresses ($X_{ij} = -\partial G/\partial x_{ij} = 0$), one gets nine indeterminated equations of g unknowns P_i and x_{ij} (Zero discriminant). One has to set one of them, for instance magnitude and direction of remanent polarization P_i, and express x_{ij} as functions of P_i. We choose $P_3 = P \neq 0$, $P_1 = P_2 = 0$. This gives :

$$x_4 = x_5 = x_6 = 0$$
$$x_i = -\frac{1}{4} P_3^2 \, s_{ij} \cdot \gamma_{i3}$$
$$\eta_3 + b_{33} P_3^2 \left[\overline{\gamma_{i3} \cdot x_i} \right] = 0 \quad \Big\} \quad i,j = 1,2,3$$
$$[s_{ij}] = [c_{ij}]^{-1}$$

The distorted symetry mmm is now $mm2$.
The total derivatives dE_i and dX_{ij} allow to compute :
- susceptibilities and compliances tensor components

$$\chi_1^x = \Delta_1 \chi_1^x \qquad \chi_1^x = \left[\eta_1 + \frac{1}{2} \overline{\gamma_{i1} \cdot x_i} \right]^{-1}$$
$$s_{55}^E = \Delta_1 s_{55}^P \qquad \Delta_1 = \left[1 - d_{15}^2 \, c_{55}^P / \chi_1^x \right]^{-1}$$
$$\chi_2^x = \Delta_2 \chi_2^x \qquad \chi_2^x = \left[\eta_2 + \frac{1}{2} \overline{\gamma_{2i} \cdot x_i} \right]^{-1}$$
$$s_{44}^E = \Delta_2 s_{44}^P \qquad \Delta_2 = \left[1 - d_{24}^2 \, c_{44}^P / \chi_2^x \right]^{-1}$$
$$\chi_3^x = \Delta_3 \chi_3^x \qquad \chi_3^x = \left[2 b_{33} P_3^2 \right]^{-1}$$
$$s_{ij}^E = \Delta_3 s_{ij}^P \qquad \Delta_3 = \left[1 - \frac{d_{3i} \, d_{3k} \, c_{ik}^P}{\chi_3^x} \right]^{-1} \quad i,j,k = 1,2,3$$

The differences $\Delta \chi = \chi^x - \chi^x$ and $\Delta s = s^E - s^P$ can be neglected, since in PVF$_2$ $\Delta \chi / \chi \# \Delta s / s \# 2.10^{-3}$.
- Piezoelectric strain components

$$d_{15} = -\frac{1}{4} \gamma_{55} \chi_1 s_{55} P_3 \qquad d_{24} = -\frac{1}{4} \gamma_{44} \chi_2 s_{44} P_3$$
$$d_{3j} = -\frac{1}{2} P_3 \chi_3 (\overline{\gamma_{i3} \cdot s_{ij}}) \qquad i,j = 1,2,3$$

- Pyroelectric coefficient of the free dielectric, with two components :
both in the P_3 direction

$$p_3^{Ex} = p_3^{Ex} + d_{3i} \, c_{ij} \alpha_i \qquad p_3^{Ex} = -P_3 \chi_3 \frac{\partial \eta_3}{\partial T}$$

The first term is primary pyroelectricity, and the other is the contribution of volume change (secondary effect).
- Thermal expansion :

$$\alpha_j^E = -s_{ij} \frac{\partial c_{jk}}{\partial T} x_k + d_{3j} \frac{P_3^x}{\chi_3}$$

- Strain-field electrostriction components, defined as :

$$K_{ijkl} = \frac{\partial^2 x_{ij}}{\partial E_k \partial E_l} = \frac{\partial d_{ijk}}{\partial E_l} = \frac{\partial \chi_{il}}{\partial X_{jk}}$$

then :

$$K_{3i} = -\frac{1}{2} \chi_3^2 \Delta_{ij} \cdot \gamma_{j3}$$

$$= 2 \frac{x_i}{\mathcal{E}_3^2} = \frac{d_{3i}}{\mathcal{E}_3}$$

where $\mathcal{E}_3$ is the equivalent internal field $\mathcal{E}_3 = P_3/\chi_3$
Using the values $P_3 = 6.5 \cdot 10^{-2}$ C.m^{-2} and $\chi_{33} = 12 \varepsilon_0$
we get $\mathcal{E}_3 = 5.4 \cdot 10^8$ V.m^{-1}, $b_{33} \varepsilon_0 = 19.7$ C^{-2} m^4.
From the set of values measured by Kepler in Kureha Piezofilm (4)

$d_{31} = 21.4 \qquad d_{32} = 2.3 \qquad d_{33} = -31.5 \quad (pC.N^{-1})$

$\Delta_{11} = 4 \qquad \Delta_{22} = -1.57 \quad (10^{-10} N\, m^{-2})$

$\alpha_1 = \alpha_2 = 0.13 \quad,\quad \alpha_3 = 1.24 \quad (10^{-4} K^{-1})$

we get :

$K_{31} = 3.9 \qquad K_{32} = 4.26 \qquad K_{33} = -5.8 \quad (10^{-20} m^2 V^{-2})$

$x_1 = 5.8 \qquad x_2 = 0.62 \qquad x_3 = -8.5 \quad (10^{-4})$

$\gamma_{31}\varepsilon_0 = 0.4 \qquad \gamma_{32}\varepsilon_0 = 0.1 \qquad \gamma_{33}\varepsilon_0 = -0.57$

The computed secondary effect is $p_3^{(2)} = 1.48 \cdot 10^{-5}$ C.m^{-2} K^{-1} ; if we assume that the difference with measured effect ($p_3^{(m)} = 2.74 \cdot 10^{-5}$ C.m^{-2} K^{-1}) is due to primary effect, then : $\partial\gamma_3\varepsilon_0/\partial T = 1.6 \cdot 10^{-5}$ K^{-1}. The last computed values are the stored energies in phase I PVF$_2$:

$$\begin{aligned} G-G_0 &= -1.086 \cdot 10^7 \text{ Jm}^{-3} \quad \text{(electrical)} \\ & +1.02 \cdot 10^5 \text{ Jm}^{-3} \quad \text{(mechanical)} \\ & +8.63 \cdot 10^5 \text{ Jm}^{-3} \quad \text{(electromechanical)} \end{aligned}$$

The total stored energy is comparable to that stored in oxygen octaedra ferroelectrics : although the remanent polarization is smaller, permittivities and stiffnesses are much lower.

Remanent Polarization

We assume first that the remanent polarization results from electrostatic and cooperative dipolar interactions ; the local field is expressed as $E_\ell = E + \beta P$, where E is the external field and β the Lorentz factor. P is given by the contributions of the N dipoles with moment μ, which make the angle θ with E_ℓ directions : $P = N \mu \langle \cos\theta \rangle$

If we assume that before being aligned, dipoles are free to rotate in the full space (n=3), or in a plane (n=2), or can take only two antiparallel directions, (n=1), $\langle \cos\theta \rangle$ is given by the relations :

$$n = 1 \qquad \langle \cos\theta \rangle = \text{th}\, u \qquad\qquad u = \mu \frac{E + \beta P}{kT}$$

$$n = 2 \quad \langle \cos\theta \rangle = \frac{\sum_1^\infty \frac{2p}{2^{2p}} \frac{u^{2p-1}}{(p!)^2}}{\sum_0^\infty \frac{1}{2^{2p}} \frac{u^{2p}}{(p!)^2}}$$

$$n = 3 \quad \langle \cos\theta \rangle = \coth u - \frac{1}{u}$$

In any case, such model predicts the existence of a spontaneous polarizarion P which disappears above a critical temperature Tc, given by :

$$n k T_c = N\mu^2 \beta \qquad P \# N\mu \frac{T}{T_c} \left[\frac{n+2}{n} \frac{T_c - T}{T} \right]^{1/2}, (T \# T_c)$$

The dielectric permittivity follows a Curie Law :

$$\varepsilon_r = 1 + \frac{C}{T_c - T} \qquad C = \frac{T_c}{2\beta\varepsilon_0}$$

We have observed such a behaviour in PVF_2 between 110° C and the melting temperature Tm = 170°C. Therefore, we consider in the following the melting temperature as the critical temperature : Tm = Tc. We obtain :

$$n/\beta\varepsilon_0 = N\mu^2/\varepsilon_0 k T_c = 15.4$$

By introducing the function $\mathcal{L}(u) = P/N\mu$, we can derive the general expressions of partial derivatives of P, which allow to compute permittivities, piezoelectric and pyroelectric effects :

$$\chi^x = \frac{\partial P}{\partial E} = \frac{1-\Delta}{\beta\Delta} \qquad p^x = -\frac{E\varepsilon}{T}\chi^x$$

$$\frac{\partial P}{\partial N} = \frac{P}{N\Delta} \qquad \frac{\partial P}{\partial \mu} = \frac{E}{\mu\beta}\frac{1-\Delta}{\Delta} + \frac{P}{\mu}\frac{2-\Delta}{\Delta}$$

where : $\Delta = 1 - n \frac{T_c}{T} \frac{d\mathcal{L}(u)}{du}$

$$\Delta \simeq 0{,}6 \ (n = 1,2,3 \text{ at room temperature})$$

The free dielectric pyroelectric coefficient is :

$$p^{xE} = p_3^{xE} - \frac{P}{\Delta}\alpha_v = -\frac{P}{\Delta}\left[\frac{1-\Delta}{T} + \alpha_v\right] + \frac{\partial P}{\partial \mu}\frac{\partial \mu}{\partial T}$$

The first term expresses the primary effect and the second the secondary effect, where α_v is the volume thermal expansion coefficient. The third term is a contribution to primary effect due to temperature change of the elementary dipole moment. In the frame of a rigid dipoles model, we will neglect this last term. The piezoelectric strain constant d_{33} is :

$$d_{33}^T = \frac{dP}{dX_3} = -\frac{P}{\Delta}\Delta_v + \frac{dP}{\partial \mu}\frac{\partial \mu}{\partial X_3}$$

where Δ_v is the volume compressibility and $\partial \mu/\partial X_3 = 0$ for rigid dipoles.
Using the average values $\alpha_v = 1.7 \ 10^{-4} \ K^{-1}$, $\Delta_v = 3(1-2\nu)/e$
$\nu = 0.35$, $e = 5.4 \ 10^9 \ N^{-1} m^2$, $\Delta_v = 3.6 \ 10^{-10} \ N \ m^{-2}$, we are able to compute P, ε_r, p_3^x and d_{33}^T for the single crystal, in the cases n = 1,2,3 at room temperature.

	n = 1	n = 2	n = 3	measured
P^c (10^{-2} C.m^{-2})	10.6	8.9	8.4	3.5 to 6.5.
ε_r^c	10.8	7.3	5.1	10 to 13
p_3^{xc} (10^{-4} Cm^{-2}K^{-1})	-2.6	-2.7	-2.5	-0.2 to -0.3
d_{33}^c (10^{-12} CN^{-1})	-63	-58	-54	-10 to -33

All these coefficients are larger than the measured ones, except for dielectric constant, which is smaller in the cases n = 2 and n = 3. It must be noted that in the computed p_3^{xc}, the primary effect account for 90% of the total effect ; this is in contradiction with experiences which show that the secondary effect is responsible for half of the measured value (5). The computed values have to be corrected to take into account the amorphous phase. We use the model proposed by Takayanagi (6), which allows averaging properties of the crystalline and amorphous phase.

From upper to lower electrodes, the polymer is divided into crystalline regions with permittivity ε^c and compliance s^c and amorphous regions (ε^a, s^a), as represented in the following figure.

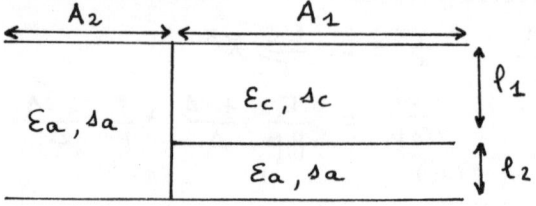

The average values are given by the relations :

$$\varepsilon = \varepsilon_a \left[1 - \phi \frac{\varepsilon_a - \varepsilon_c}{\alpha \varepsilon_c + (1-\alpha) \varepsilon_a} \right] \quad P = \phi P_c \frac{\varepsilon_a}{\alpha \varepsilon_c + (1-\alpha) \varepsilon_a}$$

$$p = p^c \phi \frac{\varepsilon_a}{\alpha \varepsilon_c + (1-\alpha) \varepsilon_a} \quad d = d^c \phi \frac{\varepsilon_a}{\alpha \varepsilon_c + (1-\alpha) \varepsilon_a} \cdot \frac{s_c}{s_c + \phi(\frac{1}{\alpha+\phi} - 1)(s_a - s_c)}$$

where ϕ is the crystallinity, and $\alpha = \frac{A_1}{A_1 + A_2} - \phi$ a "shape" factor.
We assume $s_a/s_c = 10$, $\varepsilon_a = 25$, $\phi = 0.5$ and $\alpha = 0.2$, which corresponds to a lamellar configuration of the crystallites.
The average computed values are :

	n = 1	n = 2	n = 3
P $(10^{-2} C\, m^{-2})$	5.9	5.2	5
ε_r	17	14,7	13
p^x $(10^{-4} C\, m^{-2} K^{-1})$	-1,5	-1,6	-1,5
d_{33} $(10^{-12} C\, N^{-1})$	17	16	15

These values are in qualitative agreement with measured values, excepted in the cases of ε_r (n = 1, n = 2) and p^x which is 6 times too large. As we noticed earlier, the dominant computed term is primary effect, which is not confirmed by experience. We have therefore to assume that in PVF$_2$, dipoles have less rotation freedom than provided by electrostatic interactions alone.

The rotation freedom is represented by the parameter Δ : Δ =0 at the critical temperature, Δ = 1 for frozen dipoles. In order to account for experimental results, Δ has to be larger than the value Δ = 0.6 computed in the case of electrostatic interactions ; we have found that the most appropriate value of the adjustable parameter Δ is Δ = 0.85. In those conditions, we obtain :

	n = 1	n = 2	n = 3
P^c $(10^{-2} C\cdot m^{-2})$	10	8	7
P "	6	4,9	4,3
ε_r^c	9.7	2.4	1.9
ε_r	12.2.	11.2	10.8
p^{xc} $(10^{-5} C\, m^{-2} K^{-1})$	-8	-6.4	-5.6
p^x "	-4.8	-3.9	-3.6
d_{33}^c $(10^{-12} C\, N^{-1})$	-42	-34	-30
d_{33} "	-12	-10	-9

These results suggest that the case n = 2 (rotation perpendicular to chain axis only) would be the most representative of stretched phase I PVF$_2$. The case n = 3 could be applied to unstretched PVF$_2$-PTFE. The value $\Delta > 0.6$ suggests that in PVF$_2$, ferroelectricity is improper, i.e. polarization stability is not due to electrostatic interactions only, but to mechanical interactions, as for instance in the case of ferroelastics-ferroelectrics (2) or steric hindrances, encountered in chiral liquid crystals (7).

A different assumption would be the existence of a permanent internal electric field, due to embedded charges, which stabilizes the polarization orientation. A more accurate microscopic model than the proposed model will allow to compute the coefficients of the free energy expansion used in the first part, and then to obtain exact expressions of piezoelectric and pyroelectric effects in semi-crystalline ferroelectric polymers.

In conclusion, we point out that ferroelectricity in PVF_2 can't be compared with properties of ferroelectric ceramics; the amorphous phase plays a major role in supplying high compliance and volume thermal expansion, while in ferroelectric ceramics, the intergrain phase occupies a very small relative volume, with compliance comparable to the compliance of the grains. Moreover, grains in a ferroelectric ceramic have to be both piezoelectric and pyroelectric in order to induce these properties in a macroscopic ceramic sample; in PVF_2, piezoelectricity and pyroelectricity of the lamellar crystallites doe'nt play the fundamental role in the properties of the macroscopic films : their compliance and dielectric constant are so small at room temperature compared to that of the amorphous phase, that, in a simplified interpretation, they would contribute to piezoelectric, pyroelectric and mechanical properties only by the value and direction of the remanent polarization, and shape of the crystallites.

1) A.F. Devonshire, Phil. Mag. 40, 1040 (1949) ; 52, 1065 (1951)
2) T. Mitsui, I. Tatsuzaki and E. Nakamura, An Introduction to Physics of ferroelectrics, Gordon and Breach 1976 -
3) J.F. Nye, Physical Properties of Crystals. Clarendon Press 1957
4) R.G. Kepler - private communication -
5) R.G. Kepler - Proc. 175th Meeting of the Amer. Chem. Soc. Anaheim Cal. 1978
6) M. Takayanagi, K. Imada and T. Kajiyam. J. Poly. Sci. 15, 263 (1966)
7) R.B. Meyer, L. Liebert, L. Stzelecki and P. Keller, J. Phys. Letters, 36, 169 (1975) -
8) F. Micheron, Rev. Tech. Thomson-CSF - To be published, Sept. 1978 -

POLARIZATION AND DEPOLARIZATION PROCESSES IN POLYVINYLIDENE FLUORIDE

T. Furukawa and E. Fukada

The Institute of Physical and Chemical Research, Wako, Saitama

Polyvinylidene fluoride (PVDF) exhibits strong piezoelectric activity if it has been poled. The poling produces the residual polarization which has been considered to be the origin of piezoelectricity. In the present report, we show the isothermal polarization and depolarization processes of PVDF studied by monitoring the time dependences of the piezoelectric constant e_{31} and the charge density Q after application and removal of high dc electric field. The piezoelectric film of PVDF has usually been produced anisothermally, that is, by applying high dc electric field at an elevated temperature and turning it off at room temperature. Such an anisothermal poling may produce various kinds of residual polarizations in the sample. The study of isothermal polarization and depolarization processes is expected to extract one of the most important and characteristic mechanisms of the piezoelectricity of PVDF.

The PVDF used was obtained from Kureha Chemicals Company. The 100 μm thick sheet were stretched to 4.5 times their original length at 80°C to obtain 50 μm thick samples. After annealing at 140°C for 30 min. under the condition of fixed length, gold electrodes were evaporated onto both surfaces of the samples.

Figure 1 shows the experimental set-up. We apply to the sample in the thermostat the high dc voltage and the small sinusoidal strain of 10 Hz, S_ω, simultaneously. The currents flowing through the sample consist of dc component I_0 due to dc conduction and dielectric polarization, and ac component I_ω due to electrostriction and piezoelectricity. After converting current

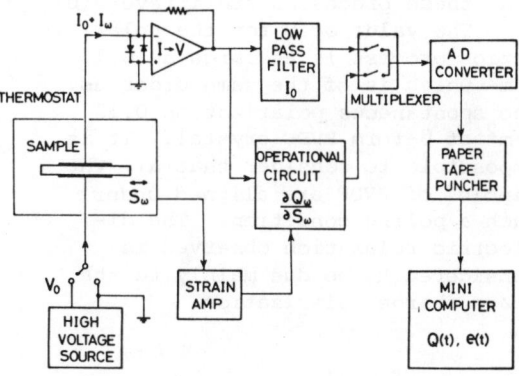

Fig. 1 Experimental set-up for the simultaneous measurement of e(t) and Q(t).

to voltage, the dc component is selected by a low pass filter. The ac component is selected by an operational circuit to obtain the piezoelectric constant. The results are led every one second to a minicomputer which calculates $Q(t)$ and $e(t)$ according to the following equations.

$$Q(t) = \int_0^t I_0(t)dt \quad (1) \qquad e(t) = \partial(\int_0^t I_\omega(t)dt)/\partial S_\omega \quad (2)$$

Typical isothermal polarization and depolarization processes are indicated by circles in Figs. 2 and 3. The poling field E_p of 3×10^7 V/m was applied at T_p of 90°C for t_p of 10^4 sec. Then E_p was turned off while T_p was kept constant. It is found that both $e(t)$ and $Q(t)$ are fitted to the following relaxation functions.

$$e(t) = e_\infty - \Delta e \cdot \exp(-(t/\tau)^\beta) \quad (3)$$

$$Q(t) = Q_\infty - \Delta Q \cdot \exp(-(t/\tau)^\beta) + Q_c \cdot t \quad (4)$$

Here τ is the relaxation time, β is a parameter expressing the distribution of relaxation times, Δe and ΔQ are the relaxation strengths, e_∞ and Q_∞ are the equilibrium values and Q_c is the conduction current. The derivatives $\partial e/\partial \log t$ and $\partial Q/\partial \log t$ also provide informations about the relaxation time τ (peak time) and the relaxation strengths Δe and ΔQ (peak area). Comparison between polarization and depolarization processes finds that the relaxation times for $e(t)$ and $Q(t)$ are not much different. On the other hand, the relaxation strengths Δe and ΔQ for the polarization process are much larger than those for the depolarization process which clearly suggests that these processes are irreversible.

The value of Q for the polarization process is as large as 0.15 C/m² which is of the same order as the spontaneous polarization 0.13 C/m² of β-form PVDF crystal. It is impossible to consider that all the dipoles of PVDF are aligned under such a poling condition. The dielectric relaxation observed is considered to be due mainly to the space charge polarization.

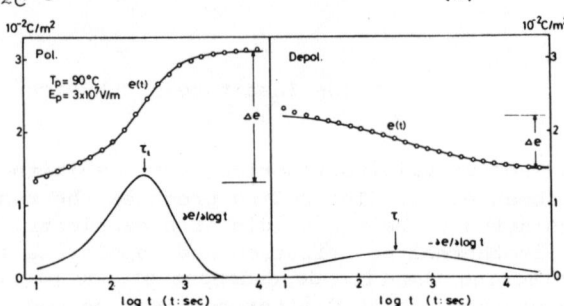

Fig. 2 Time dependence of e. Solid lines are drawn using the parameters in Eq. (3) τ =300 sec., β=0.98, Δe=1.8x10⁻² C/m², e_∞ =3.1x10⁻² for the polarization process and τ=350 sec., β=0.52, Δe=-0.7x10⁻² C/m² for the depolarization process.

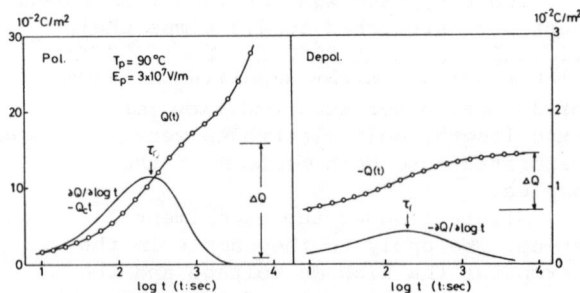

Fig. 3 Time dependence of Q. Solid lines are drawn using the parameters in Eq. (4) τ=250 sec., β=0.92, ΔQ=15x10⁻² C/m², Q_∞ =16x10⁻² C/m², Q_c=24x10⁻⁶ C/m² sec. for the polarization process and τ=180 sec., β=0.66, ΔQ=-0.7x10⁻² C/m², Q_∞ =-1.4x10⁻² C/m² as Q_c=0 for the depolarization process.

Figure 4 shows the dependence of the depolarization process on the poling time t_p. Points 1, 2, 3 and 4 indicate the values of e(t) after 10, 10^2, 10^3 and 10^4 sec. polarization respectively. Curves 1, 2, 3 and 4 show the decay of e(t) starting from these points. It is found that the piezoelectricity produced before the relaxation time τ of about 300 sec. almost vanishes due to isothermal depolarization (1,2), while the piezoelectricity produced after τ can remain unvanished (3,4).

Figure 5 shows the E_p dependence of e(t) during the polarization process at 90°C. The value of e increases with increasing E_p. The more interesting is that the relaxation time significantly decreases with increasing E_p. In other

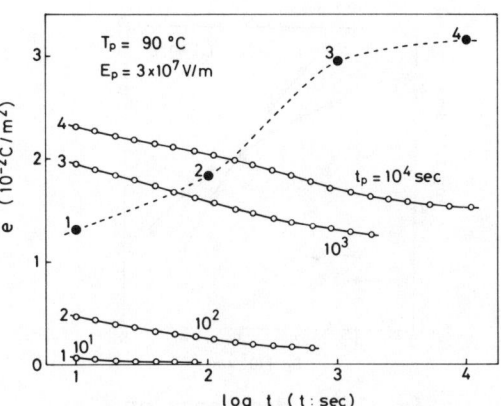

Fig. 4 t_p dependence of the depolarization process of e(t).

words, the piezoelectric structure of PVDF is built up faster at higher poling fields. In Fig. 6, ∂Q/∂log t is plotted against log t for various E_p at 90°C. The dielectric relaxation time estimated from peak time decreases with increasing E_p. The dielectric relaxation strength estimated from peak area rather decreases with increasing E_p which suggests that only part of the dielectric polarization is responsible for piezoelectricity, because the piezoelectric relaxation strength increases with increasing E_p. As shown in Fig. 7, the relaxation time for e shows stronger E_p dependence than that for Q.

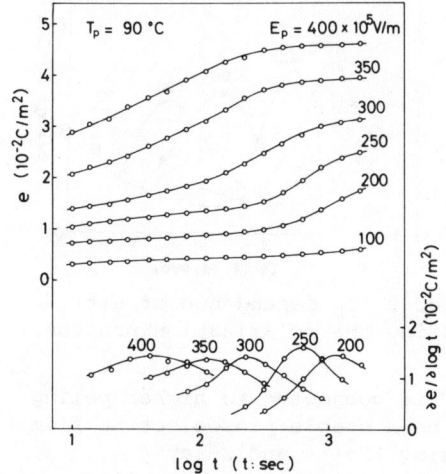

Fig. 5 E_p dependence of e(t) during the polarization process.

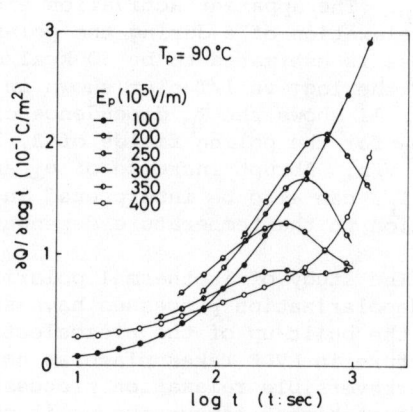

Fig. 6 E_p dependence of ∂Q/∂log t during the polarization process.

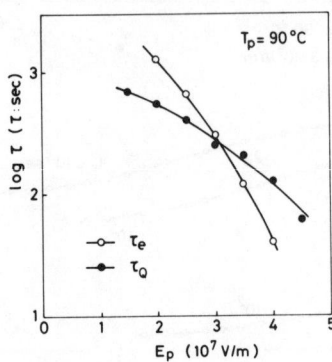

Fig. 7 E_p depenence of $\log \tau_e$ and $\log \tau_Q$.

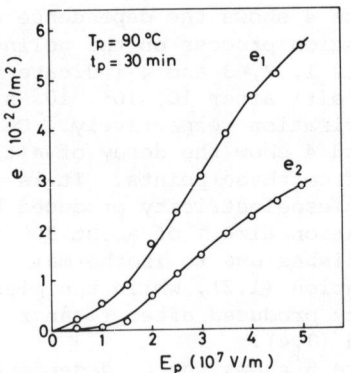

Fig. 8 E_p dependence of e_1 and e_2.

Figure 8 shows the poling characteristics of PVDF. Here e_1 is the piezoelectric constant after 30 min. polarization and e_2 is after the 30 min. isothermal depolarization. Their nonlinear E_p dependence can be interpreted as due to the finding that the relaxation times for the poling fields below 2×10^7 V/m at 90°C are larger than 30 min.

Figure 9 shows the T_p dependence of $e(t)$ at E_p of 3×10^7 V/m. It is found that the relaxation time decreases with increasing T_p. The apparent activation energy for the relaxation of e during the polarization process is estimated to be 30 kcal/mol using the $\log \tau$ vs $1/T$ plot shown in Fig. 10. Figure 11 shows the T_p dependence of e_1 and e_2 for the poling fields of 2, 3 and 4×10^7 V/m. Abrupt increase of e_1 and e_2 with T_p. can also be interpreted in relation to the temperature dependence of τ.

The study of isothermal polarization and depolarization processes have shown that the built-up of the piezoelectric structure in PVDF takes place as nonlinear and irreversible relaxation processes. If PVDF is subjected to higher poling fields at higher temperatures, it turns out to be a stable piezoelectric film within shorter period. Recent works on PVDF using IR[1,2] and X-ray[3] techniques have shown that the dipoles of PVDF can align along the poling field. The relaxation of the piezoelectric constant e caused by high dc electric fields is due presumably to the rotation of PVDF dipoles. The

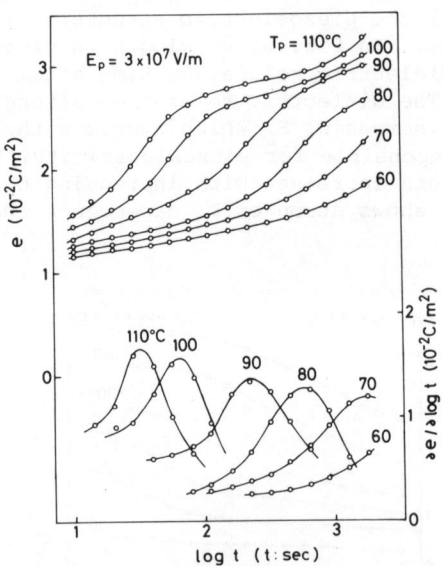

Fig. 9 T_p dependence of $e(t)$ during the polarization process.

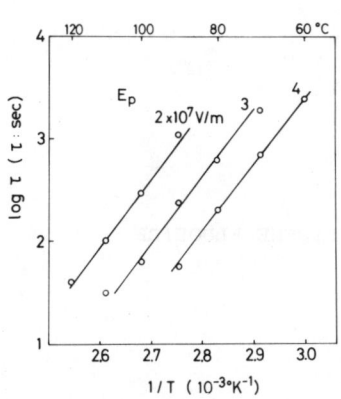

Fig. 10 $1/T_p$ dependence of $\log \tau_e$ at E_p of 2, 3 and 4×10^7 V/m.

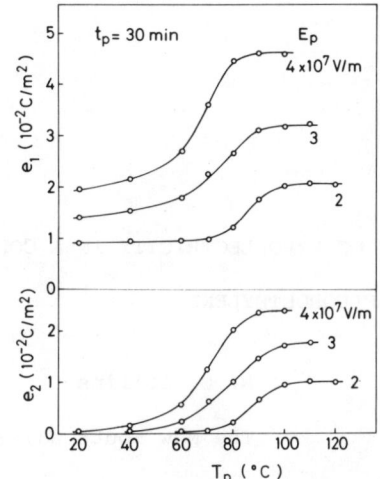

Fig. 11 T_p dependence of e_1 and e_2 at E_p of 2, 3 and 4×10^7 V/m.

dielectric relaxation simultaneously observed may include both space charge polarization and dipole orientation. The space charge polarization is considered to influence the dipole orientation especially at lower fields and higher temperatures because it generate inhomogeneous internal field in the sample. Various informations like the relaxation time and relaxation strength during the polarization and depolarization processes as a function of temperature and electric field will be the typical characteristics about the rotation of PVDF dipoles.

References
1) M. Date and E. Fukada, Rep. Progr. Polym. Phys. Jpn., 339, 20 (1977).
2) D. Naegle and D. Y. Yoon, Appl. Phys. Lett., 132, 33 (1978).
3) R. G. Kepler and R. A. Anderson, J. Appl. Phys., 1232, 49 (1978).

TWO TYPES OF PYROELECTRICITY IN A COPOLYMER OF VINYLIDENE FLUORIDE
AND TETRAFLUOROETHYLENE

R. E. Collins

The New South Wales Institute of Technology

Broadway, NSW, Australia   2007

M. G. Broadhurst, G. T. Davis, and A. S. DeReggi

National Bureau of Standards

Washington, D. C.   20234

Semicrystalline polymers like poly(vinyl fluoride) (PVF), poly(vinylidene fluoride) (PVDF), and copolymers of the latter can be made to exhibit large piezoelectric and pyroelectric coefficients by the temporary application of high electric fields [1,2]. Previous results [3] for a random copolymer of 73 weight per cent vinylidene fluoride and 27% tetrafluoroethylene showed an unusually strong dependence of pyroelectric coefficient on poling field when compared with PVDF homopolymer. Furthermore, a nearly constant value of 2.2 nC/cm^2K was obtained for all applied voltages corresponding to uniform electric fields in excess of 300 kV/cm for all temperatures of poling between 0° and 60° C. This paper presents measurements which show that such films are not uniformly poled across their thickness and the response previously reported arises from two separate effects. The first is a fast pyroelectric response arising from the reversible temperature dependence of permanent polarization within the poled portion of the film. The second is a slower time dependent response attributed to reversible motion of real charge through the bulk of the polymer. The initial response depends upon the average value of polarization while the total response depends only upon the maximum value of polarization and is independent of the polarization distribution. Measurements of pyroelectric response were made on two different time scales. One at times less than the thermal equalization time and the other at times after thermal equalization.

Pyroelectric measurements on the longer of two time scales were made by using an operational amplifier to measure the charge generated as a function of time (order of minutes) following a stepwise change in temperature. The polymer film was immersed in a stirred vessel of hexane and the step-change in temperature was accomplished bh adding a small volume of hexane of a different temperature. Results for samples poled under two different poling conditions and then subjected to a 1°C temperature change at 23°C are shown in Figure 1. The lower curve was obtained from a film poled at 300kV/cm for 30 minutes at 23°C and the upper curve was obtained from a film poled at 300 kV/cm for 30 minutes at 100°C. Data points were taken from chart paper traces and the lines correspond to exponential single relaxation time constants of 0.61 and 2.83 minutes for the samples poled at 23°C and at 100°C respectively. If the measurement temperature is reduced to 9.6°C, one obtains the lower curve in Figure 2 where the relaxation time has increased from 0.61 to 1.17 minutes.

Measurements on a shorter time scale were made using a thermal pulse method which has been described previously [4]. The technique consists of the absorption of a pulse of light of microseconds duration by an 80 nm thick evaporated metal electrode on one surface of a polymer film typically 10 to 50 µm in thickness. The electrical response (short circuit current or open-circuit voltage) from the film is measured on a time scale of milliseconds as the heat near the pulsed surface becomes uniformly distributed through the thickness of the sample. Analysis of this transient response yields the Fourier coefficients of the polarization (or charge) distribution across the thickness of the film [5]. Since the polymer is metallized on both surfaces, it can be pulsed from either side. A visual comparison of the transient responses resulting from the pulsing of opposite surfaces leads directly to a qualitative assessment of uniformity of polarization. Also, the coefficients can be determined from the deconvolution of two different measurements from the same distribution. The temperature dependence of the resistance of the thin metal electrode can be used as a probe of temperature history on either electrode and thus serves as a check on the boundary conditions for the heat flow.

Typical results obtained from the thermal pulsing are shown in Figure 3 for the film which was poled at 23°C and referred to in Figures 1 and 2. Curve (a) is the charge transferred between electrodes when heated from the electrode which was positive during poling. Curve (b) is the response obtained when pulsed from the negative side and curve (c) is a trace of the variation in resistance of the pulsed electrode. A uniformly poled sample would be expected to show the initial charge transfer and then remain unchanged with time when pulsed from either side as a constant quantity of heat is merely redistributed throughout the sample. However, one can see from the different response from opposite sides that the distribution of polarization is highly asymmetric and that the polarization is larger near the positive electrode. In general, both permanent polarization and trapped real charge contribute to the signal produced since expansion in the sample is non-uniform during thermal equalization. When the real charge density, $\rho(x)$, is equal and opposite to the gradient of permanent polarization, $dP(x)/dx$ [5,6], the polarization distribution for the sample poled at 23°C is found to be as displayed in Figure 4.

The time-dependent generation of charge following a step-change in temperature for times long compared with thermal equilibration can be explained in terms of the non-uniform polarization. Consider a non-uniformly poled polymer and for simplicity assume it is uniformly poled to a value P over part of its thickness, $0<x<t$ and unpoled over the remainder, $t<x<d$. The sample has electrodes at surfaces $x = 0$ and $x = d$ and the electrodes are short-circuited at all times which is shown schematically in Figure 5. The polymer is assumed to be electrically neutral at sufficiently long times after poling. In the case of the copolymer where the room temperature resistivity ρ is 4×10^{13} ohm-cm and the relative permittivity ε is 14 at .01 Hz, the Maxwell relaxation time, $\rho\varepsilon\varepsilon_o$, is about 50 seconds. [$\varepsilon_o$ is the permittivity of vacuum]. The polymer then contains the following virtual and real charges:

(a) Polarization charge density $+ P$ per unit area just inside the polymer at $x = 0$ and $-P$ per unit area at $x = t$.

(b) Real charge density $-Q$ and $+Q$ per unit area coincident with the polarization charges at $x = 0$ and $x = t$ respectively. At equilibrium, $|Q| = |P|$.

Assume that the polymer is subjected to a slight decrease in temperature, ΔT, which will result in an increase in polarization. Immediately after thermal equilibrium, the polymer contains non-zero charge as follows:

(a) Virtual charge $+(P + \Delta P) = + P(1 + \alpha \Delta T)$ per unit area just inside the polymer at $x = 0$ and $-(P + \Delta P)$ per unit area at $x = t$ where $\alpha_p = -P^{-1} dP/dT$.

(b) Real charge $(Q + \Delta Q)^-$ and ΔQ^+ per unit area at the polymer-metal interfaces at $x = 0$ and $x = d$, respectively. A charge ΔQ has flowed through the external circuit in order to maintain the electrodes at the same potential.

Since the electrodes are at the same potential, it is a straightforward matter to show that $\Delta Q = \Delta P(t/d) = P\alpha_p \Delta T (t/d)$. The instantaneous response to a temperature change is therefore reduced from that expected from a uniformly poled sample by the fraction of the sample which is poled.

The electric field within the sample is now no longer zero. Real charge may therefore move under the influence of this field at a rate determined by the electrical time constant, $\rho\varepsilon\varepsilon_o$, of the material. The electric field in the polymer immediately after the temperature change is shown schematically in Figure 5. In the region $t<x<d$, the field is constant and given by $E = (\alpha_p P \Delta T)(t/\varepsilon\varepsilon_o d)$. The electric field in the poled region $0<x<t$ is in the opposite direction, and is given by $E = (\alpha_p P \Delta T)(d-t)/\varepsilon\varepsilon_o d)$. It is of interest to examine in detail the electric field at the metal-polymer interface, $x = 0$. The polarization charge which is generated as a result of the temperature change is within the polymer molecules. The real charge is located on the surface of the metal electrode in contact with the polymer. These two charge planes are spatially separated a small distance, since the poled crystallites are almost certainly not in intimate contact with the surface of the polymer except at a few points. The electric field at the metal electrode is therefore opposite in sign to the field just inside the polymer. If positive charges are required to compensate the polarization at $x = t$, they must enter the polymer at $x = d$ and thus flow through the external circuit because the field is in the opposite direction at $x = 0$. Charge motion through the unpoled part of the polymer continues until the polarization charge at $x = t$ is completely compensated. This requires a total charge of $\alpha_p P \Delta T$. This model therefore predicts that the total charge flow between electrodes following a rapid temperature change consists of an immediate charge transfer, followed by a slow charge flow

in the same direction. The latter charge flow takes place at a rate determined by the electrical time constant of the material. The total charge that flows is dependent only on the level of polarization in the poled region, and is independent of the fraction of sample that is poled.

Acknowldegement

We gratefully acknowlege the financial support of the Office of Naval Research.

References

1. H. Kawai, Japan. J. Appl. Phys., 8, 975 (1969).
2. N. Murayama, T. Oikawa, T. Katto, and K. Nakamura, J. Polym. Sci.: Polym. Phys. Ed., 13, 1033 (1975).
3. G. T. Davis and M. G. Broadhurst in International Symposium on Electrets and Dielectrics, M. S. de Campos, Ed. (Academia Brasileira de Ciencias, Rio de Janeiro, 1977) p. 299.
4. R. E. Collins, Rev. Sci. Instrum., 48, 83 (1977).
5. A. S. DeReggi, C. M. Guttman, F. I. Mopsik, G. T. Davis and M. G. Broadhurst, Phys. Rev. Letters, 40, 413 (1978).
6. R. E. Collins, J. Appl. Phys., 47, 4804 (1976).

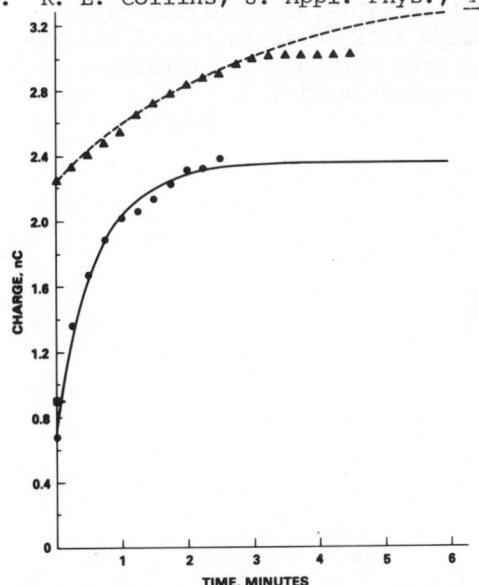

Fig. 1 Pyroelectric charge generated from PVDF-TFE copolymer films of 1.27 cm² area following a 1° step change in temperature starting a 23°C. Upper curve for sample poled at 300 kV/cm for 30 min. at 100°C, lower curve for film poled at 300 kV/cm, 30 min. at 23°C.

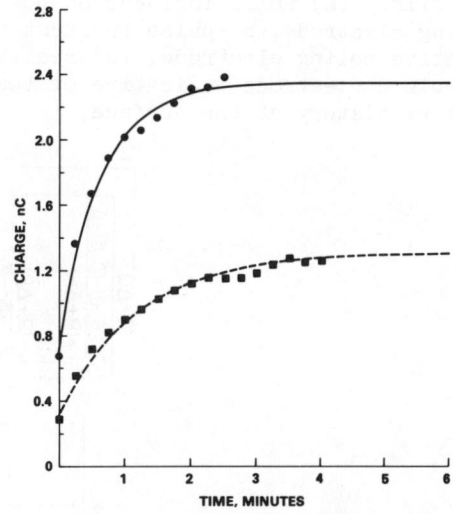

Fig. 2 Pyroelectric charge generated from PVDF-TFE copolymer film of 1.27 cm² area following a 1° step change in temperature. Upper curve: initially at 23°C, lower curve: initially at 9.6°C. Sample poled at 23°C as in Fig. 1.

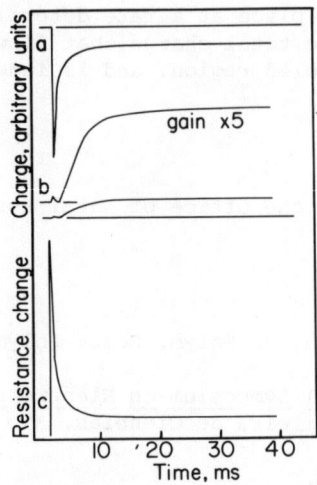

Fig. 3 Transient response from PVDF-TFE copolymer film following a thermal pulse of microseconds duration on one side of the film. (a) Pulse incident on positive poling electrode, (b) pulse incident on negative poling electrode, (c) resistance of pulsed electrode indicative of temperature history at the surface.

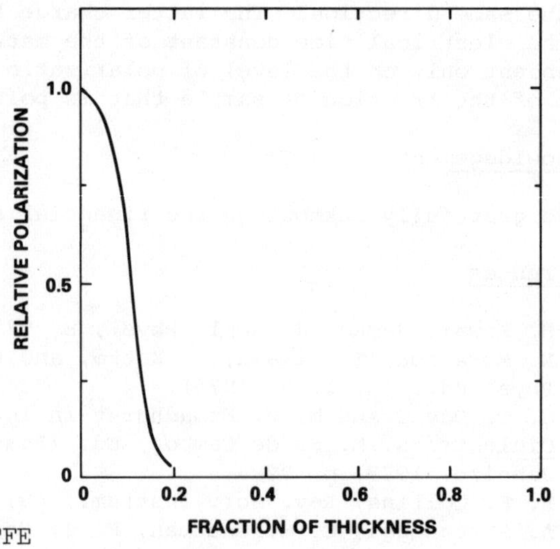

Fig. 4 Polarization distribution within PVDF-TFE copolymer film poled at 23°C as in Fig. 1 obtained from analysis of transient response in Fig. 3.

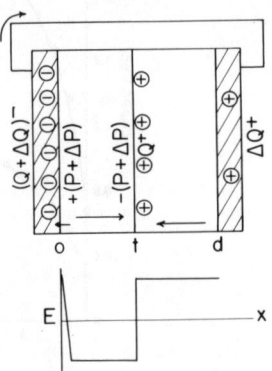

Fig. 5 Schematic diagram of non-uniformly poled polymer film illustrating distribution of polarization charge, real charge, and electric field immediately following a uniform decrease in temperature.

USE OF PIEZOELECTRIC POLYMERS TO PREVENT MARINE FOULING

Mireille Latour & Olivier Guelorget

Université des Sciences et Techniques du Languedoc
Montpellier, France.

and

Preston V. Murphy

Lectret SA, Geneva, Switzerland

INTRODUCTION

Studies on the use of high frequency sound and vibration to reduce marine fouling have been carried out in various parts of the world for the past 25 years. Some of the early work was carried out by industrial groups and technical details were not furnished.(1) Work sponsored by the U.S. government in the 1950's gave mixed results. No reduction in fouling of radar domes excited at 26KHz was observed.(2) nor was fouling reduced on test panels exposed to sound levels of 10^4 Pa.(3) However, 16 ft aluminum boats vibrated at 25 to 55KHz with 25 watt input were relatively free of fouling(4).

It has been found that the larvae of certain fouling species (Balanus, Mytilus) can be injured or killed by high frequency sound fields which exceed 10^6 Pa in intensity(5). Such high intensity sound has been used to prevent fouling on the glass windows of oceanographic sensors(6). Reports from the USSR show that low intensity vibrations can prevent the adhesion of marine larvae to ship hulls. Vibration of 0.05 to 0.5 watt/M^2 intensity was adequate to prevent fouling, however it was important that the vibration not be interrupted(7). A number of ships in the Soviet fleet have been equiped with hull vibrators for over ten years. Antifouling has been effective on vibrating surfaces, but at welded areas of the bulkheads and framing where there was no vibration heavy fouling resulted(8).

We have explored the feasibility of using piezoelectric poly (vinylidene fluoride) (PVF_2) to vibrate the surface of hulls or marine installations, and have studied the effect of the vibration on fouling. The surface might either be covered with a vibrating polymer "skin" (for massive structures), or bonded with strips of polymer. We have used polymer strips and have studied the efficiency of vibration, the long term stability of output, and the effects on marine fouling.

EXPERIMENTAL

Strips of PVF_2 film (Kureha Chemical 1cm x 20cm x 0.003cm) were bonded to test panels (plexiglass, aluminum, and steel of 2000 cm^2) and to the hull of a 2 meter boat. One film was attached to each panel and three to the boat. A marine epoxy was used to ensure rigid bonding and electrical insulation. The films were metallized both sides and excited by a 5 to 20 volts signal at 5, 15, or 19KHz. Test and non-vibrating control panels were painted with a white marine paint containing no antifoulant.

The piezofilm changes dimension in the length mode on excitation due to the predominant d_{31} piezoelectric coefficient. This oscillation is transformed to surface vibration normal to the film as shown in Fig 1. The vibration of test panels was measured as a function of frequency using a Bruel & Kjaer 4343 accelerometer and 2651 charge amplifier. Acoustic output was measured with an 8103 Hydrophone. Vibration and acoustic spectra were obtained with a Tektronix 5L4N spectrum analyzer and storage oscilloscope.

RESULTS - ELECTROACOUSTIC

The surface vibration of test panels for 7V RMS input ranged from about 0.003 to 0.1G depending on panel material, thickness, distance from the transducer, and frequency. The energy dissipation varied from .15 to .27 watt with increase in frequency from 5 to 30KHz. A typical vibration spectrum for a 2mm thick steel panel in water is shown in Fig 2. Many resonances are noted; the frequency of resonance changes with change of measurement position, as well as panel material and configuration. The acoustic output for the same panel measured in a 100 liter reservoir is shown in Fig 3. Other panels typically gave 0.2 to 3 Pa. Vibration output and piezoelectric coefficient decreased by 10% after 6 months immersion.

RESULTS - ANTIFOULING

Marine fouling studies were carried out in the Etang du Prevost on the French Mediterranean coast. This region has a very high fouling rate and the species have been well documented. Principal animal and vegetal species are shown in Fig 4.

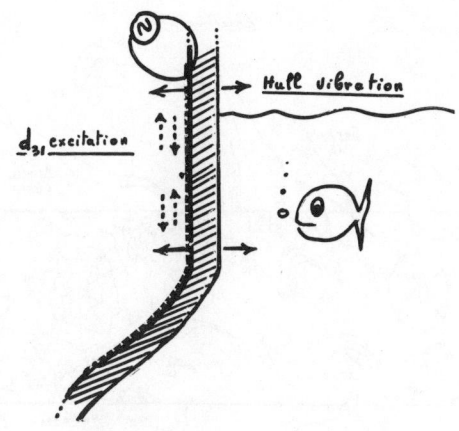

Figure 1   Schematic Diagram of Hull Vibration

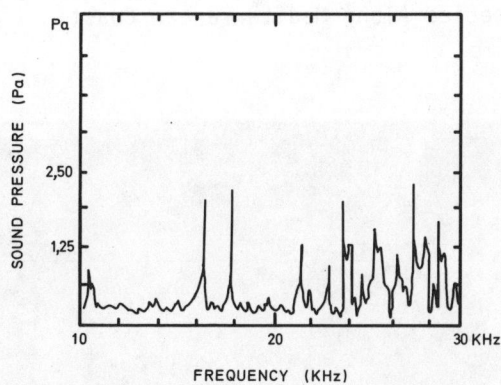

Figure 2   Acoustic Spectrum for Steel Test Panel

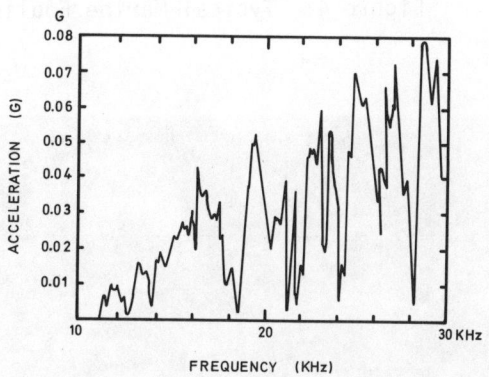

Figure 3   Vibration Spectrum for Steel Test Panel

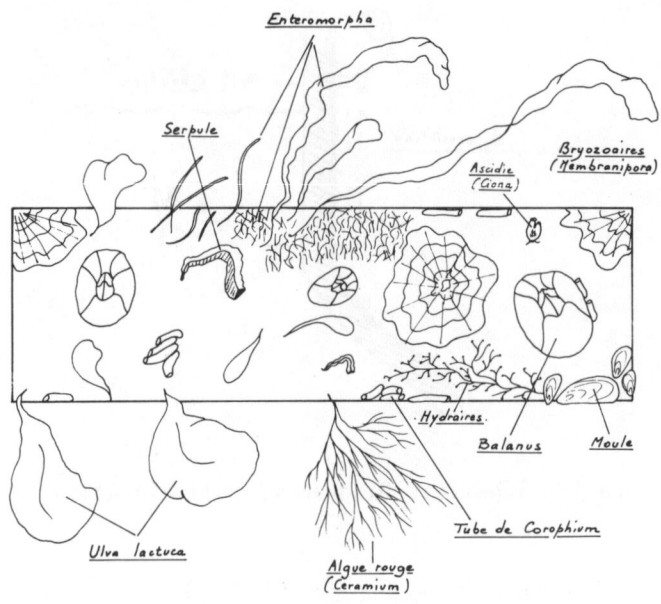

Figure 4   Typical Marine Fouling Species Along Mediterranean Coast

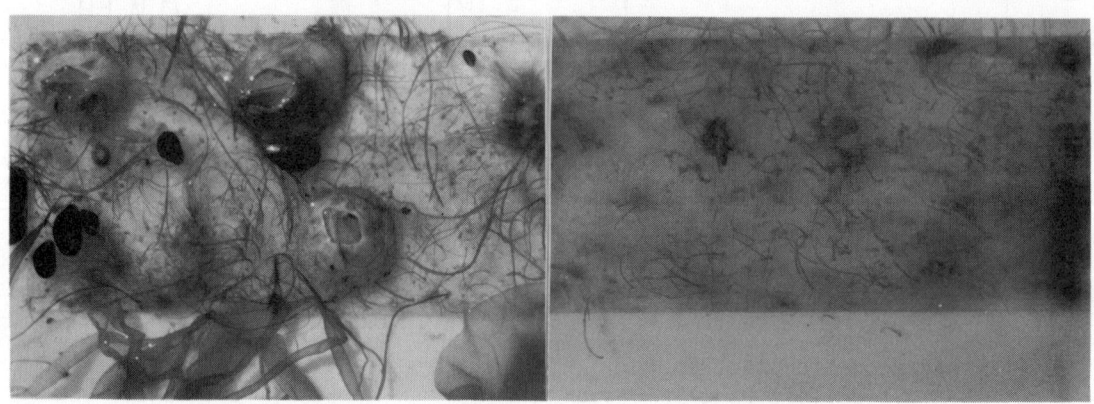

Figure 5   Control Slide After 2 Month Immersion

Figure 6   Test Slide After 3½ Month Immersion with Vibration

Test and control surfaces were covered with a series of glass microscope slides which were removed periodically and examined quantitatively for fouling. No significant difference was found between the glass slides and the painted surfaces. The control surfaces were covered with the more important species (balanus, bryozoaires, etc. as shown in Fig 5) in 1-2 months when water temperatures exceeded 17°C; subsequent attachment and growth of other species produced several gram/cm^2 of fouling in 3 - 4 months. The test panels and boat which vibrated continuously for 2 to 6 months showed a thin coverage of algae which could be wiped-off easily Fig 6. Sometimes a balanus or enteromorph was observed, however the fouling of vibrating test surfaces was one to two orders of magnitude less than stationary controls.

REFERENCES

1. "Shaking up Barnacles", Fisheries Newsletter Australia 14(12) : 13(1955).

2. "Final Report on Tests of Ultrasonic Inhibition of Marine Growth", Document AD 226998 (1959).

3. Berkowitz, H., W.B. Birch, T.T. Dietz and D.J. Zinn. "Acoustic Fouling Project", Office of Naval research, Contract Nonr-396(06) NR 163-312 (1957).

4. Waldvogel, C.W. and J.W. Pieczynski. "A research Program for Marine Growth Prevention by Ultrasonics", Glenn L. Martin Company, Baltimore, MD. Report ER 10764; PB 147155; ASTIA Document AD 219982. May 1959.

5. Suzuki, Hiroshi and Kenjiro Konno. "Basic Studies on the Antifouling by Ultrasonic Waves for Ship's Bottom Fouling Organisms. 1. Influences of Ultrasonic Waves on the Larvae of Barnacles, Balanus Amphitrite. Hawaiienses, and Mussels, Mytilus Edulis", Journal of the Tokyo University of Fisheries, 56(1-2) : 31-48(1970).

6. Kohler, W. and K.B. Sahm. "Investigations into the Use of Ultrasonics to Prevent Marine Fouling", 3rd International Conference and Exhibition for Ocean Engineering and Marine Sciences, Dusseldorf, Germany. June 1976.

7. Askel Band, A.M., "Ultrasonic Protection of Ships From Fouling", Transactions of the Oceanographic Commission USSR, 13 : 7-9 (1959)

8. Transactions, Technical Operations of the Maritime Fleet,. Thermochemical Studies. Control of Corrosion and Fouling. Central Scientific Research Institute of the Maritime Fleet No. 160, 1972", Document AD 778380 (1974).

IV
CHARGE GENERATION AND TRANSPORT

ELECTRON TRANSPORT AND BREAKDOWN IN POLYMERS AND SINGLE CRYSTAL OF LONG CHAIN HYDROCARBONS

Yoshio Inuishi

1978 International Workshop on Electric Charge in Dielectrics

Osaka University, Yamada-Kami, Suita, Osaka, Japan.

I. Introduction

Electronic transport in polymers has been investigated by many workers for long years both from scientific interest in transport in organic amorphous system and from engineering needs for electrical insulation. However, due to various experimental difficulties associated with high resistivity, short carrier life time and complicated molecular structures, satisfactory consistent results have not been established. Recent extensive applications of electronic properties of polymers such as electrets, electrostatic photography, and higher design field for electric power apparatus necessitate urgently to establish fundamental understandings of electronic processes in polymers uptill high field. In this paper our measurement[1)2)3)4)] of electronic carrier mobility, Schubweg or life time, and quantum yield for free electron production by the time of flight method using short electron beam pulse (100 ns) will be reviewed.

Based on these results, the mechanism of dielectric breakdown in PE (polyethylene) will be discussed. Electronic conduction and breakdown in the single crystal of long chain hydrocarbon $C_{36}H_{74}$ (hexatriacontane)[5)] will be also presented for comparison.

2. Time of flight measurement in polymers containing π electrons[1)2)]

As described in our previous papers,[1)4)] short pulse (100 ns) of electron beam with acceleration voltage ranging from 8 to 15KV was injected from one side of the samples in a vacuum cryostat with thin evaporated gold electrodes in both surfaces to apply d.c. bias. As shown in Fig. 1, bombarding primary electron produces many secondary electron-hole pairs within its penetration depth (1~3μm). Some of secondary carriers with appropriate sign of charge escape from these geminate recombination region due to the presence of bias field and start to migrate towards counter electrode through sample. These free carriers, however, will be trapped at deep traps after travelling their life time τ in the average or Schubweg $\omega = \mu E \tau$, where μ, E represent carrier mobility and applied field respectively. In this method we can distinguish electron from hole transport by changing the polarity of d.c. bias, when the penetration depth x_0 is much smaller than the sample thickness d. However, one has to be carefull about the space charge effect due to trapped primary electrons which was minimized in our case by limiting the charge/pulse and pulse repetition rate 0.1pC/pulse and 1/minutes, respectively. As shown in Fig.1 the drift of free carriers induces current i(t) in the outside circuit which is integrated to induced charge $Q(t) = \int i\, dt$ by large CR time constant of the preamplifier input If Schubweg w > sample thickness d, both current and charge wave form should show clear kink at time T corresponding to the transit time of carriers through sample thickness d as shown in Fig.2 by curve (a). The carrier mobility μ, then, is given by

$$T = d/\mu E \qquad (1)$$

The maximum induced charge $Q_m(T)$ at time T can be given by

$$Q_m(T) = \frac{ne\mu E}{d} [1 - \exp(-d/\mu E \tau)] \qquad (2)$$

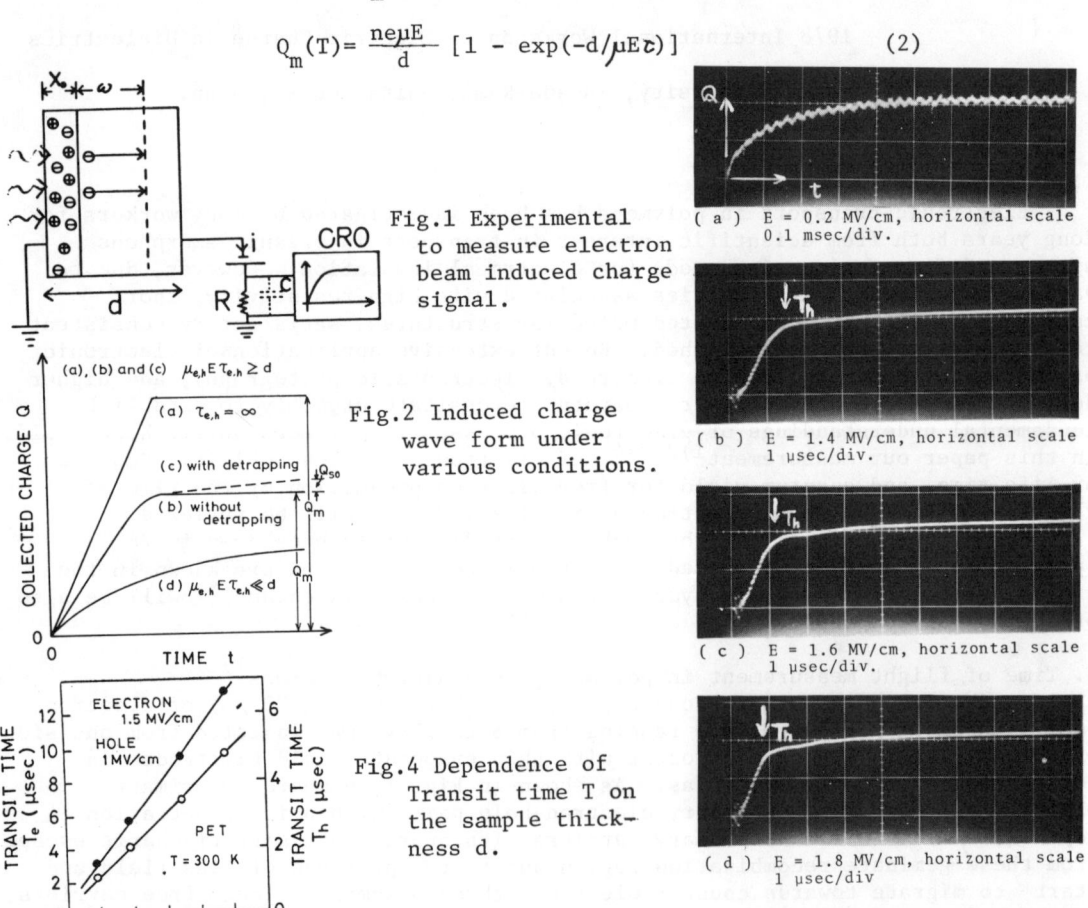

Fig.1 Experimental to measure electron beam induced charge signal.

Fig.2 Induced charge wave form under various conditions.

Fig.4 Dependence of Transit time T on the sample thickness d.

(a) E = 0.2 MV/cm, horizontal scale 0.1 msec/div.

(b) E = 1.4 MV/cm, horizontal scale 1 μsec/div.

(c) E = 1.6 MV/cm, horizontal scale 1 μsec/div.

(d) E = 1.8 MV/cm, horizontal scale 1 μsec/div

Fig.3 Examples of the charge waveform induced by pulsed electron beams.

, where $n = \eta N_p$ means number of free carriers. η and N_p are quantum yield of free carrier production and number of impinging electron per pulse, respectively. Accordingly from the experimental relation between $Q_m(T)$ and E(Hecht curve), we can determine Schubweg $w = \mu E \tau$ or $\mu \tau$ and quantum yield η. In case $w < d$, the charge wave form smeared out and does not show kink as curve (d).

Figure 3 shows several examples of the charge waveform Q(T) in PET (polyethylene terephthalate) thin film. Although at lower field ($w < d$) charge wave form does not show a clear kink, at higher field (E$\geq$1MV/cm), where $w > d$ clear kink corresponding to transit time T were observed in accordance with theoretical prediction of Fig.2. Thus obtained transit time T in PET turned out to be proportional to the sample thickness as shown in Fig.4, and insensitive to the acceleration voltage at a constant applied field, supporting the fact that we are observing the transit time through sample. T decreases with increasing field as shown in Fig.5.

From the slope of curves like Fig.5, mobilities of electron and hole in PET are estimated to be 2×10^{-5} cm^2/V·sec and 1×10^{-4} cm^2/V·sec, respectively. As shown in Fig.6 temperature dependence of the mobility in PET thus obtained can be expressed by Arrhenius type relation

$$\mu = \mu_0 \exp(-\Delta E/kT) \qquad (3)$$

, where pre-exponential mobility are 4.8cm^2/V·sec and 27cm^2/V·sec, and ΔE are 0.32eV and 0.3eV for electon

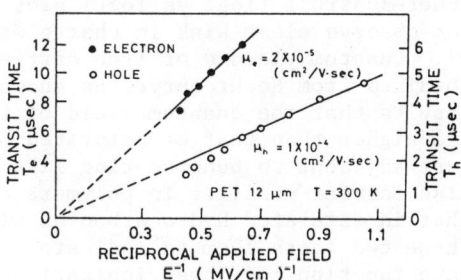

Fig.5 Dependence of the transit time T on the reciprocal of the applied field E.

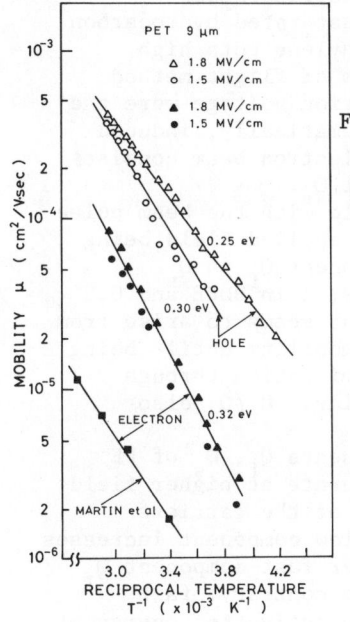

Fig.6 Dependence of the drift mobility u on the inverse absolute temperature.

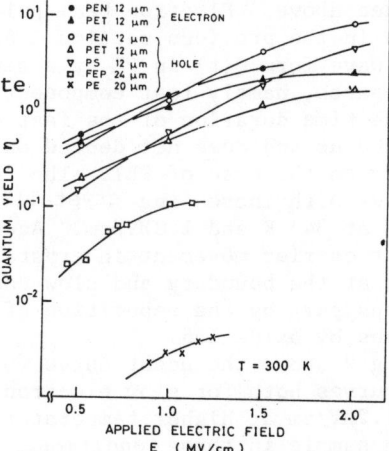

Fig.7 Field dependence of the quantum yield in various polymers.

and holes, respectively. These mobilities can be interpreted as the thermally activated hopping motion of electron and holes or hopping small polaron[1]. Carrier mobility (cm^2/V.sec) in PS (polystyrene), PEN(polyethylene naphthalate) are obtained similarly as 1×10^{-4}(electron) and 7×10^{-5}(hole) in PS, 2×10^{-4} (electron) and 6×10^{-5}(hole) in PEN, respectively.

From the Hecht curve ($Q_m(T)$ vs.E) of equation (2) $\mu\tau$ value in PET were estimated to be $10^{-7} cm^2/V$ both for electron and hole, which gives the life times of electron and hole to be 35μs and 7 μs, respectively.

The temperature dependence of the mobility in PS at lower field obtained from Scher-Montroll plot in PS also obeys to Arrehnius relation[4]. μ_c and ΔE are $120 cm^2$/V.sec and 0.4eV, respectively. It turned out that the carrier mobility in PS obtained from the kink of charge wave form at higher field (> 1.2MV/cm) shows fairly good agreement with the results obtained from Scher-Montroll (logi vs logt) plot at lower field (< 1MV/cm), where one can not observe clear kink in charge waveform.

Quantum yield η of free carriers per impinging primary electron were obtained from Hecht curves as shown in Fig.7. It is concluded from these results that the quantum yield of the polymers containing π electron system are higher than that of saturated hydrocarbon and that the presence of oxygen just adjacent to benzene ring like in PPO kills quantum yield efficiently. Also carrier mobility in polymers containing π electron is much higher than that in saturated hydrocarbon as will be mentioned in the next section. These two facts seem to indicate larger intermolecular overlapping of electron wave function and lower ionization energy in π electron systems.

3. Time of flight measurement in polyethylene[3,4] and hexatriacontane single crystal[5], as the most fundamental example of saturated hydrocarbon.

We have investigated the carrier mobility in polyethylene both high density H.D.P.E. and low density L.D.P.E.[3,4] by the time of flight method mentioned above. Electron beam pulse width and acceleration voltage were the same as in the previous section. As shown in Fig.8. schematically, induced charge wave form Q(t) in PE by a single shot of 100 ns electron beam consists of two parts, namely fast component Q_f and slow component Q_s.

The time duration of the fast component is comparable with the beam pulse width 100 ns and does not depend on sample thickness nor applied field being contrary to the case of PET. The ratio of the fast component Q_f to Q_s increases with increasing crystallinity as shown in Fig.9; 1 in HDPE and 0.2 in LDPE at 343 K and 1.8MV/cm. Accordingly fast component seems to arise from the fast carrier movement in crystalline part with high mobility until being trapped at the boundary and slow component from carrier migration through amorphous part by the repetition of trapping and detrapping. Q_f/Q_s also decreases by oxidation.

Fig.9 shows the Hecht curve (Q vs. E) for two components Q_f, Q_s of PE. These curves both for slow electron and hole tend to saturate at higher field above 1.2MV/cm at higher temperature, suggesting transit of the carriers through sample in these conditions. As shown in Fig.10, slow component increases with increasing temperature becoming more predominant over fast component Q_f above glass trasition temperature (-21 ~ -24°C). On the contrary, fast component Q_f is much less temperature-sensitive with the activation energy of 0.08eV.

The activation energy of Q_s from the slope of the curve in region II in Fig.10 gives 0.3eV, which is in good agreement with the trap depth in amorphous part of PE obtained from the simultaneous measurement of TSL[6)7)] and TSC[8)].

Region I of slow component Q_s in Fig.10 also corresponds to the field saturation region in Hecht curve in Fig.9. These facts indicate that almost all slow carriers transit between electrodes within life time at high field ($>$1.2MV/cm) and high temperature (340K). However, induced charge wave form $Q_s(t)$ even in these conditions does not show clear kink corresponding to a transit time as in the case of PET. These behaviors may be explained in terms of the wide distribution of hopping time suggested by Scher and Montroll[9)], occur in the amorphous part of PE. Fig.11 shows an example of Scher-Montroll plot of ln i(t) versus ln(t) derived from the charge wave form of slow component. One can observe clear kinks at T_r corresponding to apparent transit time of charge front. However, it should be noted that "the universality" of the curve or unique $\ln \frac{i(t)}{i(T_r)}$ vs. $\ln t/T_r$ is not fully satisfied as shown in Fig.12, also that these experimental curves can be fitted by the linear combinations of simple exponential decay such as $i(t) = I_i C_i \exp(-t/\tau_i)$ as shown in Fig.11.

In order to consider kink point in the Scher-Montroll plot T_r as transit time, one has to check the effect of the sample thickness. For example as seen in Fig.13, T_r is proportional to the sample thickness, and decreases with increasing field. Accordingly T_r can be thought to be a kind of transit time of charge front of the slow component. Apparent mobilities obtained from T_r for slow electron and hole are shown in Fig.14; 5×10^{-7} cm^2/V·sec and 3.2×10^{-7} cm^2/V·sec for electron and hole respectively at 343K and 1.2MV/cm. This result of the mobility can be also given by the Arrhenius type relation as already shown by eq.(3), where in this case μ_0 are 2×10^{-2} cm^2/V·sec and 6×10^{-3} cm^2/Vsec for electron and hole respectively, and ΔE is 0.3eV for both electron and hole. It should be noted that μ_0 is much smaller than for π electron system such as PET and PS mentioned before possibly due to the smaller overlapping of wave function in saturated hydrocarbons.

In order to clarify the fast electron transport in crystalline part of PE, prelimenary researches of conduction and time of flight measurement were done for single crystals of long chain hydrocarbon $C_{36}H_{74}$[5)].

In induced charge wave form by time of flight method, only fast component Q_f with high mobility 10^{-2} cm^2/V·sec is observed supporting the idea that fast component in PE originates from the transport in crystalline part. The relation between d.c. current density and applied field is well fit by Schottky plot.

The breakdown field of $C_{36}H_{74}$ single crystals were temperature insensitive uptill metling point as shown in Fig.15, and its magnitude (1MV/cm) is much smaller than that of PE films (4~6MV/cm) shown in Fig.16. These facts seem to indicate that breakdown occurs through impact ionization of free carrier at extended state like in alkali-halide[10)].

The breakdown strength of HDPE with larger crystalline part is smaller than that of LDPE with larger amorphous part at lower temperature below glass transition. This means that also in PE crystalline part with higher electronic mobility has lower breakdown strength due to avalanche than amorphous part with lower mobility and that accordingly breakdown starts from crystalline part at lower temperature. Breakdown strength by d.c. voltage is higher than that by pulse voltage in these temperature region especially in LDPE. This may be due to the space charge built up at the trap located in crystalline-amorphous boundary and resulting reduction of the electric field in crystalline part where breakdown initiates. However these space charge effects due to trapped carriers seem to disappear above glass transition temperature where transport in amorphous parts predominates in accordance with the results of TSC. On the other hand, at higher temperature breakdown strength of LDPE is lower than that of HDPE indicating that breakdown starts in amorphous parts as the results of some instability in electron temperature at localized state as suggested by Fröhlich[11].

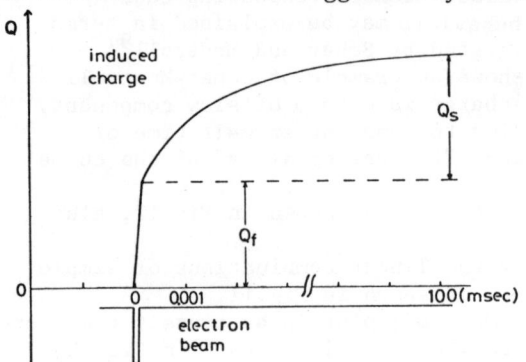

Fig.8 A typical waveform of the induced charge due to a single shot of electron beam in PE.

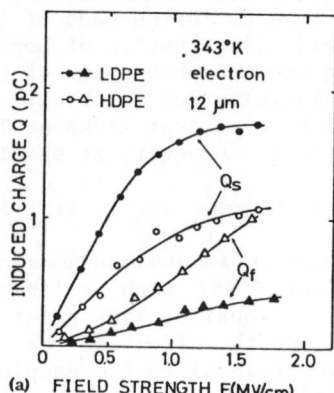

Fig.9 Dependence of the induced charge Q_s and Q_f on the applied field for electron.

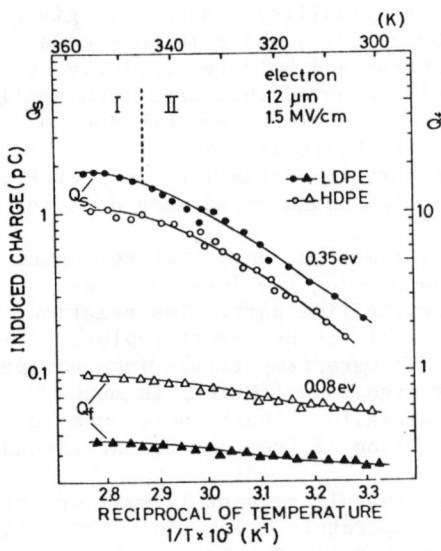

Fig.10 The temperature dependence of the induced charge Q_s and Q_f in PE.

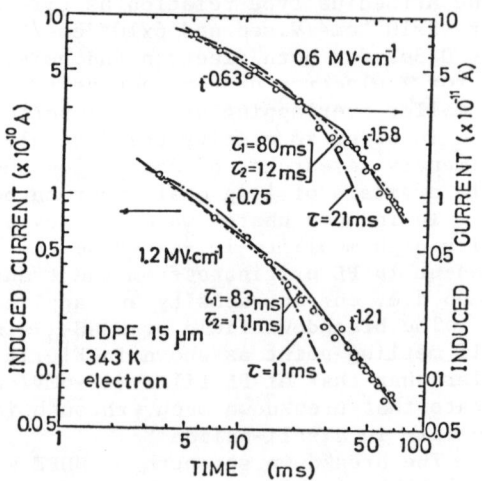

$I(t) = 0.43 \cdot e^{-\frac{1}{83}t} + 2.0 \cdot e^{-\frac{1}{10}t}$ (1.2 MV/cm)

Fig.11 Scher-Montroll plot of electron beam induced current in PE and fitting by the sum of two exponential functions.

CHARGE GENERATION AND TRANSPORT

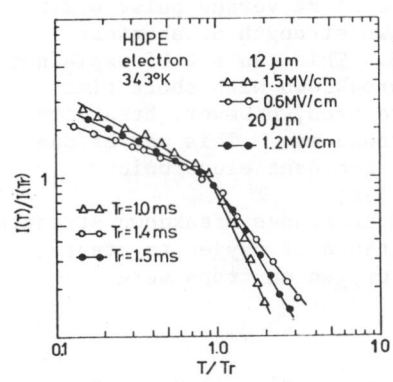

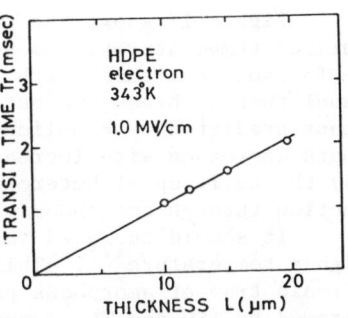

Fig.12 Dependence of $\log i(t)/i(T_r)$ on the normalized time t/T_r for electron in HDPE for various field strengths and electrode distances.

Fig.13 Dependence of T_r on the electrode distance L.

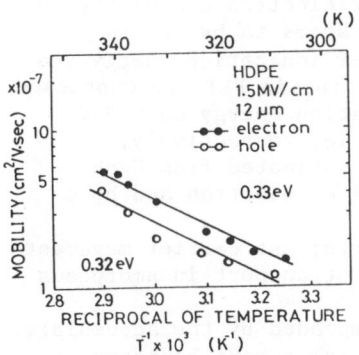

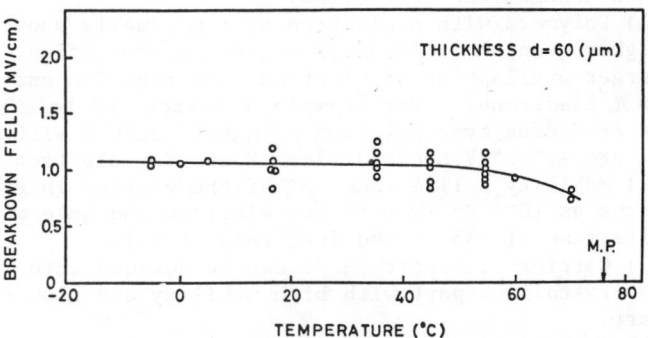

Fig.14 Temperature dependence of the apparent electron and hole mobilities in HDPE.

Fig.15 Temperature dependence of breakdown voltage of hexatriacontane single crystal.

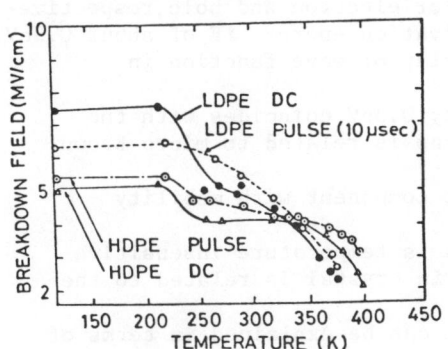

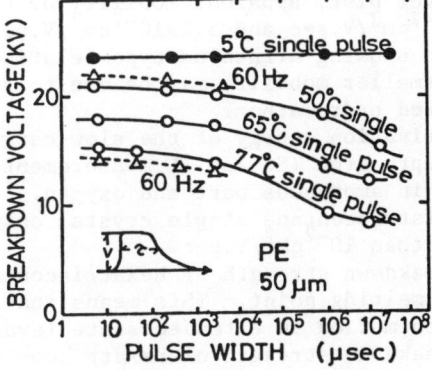

Fig.16 Temperature dependence of dielectric breakdown strength of HDPE and LDPE.

Fig.17 Dependence of breakdown voltage on pulse width at various temperatures.

Figure 17 shows the pulse breakdown strength of PE versus pulse width at various temperatures. As seen there, even breakdown strength of shortest pulse (5 μsec) decreases with increasing temperature. This can not be explained by usual thermal breakdown but by electron thermal breakdown with short time constant predicted by Fröhlich[11]. At higher temperatures, however, breakdown strength decreases with increasing pulse width at around ms. This may be due to (1) the built up of hetero-space charge by the predominant electronic conduction through amorphous part for (2) heating effect.

It should be noted that the oxidation of PE increases breakdown strength at higher temperature[12]. This indicates that importance of oxygen to create electronic trap at amorphous part. These effects of oxygen on traps were confiremed by TSC and TL measurements[7)8)].

Conclusion

From the time of flight measurement carrier mobilities in various polymers were investigated.
Main conclusions were following.
(1) Polymers with π electron system usually show higher electronic mobility and higher quantum yield of free carrier production. This seems to be due to larger overlapping of electron wave function and smaller ionization energy due to π electrons. For example, electron and hole mobilities in PET are expressed by Arrhenius type relation $\mu = \mu_0 \exp(-\Delta E/kT)$ with activation energy ca 0.3eV. μ_0 are $4.8 cm^2/V.sec$ and $27 cm^2/V.sec$ for electron and hole, respectively.
(2) Mobility x life time $\mu\tau$ of the carrier in PET was estimated from Hecht curve as 10^{-7} cm^2/V both for electron and hole which gives electron and hole life time of 35 μs and 7μs, respectively.
(3) Carrier transport in PE can be devided into two parts; fast carrier movement in crystalline part with high mobility and slow carrier transport in amorphous part.
(4) Fast carrier movement in the crystalline part is impeded by the accumulate of trapped space charge due to the traps at crystalline-amorphous boundary.
(5) Charge waveform of slow part in PE does not show clear kink as in the case of PET. However, log-log plot of Scher-Montroll plot shows clear kink. From the later plot, apparent mobility of slow carrier in PE is estimated to be $5.6 \times 10^{-7} cm^2/V.sec$ and $3.2 \times 10^{-7} cm^2/V.sec$ at 343K for electron and hole, respectively, also showing Arrhenius type relation with activation energy ΔE of about 0.3eV. These smaller mobility may be due to smaller overlap of wave function in saturated hydrocarbon.
(6) Activation energy of the slow carrier mobility 0.3eV coincides with the trap depth from TSC and TL measurements. This trap is related to micro-Brown motion in amorphous part and oxygen.
(7) Hexatriacontane single crystal only shows fast component with mobility larger than $10^{-2} cm^2/V.sec$.
(8) Breakdown strength of hexatriacontane crystal is temperature insensitive uptill melting point. This means that breakdown in crystal is related to the electron motion in extended state (avalanche).
(9) Breakdown strength of hexatriacontane crystal can be explained in terms of lower breakdown strength and higher mobility in crystalline part at lower temperature below glass transition. However, at higher temperature breakdown

strength at amorphous part becomes lower due to electron-thermal breakdown of hopping electron. Also space charge effects at the crystal-amorphous boundary play important role especially to increase d.c. breakdown strength at lower temperature especially in LDPE.

Acknowledgement

The author would like to express his sincere gratitude towards all the members of my laboratory, especially Prof.K.Yoshino, Dr.K.Hayashi, Mr.T.Nishitani, Mr.J.Kyokane and Mr.S.Harada, also the pioneering effort by Prof.K.Amakawa of Himeji Institute of Technology is deeply appreciated.

References
1) K.Hayashi, K.Yoshino and Y.Inuishi, Japan. J. Appl. Phys. Vol.12 (1973) 754
2) K.Hayashi, K.Yoshino and Y.Inuishi, Japan. J. Appl. Phys. Vol.14 (1975) 39
3) T.Nishitani, K.Yoshino and Y.Inuishi, Japan. J. Appl. Phys. Vol.15 (1976) 931
4) K.Yoshino, J.Kyokane, T.Nishitani and Y.Inuishi, J. Appl. Phys. Vol.49 (1978) 4849
5) K.Yoshino, S.Harada, J.Kyokane and Y.Inuishi, Tech. Repts. Osaka Univ. Vol.29 in press, Japan J. Appl. Phys. submitted. J.Phys.D, submitted.
6) T.Nishitani, K.Yoshino and Y.Inuishi, Japan. J. Appl. Phys. Vol.14 (1975) 721
7) T.Nishitani, K.Yoshino and Y.Inuishi, J. Ist. Electr. Eng. Japan. Vol.96-A (1976) 381 (in Japanese)
8) K.Amakawa and Y.Inuishi, J. Ist. Electr. Eng. Japan. Vol.93-A (1973) 533
9) H.Scher and E.W.Montroll, Phys.Rev. Vol.B12 (1975) 2455
10) Y.Inuishi and Y.Sung, J. Ist. Electr. Eng. Japan. Vol.82 (1962) 1909 (in Japanese)
11) H.Fröhlich, Proc.Roy. Soc. Vol.A160 (1937) 230
12) K.Amakawa, T.Moriuchi, T.Yoshida and Y.Inuishi, J. Ist. Electr. Eng. Japan Vol.84 (1964) 129 (in Japanese)

ELECTRIC CHARGES IN NON-POLAR LIQUIDS[*]

Werner F. Schmidt

Bereich Strahlenchemie, Hahn-Meitner-Institut für Kern-

forschung Berlin GmbH, 1-Berlin 39, Germany

The physico-chemical properties of excess charge carriers produced by high energy radiation in liquid hydrocarbons and related compounds are reviewed. Special emphasis is laid on the discussion of the mobility of electrons, holes, and ions as a function of temperature and electric field strength. In the second part a short description of the energy levels of the conducting carriers is given. Finally, some applications of non-polar liquids in radiation detectors are described.

Electric charges in non-polar liquids have been the object of experimental and theoretical studies since many decades. Numerous investigations have been carried out from a fundamental point of view as well as stimulated by practical problems of application of liquid insulators. In many of these investigations liquids or liquid mixtures were used, which contained solid, gaseous and liquid impurities so that the results obtained were characteristic for the particular sample rather than for the liquid itself. Several attempts have been made over the years to develop a consistent picture of conduction and breakdown in non-polar liquids. The proposed models apply, however, to specific sets of data only. Several books have been published in which certain aspects or conduction mechanisms were stressed [1 - 5].

Defined charge carriers were first observed in liquefied rare gases since purification of these compounds is relatively easy. Quite a body of information on electrons and ions in these liquids has been accumulated and the salient features have been summarized in several reviews [6-9].

The main emphasis of this review will be on the properties of charge carriers in liquid hydrocarbons and related compounds. These liquids can now be prepared with such a degree of purity that electronic conduction processes become amenable to observation.

---

*) Dedicated to Prof. Dr. Werner Stein, Berlin, on occasion of his 65th birthday

Since these liquids are perfect insulators in the pure state charge carriers have to be introduced by the application of external agents.

Charge carriers may be produced in the liquid by
a) ionization of the liquid by high energy x-rays or electrons
b) injection of electrons or holes from metal or semiconductor electrodes in the liquid under the action of light
c) photo ionization of the liquid by single or multi photon absorption or of solutes with a low ionization potential
d) injection of charge carriers from diodes or gas discharges or by field emission or field ionization on tips or sharp edges.

The detection generally incorporates either conductivity or optical absorption measurements.

Properties or processes observed are
a) yield and mobility of electrons, holes, and ions as a function of temperature, radiation dose, and electric field strength,
b) recombination,
c) attachment to electronegative molecules,
d) scattering by neutral molecules,
e) energy of the electronic conduction level,
f) photo ionization threshold for charge carrier pair production,
g) threshold for laser induced breakdown,
h) absorption spectra of ions or trapped electrons,
i) photo detachment of electrons from negative ions.

The literature on these topics has increased considerably during recent years and limitation in space precludes a detailed description and discussion.

The physico-chemical properties of electrons in fluids have been the subject of a recent international conference [10] and this writer has reviewed the properties of excess electrons in non-polar liquids [11,12] and the electrical conduction phenomena occurring in irradiated liquids [13] not too long ago.

Here an overview of the dynamics and energetics of positive and negative charge carriers in non-polar liquids is given illustrated by a few selected examples.

The most common way of producing excess charge carriers is by irradiation with high energy x-rays or electrons. For kinetic measurements the radiation is administered as a short pulse and the change of electrical or optical properties is followed as a function of time. Under the action of the high energy radiation several kinds of charge carriers are produced which may undergo reactions so that the carriers present at the end of the pulse are not necessarily those generated initially. Some of these processes are summarized in chart I. From the measurement of the conductivity the product of charge carrier concentration and mobility is obtained. The decrease of the conductivity after the radiation pulse may be due to a decrease in concentration of charge carriers as for instance in the recombination process or due to a change in mobility for instance when electron attachment occurs and mobile electrons are converted into slow ions. By proper choice of the experimental conditions the conductivity in a certain time interval may be related to a particular carrier or to a particular transport process only.

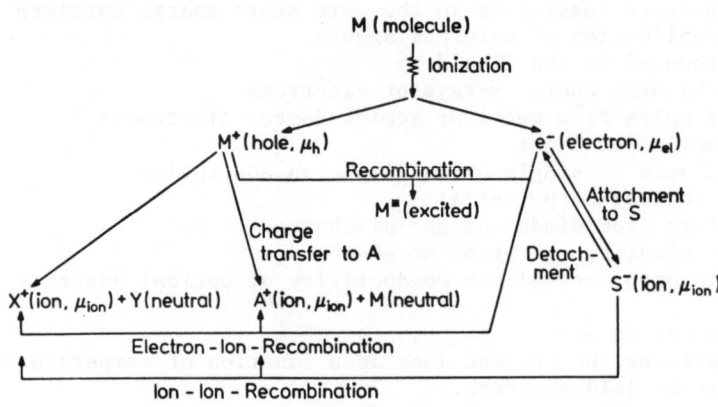

Chart I: Processes following ionization

Changes in conductivity can be detected by observing the voltage drop across a resistor in series with the conductivity cell [14,15] or by measuring the change in absorbed power of a microwave traversing the liquid [16]. Many different types of conductivity cells have been employed by different authors and a recent development from this laboratory, a 50 Ω coaxial cell with a response time of 0.5 ns is shown in fig. 1. A typical signal obtained for electron drift in liquid ethane under conditions of no volume recombination is given in fig. 2.

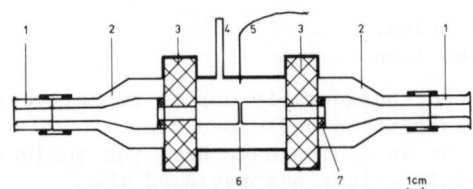

Fig. 1. Schematic of measurement cell. 1: General radio connector; 2: coaxial tapers; 3: macor ceramic; 4: cell inlet; 5: thermocouple; 6: electrode gap; 7: spacer.

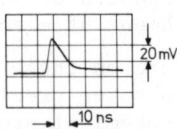

Fig. 2. Oscilloscope trace of the ionization current produced by a 2 ns pulse of X-rays. Time scale: 10 ns/division; applied voltage: 3000 V; gap distance 0.026 cm; vertical scale: 20 mV/div.

The linear decay is due to the drift of excess electrons produced by a 2.5 ns x-ray pulse to the anode.

Electron mobility measurements have been carried out by several groups [cf. 10] and values ranging from 10^{-3} to 10^3 cm^2/Vs have been reported for more than one hundred different liquids. A large variation of electron mobility with molecular structure was observed. Liquids composed of molecules with a small anisotropy of their electric polarizability exhibited greater values for the electron mobility than liquids of molecules with a large anisotropy. As an example the electron mobilities of hexane isomers are given in table 1.

Table 1: Electron mobilities in hexane isomers at 22°C [from ref. 18]

n-hexane	0.09
2 methyl pentane	0.29
3 methyl pentane	0.22
2,3 dimethyl butane	1.1
2,2 dimethyl butane	12

The electron mobility is in all liquids a function of the temperature. The degree of the temperature influence is a function of the absolute value of the mobility.

In the range of electron mobilities < 1 cm²/Vs the electron transport is thermally activated and Arrhenius type dependencies of the mobility on the temperature are observed with activation energies between 0.1 and 0.2 eV. At higher electron mobilities the influence of the temperature depends on the molecular structure. In liquid methane for instance a variation of the electron mobility with temperature as shown in fig. 3 is observed. A similar dependence was found for tetramethyl silane, neopentane, and neohexane [20,21]. In normal alkanes no such variation was detected, the electron mobility increases with temperature up to the critical temperature, as is shown for the case of liquid ethane in fig. 4.

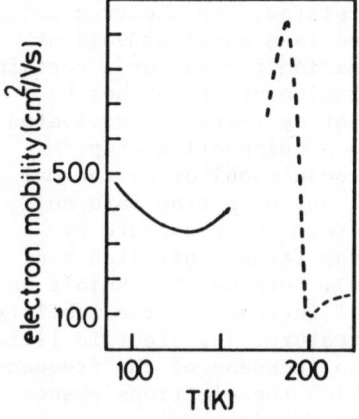

Fig. 3: Temperature dependence of the electron mobility in liquid methane. Solid line pressure lower than 13 bar, broken line pressure 55 bar [after ref. 19].

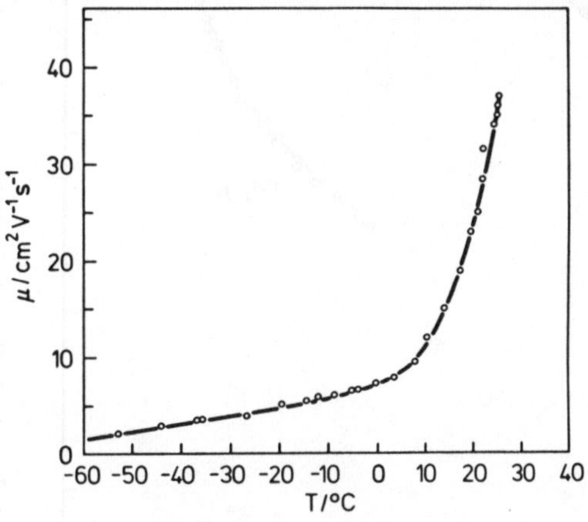

Fig. 4: Electron mobility in liquid ethane [after ref. 22 and unpublished results]

The influence of the electric field strength on the electron transport depends on the magnitude of the electron mobility. For liquids exhibiting a large electron mobility >10 cm²/Vs the drift velocity increases proportional to the electric field strength F at low values followed by a sublinear dependence at higher values of the field strength. For liquid methane, tetramethyl silane, and neopentane the drift velocity v_d was found to be proportional to $F^{0.5}$ [15,18]. The field strength F_c at which the transition from $v_d \propto F$ to $v_d \propto F^{0.5}$ occurs is related to the mean fractional energy loss which the electron suffers in scattering processes. This field dependence is indicative for the electron transport proceeding in extended states. The velocity $\mu_{el} \cdot F_c$ is comparable to the velocity of sound in the liquid. It is assumed that for $v_d > \mu_{el} \cdot F_c$ the electrons are no longer in thermal equilibrium with the molecules of the liquid. The mean electron energy is larger than k_BT. A theory of hot electron effects in molecular liquids is, however, not available.

For liquids exhibiting a low electron mobility < 1 cm²/vs the proportionality region of v_d and F is followed by a superlinear dependence. This effect has been observed for liquid ethane, propane, n-butane, n-pentane, and cyclopentane [18,23].

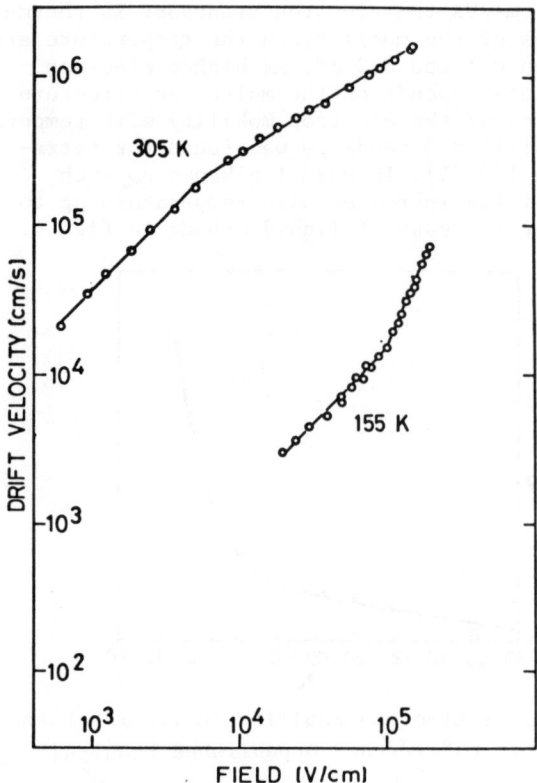

Fig. 5: Electron drift velocity in liquid ethane as a function of the electric field strength [after ref. 23 and unpublished results]

Low electron drift mobilities and a superlinear field dependence can be rationalized by assuming that the electron transport proceeds via localized states. The electron is immobilized in a physical void or by polarization forces for a certain time. Migration occurs either by tunneling or by thermally activated hopping to a neighboring trap (hopping conduction) or by thermal activation out of a trap into an extended state and recapture by another trap (trap controlled band conduction). Both models exhibit an exponential increase of the mobility with temperature. The electric field leads to an increase of the frequency with which the electrons change their trapping sites.

The field effect on the mobility is an indication of the particular transport mechanism. In liquid ethane even a transition from electron transport via localized states at low temperatures to electron transport in extended states at higher temperatures was observed. In fig. 5 these two modes of electron transport are exemplified.

The intermediate range of electron mobilities between 1 and 10 cm^2/Vs has not been studied in great detail. Theoretical considerations suggest that diffusive motion may take place [24].

From steady state [25] and pulse radiolysis studies [26,27] indirect evidence for positive charge carriers with a mobility at least ten times that of ionic species in liquid cyclohexane was obtained. The preparation of extremely pure samples allowed the direct observation of hole transport in cyclohexane, methyl cyclohexane, and trans decalin [28-30]. All the data available are summarized in table 2. The hole transport is little influenced by variation of the temperature, if at all there seems to be a slight decrease of the mobility with increasing temperature. No fast positive charge carrier was observed in isooctane. This may be either due to a higher concentration of impurities present in the particular samples or due to the fact, that the excited parent ion decomposes immediately into a different ion by splitting off a molecular group thus inhibiting resonance charge migration.

Table 2: Hole mobility $\mu \cdot 10^3$ cm^2/Vs as a function of temperature [28-30, and J. Warman private comm. 1978]

liquid	\-26	\-15	2	19	22	38	56	78	82	108
cyclohexane				10.5		8.4	9.0	9.7		
methylcyclohexane					5.8					
trans decalin	14.4	8.2	10.2		9.0	8.8	7.2		7.9	7.7

Charge transfer from holes to molecules with a lower ionization potential leads to the formation of positive ions and rate constants measured for this process are of the order 10^{10} M^{-1}s^{-1} [29].

Volume recombination of positive ions and electrons in liquid hydrocarbons was found to follow Debye's equation

$$k_r = \frac{4\pi e}{\epsilon} (\mu_{el} + \mu_{ion}) \quad (1)$$

which indicates that this process is diffusion controlled [31]. In eq. 1 k_r is the recombination coefficient, $\mu_{el} \gg \mu_{ion}$ are the electron and the ion mobility respectively, e is the elementary charge and ϵ is the relative dielectric constant of the liquid.

Electron attachment to electronegative molecules leads to the formation of negative ions. The reaction rate is a function of the solvent and the solute [cf. 11]. Electron detachment under the action of temperature or light has been observed [32,33].

Although ionic charge carriers in non-polar liquids have been studied for many decades their identities remained obscure. Except in experiments where the conductivity due to dissolved electrolytes was measured the mobilities obtained by different investigators for "parent" positive ions or negative ions varied considerably for the same liquid. Since in normal purity liquids the concentration of impurities is always much higher than the concentration of charge carriers, charge transfer and attachment processes lead to the formation of the energetically most stable ions, the properties of which are determined by the specific impurities rather than by the liquid. Defined negative ions S$^-$ are expected to be formed when an electron scavenger S is added to a pure liquid. The observed mobilities are, however, very unspecific with respect to the added solute. Methyl halides and oxygen for instance gave negative ions of the same mobility in isooctane in the temperature interval 213 to 370 K [34]. Sometimes even several negative or positive ions are observed [cf. 11]. By variation of the temperature and thus the viscosity of the liquid some authors found Walden's rule $\mu \eta$ =const to be obeyed while in other cases $\mu \eta^n$ = const with n = 3/2 was observed.

The assumption of Walden's rule allows the estimation of the Stokes' radius r by eq. 2

$$\mu = \frac{e}{6\pi\eta r} \quad (2)$$

For the positive ions in liquid methane and ethane the radii estimated are larger than the hard core radii of the neutral molecules indicating a certain degree of solvation [15,23]. The range of the polarization forces is small and can be estimated from the potential of the ion-induced dipol interaction

$$V(r) = \frac{e\alpha}{2\epsilon^2 r^4} \quad (3)$$

and by assuming

$$e\, V(r) = k_B T \quad (4)$$

where e is the electronic charge, α is the polarizability of the molecules, T is the absolute temperature and k_B is the Boltzmann constant. This estimate leads to the conclusion, that a solvation shell of nearest neighbor molecules is bound to the positive ion.

For ions of larger molecules the range of the polarization forces extends practically over the molecule only and the Stokes' radius estimated from the mobility agrees with the radius of the molecule [35].

No influence of the electric field strength on the ionic mobility up to 500 kV/cm was observed as demonstrated for negative ions in n-hexane [36].

Introduction of an ion into a non-polar liquid leads to polarization of the molecules surrounding the ion and the polarization energy for an ion of radius r can be estimated by means of Born's equation [37]

$$P_{ion} = -\frac{e^2}{2r}(1 - 1/\epsilon) \quad (5)$$

The energy of the electronic conduction level V_o can be measured by photo effect on a metal electrode. V_o is obtained as the difference of the work functions of the photo cathode in vacuum and in the liquid (eq. 6).

$$V_o = \phi_{liq} - \phi_{vac} \quad (6)$$

Since photo injection occurs adiabatically V_o refers to the extended electron state[38] and it is also called the electron affinity of the liquid. V_o depends on the molecular structure of the liquid and on the temperature [39-42]. Some values reported for different hydrocarbons and liquefied rare gases are compiled in table 3. In the case of liquefied rare gases from the V_o data it was concluded that in LHe and LNe a localized electron state would be energetically most stable while in LAr, LKr, and LXe the excess electron state would be extended [43]. These predictions were corroborated by the electron mobilities measured for these liquids (cf. table 3). For liquid hydrocarbons no such clear correlation between V_o and mobility was found.

Ionic polarization and electron affinity lead to a reduction of the ionization threshold E_1 of the molecules in the liquid phase as given by eq. 7.

Table 3: V_o - values and electron mobilities for non-polar liquids
(data compiled from various sources, cf. ref 12)

Liquid	T [K]	V_o [eV]	μ_{el} [cm^2/Vs]
methane	100	-0.25	400
ethane	100	+0.20	$1.3 \cdot 10^{-3}$
tetramethyl silane	295	-0.57	90
neopentane	295	-0.43	70
isooctane	295	-0.18	7
cyclopentane	295	-0.28	1.1
n-pentane	295	0.0	0.15
n-hexane	295	+0.04	0.09
helium	4.2	+1.05	$2 \cdot 10^{-2}$
neon	25	+0.67	$1 \cdot 10^{-3}$
argon	84	-0.20	475
krypton	116	-0.40	1200-1800
xenon	161	-0.67	2000

$$E_1 = E_g + V_o + P_+ \qquad (7)$$

E_g denotes the ionization energy in the gas phase. Recently direct photo-ionization measurements were carried out with liquid tetramethyl silane by using the synchrotron radiation from an electron storage ring as a light source [44]. The measurement yielded a value of E_1 = 8.1 eV while the gas phase value is E_g = 9.79 eV. Taking V_o = -0.57 a value of P_+ = - 1.12 eV is obtained. From Born's equation a radius of r = 2.95 A follows which is comparable to the hard core radius of the molecule. The photo conductivity as a function of the photon energy and the energy levels are shown in fig. 6.

Non-polar liquids have found application in liquid ionization chambers, especially for the determination of the quality factor of a particular type of radiation or of mixed radiation fields [45]. In the field of elementary particle physics ionization chambers filled with liquid argon are used as instruments for the measurement of the ionization density of high energy particles and as calorimeter for the measurement of the total particle energy [46-49].

Another interesting application has been reported in the field of diagnostic radiology, where a liquid ionization chamber filled with tetramethyl tin was used to obtain x-ray images of an exposed body [50]. TMT is especially suited because of its high average atomic number. One electrode of the ionization chamber is covered with an insulating foil and ions formed by the radiation in the liquid are drawn towards their respective electrodes (cf. fig. 7). At the insulating foil they form a surface charge pattern corresponding to the degree of ionization of the liquid. After removal of the insulating sheet from the chamber the surface charge pattern is developed by powder which adheres to the areas carrying charge.

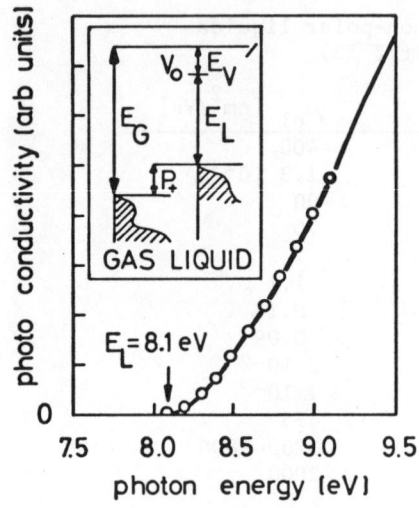 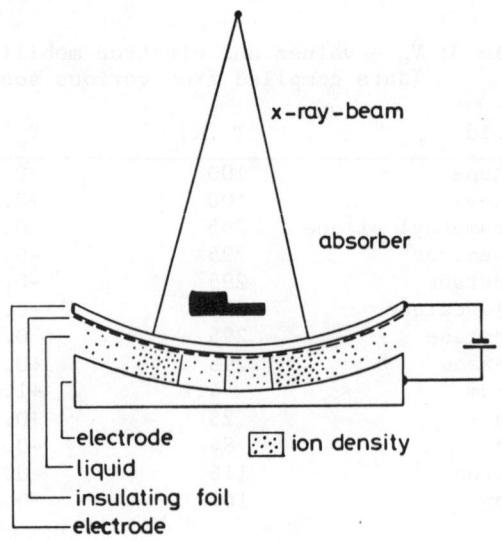

Fig. 6: Photo conductivity and energy levels of tetramethyl silane

Fig. 7: Principle of liquid ionography (after ref. 50)

Although considerable advances in the understanding of the physical behavior of excess charges, and especially of electrons, in non-polar liquids have been made during the last decade, more experimental work directed towards the exploitation of other sources of physical information is needed in order to reach a state of theoretical understanding und description which has been achieved for the problem of charge transport in solids.

Acknowledgement: This paper was prepared during a three months visit of the author with Osaka University, Japan. The author wishes to express his sincere thanks to Professors Y. Inuishi and K. Yoshino and to the other members of the Department of Electrical Engineering for the hospitality extended to him during his stay. The financial support of this visit by the Japan Society for the Promotion of Science is gratefully acknowledged.

References

[1] A. Nikuradse, "Das flüssige Dielektrikum", Springer Verlag, Berlin 1934
[2] A. Gemant, "Ions in Hydrocarbons", Interscience Publ. 1962
[3] J.A. Kok, "Der elektrische Durchschlag in flüssigen Isolierstoffen", Philips Technische Bibliothek 1963
[4] I. Adamczewski, "Ionization, Conductivity and Breakdown in Dielectric Liquids", Taylor and Francis Ltd., London 1969

[5] A.A. Zaky and R. Hawley, "Conduction and Breakdown in Mineral Oil", Peter Peregrinus Ltd. London 1973
[6] J. Jortner and S.A. Rice in "Solvated Electron", Advances in Chemistry Series 50, ACS Washington DC 1965
[7] S.A. Rice, Accounts of Chem. Res. 1, 81 (1968)
[8] J. Jortner, Actions Chim. Biol. Radiat. 14, 7 (1970)
[9] S.A. Rice and J. Jortner in "Progress in Dielectrics", Vol 6, Academic Press New York 1965
[10] International Conference on Electrons in Fluids, Canad. J. Chem. 55 (1977) No. 11
[11] W.F. Schmidt, chapter 7 in "Electron-solvent and anion-solvent interactions" L. Kevan and B.C. Webster (Editors), Elsevier 1976
[]2] W.F. Schmidt, chapter 9 in "Photoconductivity and related phenomena", J. Mort and D. Pai (Editors), Elsevier 1976
[13] A. Hummel and W.F. Schmidt, Radiat. Res. Rev. 5, 199 (1974)
[14] W.F. Schmidt and A.O. Allen, J. Chem. Phys. 52, 4788 (1970)
[15] G. Bakale and W.F. Schmidt, Z. Naturforsch. 28a, 511 (1973)
[16] J.M. Warman, P.P. Infelta, M.P. de Haas, and A. Hummel, Can. J. Chem. 55, 2249 (1977)
[17] W. Döldissen, G. Bakale and W.F. Schmidt, Chem. Phys. Lett. 56, 347 (1978)
[18] W.F. Schmidt, Can. J. Chem. 55, 2197 (1977)
[19] J.M.L. Engels and A.J.M. van Kimmenade, Phys. Lett. 59A, 43 (1976)
[20] J.P. Dodelet and G.R. Freeman, J. Chem. Phys. 65, 3376 (1976)
[21] N.E. Cipollini and A.O. Allen, J. Chem. Phys. 67, 131 (1977)
[22] W. Döldissen, G. Bakale and W.F. Schmidt, Proc. 6th Intern. Conf. Cond. Breakdown Diel. Liqu., Rouen France, July 1978, p 197
[23] W.F. Schmidt, G. Bakale and U. Sowada, J. Chem. Phys. 61, 5275 (1974)
[24] M.H. Cohen, Can. J. Chem. 55, 1906 (1977)
[25] S.J. Rzad, R.H. Schuler and A. Hummel, J. Chem. Phys. 51, 1369 (1969)
[26] A. Hummel and L.H. Luthjens, J. Chem. Phys. 59, 654 (1973)
[27] E. Zador, J.M. Warman and A. Hummel, Chem. Phys. Lett. 23, 363 (1973)
[28] M.P. de Haas, J.M. Warman and A. Hummel, Proc. 5th Intern. Conf. Cond. Breakdown Diel. Liqu., Noordwijkerhout July 1975, p. 15
[29] M.P. de Haas, J.M. Warman, P.P. Infelta and A. Hummel, Chem. Phys. Lett. 31, 382 (1975)
[30] M.P. de Haas, A. Hummel, P.P. Infelta and J.M. Warman, J. Chem. Phys. 65, 5019 (1976)
[31] T. Wada, K. Shinsaka, H. Namba and Y. Hatano, Can. J. Chem. 55, 2144 (1977)
[32] R.A. Holroyd, Ber. Bunsen Ges. physik. Chem. 81, 298 (1977)
[33] K.V. Lukin and B.S. Yakovlev, Chem. Phys. Lett. 42, 307 (1976)
[34] A.O. Allen, M.P. de Haas and A. Hummel, J. Chem. Phys. 64, 2587 (1976)
[35] G. J. Reinis and P.J. Cressman, Proc. 6th Intern. Conf. Cond. Breakdown Diel. Liqu., July 1978, Rouen France, p. 235
[36] P. Chong and Y. Inuishi, Technol. Rep. Osaka Univ. 10, 545 (1960)
[37] M. Born, Z. Physik 1, 45 (1920)
[38] J. Lekner, B. Halpern, S.A. Rice and R. Gomer, Phys. Rev. 156, 351 (1967)
[39] R.A. Holroyd and M. Allen, J. Chem. Phys. 54, 5014 (1971)
[40] R.A. Holroyd and R.L. Russel, J. Phys. Chem. 78, 2128 (1974)

[41] W. Tauchert, H. Jungblut and W.F. Schmidt, Can. J. Chem. $\underline{55}$, 1860 ($\underline{1977}$)
[42] R. Schiller, Sz. Vass and J. Mandics, Int. J. Radiat. Phys. Chem. $\underline{5}$, 491 (1973)
[43] B.E. Springett, J. Jortner and M.H. Cohen, J. Chem. Phys. $\underline{48}$, 2720 ($\underline{1968}$)
[44] W.F. Schmidt, W. Düldissen, U. Hahn and E.E. Koch, Z. Naturforsch. in press
[45] J. Mathieu, Thesis, University of Toulouse 1968
[46] G. Knies and D. Neuffer, Nucl. Instrum. Methods $\underline{120}$,1 (1974)
[47] R.A. Muller et. al. Phys. Rev. Lett. $\underline{27}$, 532 (1971)
[48] J. Prunier, R. Allemand, M. Laval and G. Thomas, Nucl. Instrum. Methods $\underline{109}$, 257 (1973)
[49] J. Engler et. al. Nucl. Instrum. Methods $\underline{120}$, 157 (1974)
[50] A. Fenster and H.E. Johns, Med. Phys. $\underline{1}$, 262 (1974)

PULSED PHOTOCONDUCTION IN SOME POLYMERS

H. Sasabe, K. Kamisako and K. Yatsuhashi

Department of Electronic Engineering, Faculty of Technology,
Tokyo University of Agriculture & Technology,
Koganei, Tokyo 184, Japan.

§1. INTRODUCTION

Recently the electrical and optical properties of electronically conducting polymers have been extensively investigated[1-3]. The coupled properties of these are well known as photoconductive and photodielectric effects: the former is concerned with the transport of photogenerated carriers in materials and the latter with the change of molecular refractive indices due to photoexcitation. In the previous papers[3] we tried to determine energy schemes of the carrier transport in pyrolized polymethacrylonitrile and polyacrylonitrile, poly-γ-ethyl-D-glutamate and polyethylene-2,6-naphthalate by means of a steady state and/or a pulsed photoconduction technique, and to study the refractive index change in polyacrylonitrile in relation to the polymer structures.

Under the excitation with a pulse-like high intensity light (*e.g.*, a Xe flash) is frequently observed a photocurrent through the bulk of polymers. From a transient action spectrum (*i.e.*, the change of photoconductivity or photocurrent with time), one can obtain quantitative information on carrier species, carrier mobilities and trapping levels in materials. Thus the pulsed photoconductivity measurement is so promising as to elucidate carrier transport mechanisms in polymeric materials. Since the electronic carrier mobility μ in most polymers is considerably small ($\mu \simeq 10^{-7} \sim 10^{-3}$ cm^2/Vs)[1-3], a transit time might be determined from the transient due to an illumination of mechanically chopped light instead of light flash. In the case of successive illumination with light pulses, however, the accumulation of space charges due to deep trapping should be taken into account. The primary purpose of this paper is to develop an analytical method of a transient action spectrum excited with a chopped light pulse, and secondly to discuss the effects of space charges.

§2. EXPERIMENTAL

Three types of polymer samples with different bonding structures are used;
(1) pyrolized polyacrylonitrile (PAN) which has conjugated double bonds along the backbone chain and hence the possibility of the intrachain carrier transport, (2) stretched polyethylene-2,6-naphthalate (PEN) which has a fairly well stacking of

naphthalene rings among adjacent chains and hence the possibility of the interchain carrier transport (*i.e.*, perpendicular to the direction of stretching) and (3) poly-γ-ethyl-D-glutamate (PEDG) which has an α-helix conformation and hence the possibility of the carrier transport along the α-helix axis with the aid of intrachain hydrogen bonds.

Table I. Parameters of Polyacrylonitrile.

Sample	Heat Treatment Condition	Absorption Edge	E_p	a
PAN-1	as received	267 nm	4.7 eV	7.0 Å
PAN-2	180°C, 2h	428	2.9	10.6
PAN-3	" , 4h	462	2.7	11.3
PAN-4	" , 8h	504	2.5	12.3
PAN-5	" , 18h	521	2.4	12.7

PAN samples were heat treated *in vacuo* at 180°C for 2, 4, 8 and 18 hours. After the analysis of Eley et al[4], the mobile length of π electrons a can be estimated from the optical gap E_p determined by the absorption edge of the sample. In Table I are listed the values of E_p and a for various PAN samples. PEN sample was provided by Dr. I.Ouchi (Teijin Co.) in the form of biaxially stretched film (PEN-B). The quenched sample (PEN-Q) was prepared by means of a melt-quenching method. The draw ratio of the uniaxially stretched sample (PEN-U) is *ca*. 4.0. The birefringence Δ (=$n_\parallel - n_\perp$) of these samples was measured by using an Abbe's refractometer. The results are: Δ(PEN-Q)=0.000, Δ(PEN-U)=0.168 and Δ(PEN-B)=0.003.

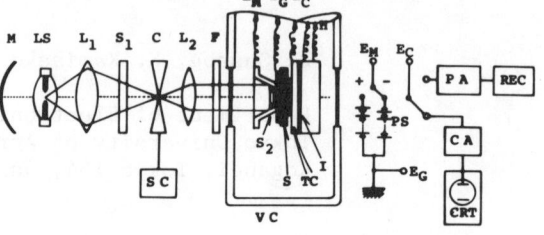

M : Mirror
LS : Light Source
L_1, L_2 : Collimating Lens
S_1, S_2 : Slit
C : Chopper
SC : Speed Controller
F : Filter
S : Sample
TC : Thermocouple
VC : Vacuum Chamber
E_M : Main Electrode
E_C : Counter Electrode
E_G : Guard Electrode
H : Heater
I : Insulator
PS : dc Power Supply
PA : Picoammeter
REC : Chart Recorder
CA : Current Amplifier
CRT : Storage Oscilloscope

Fig. 1. Diagram of measuring system.

A PEDG film was prepared by means of a solvent casting technique on the glass plate. The solvent used was a mixture of ethylacetate and methyethylketone (7:3 in weight) and hence a helix solvent.

Electrodes were formed on both sides of the film sample by a vacuum deposition of Au. The electrode [E_M] to be exposed to the light is semi-transparent (*ca*. 300 Å thick) and guarded by the guard electrode [E_G]. Fig. 1 represents schematically the arrangement of photocurrent measurements. The exciting light was supplied by a ultra-high pressurized mercury lamp (250W)[LS] and chopped mechanically by a chopper [C]. The chopper speed was *ca*. 30 rps and the crossing time of the edge of a chopper blade over the slit [S_2] was less than 0.1 ms. The light was limited in the range of wavelength in 240-380 nm by using a cut filter (TOSHIBA, UVD-25)[F]. In this range the light was absorbed by the evaporated Au layer by the amount *ca*. 90% and by the surface layer of the sample. The electrode E_M was potentialized negatively or positively by the dc power supply [PS]. The transient photocurrent signals were amplified by the high-gain current amplifier (KEITHLEY 427)[CA] and displayed on the storage oscilloscope (SONY-TEKTRONIX 7623A)[CRT], and/or by the picoammeter (TAKEDA TR-8641)[PA] and recorded on the chart [REC]. Temperature and applied field dependences of photocurrent were measured in the ranges of 20°C-180°C and 10^2-10^5 V/cm. The photocurrent measurement was carried out under the vacuum of 2×10^{-6} Torr.

§3. RESULTS AND DISCUSSION

3.1. Transient Photocurrent by Single Chopped Light Pulse:

In Fig. 2 are summarized schematically the pulsed photoresponse curves (transient photocurrents) depending on the applied field F, polarity P and temperature T. Curve#1 is most typical form of the transient. In case of curve#2 a rising time constant might be much shorter than the time constant of the measuring system. The accumulation of space charges causes a gradual decrease of photocurrent during the light excitation, as shown in curve#3 observed in the low T and F regions and might indicate the effect of carrier trapping due to the deep traps. At very low F curve#5 is occasionally observed. This might be caused by the large background field due to, say, residual charges bound in the vicinity of the surface. In case of curve#6 the current flows in the opposite direction to the applied field, which is inexplainable and still open to speculation.

3.2. Drift Mobility:

Photo 1 shows an example of the transient for PEN-Q. The apparent transit time t_T' determined by the deflection point (an arrow in Photo 1) gives the drift mobility μ' as $\mu' = l^2 / V t_T'$, where l is the sample thickness and V the applied voltage. Temperature dependences of μ' for PEN-B, PAN-2 and PAN-5 are shown in Fig. 3. In case of PEN, μ' changes with increasing T following the Arrhenius equation, $\mu' = \mu_0' \exp(-\Delta E/kT)$ where ΔE is the activation energy and k the Boltzmann constant, below T_g (=113°C), but decreases slightly with T above T_g. This might indicate that above T_g the interchain stacking of naphthalene rings are disturbed by the segmental motion of main chains. In case of PAN, μ' decreases with increasing T. This peculiar behavior might be caused by the fact that the effect of carrier trapping into deep levels, which has a longer time constant, is overlapped and makes the apparent transit time t_T' larger than the true transit time t_T. The change

Fig. 2. Typical Transients.

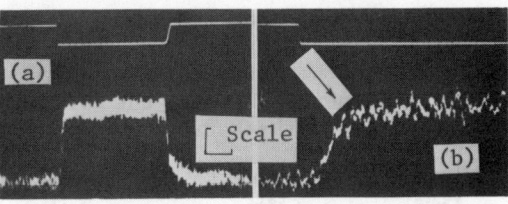

Photo 1. Photocurrent trace of PEN-Q. (91°C, +270V, 10^{-10} A/div): (a) 2 ms/div and (b) 0.2 ms/div.

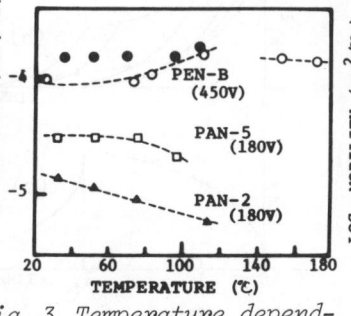

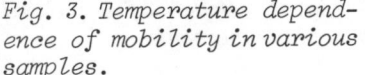

Fig. 3. Temperature dependence of mobility in various samples.

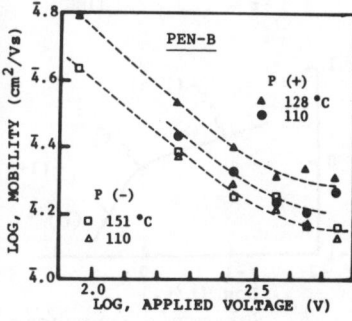

Fig. 4. Field dependence of mobility in samples.

of μ' with F is indicated in Fig. 4 for PEN-B. The tendency of μ' to decrease with increasing F is commonly observed in amorphous materials as chalcogenide glasses[5] and polyvinylcarbazole[6].

3.3 An Analysis of Curve#1: In the light flash experiments with shorter wavelength than the optical absorption edge of a sample is given a typical transient as Fig. 5(a), and the transit time is determined by the break at which the current passes from a plateau region to a long tail region. According to Scher and Montroll[7] the transient spectrum for most amorphous materials can be redrawn in the reduced form of $log[I(t)/I(t_T)]$ vs. $log[t/t_T]$ (Fig. 5(b)). They derived the spectrum theoretically with the assumption of the time dependent random walk of carriers and that the distribution of a carrier packet injected into the surface of the sample follows an $i^2 Erf(z)$. In the chopped light experiments, on the other hand, a typical transient is given as curve#1 in Fig. 1. This behavior may suggest that a chopped light pulse consists of a large number of light impulses and that an observed photocurrent is given by a superposition of successive transient photo-currents due to corresponding light impulses. Here the accumulation of space charges due to successive excitation should be taken into consideration. Then it is assumed that a chopped light pulse consists of N successive light impulses as shown in Fig. 6(a) and that a current due to the carrier packet injected into the sample is given, after Scher and Montroll, as

$$\tilde{J}_0(t) = (0.1)^{-(1-\alpha)}, \text{ for } 0 \leq t \leq 0.1,$$
$$= \tilde{t}^{-(1-\alpha)}, \text{ for } 0.1 < t \leq 1,$$
$$= \tilde{t}^{-(1+\alpha)}, \text{ for } 1 < t,$$

where α is a constant depending on the nature of the material. Space charges generated by successive excitation reduce the macroscopic applied field across the sample, and hence suppress the following carrier injection, that is the k-th transient photocurrent $\tilde{J}_k(\tilde{t})$ is

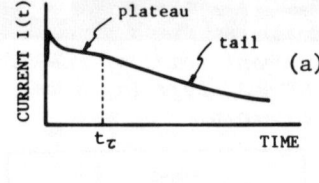

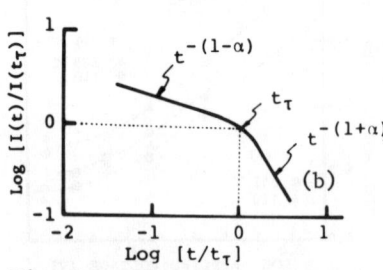

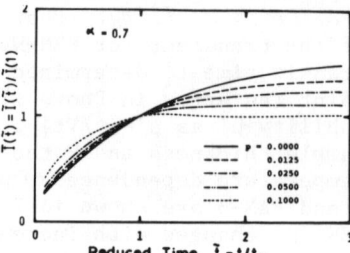

Fig. 7. *An example of the calculated curve for α= 0.7, n=20 and N=60.*

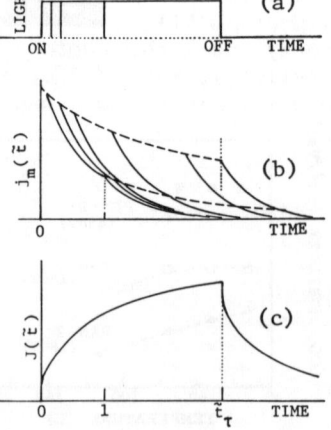

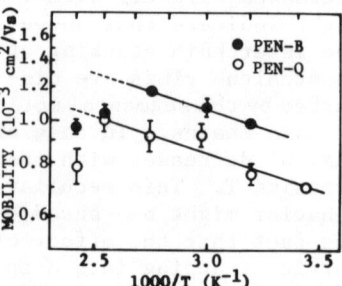

Fig. 5. *Schematic representation of photocurrent by light flash experiments (Ref. 7).*

Fig. 6. *Model of a pulsed photocurrent.*

Fig. 8. *Temperature dependence of mobility in PEN-B and PEN-Q.*

given by $\tilde{j}_k(\tilde{t})=a_k\tilde{j}_0(\tilde{t})$, where $\tilde{t}=t/t_T$ and $\tilde{j}_k(\tilde{t})=\tilde{j}_k(t)/\tilde{j}_k(t_T)$ are respectively the reduced time and current, $\tilde{j}_0(\tilde{t})$ is the 0-th photocurrent due to the first light impulse and a_k is a constant (<1) characterized with space charge effects. Let p be a measure of space charge effect reducing a macroscopic field, $\Delta\tilde{t}$ be a time interval between successive light impulses, and n be a number of light impulses during a time interval of $\tilde{t}=1$. The reduced current $\tilde{j}_m(0)$ is given as follows:

Table II. Parameters α and p.

Sample	α	p
PEN-Q	0.65	0.02
PEN-U	0.80	0.02
PEN-B	0.70	0.03

$$\tilde{j}_m(0)=\tilde{j}_0(0)-p[\tilde{j}_{m-1}(\Delta\tilde{t})+\tilde{j}_{m-2}(2\Delta\tilde{t})+\cdots+\tilde{j}_0(m\Delta\tilde{t})]=a_m\tilde{j}_0(0)$$

$$a_m=1-(p/\tilde{j}_0(0))\sum_{k=0}^{m-1}\tilde{j}_k((m-k)\Delta\tilde{t})$$

$$=1-(p/\tilde{j}_0(0))\sum_{k=0}^{m-1}a_k\tilde{j}_0((m-k)\Delta\tilde{t}).$$

The total photocurrent $\tilde{I}_m$ is given by (Fig. 6(c)),

$$\tilde{I}_m=\sum_{k=0}^{m}\tilde{j}_k((m-k)\Delta\tilde{t})=\sum_{k=0}^{m}a_k\tilde{j}_0((m-k)\Delta\tilde{t}) \text{ for } 1\leq m\leq N,$$

$$=\sum_{k=0}^{N}a_k\tilde{j}_0((m-k)\Delta\tilde{t}) \text{ for } N+1\leq m.$$

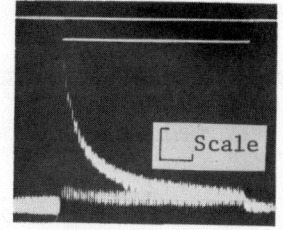

Photo 2. Trace for successive illumination: PEN-Q, 91°C, +270V, 10^{-10} A/div, 0.2 s/div.

These equations can be calculated by using a programmable calculator as a function of α and p. Fig. 7 shows an example of the reduced transient at α=0.7 for n=20 and N=60. A calculated curve was fitted to the observed transient on the graph and the best fit parameters α and p were determined. Results for PEN are listed in Table II. On the fitting procedure, the transit time is determined easily. Then the mobility μ is estimated as a function of T and F. Fig. 8 shows an Arrhenius plot of μ for PEN-B and PEN-Q. The values of ΔE for both samples are nearly equal with each other and estimated as 0.03 eV. Taking that the small value of α might correspond to a variety of trapping levels into consideration, the fact α(PEN-Q) < α(PEN-B) indicates that in case of PEN the stacking of interchain naphthalene rings is enhanced by stretching.

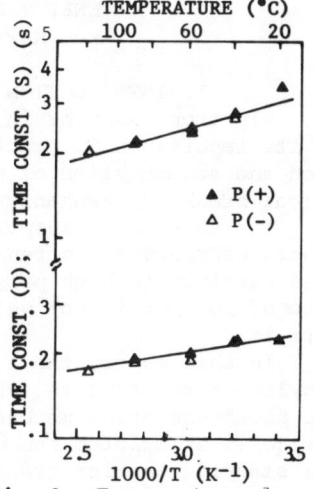

Fig. 9. Temperature dependence of trapping times for PEN-B.

3.4. *Effect of Traps:* As shown in Photo 2 the height of pulsed photocurrent $I_p(t)$ decreases exponentially with successive light pulse excitation. Assuming that this is caused by carrier trapping into shallow (s) and/or deep (d) traps and hence $I_p(t)=I_{ps}exp(-t/\tau_s)+I_{pd}exp(-t/\tau_d)$, then both time constants τ_s and τ_d can be estimated from the $logI_p(t)$ vs. $logt$ plot. τ is a trapping time and generally expressed in the form $\tau=\tau_0 exp(-\Delta E/kT)$. Here ΔE corresponds to the depth of trap level. Fig. 9 shows Arrhenius plots of τ_s and τ_d for PEN-B, the slopes of which give ΔE_s=0.02 and ΔE_d=0.04 eV, respectively.

REFERENCES: /1/e.g.,L.I.Boguslavskii et al,"Organic Semicond.&Biopolym." ,Plenum, 1970, /2/H.Martin et al, Sol.State Comm.,7,783('69),/3/H.Sasabe et al,Repts.Progr. Polym.Phys.Jpn,17,631('74),18,401('75),20,359,625('77),21,in print('78), /4/D.D. Eley et al,Trans.Faraday Soc.,49,79('53), /5/A.E.Owen et al,Phys.Chem.Glasses,17, 174('76), /6/P.J.Reucroft et al,J.Non-Cryst.Sol.,17,71('75), /7/H.Scher & E.W. Montroll,Phys.Rev. ,B12,2455('75).

FIELD-CONTROLLED PHOTOGENERATION AND TRAPPING OF CARRIERS

IN POLYETHYLENE-TEREPHTHALATE

Daisuke Ito

TOSHIBA RESEARCH AND DEVELOPMENT CENTER

ENERGY SCIENCE & TECHNOLOGY LAB., 4-1, Ukishima-cho, Kawasaki-ku, Kawasaki, 210, Japan

1. INTRODUCTION

Electron behavior in polymeric insulators is not well understood in spite of its importance in practical use. Especially, high field electronic conduction and accumulation of space charge are important for the investigation of electrical breakdown mechanism in polymeric insulators.

To study the transport of electronic carriers and trapping phenomena, electronic carriers are often generated by ultraviolet light irradiation, because free carriers in high polymers are very few. However, the photogeneration mechanism of polymer is not well known and it is also not well known what the trapping site is.

In this paper, the field dependence of photogeneration of carriers in polyethylene terephthalate (PET) was measured by u.v. light irradiation for studying the photogeneration mechanism. And the temperature dependence of tan δ, namely thermally stimulated tan δ, was simultaneously measured with thermoluminescence for studying carrier trapping phenomena.

Study of field controlled photogeneration indicates that the carriers are liberated by u.v. light from traps, which have been filled with injected electrons from the cathode. Study of carrier trapping phenomena indicates that the trapping centers and luminescence centers are attributed to carbonyl groups.

2. EXPERIMENTAL PROCEDURE AND RESULTS

Measurements were made on films of PET (6-9μm thick) with evaporated semi-transparent electrodes of gold. Carriers were generated by irradiation of u.v. light under a high DC electric field in vaccum (5×10^{-6}Torr).

When a light pulse as short as 300nsec from a Xe lamp was irradiated on the sample, a voltage was induced across a resister (1Mohm) that was connected in series with the sample. Fig. 1 shows the voltage waveforms indicating occurrence of carrier generation and transport in PET. These waveforms show that schubweg of negative carrier is very short at low temperature. Mobility of negative carriers can be measured by time of flight method. Q_{max} is defined from this voltage value multiplied by the sample capacitance as shown in Fig. 2. Positive induced signal was very small and was one-hundredth as large as the negative one. Field dependence of negative induced charge Q_{max} at room temperature is also shown in Fig. 2. The u.v. light whose wavelength was shorter than 2500Å mainly contributed to the carrier generation judging from the experiments using glass filters for the radiation.

Mobility of negative carriers depended on electric field, and the magnitude is about 10^{-4}-10^{-5} cm^2/Vsec above 1 MV/cm.[1] Super-linear field dependence of drift velocity is shown in Fig. 3. The drift velocity can be expressed by a hyperbolic sine function of applied field as also shown in Fig. 3. Jumping distance of electrons can be estimated to be 16Å from the slope. Above the glass transition temperature, the jumping distance decreases to 5Å apparently.

To study the carrier trapping phenomena, thermally stimulated current (TSC) and thermoluminescence (TL) were measured simultaneously or tan δ at 1KHz and TL were measured simultaneously. Experimental apparatus is shown in Fig. 4.

When DC high voltage was applied or removed from the sample electroluminescence (EL) was observed. EL was also observed by K. Kaneto et al. about PET.[2]

When the sample was excited by continuous u.v. light from Hg lamp under the field strength of 1MV/cm at -190°C photoelectroluminescence (PEL) was observed. The visible spectrum of PEL ranged from 4000Å to 6000Å. The PEL was not observed above the room temperature.

After 30 minutes of irradiation, the u.v. light and the applied field were removed. The TL measurement was made by warming up the sample at a rate of 5°C/min. Simultaneously, either the TSC was measured by a vibrating reed electrometer, or the dielectric properties were measured by an automatic RLC bridge.

The difference in the dielectric properties between a virgin sample, to which neither any field nor u.v. light had been applied and a sample which had been heated up to 150°C after application of electric field at room temperature is shown in Fig. 5. On the other hand, there was no significant difference between the virgin sample and a sample to which 1MV/cm field had been applied at -190°C. The dielectric properties of a sample which had been heated up to 150°C after the irradiation of u.v. light under the field at -190°C is shown in Fig. 6 with the virgin sample. In both cases new tan δ peaks β_2 [at about -90°C] and β_1 [at about -50°C] appeared in low temperature region. These new peaks could not be vanished by warming up the samples up to 150°C.

On the other hand, just after the irradiation, the height of β_2 peak decreased on subsequent warming and simultaneously the TL was observed as shown in Fig. 7.

The TSC spectra consisted of two peaks with temperature maxima of about -140°C and 125°C as shown in Fig. 8. The low temperature peak of TSC was characterized by the coincidence with the TL peak. The activation energy which was measured by the initial rise of TSC or TL was about 0.2eV.

3. DISCUSSION

The foregoing results are interpreted by the field emission of electrons from cathode and by detrapping of electrons due to light excitation from CO-groups as deep traps.

The mobility value indicates that the negative carriers are attributed to electrons. The detection of very small amount of positive signal and wave length of the irradiated light suggest that inter band transition does not occur.

The EL suggests that there exists the field emission of electrons from the cathode, so the carrier generation mechanism can be explained by detrapping of the injection electrons from deep traps by the u.v. excitation.

Fowler Nordheim plot of the induced charge fits Eq. (1) with a constant value of $B = 0.18 \times 10^6$ V/cm as shown in Fig. 9.

$$J = AE^2 \exp[-B/E] \qquad (1)$$

where J is the current density, E is the field strength and A is a constant.

Mechanical measurement of local mode relaxation processes at 1 Hz shows that relaxation of CO-groups occurs at about -105°C and about -70°C.[3] The new tan δ peaks at about -90°C and about -50°C at 1KHz, which were made to appear by the u.v. light irradiation or field application at room temperature, can be interpreted as corresponding to the relaxation processes of the CO-groups. Namely, it can be explained that the injected electrons are trapped by the CO-groups and form new dipoles which make the new dielectric loss peaks β_2 and β_1.

At low temperature, the schubweg of injected electron from the cathode is so short that the CO-groups which are not near the cathode can not be filled with the injected electrons. Therefore the tan δ peaks are not increased at low temperature without the irradiation of u.v. light. Decreasing of β_2 peak whenever the TL is observed can be interpreted that the electrons trapped by CO-groups are partly detrapped by the u.v. light exitation. When the sample is warmed up, electrons which are liberated from some shallow traps are retrapped by the CO-groups and TL occurs simultaneously. The trap depth of the shallow traps can be estimated at about 0.2eV by initial rise of TSC or TL peak. The PEL also can be explained by the same mechanism. No significant difference was observed between the TL spectrum of PET measured by Y. Takai et al.[4] and the TL spectrum of polyethylene measured by A. Charlesby et al.[5] Both of these TL spectra consist of fluorescence and phosphorescence components. The same phosphorescence spectrum is observed in pentatriacontane. These results also show that the luminescence centers are assigned as CO-groups. In the case of saturated hydrocarbon, origin of carbonyl groups is probably oxidation products.

The depth of the trap due to carbonyl groups can be estimated at deeper than 2 to 3 eV from the wavelength of the luminescence. The existence of deep traps is also shown by the experimental results that β_2 and β_1 peaks could not be vanished by warming up the samples up to 150°C. β_2 peak could be vanished only by u.v. light irradiation.

4. CONCLUSION

From the preceding discussion, electrons which are injected from the cathode by field emission are trapped by deep traps due to carbonyl groups of PET. The mechanism of photogeneration of carrier under high field is explained by the detrapping of the electrons from the deep traps by u.v. light irradiation. The depth of the deep traps is estimated at 2 to 3 eV at least.

The luminescence centers responsible for TL or PEL are also carbonyl groups. The TL mechanism is explained as follows: When the sample is warmed up, electrons which are liberated from shallow traps are retrapped by carbonyl groups and TL occurs simultaneously. The depth of the shallow traps is estimated at about 0.2eV.

REFERENCES

1) D. Ito and T. Nakakita: The Transactions of the Institute of Electrical Engineers of Japan, 98-A, 451, (1978).
2) K. Kaneto et al.: Japan. J. Appl. Phys. 13, 1023 (1974).
3) K. H. Illers and H. Breuer: J. Colloid Sci. 18, 1 (1963).
4) Y. Takai et al.: J. Polymer Sci. Polym, Phys. Ed., 16, to be published.
5) A. Charlesby and R. H. Partridge: Proc. Roy. Soc. A283, 312 (1964).

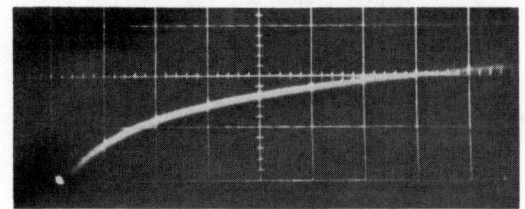
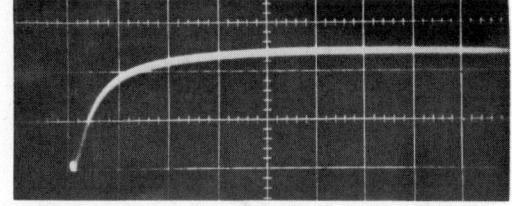

Fig. 1
a) Negative charge waveform at low temperature (-100°C), PET 6μm, 5μsec/div., 5mV/div., 2MV/cm.

b) Negative charge waveform at R.T., PET 6μm, 5μsec/div., 2MV/cm, 100mV/div.

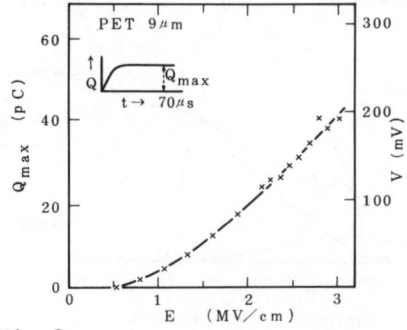

Fig. 2 Field dependence of Q_{max}.

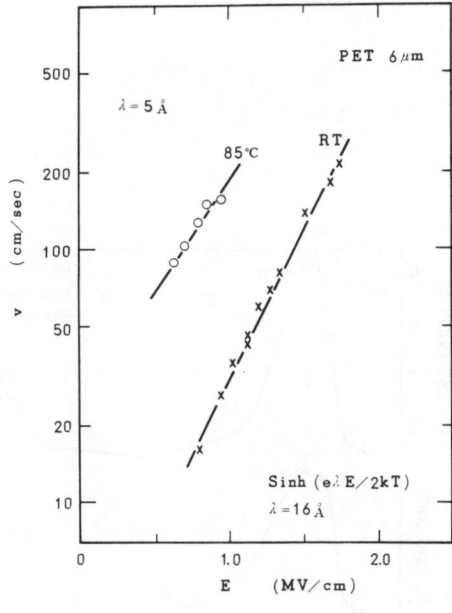

Fig. 3 Field dependence of drift velocity.

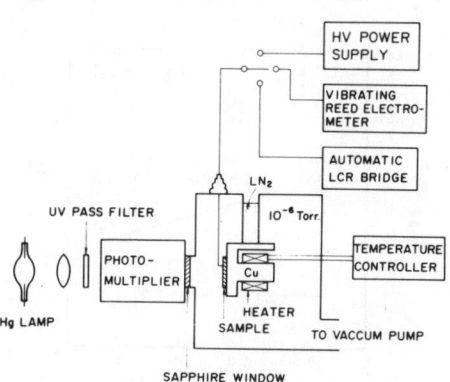

Fig. 4 Experimental apparatus

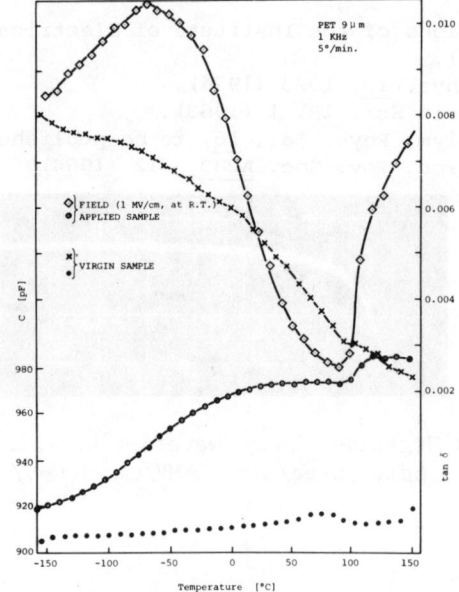

Fig.5 High field application effect on dielectric relaxation processes of PET.

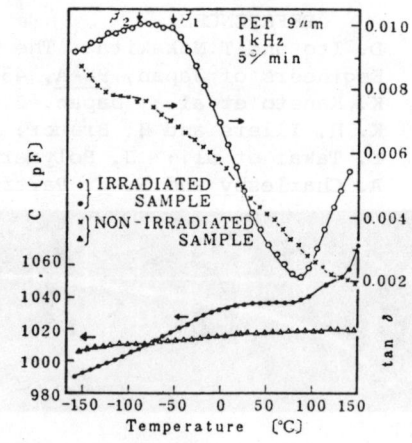

Fig.6 UV light radiation effect on dielectric relaxation processes of PET.

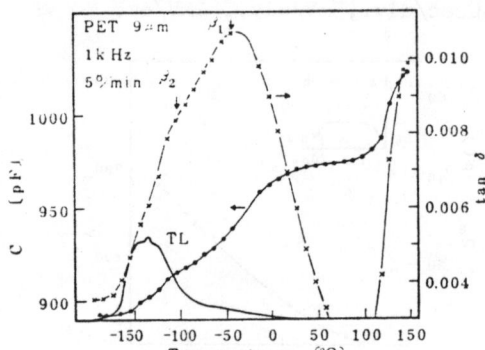

Fig.7 Thermoluminescence and dielectric relaxation processes of PET.

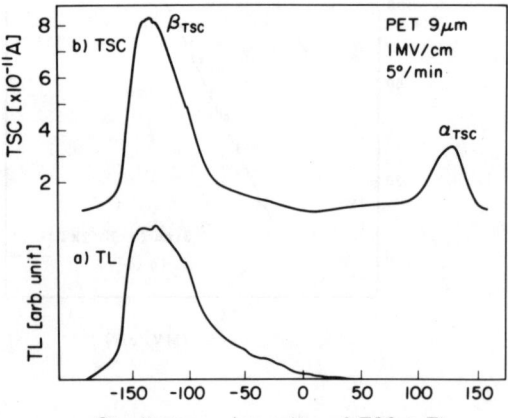

Fig.8 Simultaneous observation of TSC & TL

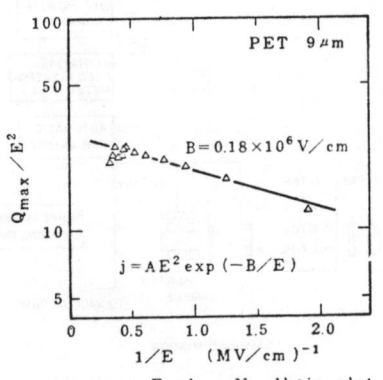

Fig.9 Fowler–Nordheim plot.

ON THE SPACE CHARGE INJECTED FROM ELECTRODE INTO

POLYMERIC MATERIALS AT LOW TEMPERATURE

Noriyuki Shimizu, Hiroyuki Katsukawa, Masaru Miyauchi

Masamitsu Kosaki and Kenji Horii

Nagoya University, Furo-cho, Chikusa-ku, Nagoya, Japan

1. Introduction
   The space charge problems in solid dielectrics are very important. Especially, under a nonuniform electric field the space charge influences strongly the starting voltage of local breakdown, so called treeing. The authors have been studying the treeing phenomena in polyethylene at low temperature. As a result, it was clarified that the homo-space charge injected from the needle electrode reduces the intense nonuniform electric field.[1,2] However, the long term effect of the space charge is still not clear.
   In this paper the authors aimed to discuss the effect of space charge on polymer material under a long period application of ac electric stress. Firstly the luminescence phenomenon is described. The luminescence is considered to be caused by the recombination between positive charge and negative one. Nextly, the degeneration of polyethylene found after ac voltage application of a long period is discussed. The detailed studies using the experimental method of electroluminescence shows that this degeneration is the change of polymer molecular structure caused by the collision with the injected charge.

2. Sample and Experimental
   The experiments were carried out mainly in liquid nitrogen(77K) using the treeing test sample of low density polyethylene ($0.918g/cm^3$; PE). A steel needle or a semiconducting polyethylene film was used as a needle electrode. The outlines of samples are shown in Fig.1.
   The sample without any defect like a void near the needle tip was mounted in a liqud nitrogen tank with a high voltage connection to the needle electrode. The light emitted from the sample was detected by a photomultiplier through a microscope. A spectrum measurement was done using color filters. Simultaneously with the light observation the partial discharge was measured by a tuning type partial discharge detector of a high sensitivity(0.06pC).
   After the voltage application the sample was examined by the microscope. No trace of a tree was detected in the sample provided for the luminescence ex-

periments after the voltage application unless otherwise mentioned.

3. Electroluminescence

Here, the electroluminescence is discussed briefly. The detail on the electroluminescence was reported else-where by the authors.[3] When ac voltage was applied to the needle electrode of a sample the luminescence phenomenon was observed at the electrode tip as shown in Fig.2. It can be said that this was entirely different from the light of the partial discharge from the following.
(1) This luminescence was detected in the viod free sample and no partial discharge pulse could be detected in the luminescence experiments.
(2) The light of the partial discharge in the sample with a void was completely different from this luminescence in terms of the spectrum, the intensity and the wave form of the photomultiplier response.

The wave form of the luminescence for ac 60Hz is shown in Fig.3. Figure 3 (a) was the case of the steel needle sample. The light is emitted by 120Hz and a crest of the light wave precedes that of the voltage wave by $\pi/4$. The intensity corresponding to the negative and positive half cycle is almost equal. On the other hand Fig.3(b) shows the case of the semiconducting PE electrode sample. The intensity of negative half cycle is larger than that of positive: The polarity difference is apparent.

The spectrum of luminescence was not affected by the electrode materials. Moreover a mitute study using a monochromator revealed that this luminescence had a broad spectrum in the range of visible light as was seen in that of the thermo-luminescence of PE. Those results indicate that this luminescence occured in the PE bulk instead of at its surface. The fairly large dimension of the luminous region in Fig.2 may support this.

The polarity difference of luminescence affected by the electrode materials suggests that this luminescence is caused by the injected charge. If the injected charge had not played any role, the electrode material could not have any effect on the luminescence in the bulk. From more detailed investigation the authors concluded this luminescence to be the electroluminescence caused by the recombination between injected positive and negative charges. This luminescence was observed in the other polymer than PE, and further not only for the ac voltage but also for the impulse voltage.

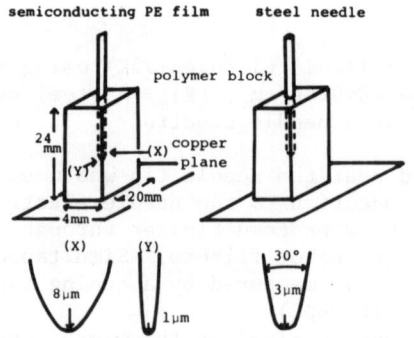

Fig.1 Samples used in experiments.

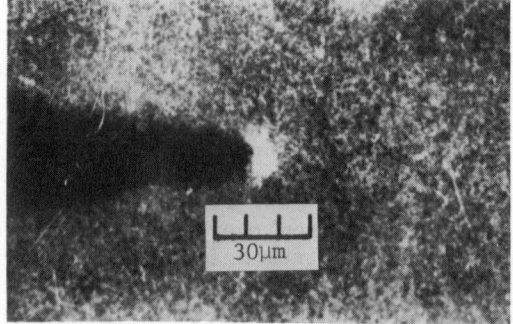

Fig.2 Electroluminescence at the tip of semiconducting PE electrode. (20kVrms)

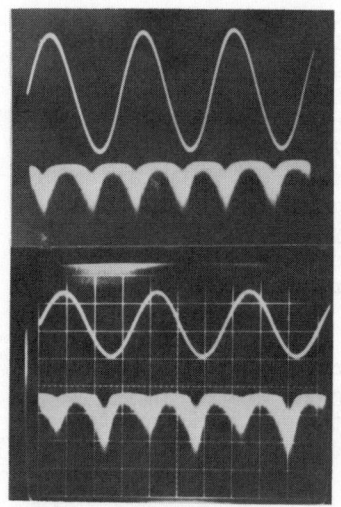

Fig.3 Wave form of luminescence for ac 60Hz applied voltage. (source;ref.3 Reproduced by kind permission of IEEE)

(a) steel needle electrode (14kVrms)

(b) semiconducting PE electrode (22kVrms)

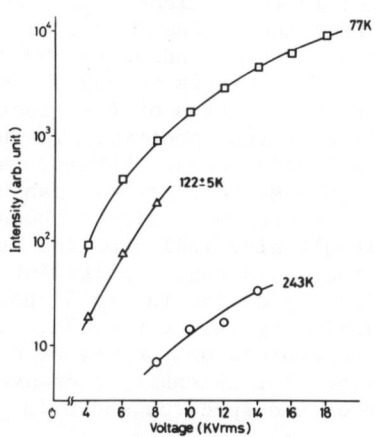

Fig.4 Temperature dependence of intensity of electroluminescence.

Figure 4 shows the temperature dependence of the intensity of the electroluminescence. In this experiment the liquid freon-12(243K) and the cold nitrogen gas (122±5K) were used besides the liquid nitrogen(77K). It can be seen that the intensity decreased with temperature. This result also supports the electroluminescence caused by the recombination of charges. The decrease may be explained by that the probability of the radiative recombination decreases with temperature. It can be understood that this electroluminescence is related closely to the structure of polymer molecules from above consideration.

4. Degeneration of Polymer

In a sample subjected to ac voltage for a long period at 77K the degenerate region was found at the needle tip.[3] Figure 5 shows this, which was obtained from the semiconducting PE electrode sample subjected to 48kVrms for 30 minutes. This degenerate region was dyed by methylene blue more clearly than the other region. Figure 5 was taken after dying. The degenerate region is not an empty void, because no partial discharge was detected in the sample containing only the degenerate region. Moreover the degenerate region remained after melting of sample by heating. This indicates that the degenerate region is not the effect of remaining space cahrge in polyethylene but the change of material structure. However, it was difficult to apply the physical- chemical analysis method to the degenerate region, because it was very small as 10μm in diameter.

The degenerate region is very close to the luminous region in terms of dimension, shape and position as can be seen in Fig.2 and Fig.5. The luminous region is considered to be the injection region of the space charge, so that this result indicates the degeneration took place just in the injection region. Moreover, for the dc voltage which did not give the electroluminescence the degenerate region did not appear. Those results suggest that the degeneration was related to the injected charge.

Figure 6 shows the temporal variation of the intensity of the electroluminescence in the semiconducting PE electrode sample. In Fig.6 the broken line shows that the intensity decreased with time. The authors considered this decrease was corresponding to the change of the morecular structure in the luminous region. The results to support this consideration was obtained from the experiment on the spectrum of the electroluminescence. Figure 7 shows the temporal variation of the spectrum. The broken line shows the spectrum at the start of the application of ac 20kVrms, the solid line shows the spectrum 5 hours later under the same voltage. It can be seen that the intensity in the range below 500nm decreased with time although that in the range above 550nm increased. This result also indicates the change of polymer molecular structure with time under the ac voltage application.

The solid line in Fig.6 shows the temporal variation of the intensity in the case including the heat cycle. That is , the sample was restored from 77K into room temperature on the way of measurement and immersed again into the liquid nitrogen. The intensity decreased suddenly after the heat cycle. Moreover the change of the spectrum shown in Fig.7 emphasized by the heat cycle. Those re-

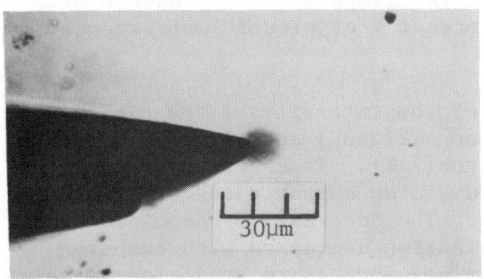

Fig.5 Degenerate region at semiconducting PE electrode tip. (48kVrms, 30minutes)

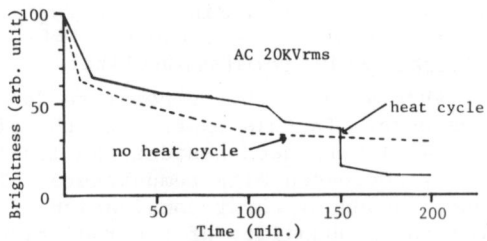

Fig.6 Temporal variation of intensity of electroluminescence in semiconducting PE electrode sample. (20kVrms)

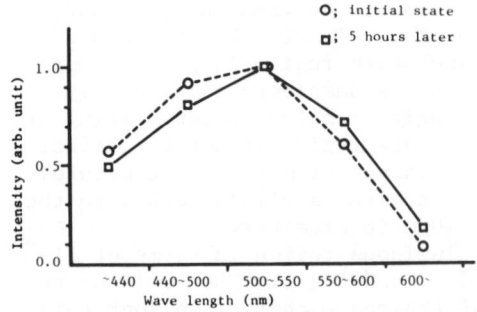

Fig.7 Temporal variation of spectrum of electroluminescence in semiconducting PE electrode sample. (20kVmrs)

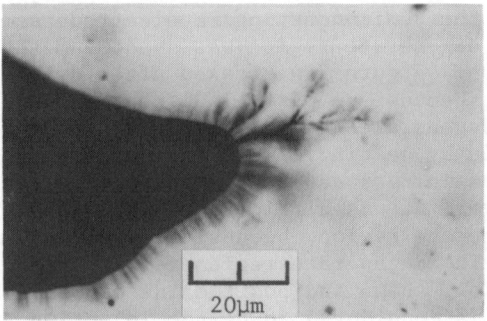

Fig.8 Tree started from degenerate region. (50kVrms, 400minutes)

sults suggest that the change of the molecular structure progressed drastically by the application of the heat cycle.

From above results the authors concluded that the degenerate region is the region in which the chain of polymer molecule was cut by the collision with the space charge injected from the electrode.

5. Tree Starting from Degenerate Region

The tree channel started from the inside of the degenerate region in the semiconducting PE electrode sample as shown in Fig.8. Figure 8 was obtained by the application of ac 50kVrms for 400 minutes. It can be seen the degenerate region is the starting point of the tree, in other words the degeneration is the pre-stage of the formation of the tree channel.

This result is very important practically because it described that the injection of the space charge can cause the initiation of the tree and consequently trigger the total breakdown of the polymer insulation material through the degeneration. The process from the degeneration to the tree starting is currently investigated, but it is probable that the void is eventually formed in the degenerate region and the partial discharge start, it causes the formation of the tree channel.

6. Conclusion

The effect of the injected charge on the polymer material was studied under the ac electric stress. The experiments carried out at the liquid nitrogen temperature using the treeing test sample led the authors to the following conclusions.
(1) The electroluminescence was found at the needle tip. This is not the light from the partial discharge but the luminescence caused by the recombination of the injected charges.
(2) The degenerate region was found at the needle tip. This degeneration is considered to be a change in the molecular structure of polymer caused by the collision with the injected charges.
(3) The tree started from the degenerate region. This indicates that the injection of the space charge can affect the life-time of the polymer insulation material through the mechanism of the degeneration.

References
[1] N.Shimizu, M.Kosaki & K.Horii; J. Appl. Phys. 48, 2191 (1977)
[2] M.Kosaki, N.Shimizu & K.Horii; IEEE Trans. on Electr. Insul. EI-12, 40 (1977)
[3] N.Shimizu, H.Katsukawa, M.Kosaki & K.Horii; Proceedings of 1978 IEEE International Symposium on Electrical Insulation, p.212 (IEEE, New York, 1978)

HIGH-FIELD CARRIER TRANSPORT IN DIELECTRIC LIQUIDS AT PRE-BREAKDOWN

AND BREAKDOWN REGIONS

K.Yoshino,H,Fujii,R.Takahashi,I.Kaizaki,K.Hayashi*,U.Kubo*,Y.Inuishi

1978 International Workshop on Electric Charges in Dielectrics

Osaka University, Yamada-Kami, Suita, Osaka, Japan.
*Kinki University, Higashi-Osaka, Japan.

1. Introduction

The dielectric breakdown in insulating liquids has been studied for many years and recently that in cryogenic liquids has been also studied extensively for the usages as coolants and as insulators in cryogenic apparatuses. However, those breakdown mechanisms are still not completely understood. Carrier transport in dielectric liquids especially at high field regions is of great interest for the understanding of the initial process of the electrical breakdown. Field dependent drift velocity and mobility in liquids have been already reported in the range of $10^3 \sim 10^5$ V/cm by means of the time-of-flight method [1,2].

In this paper, the electrical breakdown and its time lag in cryogenic liquids (Ar, N_2 and He) are studied as functions of temperature, electrode distance and electrode configuration. Breakdown mechanism and its relation with carrier transport processes at pre-breakdown regions will be discussed.

2. Experimental

Both a Fletcher-type square wave pulse generator with pulse width of 10 ~ 100 nanosec (rise time; shorter than 2 nanosec) and a vacuum tube pulse generator with pulse width of 0.5~6 μsec were used as voltage sources. Plane-sphere (10mm$^\phi$) (nearly uniform field) and plane-needle (tip radius of 2~3 μm) electrode systems made of stainless steel were used. The electrode distance was adjusted by a micrometer within the accuracy of ± 1 μm.

High purity Ar (99.99 %), N_2 and He gases were introduced into the cryostat evacuated to 10^{-4} mmHg and then liquified. Electrodes were exchanged or polished after each set of experiments to avoid the effect of the electrode damage by the previous discharge. All experiments were done under the saturated vapour pressure.

The occurrence and the time lag of the breakdown were measured by monitoring the wave form of the applied voltage pulse with an oscilloscope Tektronix 7904. The occurrence of the breakdown was also confirmed by observing light emission of the spark between electrodes. In simplest case, N(T) trials which showed breakdown

time lag longer than T among total trials N can be related to the statistical time lag T_s and the formative time lag T_f by the next equation.
$$N(T)/N = \exp[-(T-T_f)/T_s^f] \cdots\cdots\cdots(1)$$
Both T_s and T_f are evaluated from the Laue plot ($\log N(T)/N$ vs. T) [3] of the breakdown time lag T under the application of the rectangular overvoltage.

3. Results and discussion

Figure 1 shows an example of Laue plot in Liq.Ar at various field strengths. Both T_s and T_f become shorter with increasing overvoltage. As shown in Fig.2, T_f increases with increasing electrode distance d under constant electric field. However, T_s is insensitive to d. These experimental facts suggest that T_f is closely related to the propagation time of avalanche or streamer between electrodes. By assuming a single electron avalanche as the breakdown mechanism, the drift velocity v_f and the apparent mobility μ of the propagating carrier under such high field E can be estimated by the next equations, respectively.
$$v_f = d/T_f \quad \text{and} \quad \mu = d/(T_f \cdot E) \cdots\cdots\cdots(2)$$
Thus obtained v_f and μ increase with increasing field strength as shown in Fig.3. The drift velocity of electron in Liq.Ar measured by the time-of-flight method is known to saturate to about 5×10^5 cm/sec with increasing applied field up to 100 KV/cm due to hot electron effect [2]. The increase of v_f estimated from T_f over this saturation velocity should be explained by the inelastic energy loss of electron due to impact ionization of Ar atom. Deviation from the saturation drift velocity was also observed in rare gas liquids containing molecular impurity possibly due to inelastic collision [2].

Figure 4 shows voltage dependence of T_s and T_f under non-uniform field (needle-plane configuration) in Liq.Ar. In the case of negative needle (needle is negative polarity), the breakdown characteristic is similar to that of uniform field (plane-sphere) with the same electrode distance, but in the case of positive needle, it is much different from that of uniform field. Namely, if the electrode distance is the same, the breakdown voltage of positive needle is much lower than those of uniform field and negative needle. The propagation velocity of the breakdown phenomena calculated from eq.(2) by using T_f of positive needle, is the order of 10^7 cm/sec at the apparent mean field strength of 75 KV/cm, which is 10 times larger than those in the cases of negative needle and uniform field. This extremely high propagation velocity cannot be interpreted as the hot electron drift velocity. Therefore, in this case (positive needle), development of positive streamer should play important role. Similarity of the result of negative needle with that of uniform field can be explained by taking the reduction of the tip fields of negative needle by the space charge layer due to injected electrons into consideration.

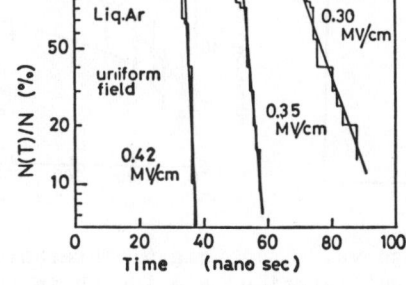

Fig.1 ; Laue plots of breakdown under the uniform field in Liq.Ar

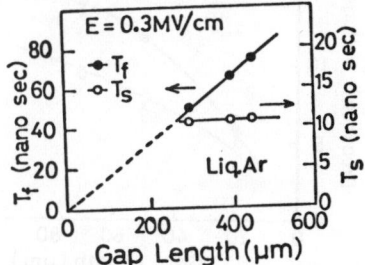

Fig.2; Dependence of T_f on gap length in Liq.Ar.

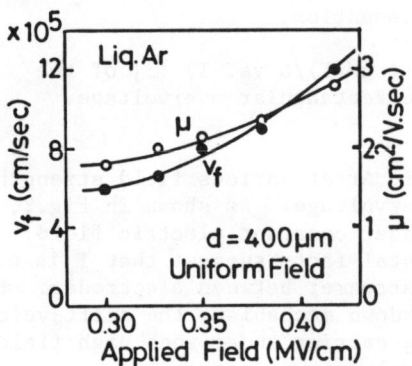

Fig.3; Dependence of v_f and μ on the electric field E in Liq.Ar.

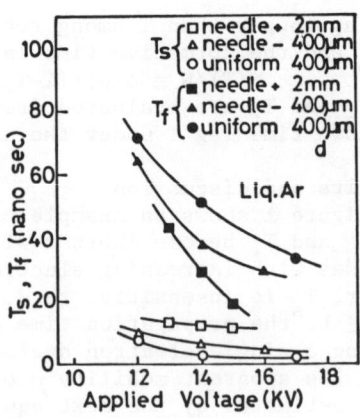

Fig.4; Dependence of T_s and T_f on applied voltage V in Liq.Ar.

In Liq.N_2, T_s decreases with increasing field remarkably. However, T_f does not depend strongly on the electrode distance. Therefore, it seems unreasonable to consider T_f as the transit time of the carriers between electrodes. These facts suggest that the electron avalanche is not the important process in the breakdown of Liq.N_2, perhaps due to (1) the lower energy input to electrons from electric field on account of low electron mobility ($\sim 10^{-3}$ cm^2/V.sec [4]) and (2) higher efficiency of the energy loss of the electron by exciting the molecular vibration of N_2 compared with Liq.Ar. These two reasons can also explain the much higher breakdown strength in Liq.N_2 (3.2 MV/cm for the 50% probability of breakdown with 50 nanosec pulse) compared with that in Liq.Ar (0.35 MV/cm).

Under non-uniform field condition, positive needle gives much lower breakdown strength on applying nanosec voltage pulse as shown in Fig.5, which is the similar characteristic with the case of Liq.Ar. However, in the case of μsec pulse, negative needle gives much lower breakdown strength as shown in Fig.6. The reason of this reversal of polarity effect with increasing pulse width is now under consideration.

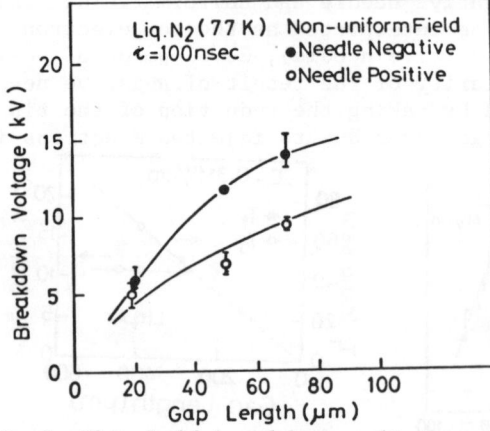

Fig.5; Threshold breakdown voltage vs. gap length in the needle-plane configuration at pulse width of 100 nsec in Liq.N_2.

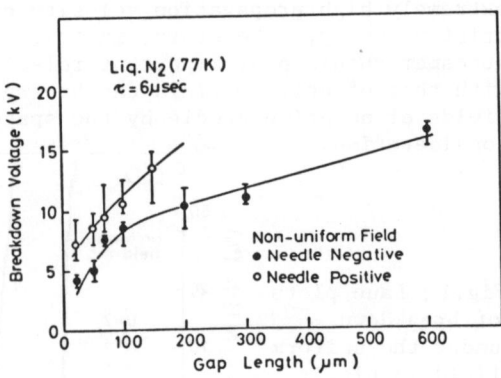

Fig.6; Threshold breakdown voltage vs. gap length in the needle-plane configuration at pulse width of 6μsec in Liq.N_2.

Both T_s and T_f obtained from Laue plot of breakdown in Liq.He, decrease with increasing applied voltage at 4.2 K as shown in Fig.7. T_f also decreases with decreasing electrode distance. These characteristics of T_f seem to indicate the propagating characteristics of breakdown phenomena between electrodes with velocity of about 5×10^5 cm/sec, which is much higher than the sound velocity in Liq.He. By the assumption of a single avalanche process, the mobility of carrier at higher field is estimated to be about 0.3 cm^2/V.sec at 4.2 K, which is larger by one order of magnitude than those of negative carrier (electron in bubble) and positive carrier (He$^+$ cluster) obtained by means of time-of-flight method at low field [5]. Accordingly the creation of quasi-free electron with higher mobility by the ejection from the electron bubble seems to be probable at extremely high fields, if the single avalanche assumption is correct.

At 2.1 K (below λ point), the mobility estimated from T_f by the single avalanche assumption is about 0.1 cm^2/V.sec, which is also larger than the low field value obtained by the time-of-flight method [5].

Breakdown field decreases slightly with decreasing temperature. However, no remarkable change was observed at λ point below which Liq.He becomes superfluid.

It should be noted that the breakdown voltage of positive needle configuration is much higher than that of negative needle and even than that of uniform field as shown in Fig.8 for the impulse of 6 μsec pulse width and also for that of 100 nanosec pulse width, contrary to the case of Liq.Ar and organic liquids. In Liq. He, the mobility of positive carrier is a little larger than that of negative carrier [5]. On the other hand, in Liq.Ar, the mobility of electron is much larger than that of positive carriers. Such a small difference of mobilities of positive and negative carriers in Liq.He may suppress the formation of positive streamer in Liq.He. However, to explain the fact that the positive needle gives much higher breakdown voltage even than the plane-sphere electrode configuration with the same electrode distance, some other effects must be taken into account.

The field strength on the plane electrode under the needle-plane electrode configuration is calculated by the simple method already reported [6]. Thus calculated field strength on plane electrode (cathode) under positive needle configuration is much smaller than the apparent mean field strength calculated by only dividing applied voltage V with electrode distance d in such case. Fig.9 shows the dependence of the field strength $E_b(d)$ at the negative plane electrode surface at the breakdown threshold on the electrode distance, which is just similar to that of breakdown field under sphere-plane electrode configuration (uniform field). This fact should indicate that, in Liq.He, negative electrode plays most important role in breakdown process. Perhaps electron injection from negative electrode (asperities on plane electrode) is determinative process for the breakdown initiation. The role of injected electron at this initial stage is now under study.

Liq.He shows intermediate breakdown strength of about 1.5 MV/cm (for the 50% probability of breakdown with 50 nanosec pulse) between those of Liq.Ar (0.35 MV/cm) and N_2 (3.2 MV/cm). This also seems reasonable, because (1) the carrier mobility in Liq.He is smaller than that in Liq.Ar but larger than that in Liq.N_2, and (2) He atom has no intra-vibrational mode to be excited, contrary to N_2 molecule. Therefore, the increase of electron energy in Liq.He is easier than in Liq.N_2 but much more difficult than in Liq.Ar.

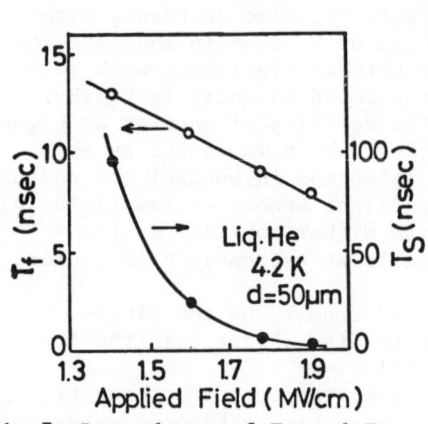

Fig.7; Dependence of T_s and T_f on the uniform applied field in Liq.He at 4.2 K.

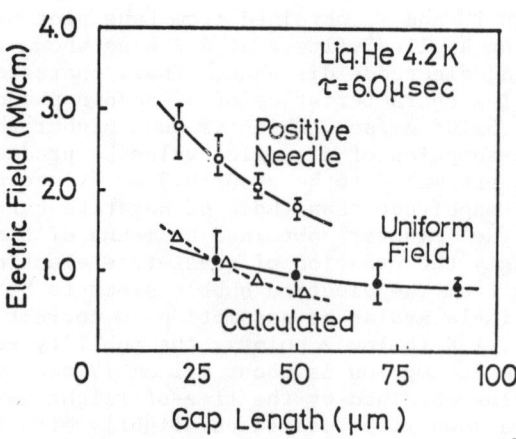

Fig.9; Calculated field strength $E_b^{NU}(d)$ on the cathode surface at breakdown threshold vs. gap length in the case of positive needle (tip radius r = 2 μm). ○:The apparent mean field(V_b/d). ●:The breakdown field under uniform field condition.

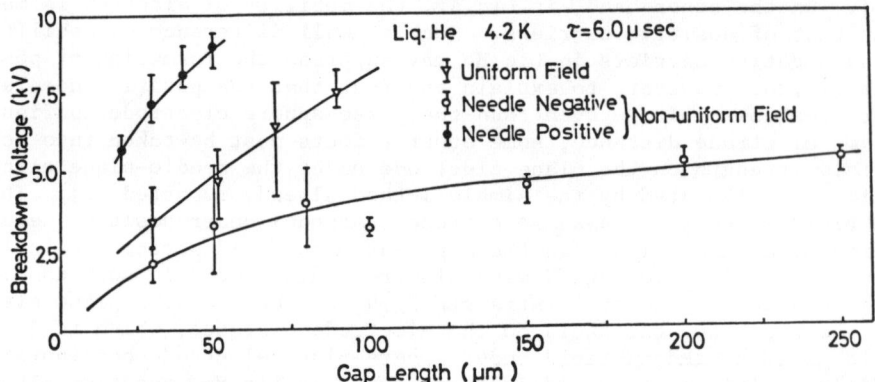

Fig.8; Dependence of the threshold breakdown voltages on the gap length under uniform and non-uniform field conditions in Liq.He at 4.2 K.

References
1) L.S.Miller, S.Howe and W.E.Spear; Phys. Rev. 166 (1968) 871.
2) K.Yoshino, U.Sowada and W.F.Schmidt; Phys. Rev. A 14 (1976) 438.
3) M.von Laue; Ann. d. Physik Lpz. 76 (1925) 261.
4) R.J.Loveland, P.G.LeComber and W.E.Spear; Phys. Rev. B 6 (1972) 3121.
5) K.W.Schwarz; Phys. Rev. A 6 (1972) 837.
6) R.Coelho and J.Debeau; J. Phys. D: Appl. Phys. 4 (1971) 1266.

ELECTRON MOBILITIES AND ELECTRON-ION RECOMBINATION

RATE CONSTANTS IN LIQUID AND SOLID NEOPENTANE

H. Namba, K. Shinsaka and Y. Hatano

Dept. of Chem., Tokyo Institute of Technology

Meguro-ku, Tokyo 152, Japan

Introduction

The behavior of electrons in nonpolar liquids, especially in liquid hydrocarbons, has been studied for these years[1]. The electron mobility μ_e has been measured for a large number of liquid hydrocarbons. In solid aliphatic hydrocarbons, however, which are good electrical insulator under the usual condition, studies of mobility are rare partly for reasons of experimental difficulty[2,3]. We already reported that μ_e and the rate constant of electron-ion recombination, k_r, could be determined by measuring and analyzing the decay of electron current induced by a few n-sec pulsed x-ray irradiation upon liquid neopentane-n-hexane mixtures and demonstrated that this method was quite useful for studying electron transport and reactivity in liquid hydrocarbons[4]. We have applied the same method to the investigation of electron behavior in solid neopentane, which is known as a typical plastic crystal.

Experimental

Sample purification and measurement technique are almost the same as a previous one[4]. The three different neopentane samples; Phillips research grade (sample A), Tokyo-Kasei extra pure (sample B) and Takachiho pure grade (sample C) were used. These samples were stirred under vacuum with concentrated H_2SO_4 for 20hr, with NaOH pellets for 10hr, with $LiAlH_4$ powder for 10hr, and with Na-K alloy for 20hr. Hydrocarbon impurities were analyzed by a gaschromatograph with a flame ionization detector after conductivity measurements. The irradiation dose in the series of experiments was so small that the amount of the radiolysis products was extremely small, less than 10^{-7}%. Main hydrocarbon impurities and their quantities were as follows, sample A (n-butane 0.06 mol%), sample B (n-butane 2.0 mol%) and sample C (n-butane 0.03 mol%, n-pentane 0.05 mol%), after the rigorous purification to remove O_2, CO_2, H_2O, etc. The quantities of other impurities were less than 1% of the main impurity. A small amount of n-butane (Takachiho research grade, purity 99.7%) was added to the Phillips neopentane to examine the effects of the hydrocarbon impurity on the conductivity in neopentane

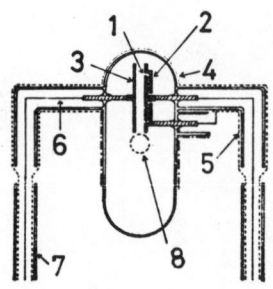

Fig.1. The cell for conductivity measurements
(1) collecting electrode
(2) guard ring
(3) high voltage electrode
(4) silver paint
(5) copper mesh shield
(6) copper wire
(7) coaxial cable
(8) side arm to the vacuum line

samples. After a final purification with Na-K alloy for 20hr, the samples were introduced into a conductance cell in vacuo.

The cell used for conductivity measurements is schematically shown in Fig.1. Circular plates of stainless steel were placed in parallel, having an interelectrode distance of 4.8mm. The collecting electrode had an area of 3.8cm^2 and was protected by a guard ring from edge effects. The electrodes were sealed in a glass envelope whose cross-section was elliptical in shape so as to prevent the cell from breaking because of volume expansion of the sample due to the phase change from solid to liquid. The samples were introduced into the cell through a side arm with several break-seals. The conductance cell was coated on the outside surface with Ag paint to shield it from electromagnetic effects.

The apparatus is schematically shown in Fig.2. The conductance cell containing a sample was set in a cooling box, whose temperature was controlled with the flow of cold nitrogen gas from a liquid nitrogen Dewar vessel with a heater in it. The temperature was detected with copper-constantan thermocouples which were connected with a thermocontroller and a recorder. Temperature was controlled within 1°C. A negative high voltage was supplied to the high voltage electrode by a Fluke 408B or a 415B high voltage power supply through a low-pass filter(8kHz). The sample between the electrodes was irradiated with a few nanosecond pulse of X-rays through a window surrounded by lead bricks of 5cm thickness. The single pulse of X-rays was generated by an impinging pulse of 0.6 MeV electrons from a Febetron 706 accelerator on 0.1mm thick tungsten foil. The electron current induced in the external circuit of the cell was observed by an oscilloscope (Tektronix 454A MOD-163D, 150 MHz) with 20 MHz filter. The oscilloscope was connected to the collecting electrode by a 50 Ω coaxial cable with a 50 Ω termination.

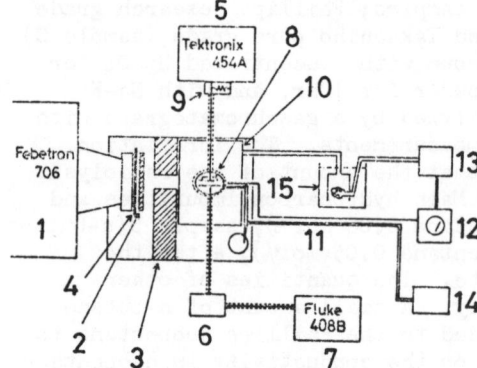

Fig.2. The schematic representation of the apparatus
(1) Febetron 706 accelerator
(2) lead plate of 0 to 8 mm thickness
(3) lead brick of 5 cm thickness
(4) tungsten foil
(5) 150 MHz oscilloscope with 20 MHz filter
(6) high voltage filter
(7) high voltage power supply
(8) conductivity measurement cell
(9) termination resistor
(10) cell holder (cooling box)
(11) thermocouples
(12) thermocontroller
(13) thyristor
(14) recorder
(15) liq. N$_2$ Dewar vessel with a heater

Results and Discussion

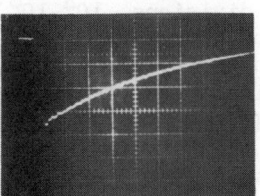

Fig.3. The conductivity decay signal in liquid neopentane (sample A) at room temperature
Ver. 50 mV/ div.
Hor. 100 ns/ div.
E= 3.1 kV/ cm

The method of analyzing the results is almost the same as that of the previous work[4]. The decay of the conductivity singnal shown in Fig.3 was observed when neopentane was irradiated with a pulse of X-rays at the higher dose per pulse. Reducing the absorbed dose per pulse, the decay of the signal becomes longer and below a particular dose the decay rate is independent of the pulse dose[4]. The electron current $I_e(t)$ observed following a pulse of X-rays is given by

$$I_e(t) = \frac{eA\mu_e E}{d} \int_0^d n_e(x,t)dx \quad [1]$$

where e is the electron charge, A the area of the collecting electrode, d the distance between electrodes, E the applied field strength, and n_e the electron concentration at the distance x from the collecting electrode. The change in the electron concentration is given by

$$-(\frac{\partial n_e}{\partial t}) = \mu_e E(\frac{\partial n_e}{\partial x}) + k_{tr}n_e + k_r n_e n_+ \quad [2]$$

The first term on the right hand side of [2] represents the contribution due to drifting of electrons to the collecting electrode in the presence of an external field. The second term is due to the conversion of high-mobility electrons to negative charge carriers of lower mobility with the rate constant k_{tr}, possibly due to attachment to residual impurities which might remain even after rigorous purification described in the experimental section. The third is due to electron-ion recombination with the rate constant k_r, where n_+ is the concentration of positive ions. Reducing the pulse dose makes the contribution of the third term of [2] to the observed decay rate less important, because both n_+ and n_e become smaller in proportion to the absorbed dose. At the lower pulse-dose where decay rate is apparently constant, the observed decay may be ascribed to only the first two terms of [2].

If only the first two terms of [2] contribute to the decay, the electron current is given by

$$I_e(t) = I_o(1 - \frac{\mu_e Et}{d})\exp(-k_{tr}t) \quad [3]$$

where I_o, the peak current intensity at the end of the pulse, is equal to $eA\mu_e E n_o$, where n_o is the initial concentration of free electrons. The most probable value for μ_e is determined, using the least square method with a computer[5], as the value at which the plot of $[I_e(t)/(1-\mu_e Et/d)]$ vs. t can be represented well by the exponential decay $[I_o \exp(-k_{tr}t)]$. The error limit of μ_e is 5% in liquid neopentane and is 15% in solid neopentane.

At the higher pulse dose, electron-ion recombination should compete with the first and the second processes of [2]. The differential equation [2] is impossible to solve, but it can be calculated numerically. We use the two methods to determine the rate constant of electron-ion recombination, k_r. The first is to calculate [2] numerically with a computer[5]. The electron drift time and space are divided into small sections (the number of divisions, 10^4-10^6) and $n_e(x,t)$ is calculated and obtained for a tentative value of k_r according to [2], where μ_e and k_{tr} are obtained from the decay of the conductivity signal of the same sample at low pulse dose, and the initial concentrations of n_+ and n_e, which are equal to n_o, are calculated from the peak current, I_o, at high dose. The electron current $I_e(t)$ is obtained by the substitution of the $n_e(x,t)$ into [1]. Comparing the calculated $I_e(t)$ with the experimental result, we determined the most probable value of k_r. The second method to obtain k_r is to use an approximate formula;

$$I_e(t) = I_o \left(\frac{1-\mu_e Et/d}{1+n_o k_r t}\right) \exp(-k_{tr} t) \quad [4]$$

The value of k_r can be so chosen as to fit [4] to the experimental result. In the actual calculation to find the value of k_r, equation [5] derived by combining [4] with [3], is used

$$1 + (n_o)_h k_r t = \left(\frac{I_e(t)}{I_o}\right)_l / \left(\frac{I_e(t)}{I_o}\right)_h \quad [5]$$

where subscripts l and h represent the low and the high pulse dose, respectively. Substituting two sets of $I_e(t)/I_o$ into [5], one can determine the value of k_r, using the value of n_o at high pulse dose. The values of k_r obtained by these two methods are in good agreement with each other.

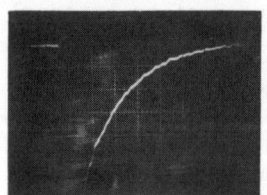

Fig.4. The conductivity decay signal in solid neopentane (sample A) at $-20\,°C$
Ver. 50 mV/ div.
Hor. 50 ns/ div.
E= 3.1 kV/ cm

The examples of the conductivity decay signals in solid neopentane are shown in Fig.4. The values of μ_e and k_r are determined in solid neopentane as well as in liquid neopentane. The temperature dependence of μ_e in different neopentane samples at the field strength of 6.3 kV/cm is shown in Fig.5. In liquid neopentane, μ_e is independent of temperature in the range of $30\,°C \sim -16\,°C$ and the effect of the impurity seems to be small. If the electron

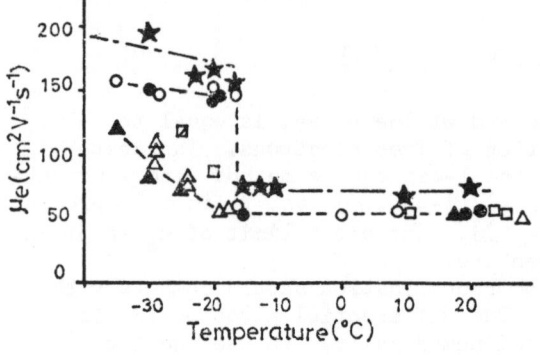

Fig.5. The temperature dependence of μ_e in different neopentane samples
A ○ n-C_4H_{10} 0.06 mol%
B ▲ n-C_4H_{10} 2.0 mol%
C ● (n-C_4H_{10} 0.03 mol%
    (n-C_5H_{12} 0.05 mol%
D □ n-C_4H_{10} 0.8 mol%
    (▨ E=7.3kV/cm)
    (sample A + n-C_4H_{10})
E △ n-C_4H_{10} 2.0 mol%
    (sample A + n-C_4H_{10})
★ TOF + conductivity[3])

mobility of the sample containing a small amount of a solute hydrocarbon obeys [6]

$$\mu_m = \mu_p^{1-x} \mu_s^{x} \qquad [6]$$

where μ_p and μ_s are the electron mobilities in pure neopentane and in pure solute hydrocarbon, respectively, and x is the mole fraction of the solute, the difference between μ_m and μ_p is less than 8% at the maximum concentration of the solute (2 mol% n-butane) in this experiment. The experimental result does not conflict with this estimation. In sample A, however, at its melting point a steep increase in the value of μ_e with decreasing temperature has been observed, whereas in the presence of a small amount of n-butane the increase in the value of μ_e has been clearly suppressed. This anomalous effect of n-butane might be closely connected with the electron scattering by the solute molecule and might be quite valuable for a better understanding of electron transport phenomena in nonpolar media.

In contrast with the anomalous effect of n-butane on μ_e, the values of k_r in solid neopentane are proportional to the value of μ_e with or without n-butane, and the proportional constant agrees with the value calculated from the reduced Debye equation[4]

$$k_r = \frac{4\pi e}{\varepsilon} \mu_e \qquad [7]$$

where ε is the dielectric constant of neopentane. The relationship between k_r and μ_e is shown in Fig.6. The value of k_r in liquid neopentane-n-hexane mixtures[4] and in some neat hydrocarbons and tetramethylsilane determined by other groups[6] are listed in the same figure. The solid line is calculated from [7]. The result means the electron-ion recombination is diffusion-controlled in condensed nonpolar hydrocarbons in which the electron mobility widely ranges from 0.1 to 150 $cm^2V^{-1}sec^{-1}$.

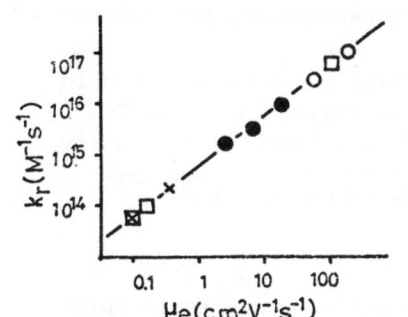

Fig.6. The variation of k_r as a function of μ_e in solid and liquid neopentane (O) and other nonpolar liquids (●[4], ×[6], □[6])

1) Int. Conf. on Electrons in Fluids, Banff, Canada (1976)
2) R. M. Minday, L. D. Schmidt and H. T. Davis, J. Chem. Phys., 54, 3112 (1971); Y. Maruyama and K. Funabashi, J. Chem. Phys., 56, 2342 (1972); G. F. Novikov and B. S. Yakovlev, Int. J. Radiat. Phys. Chem., 8, 517 (1976); J.-P. Dodelet and G. R. Freeman, Can. J. Chem., 55, 2264 (1977)
3) K. Shinsaka and G. R. Freeman, Can. J. Chem., 52, 3556 (1974)
4) T. Wada, K. Shinsaka, H. Namba and Y. Hatano, Can. J. Chem., 55, 2144 (1977)
5) HITAC M-180 system. Computer Center, Tokyo Inst. of Technology.
6) J. H. Baxendale, J. P. Keene and E. J. Rasburn, J. Chem. Soc. Faraday Trans. I, 70, 705 (1974); A. O. Allen and R. A. Holroyd, J. Phys. Chem., 78, 796 (1974)

ELECTRON TRANSPORT MECHANISMS IN DIELECTRICS: TRANSITIONS CAUSED BY DENSITY CHANGES

G. R. Freeman, J.-P. Dodelet and I. György

Department of Chemistry, University of Alberta,

Edmonton, Alberta, Canada T6G 2G2

The mobility of thermal electrons in liquid neopentane (3,3-dimethylpropane) at 300K is almost three orders of magnitude greater than that in n-pentane. The high mobility in neopentane is attributed to the persistence of the quasifree state of electrons in that liquid, whereas electrons form localized states in n-pentane. The persistence of the quasifree state is associated with the spherelike shape, and the consequent isotropic polarizability, of the neopentane molecules.[1] Molecules of the straight-chain n-pentane are not spherelike.

In the low density vapors, electrons are more mobile in n-pentane than in neopentane.[2] This reversal of behavior between the two phases is not understood. Electron transport has therefore been studied as a function of density in the fluid pentanes to determine the location of the reversal. Measurements were made for all three pentane isomers: n-$CH_3CH_2CH_2CH_2CH_3$, iso-$(CH_3)_2CHCH_2CH_3$ and neo-$(CH_3)_4C$.

EXPERIMENTAL

Highly purified pentanes were contained in heavy walled conductance cells that would withstand the pressures of the supercritical gases.[3]

Mobilities were measured by a time flight method.[3]

The PVT parameters of the pentane isomers were obtained from the recent work of Das and coworkers.[4]

RESULTS AND DISCUSSION

When comparing mobilities μ in low density gases, the data may be normalized for differences in gas densities n by multiplying the mobility by n and dividing the electric field strength by n. This is because the mobility is proportional to the length of time τ between collisions, during which the electron is accelerated by the field:

$$\mu = (e/m)\tau \quad (1)$$

where e and m are the charge and mass of the electron, respectively. Similarly, field effects depend on the product of the field strength E and the length of time τ that the field can act upon the electron without interference. When the electron and ion concentrations are very small, the time between collisions suffered by the electron is given by (2),

$$\tau = (nv\sigma)^{-1} \quad (2)$$

where σ is the collision cross section of the molecules and v is the velocity of the electron with respect to the molecules. Accordingly, one has the relationship:

$$\mu n = e/mv\sigma \quad (3)$$

If field effects became noticeable at a constant $(E\tau)_{threshold}$ one would find the threshold field strengths given by (4).

$$(E/n)_{thresh} = (E\tau)_{thresh} v\sigma \quad (4)$$

For larger values of μn one has smaller values of vσ, and one might expect field dependent mobilities to be observed at lower values of E/n. The reverse is true for the pentane isomers (Figure 1).

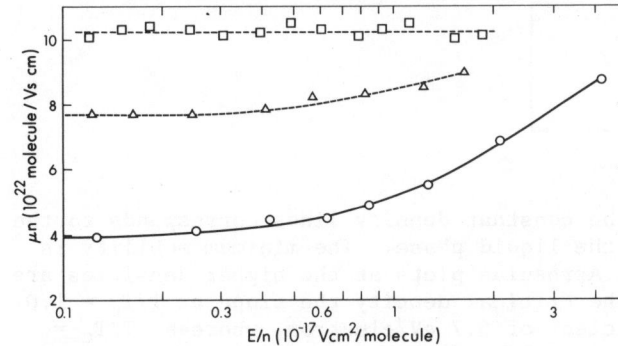

FIGURE 1. Plots of μn against E/n for electrons in low density vapors of n-pentane (□), isopentane (△) and neopentane (○). The respective values of n(10^{19} molecules/cm^3), T(K) are: 9.9, 362; 3.6, 325; 7.2, 335.

The temperature dependences of the mobilities also differ in the isomeric vapors. They vary in the order neopentane > isopentane > n-pentane and correlate with the field dependences. This behavior is attributed to a variation of σ with v. The scattering cross section can be approximated as a power function of v, for a limited range of v.[5]

$$\sigma_v = A_\alpha v^{-\alpha} \quad (5)$$

Analysis of the thermal electron mobilities in the low density vapors at temperatures in the vicinity of 400K gave α = 1.1 for n-pentane, 1.5 for isopentane and 3.2 for neopentane. Values of α > 1 for nonpolar molecules cannot be explained by classical theory. The least polar and most symmetrical molecule, neopentane, has the largest value of α. Two possible explanations are: a) a Ramsauer-Townsend effect exists for each of the isomers, but the scattering minimum is shifted to lower energies for the less spherelike molecules; or b) transient negative ion states lying near zero energy make a stronger contribution to scattering from the more spherelike molecules.

The temperature dependence of the mobility in a constant density vapor increases with increasing density. Arrhenius plots of the mobilities in isopentane vapor at a series of densities are displayed in Figure 2. The curve that

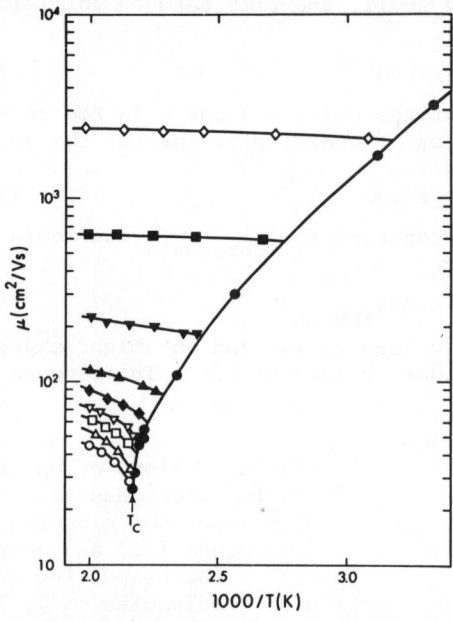

FIGURE 2. Arrhenius plots of thermal electron mobilities in isopentane vapor at different densities. $N(10^{19}$ molecules/cm$^3)$; ◇, 3.60; ■, 13.3; ▼, 39.3; ▲, 67.0; ◆, 94.4; ▽, 112; □, 138; △, 160; ○ 197 (critical). Mobilities in vapors along the coexistence curve are marked ●

forms the low temperature limit of the constant density lines corresponds to the saturated vapor in equilibrium with the liquid phase. The minimum mobility is obtained in the critical fluid. The Arrhenius plots at the higher densities are distincly curved. For example, at the critical density the slope at $T/T_c = 1.01$ is equivalent to an Arrhenius coefficient of 0.7 eV/electron, whereas $T/T_c = 1.07$ in the same gas the coefficient is 0.14 eV/electron. Similar behavior occurs in the vapors of the other pentane isomers. The density effects are due to the interaction of the electron with more than one molecule at a time. At the higher densities incipient droplets seem to be involved.

Increasing the density of n-pentane vapor above 1×10^{20} molecules/cm^3 (0.05 n_c) causes μn to decrease gently (Figure 3). In the liquid phase at densities > 1.3 n_c the mobility drops sharply.

The density effect can be interpreted in terms of an equilibrium between delocalized and localized states.

$$e^-_{del} \rightleftharpoons e^-_{loc} \qquad (6)$$

The behavior in neopentane is quite different from that in n-pentane. In neopentane vapor μn remains constant up to $n = 0.4$ n_c then increases with increasing density (Figure 3). The trend continues through the critical region, then reverses for $n > 1.8$ n_c. The initial increase is due to the formation of

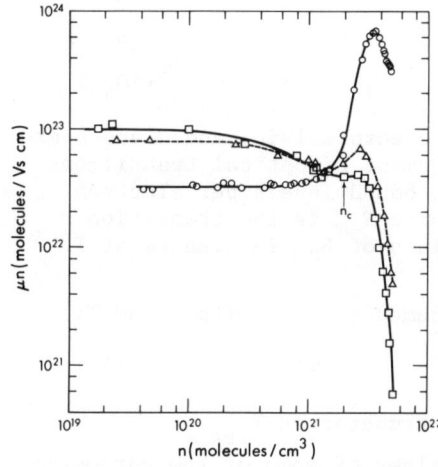

FIGURE 3. Plots of μn against fluid density n in the coexistence vapors and liquids of n-pentane (□), isopentane (△) and neopentane (○). The densities of these equilibrium fluids were changed by changing the temperature, so the temperature is highest at T_c and lowest at the two ends of a curve.

a conduction band. The decrease at the highest densities is due to scattering by the repulsive cores of the molecules.

The cross-over of the μn values in neopentane and n-pentane occurs in the gas phase at n = 0.7 n_c (Figure 3).

The behavior in isopentane is intermediate between those in neo- and n-pentane (Figure 3). Quasilocalization occurs in the high density gas, but there is a maximum in μn in the liquid at n = 1.4 n_c. There is also a hint of a maximum in the n-pentane curve at 1.3 n_c (Figure 3). Thus, as the molecules become less spherelike the maximum diminishes and shifts to lower densities.

The mobilities on the high density sides of the maxima in n-pentane and isopentane can be described quantitatively in terms of thermally activated electron transitions between localized and delocalized states, reaction (6). In fluid alkanes electron transport occurs mainly in the delocalized states.

The distribution of transition energies from localized to delocalized states is assumed to be a Gaussian centered about a value E_o, with a dispersion σ_E.

$$N(E) = \pi^{-\frac{1}{2}} \sigma_E^{-1} \exp[-(E-E_o)^2/\sigma_E^2] \tag{7}$$

The mobility μ_{hd} at high densities is given by equation (8).

$$\mu_{hd} = \mu^o \int_{-\infty}^{\infty} N(E)[1 + \exp(E/kT)]^{-1} dE \tag{8}$$

where μ^o is the mobility in the delocalized state. It is assumed that the entropy change for a transition is small and that the energy ranges of the localized and extended states overlap. For example, in neopentane localized states can lie at higher energies than the delocalized state.

It is not yet possible to determine unique values of the parameters μ^o, E_o and σ_E, but limitations can be put on them by comparison with other data. By analogy with the temperature dependences of the optical absorption parameters of solvated electrons, one may assume:[6]

$$E_o = E(0) - aT \tag{9}$$

$$\sigma_E = \sigma_E(0) + bT \tag{10}$$

where $E(0)$ and $\sigma_E(0)$ are the values extrapolated to zero Kelvin, and a and b are positive or zero. Due to the Franck-Condon requirement of optical transitions and the probability that there are only one or two bound levels per electron trap in a liquid alkane, one may take $E_o \leq E_{am}$, where E_{am} is the transition energy at the optical absorption maximum. The values of E_{am} in alkanes at $T > 300K$ are probably <0.4 eV.[7,8]

The mobility in the delocalized state was assumed to vary with n and T:[3]

$$\mu^o = \mu^o_{ref}\left(\frac{n_{ref}}{n_T}\right)^2 \frac{T_{ref}}{T} \tag{11}$$

where μ^o_{ref} is the value of μ^o at density n_{ref} and temperature T_{ref}.

Although unique values are not attainable, values of some of the parameters may be set arbitrarily to allow assessment of variations in the others from one system to another. Such calculations have been made for a number of hydrocarbons.[3,9-12] The value of the ratio E_o/σ_E at a given temperature is relatively insensitive to the assumptions invoked and therefore appears to be characteristic of the liquid. Results for 300K are summarized in Table I. The data for both the alkanes and alkenes are approximately described by equation (12).

$$\mu_{300K} \approx 40 \exp(-2.7 E_o/\sigma_E) \quad , \quad cm^2/Vs \tag{12}$$

A larger value of E_o/σ_E implies tighter trapping of the localized electrons, hence fewer electrons in the delocalized state.

In these simple alkanes the localization potentials are created by random orientations of the close-packed, irregularly shaped molecules. The potential relates to the anisotropy of polarizability of the carbon framework. The C-H bonds are nearly transparent to the electrons.

Lack of space prevents a discussion of the individual parameter values, so we simply make reference to the original articles.[3,9-12] A major remaining problem is that the temperature dependence of σ_E changes greatly from system to system in a way that we do not understand. This study will be continued at higher field strengths to obtain further information about the binding and transport of electrons in dielectric fluids.

REFERENCES

1. J.-P. Dodelet and G. R. Freeman, Can. J. Chem. 50, 2667 (1972).
2. L. G. Christophorou, Int. J. Radiat. Phys. Chem. 7, 205 (1975).
3. J.-P. Dodelet and G. R. Freeman, Can. J. Chem. 55, 2264 (1977).
4. T. R. Das, C. O. Reed, Jr. and P. T. Eubank, J. Chem. Eng. Data, 22, 3, 9, 16 (1977).
5. E. W. McDaniel, "Collision Phenomena in Ionized Gases", Wiley, New York, 1965 p.436.

6. J.-P. Dodelet, F.-Y. Jou and G. R. Freeman, J. Phys. Chem. *79*, 3876 (1975).
7. T. Shida, S. Iwata and T. Watanabe, J. Phys. Chem. *76*, 3683, 3691 (1972).
8. F.-Y. Jou and G. R. Freeman, Can. J. Chem. *54*, 3693 (1976).
9. J.-P. Dodelet and G. R. Freeman, Can. J. Chem. *55*, 2893 (1977).
10. J.-P. Dodelet, K. Shinsaka and G. R. Freeman, Can. J. Chem. *54*, 744 (1976).
11. T. G. Ryan and G. R. Freeman, J. Chem. Phys. *68*, 5144 (1978).
12. I. György and G. R. Freeman, J. Chem. Phys. (submitted).

TABLE I

Values of μ and E_o/σ_E for Thermal Electrons in Liquid Hydrocarbons at 300K.[a]

Liquid	μ (cm^2/Vs)	E_o/σ_E
alkanes		
n-pentane	0.14	2.0
3-methylpentane	0.25	1.9
isopentane	0.9	1.5
2,3-dimethylbutane	1.2	1.5
neopentane	62.	-0.2[b]
alkenes		
trans-butene-2	0.06	2.4
butene-1	0.14	2.0
isobutene	1.7	1.2
cis-butene-2	2.6	0.9

a. Data taken from references 3, 9-12.

b. The value of μ^o for neopentane is not adequately described by equation (11) This value of E_o/σ_E was obtained from a slightly different version of the model, but is included here to illustrate the suggestion that E_o can be negative.

TRANSPORT OF SPACE CHARGE IN NAPHTHALENE CRYSTALS*

M. CAMPOS
DEPARTAMENTO DE FÍSICA E CIÊNCIA DOS MATERIAIS
INSTITUTO DE FÍSICA E QUÍMICA DE SÃO CARLOS
UNIVERSIDADE DE SÃO PAULO - C.Postal - 369
13560 - SÃO CARLOS, S.P., BRASIL

INTRODUCTION

It is well known that currents considerably in excess of those predicted by Ohm's law can, under certain circumstances, be drawn through insulators when sufficiently large voltages are applied. Such currents are limited by space charge of the free and trapped carriers. It is well known also that if the abrupt increase of the current in function of the voltage is explained by using the theory of space-charge-limited current[1], the values found for the trap density are then improbably low[2]. Also for naphthalene crystals we show[3] that although the current-voltage curves agree closely with the theory for an insulator with a discrete level of traps above the Fermi level, the trap-filled limit voltage is independent of the sample thickness, rather than $V_{TFL} \alpha L^2$ as expected. To solve these problems the behavior of the current as a function of the voltage has been explained by different points of view: emptying of traps[4], filamentary regions[3], hot-electron effect[5], etc. One of the aims of the present work is to show that this behavior for naphthalene single crystals can be explained[6] using a theory based on the assumption that the insulator crystal contains centers with deep trapping levels and that hopping conduction occurs between these centers.

EXPERIMENTAL

Single crystals were grown from the melt, using C.P. naphthalene, twice sublimed. The thicknesses of the crystals varied from 0.03 to 0.4 cm and their areas ranged from 0.1 to 0.4 cm². Contacts were of silver paste which is able of injecting electrons freely up to a current density of at least 10^{-6} A/cm². No saturation effect of the contact was observed in the whole range of the measurements. The use of guard rings to prevent surface currents was found to be unnecessary. All electric fields were applied normally to the ab crystallographic plane. The measurements were carried out in the dark, in a dry air atmosphere inside a thermostated metal cell. All current-voltage curves were measured under steady-state conditions and particular care was taken to ensure that thermal equilibrium had been attained before each measurement. Isothermal decays were recorded during approximately one hour, in good agreement with the range of thermal release times, which can be calculated from the previously reported deep trap parameters[2]. Except where otherwise stated, all measurements were taken at room temperature.

THEORY

It is well known from the general theory of the space-charge-limited current that if an insulator contains traps, the current is limited by the space charge captured at the traps. If the centers which have captured electrons are sufficiently mobile or if hopping conduction can occur between these centers, the current will be mainly due to the electrons captured by the centers than due to electrons transported across the conduction band. In this condition there will be formation of a barrier at the anode and the current is limited not only by the space charge in the bulk, but also by the barrier. The concentration of the field at the anode result in a field-induced liberation of carriers from the centers and in a field emission of holes into the material, that may be due to the Schottky mechanism[7]. The voltage drop across a contact can be written as

$$V_c = \frac{1}{C_c} \int_0^t (I - I_L) dt \qquad (1)$$

where C_c is the capacitance of the contact, I is the total current I_L the leakage current, and $V_c = V - V_i$ where V is the applied voltage. The leakage current across the anode capacitance is

$$I_L = A T^2 \exp\left(-\frac{\phi}{KT}\right) \exp\left[\frac{1}{KT}\left(\frac{e^3}{\varepsilon} \frac{\int_0^t (I - I_L) dt}{d_c C_c}\right)^{1/2}\right] \qquad (2)$$

where A is the Richardson constant, ϕ is the height of the potential barrier and d_c is the thickness corresponding to the contact capacitance. The current I is

$$I = \frac{C_c d_c u_f v_i^2 S}{L^3} \qquad (3)$$

So the charge Q will be

$$Q = VC_c - \left(\frac{C_c L^3 I}{d_c u_f S}\right)^{1/2} = \int_0^t (I - I_L) dt \qquad (4)$$

Equation (4) can be reduced to

$$\int_{I_o}^{I} \frac{dI}{(I - I_L) I^{1/2}} = \frac{2t}{(C_c L^3 / u_f d_c)^{1/2}} \qquad (5)$$

If $I_L \ll I$

$$I = VC_c \frac{\tau}{(\tau + t)^2} \quad \text{where } \tau = L^3 / u_f d_c V \qquad (6)$$

Therefore, if $t \ll \tau$ we have $I = u_f \varepsilon S V^2 / L^3$

From $\frac{dQ}{dt} = \frac{VC_c \tau}{(\tau + t)^2}$ we have $Q = VC_c \frac{t}{t + \tau}$ \qquad (7)

It follows from the above expression that if $t \ll \tau$

$$Q = (u_f \varepsilon S V^2 / L^3) t \qquad (8)$$

RESULTS AND DISCUSSION

Figure 1 shows a typical experimental dependence of the charge Q accumulated after different intervals of time on the value of the current I, measured at the same time. The charge accumulated in the system was calculated from the area under the time dependence of the discharge current.

As we can see from Fig. 1 we find a linear dependence of Q as a function of $\sqrt{I}$, in good agreement with the theory.

Using the values given in Fig. 1 and $L = 0.2$ cm, $S = 0.64$ cm^2, dielectric constant 2.8 we find $C_c = 22$ pF, $d_c = 7.2 \times 10^{-3}$ cm, $u_f = 2 \times 10^{-7}$ cm^2/Vsec and $\tau \simeq 5$ hours. As the measurements were done during one hour the condition that $t < \tau$ is satisfied, so for the first minutes we have $t \ll \tau$. In Fig. 2 we can see that for the first minutes $Q \propto V^2$ as predicted by the theory.

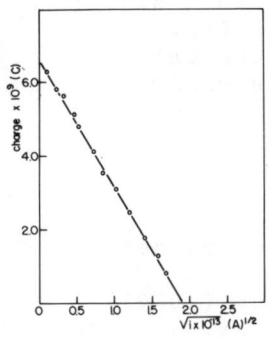

Fig. 1 - Dependence of the charge as a function of the current for a voltage of 200 V.

Figure 3 shows the dependence of the charge Q accumulated in function of time As we can see $Q \propto t$. At high voltages the total current across the system is governed by the leakage current I_L, which results from the field emission into the insulator crystal near the anode. In Fig. 4 we can see that the current depends exponentially on the square root of the voltage as was predicted by the theory. From this figure we can calculate the height of the barrier ϕ which hinders the current at the anode. The value found was $\phi = 0.8$ eV

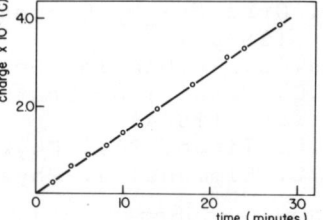

Fig. 3 - Dependence of the charge as a function of time.

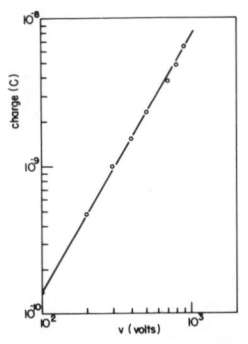

Fig. 2 - Charge calculated from a discharge current curve as a function of the voltage.

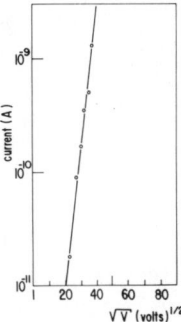

Fig. 4 - the current as a function of high values of the voltage.

CONCLUSIONS

Results of current-voltage in naphthalene with silver contacts can be he explained by using a theory of hopping. The barrier formed at the anode can be characterized by the calculations of its capacitance, thickness and height. The mobility was also calculated.

REFERENCES

(1) M.A. Lampert and P. Mark - "Current Injection in Solids" (Academic Press, New York, 1970);
(2) M. Campos - Molec. Cryst. and Liq. Cryst. $\underline{18}$, 105 (1972);
(3) J. Dresner; M. Campos; R.A. Moreno - J. Appl. Phys. $\underline{44}$, 3708, (1973);
(4) Yu. S. Ryabinkin, Sov. Phys. Solid State $\underline{6}$, 2382 (1965);
(5) J.C. Vesely; M. Shatzkes; P.J. Burkhardt, Phys. Rev. $\underline{10}$, 582 (1974)
(6) B.L. Timan, Sov. Phys. Semicondutors $\underline{7}$, 167 (1973);
(7) J.G. Simmons, J. Phys. D. Appl. Phys. $\underline{4}$, 613 (1971).

(*) Work supported by FAPESP, CNPq and FINEP.

INFLUENCE OF LOCALIZED ELECTRONIC STATES ON THE ELECTRICAL CONDUCTION IN IRON

TRIFLUORIDE THIN FILMS

A.S. Barrière, A. Lachter, M. Lascaud and J. Salardenne.

Laboratoire de Recherches en Electrotechnique et Physique du Solide
Université de Bordeaux I ; 351, cours de la Libération,
33405 Talence Cédex, France.

The iron trifluoride is anti-ferromagnetic and present a weak ferromagnetism below 364 K (1to3). These properties give a good determination of the material and allow an interpretation of its conduction mechanisms. Unfortunately, after growing single crystals, the cubic-rhomboedric transition wich occurs at 683 K causes the break up of the samples (4). However crystallized and quasi stoechiometric thin films of FeF_3 were prepared. Two methods were used (5) :
(a) sublimation of pure FeF_3 powder within a partial pressure of fluorine (10^{-5} torr) and condensation on a hot substrate (T_s = 550K)
(b) fluorination by fluorine (0.5 bar, 570K) of iron thin films obtained by evaporation under vacuum (10^{-8} torr) or r.f. sputtering on nickel substrat.

Under these conditions the layers are constituted by a pile of grains (scanning electron microscopy observation).
The films are well-crystallized and, in case "a", the X-ray diffraction studies of the residual powder showed that there was no other diffraction ray except that of the FeF_3 spectrum which was observed.
The composition of the layers and the concentration gradient of the elements along the depth were studied by nuclear analysis (backscattering and α-X reactions) (5.6). In the bulk of the samples the ratio of the number of lack of fluorine to the number of iron atoms is near to τ = 10^{-4}. The lack of fluorine is more important at the external surface of the films. It has been correlated with the presence of oxygen and attributed to water adsorption during the transport of the samples from the preparation vacuum to the experimental enclosure. The main impurities are chlorine which is essentially localized at the layer-substrate interface and potassium and calcium, distributed throughout the whole of the depth and which are probably pushed back to the grain joints. The table gives the ratio of the number of impurity atoms to the number of iron atoms for a 300Å layer prepared by sublimation under reactive atmosphere.

Impurities	K	Ca	Cl
$\tau\ °/_{oo}$	0.8 ± 0.1	1.0 ± 0.1	2.9 ± 0.05

In the films Fe^{3+} ions are only visible by Mössbauer spectroscopy (7). At liquid helium temperature, the layers are antiferromagnetic. At room temperature, two spectra are superposed : one is relative to an antiferromagnetic material, another to a super paramagnetic compound. The study of the magnetic susceptibility reveals that the Neel temperature of the films is 354K. These two results show that the samples are inhomogeneous. The dimension of the grains varies as observed by scanning electron microscopy. There diameter is between 100 Å and 800 Å. Furthermore the weak ferromagnetism, which characterized the single crystal, is observed in the layers, but it is four times lower than for powder. This decrease can be explained by the presence of weak distorsions in the lattice of FeF_3 thin films.

The optical absorption studies have been made between 290K and 4.2K and for incident radiation energies in the range from 0.3eV to 6eV (8.9).

Figure 1 gives the optical absorption coefficient α versus the photon energy $\hbar\omega$ for three different samples. The curves a and b are relative to a 5000 Å FeF_3 film prepared by reactive sublimation and a 4000 Å FeF_3 layer obtained by fluorination of iron thin film. Their surface is $4cm^2$. The curve c corresponds to a single crystal of FeF_3 where the volume is about $1mm^3$.

These results show that films have characteristics which are very close to a single crystal. In particular it can be noticed, in all cases, the presence of the peaks at 2.05, 2.69, 3.33, 3.69, 3.96 and 4.59eV which correspond to the spin transitions between the 3d states of Fe^{3+} ions. Nevertheless, for layers, perturbations in the internal periodic field of lattice are responsible for a broadening of these absorption peaks (10.11). Furthermore, an absorption edge at $\phi=1.1eV$ characterizes thin films. It probably corresponds to transition between localized electronic states in the forbidden band, due to the lack of fluorine and delocalized levels of the conduction band of the material. The magnitude of this absorption has been correlated with the difference of stoechiometry observed by nuclear analysis, which varies with the preparation conditions.

The optical gap has been evaluated at about 5.96 eV in FeF_3 thin film by extrapolation to α=o of $(\alpha\ \hbar\omega)^{1/2}$ plot in function of $\hbar\omega$ (12). The fundamental absorption is preceded by two excitonic peaks at energies of 5.27 and 5.76 eV.

The electrical study of metal-FeF_3-metal structures was made between 77K and 550K, from 10mV to 10V.

For a 5000 Å thin insulating film obtained by the (a) method, we have reported in figure 2 the variations of the current in function of the temperature for different voltages. Two domains where found corresponding to different conduction mechanisms. For T<200K, the activation energy is very weak (W≈0.005 eV). For 250K≤ T≤ 500K, W is about 0.8 eV. In all cases, for an electric field lower than $510^5 Vcm^{-1}$, the current-voltage characteristics are ohmic.

For a FeF_3 layer obtained by the (b) method the results are very similar. However, for T>250K, the activation energy is increased (0.85eV<W<0.9eV) and the conductivity is about ten times lower.

For low temperatures the conductivity probably corresponds to a tunneling

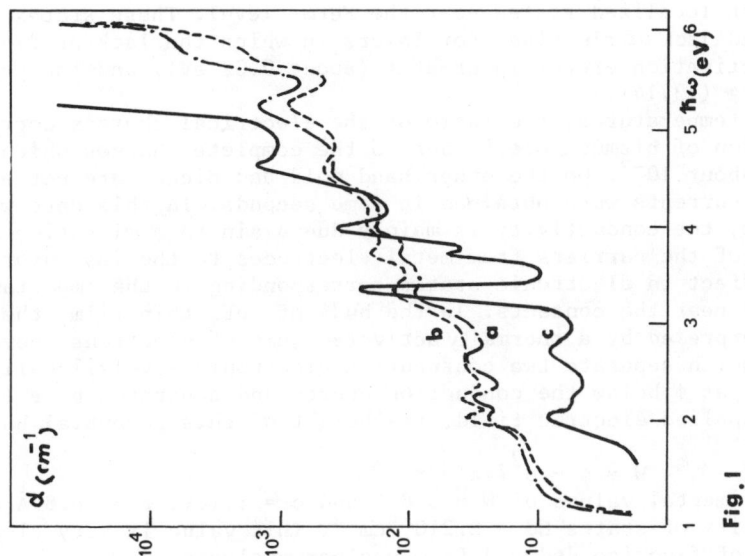

Fig. 1

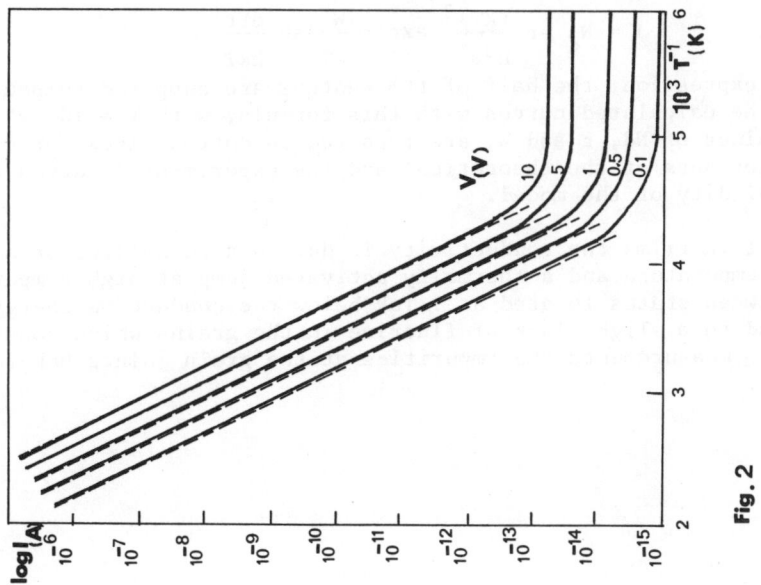

Fig. 2

process between localized states near the Fermi level. These states would be due to the observed lack of fluorine. For layers in which the lack of fluorine is increased the activation energy in greater (about 0.02 eV), and the process could be some hopping (13.14).

For high temperatures, the ratio of the electrical charges corresponding to the fluorination of bismuth electrodes to the complete charges which cross the insulator is about 10^{-3}. On the other hand gold and nickel are not attacked and the permanent currents were obtained in some seconds. In this case we feel that, in this domain, the conductivity is mainly due again to a migration of electrons. The injection of the carriers from metal electrodes to the insulator can be due to a tunnel effect in electronic states corresponding to the important lack of fluorine noted near the contacts. In the bulk of FeF_3 thin film, the conductivity can be interpreted by a thermally activated jump of electrons above the potential barrier which separate two consecutive electronic levels, localized in the forbidden band at ϕ below the conduction energy and separated by r (15).

Without applied electric field, the height of this potential barrier is given by :

$$W = \phi - e^2/\pi\varepsilon_o\varepsilon_r r$$

with the experimental values of $W \simeq 0.8 eV$ and $\phi = 1.1 eV$, $r = 76.8 \text{ Å}$ corresponding to a density of states $N_d = 2.2 \cdot 10^{18} cm^{-3}$. This value is very close to the ratio of lack of fluorine deduced from nuclear analysis.

This correspondance is agains found for films prepared by sublimation under vacuum. For these layers τ can be as high as 1%, the conductivity is multiplied by 10^4 and W is included between 0.4 and 0.5eV.

In this case we think that if ν represent the thermal vibration frequency of the electrons in a coulombic center and F the electric field in the cell is supposed constant, the current density can be given by :

$$J = N_d \; er \; \frac{(kT)^3}{h^3 \nu^2} \; \exp(-\frac{W}{kT}) sh \; \frac{eFr}{2kT}.$$

In this expression, the half of the centers are supposed compensated. On figure 2 the calculated curves with this formula, with $\nu = 10^{13} s^{-1}$ and the experimental values of Nd, r and W, are reported in dotted lines for different voltages. The closeness of the theoretical and the experimental curves seems to indicate the validity of the model.

In FeF_3 thin films the conductivity is due to a tunnelling or a hopping cess at low temperature and a thermally activated jump at high temperature c electrons between states located at 1.1eV below the conduction energy. Thes tes correspond to a slight lack of fluorine in the grains which constitute layers. For d.c. measurements the impurities at the grain joints behave lik short circuit.

References :

(1) G.K. Wertheim, H.J. Guggenheim and D.N.E. Buchanan - Solid State Com. 5(1967) 537.

(2) L.M. Levison - J. Phys. Chem. Solids, 29(1968) 1331.

(3) E.O. Wollan, H.R. Child, W.C. Koehler and M.K. Wilkinson- Phys. Rev. 112(1958) 1132.

(4) J.R. Shane and M. Kestigian - J. of Appl. Phys., vol 39, n° 2 (1968) 1027.

(5) A. Lachter, M. Lascaud, A.S. Barrière, L. Lozano, J. Portier and B. Saboya J. of Crystal growth, vol 43, 05 (1978) 621.

(6) B. Saboya, J.F. Chemin, J. Roturier, A.S. Barrière, Y. Danto and J. Salardenne - J. Appl. Phys., 8 (1975) 1008.

(7) A. Lachter, F. Mesnil and A.S. Barrière (to be published).

(8) A. Lachter, A.S. Barrière and J. Salardenne - Solid State Chem. Europ. Meeting (1978) com. III, A.17.

(9) A. Lachter, J. Salardenne and A.S. Barrière - Phys. Stat. Sol. b90 (1978).

(10) A.J. Kurtzig and H.J. Guggenheim - Appl. Phys. Letters, 16 (1970) 43.

(11) J.D. Dow and D. Redfield - Phys. Rev. B,1 (1970) 3358.

(12) J. Tauc - Opt. Prop. of Solids - North Holland Publ. co., Amsterdam (1972) 277.

(13) A. Miller and E. Abrahams - Phys. Rev. 120 (1960) 745.

(14) N.F. Mott - Phil. Mag., 19 (1969) 835.

(15) R.M. Hill - Phil . Mag., 23 (1971) 59.

DIELECTRIC STUDIES ON IONIC CHARGE CARRIER IN AMORPHOUS POLYMERS

Ichiro Murakami, Kō Saito and Hiroshi Ochiai

Department of Chemistry, Faculty of Science, Hiroshima University
Higashi Senda-machi, Hiroshima 730, Japan

As for the electric conduction of amorphous polymers, it is known that the ionic impurities in the polymers play a major role in transportation of electric charges, and the ionic jump from a trapped position to another is the fundamental mechanism of charge migration. (1),(2),(3) On the other hand, the electric charge displacement due to the ionic jump seems to cause the interfacial or space charge polarization which often hinders the observation of dielectric absorption ascribed to the molecular relaxation. However, since the behavior of ionic charge carriers closely depends not only on the structure of polymer matrixes but also on kinds and enviroments of ionic charge particles, phenomena of the interfacial polarization have not been thoroughly made clear yet. If the ionic jump is responsible essentially for the interfacial or space charge polarization, a simple relation between the electric conduction process and the dielectric relaxation process due to the interfacial polarization can be expected.

In this work, the dc conductivity and the dielectric absorption attributed to interfacial polarization, observed in lower frequency region, were measured for several amorphous polymers. From the measurements, the relationship mentioned above is examined.

Experimental
    The sample used were poly(methyl methacrylate)(MMA), partial hydrolyzate of MMA (A-MMA), silver salt of A-MMA (Ag-MMA), cupric salt of A-MMA (Cu-MMA), poly(vinyl alcohol)(PVA), PVA crosslinked by borax (B-VA), and poly(vinyl chloride)(PVC). The MMA sample was a commercial grade one and the degree of polymerization (DP) was determined to be 850. A-MMA was prepated by hydrolysis of MMA sample with 85 wt % sulfuric acid at 25ºC and the degree of hydrolysis (percentage of acid groups for monomeric units of MMA) was determined to be 3.7 %[5] A_g-MMA was the silver salt of A-MMA sample[5] Cu-MMA was the cupric salt of partial hydrolyzate of MMA sample (degree of hydrolysis, 7.5 %)[5] PVA sample was supplied commercially (DP, 2000). The degree of borax content in B-VA sample (moles of borax/base moles of PVA) was 1.33×10^{-1}[6] PVC sample was a commercial grade one

(Geon 103 EP)(DP, 1050). Samples of MMA, A-MMA, A_g-MMA, Cu-MMA, were molded into disc plates about 0.5 mm in thickness by a laboratry press at temperatures far above T_g (glass transition temperature). Samples of PVA, B-VA and PVC were made into films by solvent casting, and thickness of the films were 0.03 - 0.05 mm.

The dielectric measurement were carried out over frequency range of 10^{-4} to 10^6 Hz at various fixed temperatures below and above T_g. The apparatuses used were a inductive-ratio-arm bridge (TR-10, Ando Elect. Co.) in the range of 30 - 10^6 Hz and a resistive-ratio-arm bridge (TR-4, Ando Elect. Co.) in the range of 10^{-1} - 10 Hz. The values of dielectric loss (ε'') in the range from 10^{-4} to 10^{-2} Hz were estimated from the relaxation current (absorption current) by use of Hamon's equation. (4) The relaxation current was determined by measuring the charging and discharging current.

The electric conductivity was estimated from the stady state current of dc conduction, which was determined by extrapolating the plot of charging current against log time (sec). The charging and discharging currents were measured by a high input-resistance electrometer (PM 19A, TOA Elect. Co.) in term of the voltage drop produced by flow of charge through a multimeghom resister (10^7, 10^8, 10^9 ohm, selective). Electromotive force was provided by dry batteries. The electric field strength applied was in the range from 0.3 to 4.7 kV/cm, where the Ohmic relation was observed.

The electrodes used dc-conduction and dielectric measurements, 3-terminal electrodes system with guard electrode, were of everpolated silver.

Results and Discussion

Typical dielectric absorption spectra are shown in Figs. 1 - 6. For MMA sample, larger absorption peaks are observed in lower frequency region below 10^{-2} Hz, as shown in Fig. 1. These absorption peaks except that for 102°C are considered to differ from the α-relaxation peaks attributed to micro-Brownian motion of main chain, i.e., in the case of 116.5°C, the α-absorption is observed as a shoulder at the frequency about 10 Hz. Similar absorptions are observed in Figs. 2,3, and 5 for A-MMA, A_g-MMA and Cu-MMA, respectively. The temperature

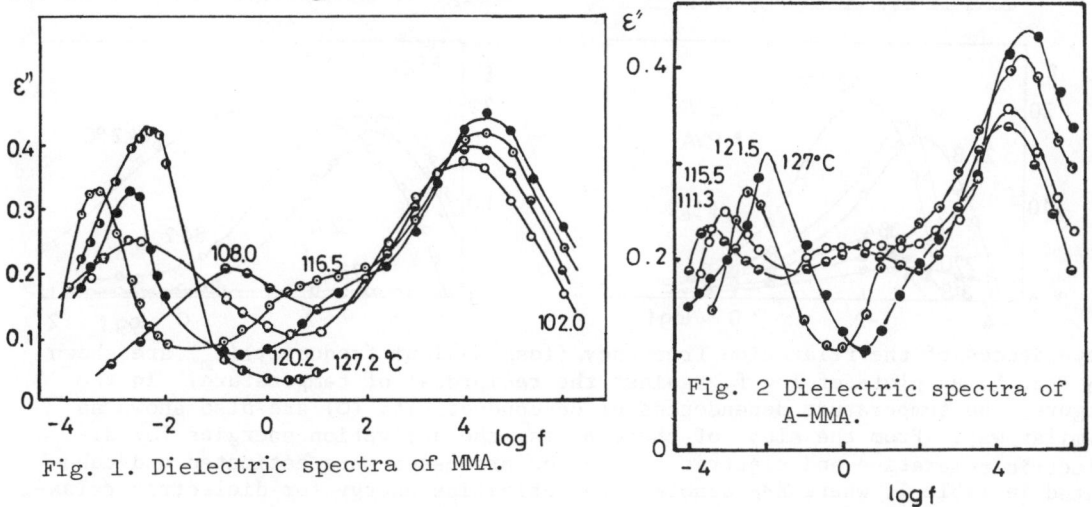

Fig. 1. Dielectric spectra of MMA.

Fig. 2 Dielectric spectra of A-MMA.

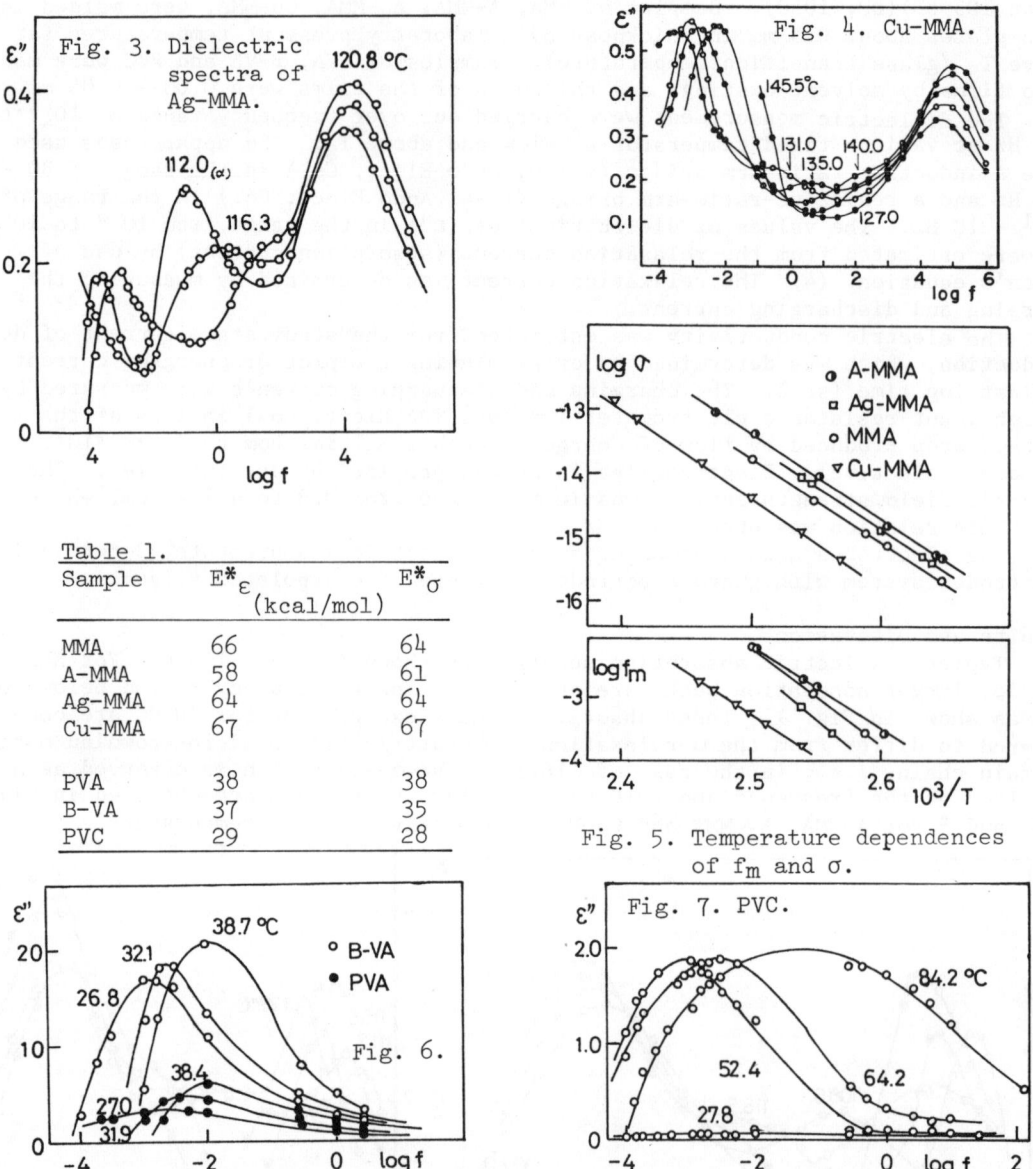

Fig. 3. Dielectric spectra of Ag-MMA.

Fig. 4. Cu-MMA.

Fig. 5. Temperature dependences of f_m and σ.

Fig. 6.

Fig. 7. PVC.

Table 1.

Sample	E^*_ε (kcal/mol)	E^*_σ
MMA	66	64
A-MMA	58	61
Ag-MMA	64	64
Cu-MMA	67	67
PVA	38	38
B-VA	37	35
PVC	29	28

dependences of the relaxation frequency (loss maximum frequency)(f_m) are shown in Fig. 5, as plots of log f_m against the reciprocal of temperature. In the figure, the temperature dependences of dc conductivity (σ) are also shown as similar way. From the slope of these plots, the activation energies for dielectric relaxation and electric conduction processes were estimated and tabulated in Table 1, where E^*_ε denotes the activation energy for dielectric relaxation process, E^*_σ, that for dc conduction process. The values of E^*_ε are in far agreement with those of E^*_σ, as shown in Table 1. Furthermore, behaviors of the absorption peak mentioned above was found to be affected by the condition of sample preparation. These facts seem to imply that the dielectric absorption observed in this study is ascribed to the interfacial or space charge polarization due to the ionic displacement in these amorphous polymers.

In the cases of PVA, B-VA and PVC samples, the same dielectric absorptions were observed at temperatures below T_g in the frequency range below 10^{-1} Hz, as shown in Figs. 6 and 7. The plots of log f_m and log σ against the reciprocal of temperature are shown in Fig. 8. The acivation energies for both dielectric relaxation and dc conduction processes are listed also in Table 1. The values of E^*_ε are found to be agree with those of E^*_σ as well as in the cases of methacrylic samples. From the fact that the dielectric absorption discussed here is observed even in the temperature region below T_g, it is considered that the mechanism of ionic migration in the glassy state is essentially the same as in the rubbery state.

The relation between the electric conduction and the dielectric relaxation process observed here can be illustrated by plotting log σ against log f_m, as shown in Fig. 9. Since the plots in this figure are represented by straight lines, the empirical equation can be derived as follows,

$$\sigma = \sigma_o (f_m)^n \qquad\qquad 1,$$

where n denotes the line slope of plot in Fig. 9, and σ_o is a constant which means the conductivity corresponding to $f_m = 1$ Hz, and these values are listed in Table 2. Refferring that the relaxation time (τ) is defined as $\tau = 2\pi f_m$, the equation 1 is modified as follows,

$$\sigma \cdot (\tau)^n = \sigma_o / (2\pi)^n = \text{constant} \qquad\qquad 2.$$

Since τ is a measure of the retardation phenomena, the equation 2 is considered to represent the modified Walden's rule; $\sigma(\eta)^m$ = constant. (7) As for the electric conduction in rubbery state of polymers, using the relaxation time of the α-relaxation process, Saito has reported the similar relationship. (8) As for the dielectric relaxation discussed here, n is found to nearly equal to unity

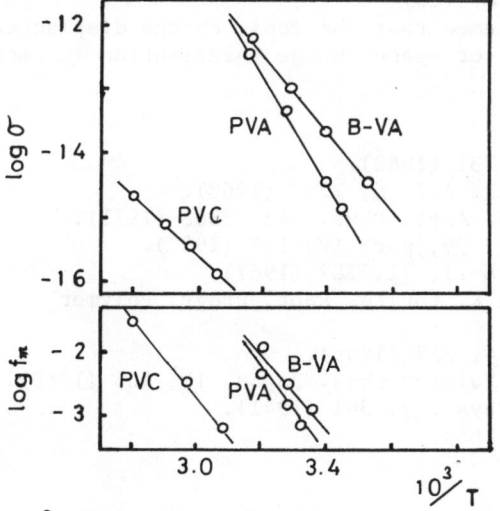

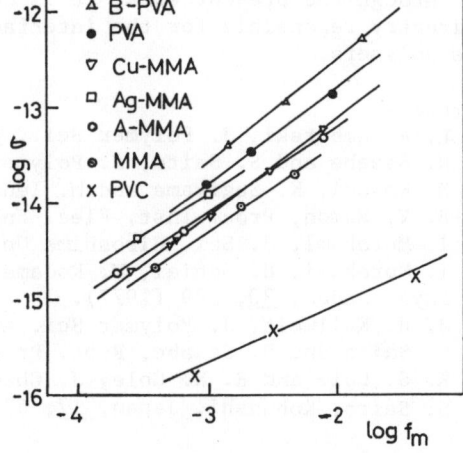

Fig. 8. Temperature dependences f_m and σ.

Fig. 9. Relation between f_m and σ.

Table 2.

Sample	$T_g(°C)$	$t_{-3}(°C)$	σ_o (ohm^{-1}cm^{-1}volt^{-1})	n	log $\sigma_{t_{-3}}$	ε_o
MMA	95	118	2.5 x 10^{-12}	0.92	−14.55	5.8 (α)
A-MMA	96	117	3.2 x 10^{-12}	1.05	−14.25	5.9 (α)
Ag-MMA	98	122	6.3 x 10^{-12}	1.00	−13.90	6.1 (α)
Cu-MMA	118	131	7.9 x 10^{-12}	1.00	−14.25	6.0 (α)
PVA	76	27	1.4 x 10^{-11}	1.00	−13.80	8.3 (β)
B-VA	79	23	3.8 x 10^{-11}	0.95	−13.55	8.4 (β)
PVC	84	55	1.4 x 10^{-14}	0.96	−15.67	4.0 (β)

as revealed in Table 2. This fact implies that the ionic jump is responsible directly for the dielectric relaxation phenomena such as space charge or interfacial polarization.

In order to see the effect of dielectric constant of medium on the mobility of ionic charge carriers, the value of σ corresponding to the temperature (t_{-3}) at which f_m becomes 1 Hz ($\sigma_{t_{-3}}$) was plotted against reciprocal of static dielectric constant (ε_o) in Fig. 10. The value of ε_o was determined by application of Cole's circular arc law (9) from the complex dielectric constants ($\varepsilon^* = \varepsilon' - j\varepsilon''$) measured at near t_{-3}. Both values of ε_o and t_{-3} are listed in Table 2, where α denotes the α-process, β, the β-process. The plot thus obtained shows approximately straight line and agrees with that observed by Saito. (10)

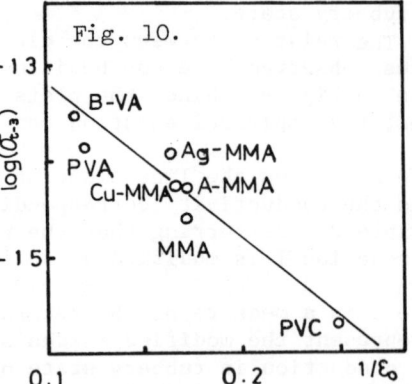

Fig. 10.

Through the present work, it is confirmed that the ionic charge displacement is directry reponsible for the interfacial or space charge polarization in amorphous polymers.

References
(1) L. E. Amborski, J. Polymer Sci., 62, 331 (1962).
(2) H. Sasabe and S. Saito, J. Polymer Sci. A-2, 6, 1297 (1968).
(3) M. Kosaki, K. Sugiyama and M. Ieda, J. Appl. Phys., 42, 3388 (1971).
(4) B. V. Hamon, Proc. Inst. Elec. Engrs., 99, part IV, 151 (1952).
(5) I. Murakami, J. Sci. Hiroshima Univ. A-II, 31, 187 (1967).
(6) I. Murakami, H. Ochiai, M. Kodama and K. Yorita, Rept. Progr. Polymer Phys. Japan, 20, 399 (1977).
(7) J. H. Kallweit, J. Polymer Sci. A-1, 4, 337 (1966).
(8) S. Saito and H. Sasabe, Rept. Progr. Polymer Phys. Japan, 12, 405 (1969).
(9) K. S. Cole and R. H. Cole, J. Chem. Phys., 9, 341 (1941).
(10) S. Saito, Kobunshi, Japan, 17, 672 (1968).

V
THERMALLY STIMULATED CURRENTS

SOME ASPECTS OF THERMALLY STIMULATED POLARIZATION

J. Gasiot, J. Vanderschueren[■], P. Parot[□], A. Linkens[■],

J.C. Manifacier, J. Jimenez-Lopez

Université des Sciences et Techniques du Languedoc,
Laboratoire Associé au C.N.R.S., C.E.E.S.
34060 MONTPELLIER - FRANCE

[■] Dept. of Physics and Chemistry, University of Liege
SartTilman, B 4000 LIEGE - BELGIUM

[□] C.E.N.G., 38042 GRENOBLE - FRANCE

Thermally Stimulated Depolarization Currents (TSDC) constitutes one of the methods for studies on dipolar relaxation, we are concerned here with the natural complement to this method, i.e., the Thermally Stimulated Polarization Currents (TSPC) - experimental arrangement is the same as for (TSDC), but we start at low temperature with a sample in a random state, then an electrical field is applied and the current measured during the heating of the sample - TSPC gives rise to a double peak, the first positive and the second negative, if as generally assumed the steady state polarization under electrical field stress decreases with the increasing temperature. Studies on TSPC for dipolar relaxation are quite scarce, the interest of this method has been developped in alkali halides by P.R. Moran (1), however no reversal currents have been observed (2). In polymers it has been shown that TSPC spectra are in good qualitative agreement with the theoretical predictions of the model involving a non negligible temperature dependence on the equilibrium polarization (3) (4). However the curves obtained for a single relaxation time associated with a Langevin dependence of steady-state polarization (5) (6) cannot quantitatively describe the TSPC spectra on polymers this end can only be reached by taking into account distributed relaxation times. It is the purpose of this paper to illustrate several characteristic features of TSPC measured in some materials. Numerical solutions have been obtained and the predictions of the theory have been tested in several experimental conditions.

Usually, the steady orientational polarization P_o formed in a dielectric by applying an external electrical field F_p at a temperature T_p is assumed to obey the familiar Langevin function :

$$P_o = \frac{s N \mu^2 F_r}{k T_p} \quad (1)$$

where F_r is the local field, N is the number of permanent dipoles per unit volume, μ is their dipole moment, k is Boltzmann's constant and s is a geometrical factor depending on the possible dipolar orientation (s = 1/3 for free rotating dipoles). Expressed in kinetic terms, this relation involves that the directing effect of the field F_r on the dipoles is counteracted by the thermal motion of the molecules. During a thermal relaxation, the rate of change of the polarization P induces a current density I obeying to the classical Debye law,

$$J = -\frac{dP}{dt}\{T(t)\} = \frac{P_o(T, F_r) - P\{T(t)\}}{\tau(T)} \quad (2)$$

the temperature is related to time by $T = \beta t + T_o$, β heating rate. The relaxation time τ is given by the Arrhenius expression

$$\tau = \tau_o \exp(E/kT) \quad (3)$$

τ_o^{-1} is the characteristic frequency factor and E is the activation energy.

Equations (1) (2) and (3) allow us to describe the current behaviour during a thermal cycle. In the simplest cases of TSDC and TSPC we can deduce analytical solutions or in more complexes cases numerical solutions could be reached. Assuming temperature-independent steady-state polarization, Mc Keever and Hughes (2) have derived an expression for the thermally stimulated polarization current density which is equivalent to the Bucci-Fieschi equation so that the TSPC peak related to a given relaxation process should be characterized by the same position, height and shape as the corresponding TSDC peak. On the other hand, it was shown by Moran and Fields (1) and Manifacier et al. (5) that, taking into account the Langevin function, the isothermal Debye equations for polarization can be extended to the TSPC case and solved analytically and that the resulting TSPC current density is then characterized by a current reversal. The later phenomenon can be easily understood qualitatively : in TSPC experiments, we are starting with randomized dipoles and thus zero polarization ; during heating, the number of oriented dipoles and thus the polarization first increases on account of the exponential temperature dependence of the relaxation time ; this process continues until an equilibrium state is reached giving rise to a peak polarization current. At higher temperatures thermal agitation counteracts the directing action of the applied field and the total polarization decreases. As a consequence, the TSPC current must change polarity, giving rise to a second peak with opposite sign.

Schematic behaviour of the current for successive thermal cycles under constant electrical field and heating or cooling rate is given fig. 1 and 2, carrying out the following steps (1) heating the sample from T_o to T_1 ($TSPC_1$) (2) cooling from the maximum temperature T_1 to T_o (3) the same as in step 1. ($TSPC_2$). The differences observed in the dielectric current behaviour in fig.1 and 2 result from the temperature dependence of the steady state polarization stated for fig. 1 and neglected in fig.2.

As pointed out previously for TSPC experiment, thermal cycling is interesting in itself and provides information not available from TSDC alone.

1) From $TSPC_1$ at the inversion temperature T_i we deduce $P_o (T_i)$ from (2)

$$P_o(T_1) = P\{T_i(t)\} = \int_{T_o}^{T_i} \frac{J\,dT}{\beta} \qquad (4)$$

2) The reversal current in TSPC at high temperature is proportionnal to dP/dT, i.e. to T^{-2} if P_o obeys to equation (1).

3) The appearence of current reversals produced by carrier injection is also possible during a TSPC process carried out with a real dielectric, but it should then be easy to distinguish between the two phenomena, either by carrying out a series of successive thermal polarization cycles or by varying experimental parameters such as magnitude and polarity of the applied field or the nature of electrodes.

4) It is important to note that the TSDC calculations, based on simplified expressions of the current density, have always been made by neglecting the temperature dependence of the steady-state polarization.

Points (1) (2) and (3) allow us an experimental check on the starting hypothesis, the results obtained must be used to improve the TSDC determinations.

Finally we must remark that the methods used for TSDC experiment in the determination of the dipole parameters are directly applicable to TSPC experiments.

We have observed reversal currents from TSPC experiments in MOS structures, in which the oxide was a borophosphosilicate (8), unfortunately the shape of the current was strongly affected by the injection phenomenon, this prevents any quantitative analysis. In the mean time we measured TSPC on some ionic materials, and never found any significant reversal current ; these results are similar to those presented by the authors working on this subject (1) (2), indicating a low temperature dependence of the steady state polarization, in opposition with the general hypothesis on this parameter (5).

In polymer we present an example of results obtained on commercial samples of radial block copolymers of styrene-butadiene with total styrene content 40 Wt-% an density of 0.95g/cc provided from Phillips petroleum. The samples were compression-molded into 1mm thick plates at 150°C, they were then provided with vacuum evaporated electrodes and stored in a vacuum over phosphorous pentaide for at least a week before measurement. Thermal cycling with electrical field applied and TSDC are shown fig.3. The $TSPC_1$ (curve 4) present a strong reversal current, the cooling with applied field clearly gives rise to the appearence of a polarization peak (curve 2) while a subsequent heating, although producing no positive polarization current, shows well, in the α-range, the trends predicted in the theoretical model (curve 3) ; it is likely that the weak positive peak expected in this case (see fig.1) is masked by the negative current appearing in the β-range and whose origin is not clear. Quantitative information from this experiment cannot be reached by looking at the particular properties of such materials. In polymer dielectric relaxation arise from the motion of individual molecular groups or from a cooperative motion of segments of the main chain ; in this case equation 1 is not suitable and a strong temperature dependence of the steady state polarization can be observed ; for example, a decrease by a factor two or more in some twenty degrees is observed in the transition range of certain methacrylic copolymers (9). Furthermore is now well established that the relaxation processes in polymers are usually characterized by a distribution of the relaxation time resulting from a distribution on τ . Such properties reduce all attemp to extract quantitative information reached only from TSDC or TSPC as long as distribution is unknown.

References

1) Moran P.R. and Fields D.E., J.A.P. 45, 1974, 1836
2) Mc Keever S.W.S. and Hughes D.M., J. Phys. D. Appl. Phys. 8, 1975, 1520-9
3) Mizutani T., Suzuoki Y. and Ieda M., J. of Appl. Phys. 48 1977, 2408-13
4) Vanderschueren J., Gasiot J., Fillard J.P., Linkens A., Parot P., European Symposium on Electric Phenomena Polymer Sciences. Pise, 29-31 Mars 1978
5) Manifacier J.C., Gasiot J., Parot P., and Fillard J.P., J. Phys. C, Solid State Phys. 11, 1978 p. 1011-15
6) Linkens A., Vanderschueren J., Parot P. and Gasiot J., Comput. Phys. Com. 13, 1978, 411-19
7) Bucci C., Fieschi R. and Guidi G., Phys. Rev. 148, 1966, 816.23
8) Parot P. (Thesis) Montpellier 1976
9) Sanno N., Nishio H., Harakami I. and Yamamura H., J. Sci. Hiroshima Univ. A39, 153 (1975)

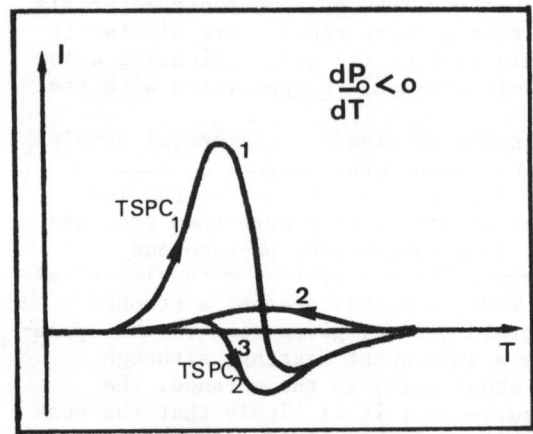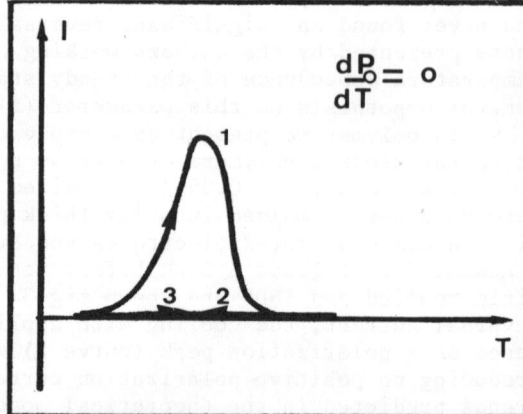

Figure -1-   Figure -2-

Schematic behaviour of dipolar relaxation current during a thermal cycle under electrical field applied.

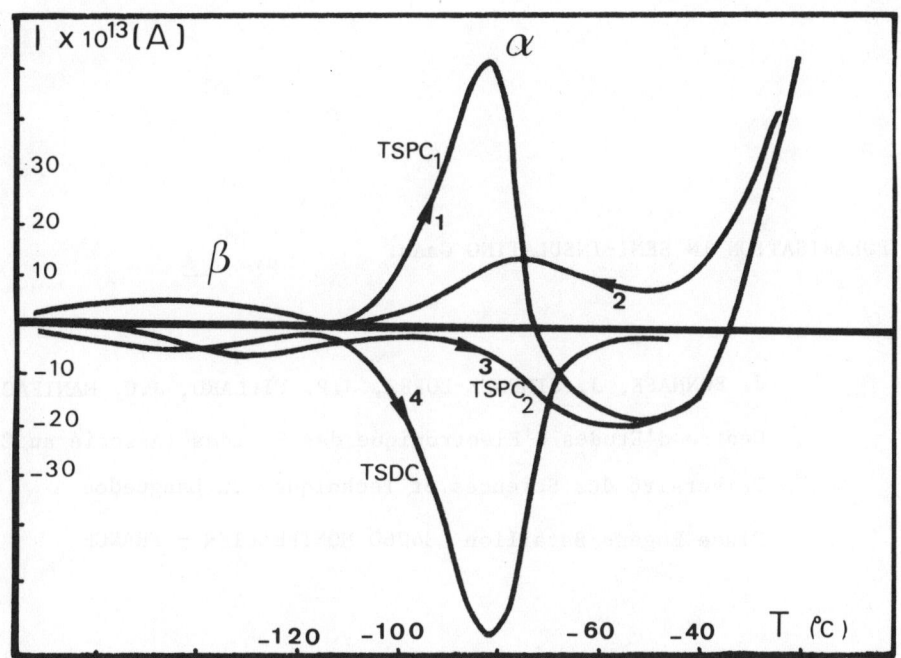

Figure -3-

TSPC (curves 1,2,3) and TSDC (curve 4) currents obtained in styrene-butadiene block copolymer (field strengh : 13kV/cm, heating rate β : 0.5/s).

INTERNAL POLARISATION IN SEMI-INSULATING GaAs.

J. BONNAFE, J. JIMENEZ-LOPEZ, J.P. FILLARD, J.C. MANIFACIER

Centre d'Etudes d'Electronique des Solides (Associé au C.N.R.S)

Université des Sciences et Techniques du Languedoc

Place Eugène Bataillon   34060 MONTPELLIER - FRANCE

Semi insulating GaAs is extensively studied for several years with the intent of competing with silicium in some wide spread applications such as fast logic transistors.

High quality materials are now available. They are obtained by compensating the originaly n type GaAs by deep acceptors such as chromium.

Difficulties yet arise in the knowledge of the exact transport properties from several origins namely :
- the presence of complex and evolutive deep traps is sensitive even at low densities
- the high resistivity introducing a dielectric relaxation time $\tau_D = \frac{\varepsilon}{\sigma}$ larger than the lifetime (1) (2), τ_o.
- the boundary conditions at the surface and in the electrical contacts (3).

Lots of work were done on GaAs with deep level spectroscopy techniques, namely Thermally Stimulated Conductivity (TSC) on bulk insulating samples or transient capacitances (photocapacitance and especially DLTS) on Schottky barriers on conducting material.

Nevertheless TSC is constitutively limited by the dark conductivity in the high energy range of the activation energies whereas DLTS is limited in the range of the high resistivity materials by the bulk serie resistance.

The purpose of this paper is to report some new preliminary results in the field of Thermally Activated Polarisation (TAP), Depolarisation (TAD) and Radiation Induced TAD (RITAD) (4).

By using such internal field techniques it is hoped to get a new glance on the electronic mechanisms relaxing the trapped charges, especially in the range of dominent dark conductivity. In some ways this method should reconcile the advantages of both the TSC and the DLTS.

The refered samples were grown by R.T.C. company. They are, tailored in a platelet shape (fig.1) typically 4 x 4 x 10.4 m.m^3 and indifferently provided with silver or insulating contacts, following the indicated configurations ; resistivities in the ohmic range, at room temperature, correspond to intrinsic material $\rho = 3.10^9$ Ωcm. The temperature range explored is 77-400K with typical heating rates 0.5K.s^{-1}.

A classical TSC spectrum obtained with a guard ring configuration is reported on fig.2. It can be observed that eliminating surface currents allows to greatly improve the sensitivity and the selectivity. It can also be observed that the merging dark current prevent finding peaks in the temperature range above 270K.

TAP and TAD experiments are pure electrical experiments ; typical curves are reported on fig.3 for a given bias voltage. TAP experiment obviously requires one (or two) insulating electrode whereas TAD can be performed indifferently with silver of insulating contacts. It can be observed on the graph that in the range of medium bias voltage (30-volts) the only TAD peak exactly correspond to the TAP one (in the reverse polarity) and follows satisfactorily the dark current rise. A similar situation is observed with inverted bias ; the low temperature activation energy is 0.42eV. In our opinion this is a quite typical situation were the theory developed by P. MULLER of internal "conduction self relaxation" can apply (5) (6) (7) (8). Let us recall briefly : suppose that we cool down, under applied bias voltage, a sample the resistivity of which is very temperature dependent. Then, at low temperature, the charges are not allowed to be transfered on the respective metallic electrode because they are strongly trapped at the interface or because the contact is blocking (insulating). Then, short circuiting the sample electrodes at low temperature will result in reversing the electric field in the bulk of the sample. This inverted field comes obviously from the sheet of opposite charges that are firmly held in place at electrodes ; it is, within a factor (1-f), opposite to the applied field (see P. MULLER).

Let us steadily heat the sample, these charges do not move but can be progressively cancelled by thermodynamical carriers coming from the bulk as long as the bulk conductivity rises with temperature ; this gives rise to a drift current in the bulk sample and of course in the external circuit. The bell shape of such "TAD" curves is due to the rising conductivity and the transient decrease of the internal field. Such a process suggested by P. MULLER and also by S.C. AGARWALL (9) has obviously nothing to do with traps (neither in the bulk nor in the contacts) but is only concerned with the conductivity which must lie in the range of $10^{-12}-10^{-14} \Omega^{-1}$cm^{-1} in order to be compatible with the temperature rate β and the frequency band pass ; in fact this theory is developped in the case of a constant reciprocal heating rate a, but it is considered that the peak is narrow enough to keep $\beta = -aT_m^2$.

When dealing with n or p type semiconductor like samples P. MULLER uses this method with a single insulating electrode in order to determine the type of the material : the polarity giving rise to a well developped TAD peak must correspond to a reverse biasing of the non insulating contact. In the present case similar peak heights are obtained on both the polarities when polarisation is applied at room temperature ; this is perfect agreement with the quasi intrinsic nature of our material. If the polarisation is applied at 250-260K, then the TAD$_+$ is greatly enhanced with respect to TAD$_-$. This is in agreement with the p type conduction in this temperature range (10) (11) due to Cu doping

(0.42eV). The low temperature side of the TAD.TAP peaks gives an activation energy E_o = 0.42eV in accordance with that of the dark direct current. The width of the TAD.TAP peaks and the temperature T_m of the maximum also agree with MULLER's formula $H = J_m ke/aP_o$ to give an activation energy in the range of 0.43eV.
It must be remarked that changing the polarity slightly changes the shape of the peak : symetrical in the (TAD+ TAP-) configuration, the peak becomes asymetrical for (TAD-TAP+). With either polarity the peak slightly broadens and shifts toward higher temperature with increasing voltage. Corresponding charge $Q = \int I.dt$ of the peak, linearly depends on voltage in agreement with theory.

Experiments were also made with identical ammeters on both electrodes : the identical values of the transient current indicates that no appreciable single space charge was relaxing (12).

This explanation of TAD relaxation as a bulk conductivity process is also supported by RITAD experiments. The sample is irradiated at low temperature under bias voltage with monochromatic (0.890μm) light in the same conditions as for TSC experiments ; the thermally stimulated relaxation is then observed under short circuit conditions. Two peaks appear on fig.4:the first one (LT) at 150K is of opposite polarity with respect to the biasing voltage (as in a classical TSD) the second one (HT) at 280K is always positive with respect to the illuminated electrode ; the TAD peak is wiped out. Roughly speaking, the integral of the RITAD LT peaks corresponds to the previous TAD one. HT peak is a power ten lower.

A satisfying explanation arise in considering that bulk thermally stimulated carriers in the low temperature range relaxes the induced internal polarisation as well as the thermal equilibrium carriers did in TAD peaks at higher temperature. Then MULLER's theory should be extended to thermally stimulated conductivity giving rise to a peak, the maximum of which is controled either by the trap exhaustion or by the internal field reduction depending on the respective values of the trapped charge and the internal initial polarisation. In the present case the good coïncidence of the RITAD LT peak with the TSC one (fig. 4) leads to the opinion that internal polarisation compensation is observed. Other traps emptying cannot be observed due to the previous cancelation of the internal field ; this also explain in the MULLER's theory why the TAD peak also disappear. The low temperature slope of the peak gives an activation energy. E_1 = 0.22eV in good agreement with the temperature shift of the corresponding TSC peak (E_1 = 0.22eV).

Then, turning to the RITAD HT positive peak, an interpretation involving a bulk process nor a space charge self relaxation cannot be invoked. This peak is exclusively dependent to the illuminated electrode which leads to think that it could be attributed to deep traps relaxing in the region of the band bending underlying the contact. Supposing a drift of carriers from the contact toward the bulk, the positive polarity should correspond to an electron emitting trap. Activation energy in the low temperature side of the peak leads to E_2 = 0.72eV, in agreement with the quasi intrinsic dark contuctivity (3) at room temperature (0.71eV). This donnor level attributed to Cr^{++} (13) was previously detected on conductive substrates or epitaxial layers by DLTS technics (14).

This HT peak appears in the only region where Current Driven TSC and Voltage Driven TSC, (CCVD) spectra deviate ; this point will be developed later.

These preliminary results will be included in a more extented and detailled paper.

REFERENCES -

1) M. ILEGEMS, H.J. QUEISSER, Phys. Rev. B 12 1443 (1975)
2) H. HENISCH, Thermally Stimulated Processes in Solids : New prospects (1976) J.P. Fillard, J. Van Turnhout Ed.
3) J.JIMENEZ-LOPEZ, J. BONNAFE, J.P. FILLARD, J. Appl. Phys. to appear
4) E.B. PODGORSAK, P.R. MORAN, Phys. Rev. B 8 3405 (1973)
5) P. MULLER, J. TELTOW, Phys. S. Sol. (a) 12 471 (1972)
6) P. MULLER, Phys. S. Sol. (a) 23 165 (1974)
7) P. MULLER, Phys. S. Sol. (a) 28 521 (1975)
8) P. MULLER, Phys. S. Sol. (a) 33 543 (1976)
9) S.C. AGARWALL, Phys. Rev. B 10 4340 (1974)
10) A. MITONNEAU, G.M. MARTIN, A. MIRCEA, Gallium Arsenide and related compounds (1976), C. Hilsum Ed.
11) G.M. MARTIN, J. HALLAIS, G. POIBLAUD, Thermally stimulated processes in solids : new prospects (1976), J.P. Fillard, J. Van Turnhout Ed.
12) A. ROSE, Concepts in photoconductivity and allied problems (1963), Willey Ed.
13) D.C. LOOK, S.S. Com. 24 825 (1977)
14) D.V. LANG, R.A. LOGAN, J. of Elect. Mat. 4 1053 (1975)
15) J. JIMENEZ-LOPEZ, J. BONNAFE, to be published

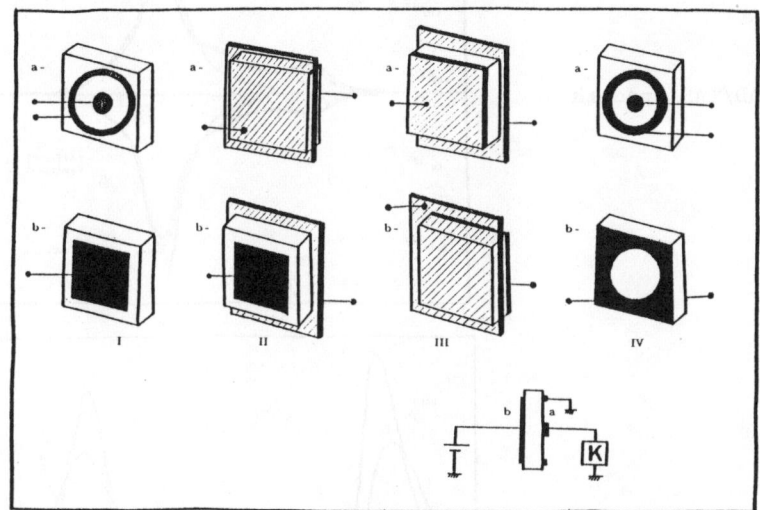

Fig.1 - Sample configurations : a) a front face, b) back face.
    I- Silver past contacts
    II- Asymetrical insulating contact
    III- Symetrical insulating contact
    IV- Silver past contact.

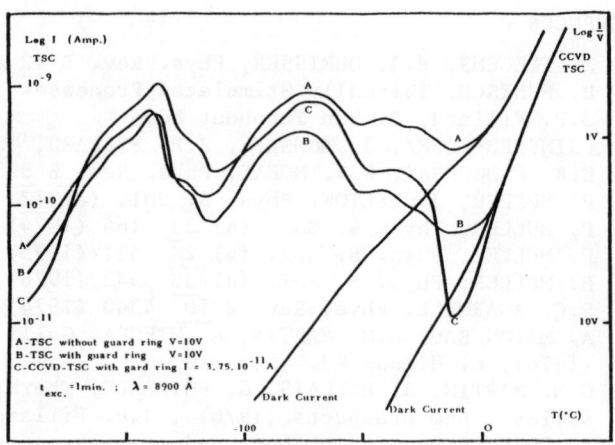

Fig.2 - TSC and CCVD/TSC spectra

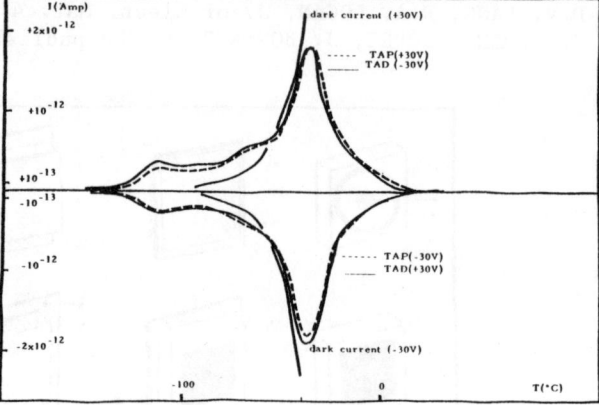

Fig.3 - TAD/TAP spectra

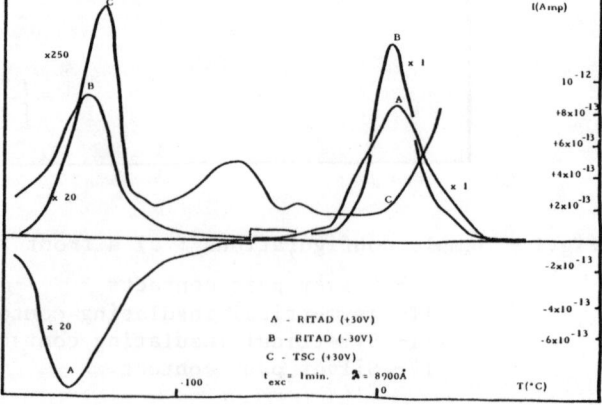

Fig.4 - RITAD/TSC spectra

THERMALLY STIMULATED DEPOLARIZATION CURRENTS IN CaWO$_4$ CRYSTALS

M. Böhm, A. Scharmann, and M. Thomae

I. Physikalisches Institut, University of Giessen

D-6300 Giessen, Fed. Rep. Germany

1. Introduction

Calcium tungstate is to be seen as a typical representative of the class of oxygen dominated phosphors. This phosphor has been investigated for a long time with respect to its attractive scintillation and laser properties. Thereby the method of luminescence is widely used as tool to study the defects. However, the results cannot be explained in the frame of an uniform model. Various other experimental techniques have to be performed to obtain more information on optical, dielectric and magnetic properties. The current paper deals with glow curves of electrical conductivity and luminescence.

2. Experimental Techniques and Results

The single crystals of calcium tungstate investigated in these studies are doped with 10^{-3} Gd and reduced in vacuum. The size of the samples are about 5 x 4 x 1.3 mm^3. Silver paint is used for making the electrodes. The samples are sandwiched between insulating foils of mica in capacitor to avoid the formation of space charges.

The depolarization measurements are performed using the ionic thermo-current method (1, 2). An external electric field of about 3×10^3 V/cm is applied to the crystal at room temperature for an interval of time. After polarization the temperature is reduced with the field still applied to the temperature of liquid air; the cooling rate is about -35 deg/min. The field is then removed and the sample is warmed at a linear rate of about 20 deg/min. During heating-up a depolarization current is detected in short circuit conditions which peaks at about 230 K (Fig. 1). This depolarization peak which exhibits typical non-symmetrical single band structure described by first order kinetics (3, 4) is supposed to be due to reorientation of microscopic dipoles.

Some other measurements yield results which can be explained quite well in terms of this model. The effect of the polarization time t_p on the current intensity is illustrated in Fig. 2. The

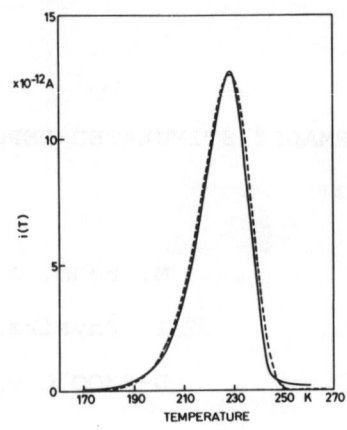

Fig. 1. Thermally stimulated depolarization current in short circuit conditions by polarizing at room temperature (———) and calculated glow curve (– – –) with the activation energy E = 0.44 eV and the pre-exponential factor τ_o = 5.7x10^{-9} s.

built-up of polarization at a fixed temperature can be described by an exponential law and a saturation behaviour is found as it is expected from relaxation processes.

The effect of the polarization temperature T_p is demonstrated in Fig. 3. The left dashed curve is obtained when the sample is pola-

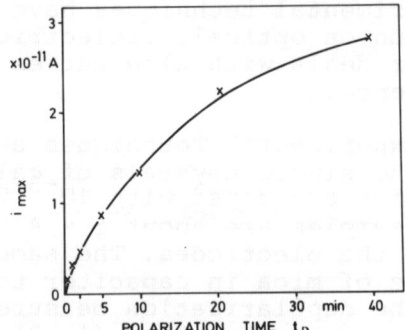

Fig. 2. Peak current of thermally stimulated depolarization in dependence on polarizing time.

rized for 30 min and cooled down with the electric field applied. The right dashed curve results from measurements where the polarization time can be neglected, i. e. the sample is cooled down immediately after the polarizing field is applied. The current values belonging to shorter polarization times are found to be shifted to higher temperatures.

A stable electrical polarization can be induced even at low temperatures by means of X-irradiation. The sample is first cooled to liquid air temperature with short-circuited electrodes and then

X-irradiated with the external field applied. A thermally stimulated depolarization current is observed during the subsequent heating in short circuit again. The integral thermoluminescence (TL) is detected simultaneously (Fig. 4).

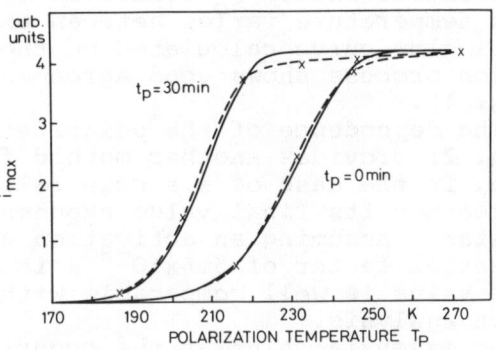

Fig. 3. Peak current of thermally stimulated depolarization versus polarization temperature at two different polarization times; experimental (– – –) and theoretical (———) curve calculated from the model of a dipolar relaxation mechanism.

3. Discussion

A paramagnetic hole centre similar to the V_k-centre in alkali halides plays a dominant role in the TL process and charge transfer (5, 6). The migration and radiative recombination of these hole centres due to thermal activation is assumed to be responsible for the first glow peak (7).

The second glow peak of TL at about 220 K is observed even in other scheelites doped with rare earth ions. In our case the Gd^{3+} located on a metal site has a paramagnetic ground state and acts

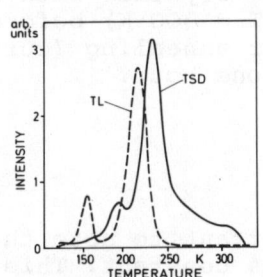

Fig. 4. Thermally stimulated depolarization (TSD) current in short circuit and thermoluminescence (TL) after both X-irradiation and polarization at liquid air temperature.

as electron trap. In the temperature region of the second glow peak an increase of Gd^{3+} concentration can be observed by means of EPR technique. Thus the luminescence emission is expected to be caused by the thermal liberation and recombination of trapped electrons.

The results of the depolarization measurements without previous X-irradiation cannot be explained by the occurrance of free charge carriers but rather by a single dipolar relaxation process (8). The glow peak of thermally stimulated depolarization (TSD) in short circuit shows a non-symmetrical shape which is typical for kinetics with first-order conditions. Presuming a monomolecular mechanism at the reorientation during heating, i. e. non-interacting dipoles there are various approximative methods for evaluating the thermal activation energy E based on the shape of the glow curve. A value of about 0.4 eV is determined in this way. Another method making use of the initial rise yields a similar value of 0.38 eV. The pre-exponential factor τ_o evaluated by means of activation energy and peak temperature varies between 6×10^{-9} s and 5×10^{-8} s. The theoretical glow curve calculated on the assumption of a single relaxation process shows good agreement with the experimental results (Fig. 1).

The dependence of the polarization upon the polarizing time (Fig. 2) provides another method for evaluating the relaxation time. In the case of a single relaxation process the polarization approaches its final value exponentially with the relaxation time constant. Assuming an activation energy of about 0.4 eV a pre-exponential factor of 5.6×10^{-8} s is found from the experimental data. This value is well comparable with others determined from glow curve analysis.

An essential hint at the occurrance of dipolar reorientation is given by the dependence of the build-up of polarization on the temperature (Fig. 3). The maximum current shows a steep decrease at polarization temperatures below the glow peak temperature. Besides this decrease depends also on both the polarization time and the cooling rate. These phenomena can be ascribed to the increase of relaxation time and the decrease of jump probability, resp. It is

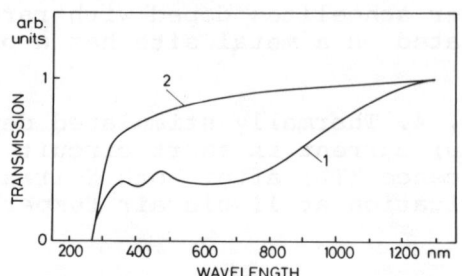

Fig. 5. Transmission spectra of $CaWO_4$ crystals with oxygen defficiency (T = 300 K) before (curve 1) and after annealing (curve 2) at 1000 °C for one hour.

important to note that the position of the depolarization peak remains constant. This fact gives further evidence for the occurrence of a dipolar relaxation process (9). The theoretical curves are calculated from the exponential law for polarization (Fig. 3). The values of activation energy and pre-exponential factor are found to be 0.44 eV and 6×10^{-9} s.

In conclusion, the results of optical absorption measurements are to be discussed in order to get some information concerning the atomic properties (Fig. 5). The broad band between 500 nm and 1200 nm which peaks at around 700 nm is assumed to arise from oxygen vacancies. Similar results are obtained from other oxygen dominated phosphors with oxygen defficiency such as $LiNbO_3$ or $Ca(NbO_3)_2$. The possible occupation with one or two electrons produces point centres similar to F- or F²-centres in alkali halides. When heating up the oxygen vacancies become mobile and their concentration is strongly reduced.

The short wavelength band at about 400 nm is likely caused by W^{5+} ions in connection with a $(WO_4)^{3-}$-complex centre. Two reduced W ions are suitable for charge compensation of the excess positive charge of oxygen vacancy. Annealing gives rise to an oxydation of these centres and the band vanishes. Concerning the current measurement the TSD signal is found to be reduced significantly by annealing procedure.

As a result of absorption measurements the electric dipole complex which participates on the relaxation process is assumed to involve an oxygen vacancy and two fivefold charged tungsten ions in the substitution arrangement (10)

$$\{Ca_{1-3y}Gd_{2y}V_y\}\{W^{6+}_{1-2x}W^{5+}_{2x}\}\{O_{4-x}V_x\}.$$

For example in $BaWO_4$ crystals fivefold charged tungsten ions are found to be localized on lattice defects. In the present case these defects might be oxygen vacancies. The electric moment is formed by two units of positive charge (relative to the perfect crystal) of one oxygen vacancy and two units of negative charge of two W^{5+} ions. The reorientation of the dipole can be explained by oxygen vacancy motion to another site within the WO_3-complex.

4. Concluding Remarks

The origin of TSD peaks due to a polarizing field during simultaneous X-irradiation is not clear up to now. Such thermally stimulated processes involve generation and trapping of free electronic charge carriers. An inhomogeneous charge distribution is probably built up within the sample due to a concentration gradient of trapped carriers. The internal electric field, then, possibly gives rise to dipolar reorientation processes during the thermal stimulation. This might be one possible explanation for the fact that the TSD signal due to X-irradiation can be correlated to that of the usual depolarization measurement. The effect of X-irradiation on the TSD signal and the correlation to the TL behaviour should yield useful information concerning the defect properties which will be published subsequently.

References

1. C. Bucci and R. Fieschi, Phys. Rev. Lett. 12(1964)16
2. C. Bucci, R. Fieschi and G. Guidi, Phys. Rev. 148(1966)816
3. M. Böhm and A. Scharmann, phys. stat. sol.(a)4(1971)99
4. M. Böhm and A. Scharmann, phys. stat. sol.(a)5(1971)563
5. G. Born, R. Grasser and A. Scharmann, phys. stat.sol.28(1968)583
6. R. Biederbick et. al., phys. stat. sol.(b)69(1975)55
7. M. Böhm et. al., J. Luminescence 17(1978)291
8. A. Scharmann et. al., J. of Electrostatics 3(1977)1
9. A. Kessler, J. Electrochem. Soc. 123 (1976)1237
10. K. Nassau and G. M. Loiacono, J. Phys. Chem. Solids 24(1963)1503

SPACE CHARGE IN Pb(Zr, Ti)O$_3$ CERAMICS

Sadayuki Takahashi

Central Res. Lab., Nippon Elect. Co., Ltd.

4-1-1, Miyazaki, Takatsu-ku, Kawasaki, Japan

1. Introduction

Pb(Zr, Ti)O$_3$ family ferroelectric ceramics exhibit high piezoelectric activity and are stable against temperature variation and elapsed time. Their electromechanical characteristics can be modified by introducing minor additives. Because of these convenient characteristics, these ceramic compositions are suitable for use as electromechanical transducers. Most of the addition effect has been considered to relate to the internal bias field modification inside ceramics.[1, 2, 3, 4, 5] Internal bias field has been assumed to be induced by space charge polarization other than ferroelectric polarization.[1, 2, 3]

This paper reports the results of investigation on internal bias field effects on aging characteristics of Pb(Zr, Ti)O$_3$ family ceramics. Different kind of Pb(Zr, Ti)O$_3$ ceramic compositions modified by additives, showing low and high aging characteristics, were prepared. Using these ceramics, the following experiments were carried out. (1) Thermally Stimulated Current (TSC) measurement to observe space charges. (2) Ferroelectric domain switching current measurement to determine the internal bias field strength. (3) Resonance frequency aging characteristics measurement under various biasing field conditions. Details are given in the followings.

2. Experimental
(2.1) Sample preparation

The following ceramic compositions were chosen as high and low aging characteristics specimens, respectively.
(1) Pb(Zr$_{0.52}$Ti$_{0.48}$)O$_3$ + 0.2 wt % metal oxide (Fe$_2$O$_3$, Ga$_2$O$_3$, In$_2$O$_3$) for high aging specimens.
(2) Pb(Zr$_{0.52}$Ti$_{0.48}$)O$_3$ + 1.0 wt % metal oxide (Ta$_2$O$_5$, Nb$_2$O$_5$, La$_2$O$_3$) for low aging specimens.
The raw materials used for the principal constituents were PbO(99.98%), ZrO$_2$ (99.99%), TiO$_2$(99.95%). The other materials used were chemically pure or reagent-grade. Specimens were prepared by the same ceramics technique as reported

previously.[1] The 0.25mm thick 10mm square plates and 0.25mm thick 3mm square plates were used for TSC and internal bias field measurements, respectively. Disks 17mm in diameter and 0.25mm thick were used for piezoelectric resonance frequency aging characteristics measurements.

(2.2) Measurements

TSC measurement was carried out in the 25°C to 600°C temperature range. The electrometer was connected across the test specimen, which was heated at a 25°C/min rate.

A ferroelectric domain switching current measurement block diagram is shown in Fig. 1. The trigonal wave cycle duration is 30 seconds and peak voltage is ±1.25 KV. Test specimens had been previously poled and aged for more than 100 days before both experiments.

In switching current measurement internal bias field effects manifest themselves in the manner shown in Fig. 2. The current curve has been displaced asymmetrically with respect to the origin by the amount of the internal bias E_i.[5]

Aging characteristics were estimated for the resonance frequency in disk radial vibration mode (~130 KHz). Specimens which had been previously poled and aged

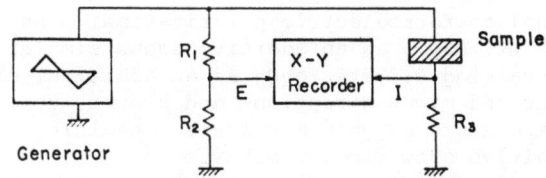

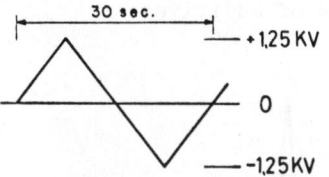

Fig. 1 Domain switching current measurement blockdiagram.

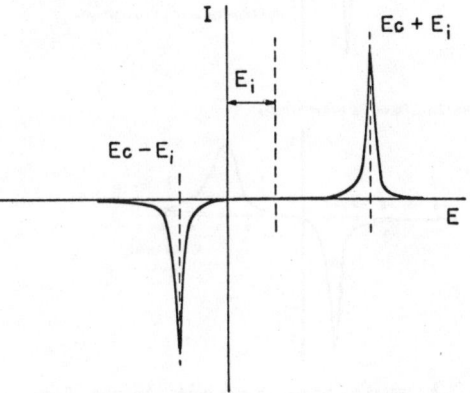

Fig. 2 Domain switching current curve for poled and aged ceramics. Used with this curve, internal bias field E_i is determined.

were switched again by applying a half cycle of the trigonal wave voltage.

3. Results and Discussion
(3.1) TSC

Figure 3 shows typical discharge current curves. In the specimen containing 1.0 wt% Ta_2O_5 as an additive, discharge current flows in one direction, attaining a maximum at the Curie point. The total charge flow in the 25°C to 600°C temperature range amounts to 45µC/cm², approximately

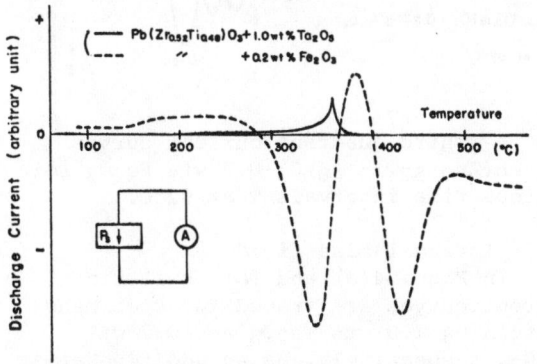

Fig. 3 Discharge current variation with temperature rise for slowly heated $Pb(Zr_{0.52}Ti_{0.48})O_3$ ceramics containing 1.0 wt% Ta_2O_5 or 0.2 wt% Fe_2O_3 as an additive.

equal to ferroelectric polarization. The specimen containing 1.0 wt% Nb_2O_5 or 1.0 wt% La_2O_3 as an additive shows similar current curves. In the specimen containing 0.2 wt% Fe_2O_3 as an additive, discharge current flows towards both plus and minus directions and amounts to more than 1000 $\mu C/cm^2$ in the same temperature range. Specimens containing 0.2 wt% Ga_2O_3 or 0.2 wt% In_2O_3 as an additive show similar behaviours.

The results indicate that an anomalous amount of charges, space charge, besides the ferroelectric polarization, exist in $Pb(Zr, Ti)O_3$ ceramics containing certain kinds of additives.

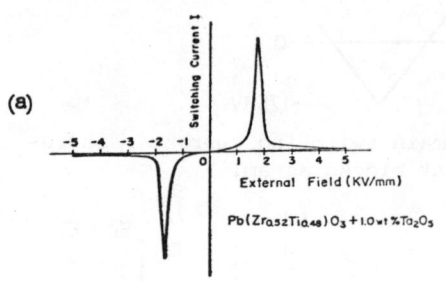

(a)

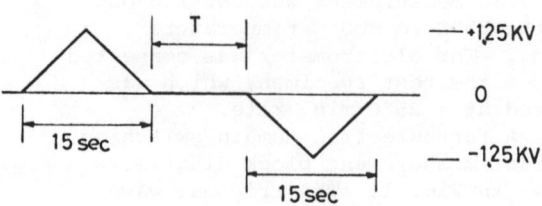

Fig. 5 Applied voltage wave form for measurement of internal bias field variation with time.

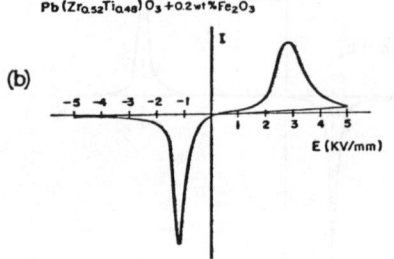

(b)

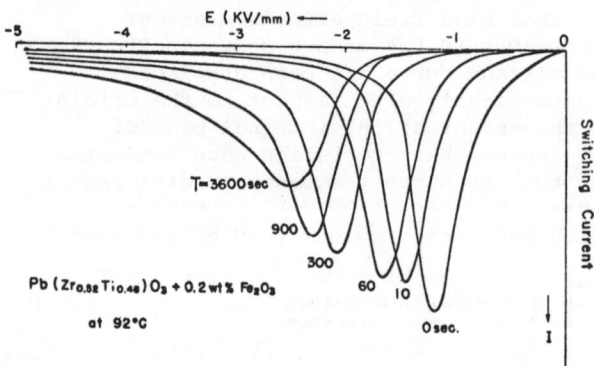

Fig. 4 Switching current curve for $Pb(Zr_{0.52}Ti_{0.48})O_3$ ceramic containing 1.0 wt% Ta_2O_5 or 0.2 wt% Fe_2O_3 as an additive at 25°C.

Table I Internal bias field strength data for all specimens investigated.

Additive	E_i (V/MM)
0.2 wt% Fe_2O_3	750
0.2 wt% Ga_2O_3	730
0.2 wt% In_2O_3	700
1.0 wt% Ta_2O_5	80
1.0 wt% Nb_2O_5	100
1.0 wt% La_2O_5	70

Fig. 6 Third quadrant current curves for $Pb(Zr_{0.52}Ti_{0.48})O_3$ + 0.2 wt% Fe_2O_3 for various time intervals T at 92°C.

(3.2) Internal bias field
In Figs. 4(a) and (b), switching current curves are traced for specimens containing 1.0 wt% Ta_2O_5 and 0.2 wt% Fe_2O_3, respectively, as an additive at 25°C. The current curve shows approximate symmetry in (a), while it shows asymmetry in (b). This is due to the difference E_i values, as explained before. In the specimen containing 1.0 wt% Ta_2O_5, $E_i \sim 80$V/mm. In the other specimen, $E_i \sim 700$V/mm.

Table I shows Ei values for all specimens investigated at room temperature. Specimens containing 0.2 wt% Fe_2O_3, 0.2 wt% Ga_2O_3 or 0.2 wt% In_2O_3 as an additive show a strong field, more than 700 V/mm. Specimens containing 1.0 wt% Ta_2O_5, 1.0 wt% Nb_2O_5 or 1.0 wt% La_2O_3 show a weak field, less than 100V /mm.

It is clear that the ceramic compositions having anomalous amount of space charges, besides ferroelectric polarization, show a strong internal bias field. This field seems to arise from the space charge polarization.

The value of Ei varies with time. This variation can be observed by applying a particular wave form voltage, as shown in Fig. 5. The current curve, appearing in the third quadrant, shift to higher field side with an increase in time interval T. This is shown in Fig. 6 for the specimen containing 0.2 wt% Fe_2O_3 at 92°C. Time variation in Ei values at various temperatures is shown in Fig. 7. Strong Ei value is retained for a fairly long time at a lower temperature, but it decreases in accordance with the familier aging law

$$Ei = A \log t + B$$

where, A depends on the temperature.(6)

(3.3) Resonance frequency aging

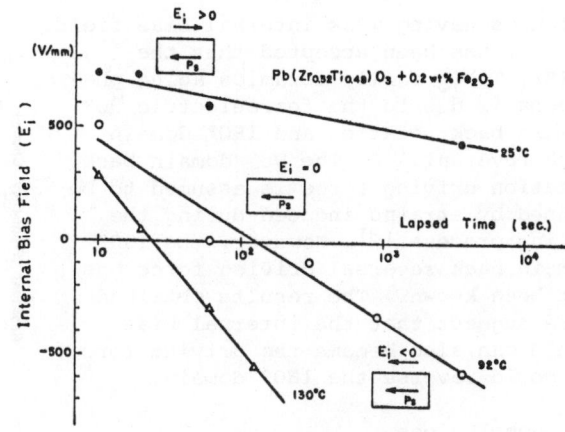

Fig. 7 Variation in internal bias field Ei with time and temperature for $Pb(Zr_{0.52}Ti_{0.48})O_3 + 0.2$ wt% Fe_2O_3 specimen.

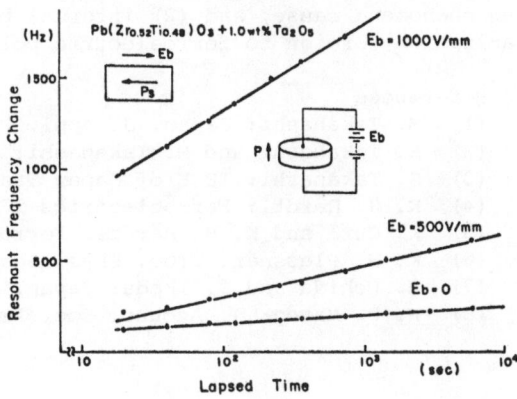

Fig. 8 Resonance frequency aging characteristics for specimen $Pb(Zr_{0.52}Ti_{0.48})O_3 + 1.0$ wt% Ta_2O_5 under various biasing field conditions at 25°C.

Figure 8 shows resonance frequency aging characteristics under various biasing conditions at 25°C. Biasing field is chosen in the opposite direction to ferroelectric polarization. It is weaker than the coercive field, to make backward domain motion slow with time. The specimen composition used is $Pb(Zr_{0.52}Ti_{0.48})O_3 + 1.0$ wt% Ta_2O_5. The ordinate indicates frequency deviation from values at 0 second. It is clear that the amount of frequency change during the aging process shows a tendency to increase with increase in bias field strength Eb.

Resonance frequency aging characteristics for all specimens investigated are shown in Fig. 9. The ordinate indicates the frequency deviation from values at 3 seconds. High aging characteristics are found for ceramics compositions having strong internal bias field. Low aging characteristics are found for ceramics having weak internal bias field.

It has been accepted that the $Pb(Zr, Ti)O_3$ family ceramics aging phenomena is due to the ferroelectric 90° domain back rotation and 180° domain back reversal.[7] The 90° domain back rotation driving force is assumed to be caused by strains induced during the poling process.[8] However, the 180° domain back reversal driving force has not been known. The results obtained here suggest that the internal bias field can also become the driving force. It could reverse the 180° domain.

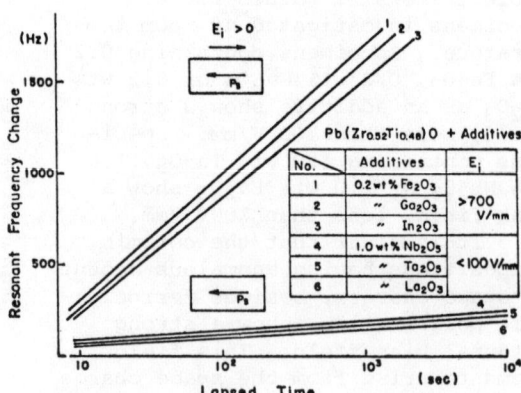

Fig. 9 Resonance frequency aging characteristics for all specimens invesigated.

4. Conclusions

The relation between $Pb(Zr, Ti)O_3$ family ceramics internal bias field and resonance frequency aging rate was studied. The results obtained here suggest that (1) internal bias field opposite to polarization direction become an aging phenomena cause, and (2) internal bias field is induced by space charge polarization foreign to ferroelectric polarization.

References
(1) M. Takahashi: Japan. J. appl. Phys. 9, 10, p.1236 (1970)
(2) S. Takahashi and M. Takahashi: Japan. J. appl. Phys. 11, 1, p.31 (1972)
(3) S. Takahashi: IECE of Japan Tras. 59-C, p.636 (1976)
(4) K. H. Härdtl: Ferroelectrics 12, p.9 (1976)
(5) K. Curl and K. H. Härdtl: Ferroelectrics 17, P.473 (1978)
(6) K. W. Plessner: Proc. Phys. Soc. A69, p.1261 (1956)
(7) N. Uchida and T. Ikeda: Japan. J. appl. Phys. 7, 10, p.1219 (1968)
(8) W. P. Mason: J. Acoust. Soc. Amer. 27, 1, p.73 (1955)

TSC STUDIES OF CHARGE STORAGE AND TRANSPORT IN ORGANIC CRYSTALS

A CRITICAL APPROACH

Anna Samoć and Marek Samoć

Institute of Organic and Physical Chemistry

Wrocław Technical University, 50-370 Wrocław,
Wyspiańskiego 27, Poland

1. Introduction

The interest in the mechanisms of transport and storage of charges in organic dielectrics arises not only from purely scientific reasons but also because of the possibilities of applications of these materials in technology. Among the possible factors influencing the processes of transport and storage of charges the major role is played by local levels capturing charge carriers -usually referred to as traps. Therefore the studies of the origin and properties of trapping levels in organic insulators are of great importance. Apart of investigations of technological materials there is also a need of performing similar studies on model substances which posess a well-defined structure and, especially, purity. This should enable one to more unambigously interpret experiments and, eventually to draw conclusions about the mechanisms of formation and action of trapping levels which may be generalized for the whole class of organic insulators.

We have performed a study of trapping levels in organic single crystals, especially in pure and doped crystals of anthracene. Several experimental techniques have been employed, among them that of thermally stimulated currents (TSC). This communication deals with the problems encountered on the interpretation of the results of TSC. Some suggestions are made concerning the possible modifications of the TSC theory which arise from the recent progress of understanding of charge transport in disordered materials.

2. Experimental data

In anthracene and similar aromatic compounds the concentrations of intrinsic charge carriers are extremely low and may be disregarded. On the other hand, it is quite easy to introduce excess charge carriers either by using the UV irradiation or from appropriate injecting contacts. We employed both techniques to achieve filling of traps, the latter technique being more convenient because of the possibility to selectively inject carriers of one sign only (homocharge) -cf. (1,2).

The TSC spectra observed for various crystals with various ways of excitation are usually rather complex. Very often individual peaks overlap, which makes the determination of trap depths difficult. In such cases one often uses the so-called "partial heating" technique (3-5) which is believed to enable one to resolve overlapping peaks due to discrete traps or to study energetic distributions of traps. The latter case is of great importance for organic solids since the studies of space-charge-limited currents (SCLC) in these materials almost invariably reveal the presence of distributions of traps being recently considered as superpositions of Gaussians (6).

It is well known that the shape of TSC peaks is sensitive to the non-discreteness of trapping levels (cf.(1,2)), and, in principle one can estimate the width of a distribution from the width of the TSC peak (7). However the appropriate expressions have been derived with the assumption that no retrapping and no recombination processes take place. It can be easily shown that this is not always true. Namely, for no retrapping and no recombination (the term recombination may also denote deep, practically irreversible trapping of carriers in the bulk of a sample) both the position and the magnitude of a TSC peak should not depend on the voltage applied to the sample (1,8). Unfortunately, very few authors of the papers dealing with TSC quote such dependences of their spectra.

The retrapping processes are of special importance while dealing with doped crystals i.e. containing, say, $10^{-7}-10^{-5}$ M/M of an impurity forming a charge carrier trap. We have studied the best known system of this kind: tetracene in anthracene. Tetracene is known to form a 0.42 eV hole trap in anthracene crystals, which causes a large decrease of the effective mobility of holes in this material. Our TSC measurements in this system reveal an enhanced peak at ca. 180 K. Partial heating measurements indicate that the trapping level formed by tetracene is not discrete. It has been also found that the position of the 180 K peak varies both with the concentration of the dopant and with the voltage applied. These evidences of fast retrapping put in doubt any closer examination of the peak shape, mainly because there are no TSC theories in which the retrapping and the energetic distribution of traps are treated simultaneously.

We have also performed electron TSC measurements on crystals of anthracene doped with phenazine. It is known (9) that phenazine forms an electron trap of ca. 0.54 eV. However, in most measurements, instead of detecting an additional peak in the electron TSC spectrum of doped crystals we have observed a marked increase of the "initial rise" activation energy of a peak which was already present in TSC spectra of nominally pure anthracene crystals (the activation energy equal to ca. 0.6 eV in pure crystals has been found to increase to 0.7-1.1 eV in doped samples).

Furthermore, while performing "partial heating" experiments on various crystals we have found that the initial rise energies often go beyond any reasonable limit i.e. reach the values which correspond to frequency factors of the order, say, 10^{20} s^{-1}, which is physically unreasonable for neutral trapping centres.

The mentioned above facts have been quoted mainly to emphasize that the present state of the TSC theory is not satisfactory and either the application of the theory to the results is doubtful from the very beginning or the results obtained from the calculations cannot be explained within the theory framework.

In subsequent sections we shall present two possible approaches to the problem of describing thermally stimulated currents in a more realistic way.

3. Multiple trapping approach

It has been already stated that the description of TSC in organic materials should take into account both the energetic distribution of traps and the existence of retrapping. Some further complications are caused by the facts that the carriers of both signs can be trapped in the sample and the trapped carriers may then act as recombination centres for the carriers of the opposite sign and that the excess charges of trapped carriers create the internal electric field varying both with distance and time.

It is clear that the common TSC approach using a set of two or three ordinary differential equations (10,11) cannot account for such a situation. On the other hand a full solution of the appropriate partial differential equations which should be used instead is virtually impossible. Thus we suggest to introduce the following simplifications to the TSC theory. Firstly, let us distinguish between the large signal and small signal experiments. The term "small signal" means simply that the voltage applied to the sample is large enough to neglect the internal field of trapped carriers. Such a regime may be easily achieved experimentally by using low excitation doses. Further, we shall deal with one-carrier TSC only which means that there are no true recombination centres (although there are some deep traps which action is similar to that of recombination centres). To simplify the model even more let us consider the continuous distribution of trap energies as a sum of k types of levels. The equations describing the TSC simplify now to

$$\frac{\partial n_i}{\partial t} = -r_i n_i + \omega_i n_c \quad (1), \qquad \frac{\partial n_c}{\partial t} = -\sum_i \frac{\partial n_i}{\partial t} - \mu F \frac{\partial n_c}{\partial x} \quad (2)$$

$$J = \mu e F \int_0^L n_c \, dx \quad (3)$$

where n_i are the concentrations of trapped carriers, n_c is the concentration of free carriers, r_i and ω_i stand for the release and capture coefficients, respectively, in particular

$$r_i = \nu_i \exp(-E_i/kT) \quad (4)$$

μ is the mobility and F stands for the electric field. Note that in the small signal regime one can suppose that the coefficients ω_i are constant because the traps are not saturated. The above equations are identical to those employed by Schmidlin (12) and Noolandi (13) to describe the transient photoconductivity. The method of Laplace transform used by the mentioned authors to solve the set (1) and (2) is, however, inapplicable to the TSC case since the coefficients r_i become variable while temperature is changed. Nevertheless, some conclusions about the behaviour of the thermally stimulated current can be drawn from the isothermal solutions of Eqs.(1) and (2). By taking as the initial condition $n_i(0,x) = f_i(x)$ we can obtain solutions valid for the isothermal decay current experiment - an isothermal analogue of TSC (cf.(1)).

Three model spatial distributions have been analyzed. They are listed below together with the Laplace transforms of the current.

i) surface charge $\qquad n_i(0,x) = \lambda_i \delta(x-0)$

$$\tilde{J}(s) = \mu eF \sum \frac{r_i^* \lambda_i}{s + r_i^*} \left[\frac{1}{a}\left(1 - e^{-a}\right) \right] \qquad (5)$$

ii) uniform distribution $\qquad n_i(0,x) = \varepsilon_i \neq f(x)$

$$\tilde{J}(s) = \mu eF \sum \frac{r_i^* \varepsilon_i}{s + r_i^*} \left[\frac{1}{a}\left(1 - \frac{1}{a} + \frac{e^{-a}}{a}\right) \right] \qquad (6)$$

iii) exponential distribution $\qquad n_i(0,x) = \gamma_i \exp(-\alpha_i x^*)$

$$\tilde{J}(s) = \mu eF \sum \frac{r_i^* \gamma_i}{s + r_i^*} \left[\frac{1}{a - \alpha_i} \left(\frac{1}{\alpha_i} + \frac{e^{-a}}{a} - \frac{e^{-\alpha_i}}{\alpha_i} - \frac{1}{a} \right) \right] \qquad (7)$$

where

$$a = s\left(1 + \sum \frac{\omega_i^*}{s + r_i^*}\right) \qquad (8)$$

and $\qquad \omega_i^* = \omega_i t_f, \quad r_i^* = r_i t_f \qquad t_f = \frac{L}{\mu F} \qquad x^* = \frac{x}{L}.$

It should be noted that in the case i) the current obtained after carrying out the inverse Laplace transformation will appear as a sum of convolutions of the functions $\exp(-r_i t)$ with the function being a solution given by Schmidlin (12) for the transient photocurrent (it involves exponential functions and modified first order Bessel functions). Inverse transforms of (5), (6) and (7) which are not written here because of their complexity allow to predict some features of the thermally stimulated current, in particular its dispersion. Considering a single trap one can note that the largest dispersion should occur for the moderate retrapping, i.e. if the carriers once released are trapped only a few times.

The largest difference between the classical description and that using partial differential equations occurs if the time range considered is comparable to the value of t_f which for the TSC means that kT^2/E_i is of the order of t_f. Such a situation is possible in low-mobility materials, e.g. polymers. However, even in materials where mobilities are high (of the order of 1 cm^2/Vs as in anthracene) similar situation may arise if we deal with the material containing apart of some deeper traps, a large density of shallower ones. Then, either a concept of shallow trapping controlled mobility may be utilized to describe the transport or one can consider simulataneously the kinetics of trapping and detrapping by both sets of traps. Note however, that in the former case the low mobility value makes the steady-state approximation $dn/dt = 0$ to be no longer applicable.

We have studied the problem of two sets of traps using an appropriate set of ordinary differential equations (8) . It has been found that if the total time of residence of carriers in shallower traps (which is approximately equal to the effective transit time) is comparable to the time of release of carriers from the deeper trap, the initial rise activation energy may reach the value equal to the sum of depths of both traps, or in other words, the sum of activation energies of the release and the transport.

This is probably the reason of surprisingly large activation energies observed in our experiments. A similar situation may take place in polymers where the activation energies reported for so-called ϱ peaks are also too high to be attributed to depths of trapping levels.

4. CTRW approach

The problem of the transport of carriers via localized states has been also recently discussed in terms of the continuous time random walk (CTRW) model of Scher and Montroll (14). It is tempting to apply this stochastic approach to the description of TSC (cf.(15)). The crucial point of adopting of the CTRW theory to the TSC is accounting for the temperature changes which produce substantial changes in the parameters describing the rate of hopping. We attempted to reach this goal by an appropriate modification of the function $\psi(t)$ describing the distribution of hopping times. Then the resulting current was calculated employing the formula

$$J \sim d\langle 1\rangle/dt \quad \text{where} \quad \langle 1\rangle \sim \mathcal{L}^{-1} \frac{\tilde{\psi}(s)}{s(1 - \tilde{\psi}(s))} \tag{9}$$

The tilde denotes the Laplace transform.

This formula holds only when the effects of the absorbing boundary are negligible so that it should allow to describe the initial portion of a TSC curve. As the starting ψ function we employed a simple exponential function and assumed the heating scheme being somewhat similar to the hyperbolical heating scheme

$$T^{-1} = T_0^{-1} - \ln(\beta t + 1) \tag{10}$$

This allowed us to modify the ψ function, however, the performed calculations (16) of the initial rise of TSC curves have showed that the shape of TSC curves depends strongly on the values of such parameters as the starting temperature and the heating rate. Therefore it should be concluded that further efforts are necessary to obtain the stochastic description of the thermally stimulated current. Another problem is, of course, that in the present theory there is no place to include the possibility of the energetic distribution of hopping sites.

References
1) A.Samoć, M.Samoć and J.Sworakowski, phys. Stat. Sol., A36, 735 (1976)
2) A.Samoć, M.Samoć, J.Sworakowski, J.M.Thomas and J.O.Williams, ibid., A37, 271 (1976)
3) W.Hoogenstraaten, Philips Res. Repts., 13, 515 (1958)
4) K.H.Nicholas and J.Woods, Brit. J. Appl. Phys., 15, 783 (1964)
5) H.Gobrecht and D.Hoffmann, J. Phys. Chem. Solids, 27, 509 (1966)
6) S.Nespurek and E.A.Silinsh, Phys. Stat. Sol., A34, 747 (1976)
7) J.G.Simmons, G.W.Taylor and M.C.Tam, Phys. Rev., B7, 3714 (1973)
8) A.Samoć and M.Samoć, Conf. Space Charge in Dielectrics, Karpacz 1977, to appear in J. Electrostatics
9) K.H.Probst and N.Karl, Phys. Stat. Sol., A27, 499 (1975)

10) R.R.Haering and E.N.Adams, Phys. Rev., 117, 451 (1960)
11) G.A.Dussel and R.H.Bube, Phys. Rev., 155, 764 (1967)
12) F.W.Schmidlin, Phys. Rev., B16, 4466 (1977)
13) J.Noolandi, Phys. Rev., B16, 4466 (1977); idem., ibid., B16, 4474 (1977);
    W.D.Lakin, L.Marks and J.Noolandi, ibid., B15, 5834 (1977)
14) H.Scher and E.W.Montroll, Phys. Rev., B12, 2455 (1975)
15) J.K.Jeszka, M.Zieliński and M.Kryszewski, this meeting
16) M.Ziółkowski, M.Sc.Thesis, Wrocław 1978, unpublished

THERMAL-DEPOLARIZATION-CURRENT STUDY OF SPACE CHARGE POLARIZATION

IN POLYMER COMPOSITES

S.Hayashi, T.Tanaka, S.Hirabayashi and K.Shibayama

Manufacturing Development Laboratory, Mitsubishi

Electric Corporation, Amagasaki, Hyogo, Japan.

1 INTRODUCTION

It has been known that dipole and space charge are able to form polarizations in polymers. Thermal-Depolarization-Current (TDC) has been measured to study the mechanism of polarizations in polymers. The mechanism of dipole orientation polarization has been studied by many investigators,[1] but about the mechanism of space charge polarization there are not so many reports. One of the reasons is that the behaviour of TDC due to space charge polarization is more complicated than that of TDC due to dipole orientation polarization in polymers. Especially in polymer composites, few works have been done on the mechanism of space charge polarization. It is important to clarify the mechanism of space charge polarization in polymers or polymer composites because the polarization is considered to be correlated with the electrical breakdown[2,3] and they are often used of insulatos.

In general, space charges tend to accumulate near the boundaries of different materials under applying electric field (Maxwell-Wagner effect). In homogeneous polymers, they are expected to accumulate near the electrodes and in heterogeneous polymers on the interfaces. In our previous papers,[4,5] we have chosen polymer composites containing mica-flakes with stratified structure as the case of heterogeneous polymers and it has found that a TDC peak due to interfacial polarization is observed in the polymer composites.

In this report, the TDC of a polymer composite containing organic papers is measured in order to make a study of interfacial polarization clearer.

Furthermore, the behaviour of d.c. conduction is observed to clarify effect of interfacial polarization on the transport of charge carriers in polymer composites containing mica-flakes.

2 EXPERIMENTAL

We chose polyamide papers (Du.Pont. Nomex 410) or mica-flakes (Okabe Mica Co.) as fillers for polymer composite samples. Nomex papers or mica mats prepared from small muscovite mica-flakes were impregnated with epoxy resin with composition of weight ratio of Epikote 828 (Shell Chemical Co.): Epikote 817 (shell Chemical Co.): catalyzer=75:25:15 for the polymer composite containing Nomex papers or Epilote 828:tetrahydrophthalic anhydride: phenyl glycidyl ether: Zn-octylate=100:104:25:2 for the polymer composite containing mica-flakes between two glass plates. The composite samples were cured at 120°C for 12h subsequently at 150°C for 8h and finally at 180°C for 4h. The composite samples containing Nomex papers were called sample Cf and unfilled epoxy resin samples were called sample R.

The composite samples containing mica-flakes were called sample Cm and unloaded epoxy resin samples were called sample R1. The volume fraction ϕ of filler ranged from 0.4 to 0.9 for sample Cf and 0.20 for sample Cm. These samples were provided with electrodes and a guard ring of vacuum evaporated aluminum.

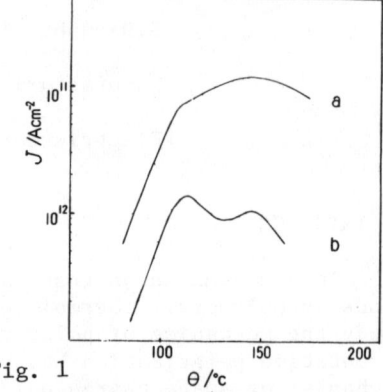

Fig. 1
TSD spectra of sample Cf, $E_p = 10$ kV cm^{-1}
a; $T_p = 165$ °C, b; $T_p = 72$ °C.

The measurement of TDC was performed in the following manner. A sample was polarized with electric field E_p at a temperature T_p for 20 minutes and subsequently cooled rapidly to room temperature before the field was removed. After short circuiting the electrodes the sample was heated at a constant rate β, 0.033 K sec^{-1} and the depolarization current was measured using vibrating reed electrometer (Takeda Riken Co.).

Dielectric constant was measured with a.c. bridge (Ando Electric Co.).

Apparent mobility was measured by the method of the polarity reversal.

3 RESULTS AND DISCUSSION

The TDC spectra of sample Cf are shown in Fig.1. The sample Cf was polarized with E_p of 10kv/cm at $T_p=165$°C for the curve a and at $T_p=72$°C for the curve b. The curve b shows two pronounced peaks at 145°C (αf) and 115°C (βc). Figure 2 shows the TDC spectrum of sample R polarized with E_p of 10kv/cm at $T_p=72$°C. A peak is observed at 105°C (βR).

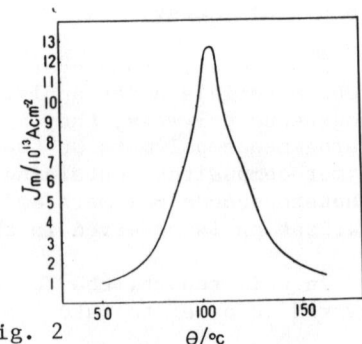

Fig. 2
TSD spectrum of sample-R, $T_p = 72$ °C
$E_p = 10$ kV cm^{-1}

It is well known that TDC due to dipole orientation polarization is directly proportional to Ep.[6] The βc and β_R peaks are considered to be due to dipole orientation polarization associated with the glass transition of epoxy resin matrix for the following reasons.
1) The maxmum currents of these peaks increase linearly with Ep.
2) These peak temperatures are nearly equal to the glass transition temperatures measured by the measurement of mechanical property (TMA).

Then, we are concerned with αf peak.
It is known that TDC arising from space charge polarization due to electrons strongly depends on the collecting electric field Ec[7],[8] which is applied during TDC measurement and that the sign of the current changes with Ec which is much smaller than Ep. The αf peak characterized by sample Cf is not considered to be attributed to the polarization of injected electrons because the sign of the peak does not change with Ec=0.5 Ep. Figure 3 shows the Ep dependence of the maxmum current Jm of the αf peaks. It is clear that the Jm of the peak increases with Ep and becomes constant at higher Ep. Supposing the peak is due to dipole orientation polarization, it is possible to apply two site hopping model[7] to analyze the dependence of Jm on Ep. The distance between two sites of the model was calculated from the dependence of Jm on Ep by curve fitting method. The value of the distance is almost 40Å at 165°C. It is too large to a length of a dipole.

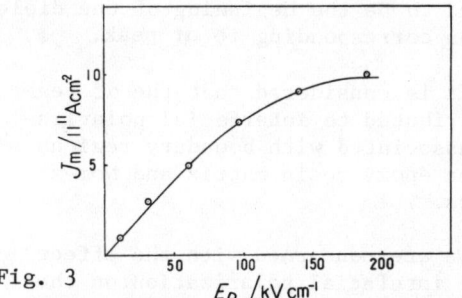

Fig. 3 Electric field dependece of αf peak strength. Sample Cf T_p = 165°C

The αf peak is considered not to be attributed to dipole orientation polarization but to space charge polarization. It is well known that space charges tend to accumulate near the boundaries of different materials[11] whose time constants (ε/σ) are not equal to each other. The conductivity of sample R is about 3.5 x

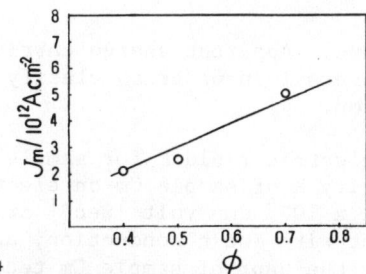

Fig. 4 αf peak strength versus filler volume ratio. E_p = 10 kV cm^{-1} T_p = 165°C

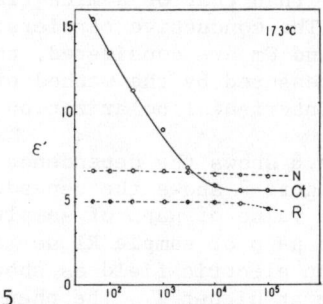

Fig. 5 Frequency dependence of ε' of sample Cf, N and R.

10^{-12} S cm^{-1} and that of Nomex film is about 4.2 × 10^{-14} S cm^{-1} at 165°C. Permittivity of sample R is about 4.9 and that of Nomex film is 6.6. There is much difference from time constant between eipoxy and Nomex. The αf peak is considered to be attributed to space charge polarization due to conductive ions in sample Cf. Figure 4 shows the dependence of the αf peak on the volume fraction φ of Nomex papers. The αf peak is considered to be associated with boundary regions between epoxy resin matrix and Nomex filler.

In general, interfacial polarization in heterogeneous mixtures causes a dielectric relaxation at low frequency range. Figure 5 shows the frequency dispersion curves of ε' of sample Cf, sample R and Nomex film at 173°C. Real part of dielectric constant increases only in the sample Cf. The increment of ε' at lower frequency is considered to be the beginning of the dielectric dispersion corresponding to αf peak.

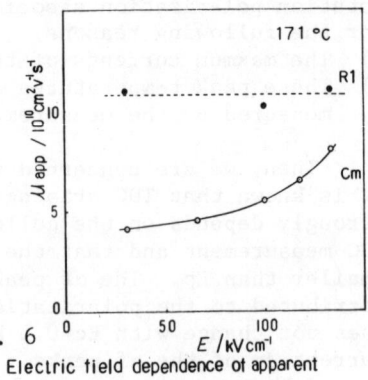

Fig. 6 Electric field dependence of apparent mobility; sample Cm(○), sample R1 (●).

It is considered that the αf peak is attributed to interfacial polarization associated with boundary regions between epoxy resin matrix and Nomex fillers.

We are concerned with the effect of the intefacial polarization on the d.c. conduction. It has been found that ions are predominant in the electric conduction of epoxy resin in the relatively low electric field.[9],[10] The conductivity K of sample Cm is much larger than that of a mica-flake at 171°C. The conductive carriers of sample R1 and Cm are considered, therefore to be ions. Apparent charge carrier μapp were measured by the method of the polarity reversal in order to clarify the effect of interfacial polarization on d.c. conduction.

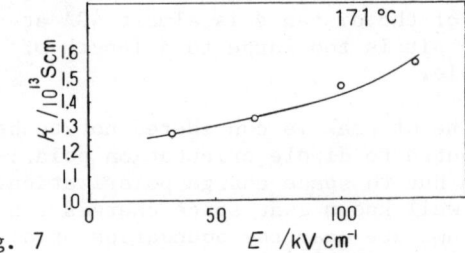

Fig. 7 Electric field dependence of conductivity, sample Cm.

Figure 6 shows the dependence of the μapp on electric field E for sample Cm and R1. Figure 7 shows the dependence of conductivity K of sample Cm on electric field. The value of μapp of sample R1 is about 1.1 × 10^{-9} cm^2 volt^{-1} Sec^{-1} at 171°C. The μapp of sample R1 seems to be consistent with ionic conduction[9] and independs on electric field as shown in Fig.6. But the μapp of sample Cm tends to increase at higher E. The phenomenon is characteristic of the composite sample Cm. The increment of μapp is considered to be caused by space charge effect of the polymer composite sample Cm. In our previous paper[4], it has been known that interfacial polarization exists in sample Cm. The dependence of the interfacial polarization P on Ep is shown in Fig.8. It is considered that the internal field due to interfacial polarization decreases external applied field. The effect of the internal field may apparently decreases the μapp of sample Cm. But, the interfacial polarization of sample Cm becomes nearly constant at higher Ep. The effect of internal field is considered to decrease relatively at higher Ep. The μapp of sample Cm probably tends, therefore, to increase at higher Ep.

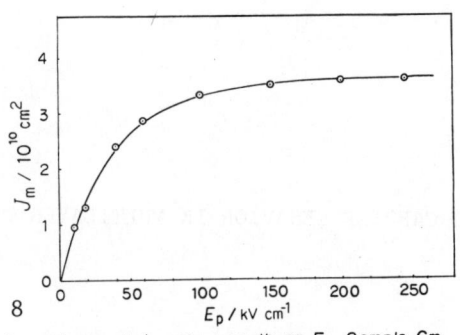

Fig. 8 Dependence of α_c peak strength on E_p. Sample Cm

4 CONCLUSION

The Thermal-Depolarization-Current of polymer-composite containing organic Nomex papers has been studied in the temperature ranging from room temperature to 180°C. A peak characteristic of the composite is observed near 145°C. The peak is considered to attributed to interfacial polarization due to ionic charge carriers at boundaries between the polymer matrix and Nomex papers.

And the effect of interfacial polarization on charge transportation has been studied by the method of the polarity reversal.

Internal field of interfacial polarization decreases apparently applied field and apparent mobility decreases. But, the internal effect reduces relativly compared with applied electric field at higher electric field, and apparent mobility increases.

References

1) J. van Turnhout, Polymer J. $\underline{2}$, 173 (1971).
2) K. Amakawa, T. Moriuchi, T. Yoshida, and Y. Inuishi, J. Inst. Electr. Eng. Jpn. $\underline{84}$, 129 (1964).
3) D.B. Watson, J. Phys. D5, 410 (1972).
4) T. Tanaka, S. Hayashi, and K. Shibayama, J. Appl. Phys. $\underline{48}$, 3478 (1978)
5) T. Tanaka, S. Hayashi, S. Hirabayashi, and K. Shibayama, J. Appl. Phys. $\underline{49}$, 2490 (1978).
6) C. Bucci, R. Fieschi, and G. Guidi, Phys. Rev. $\underline{148}$, 816 (1966).
7) G. Sawa, M. Kawade, D.C. Lee, and M. Ieda, Jpn. J. Appl. Phys. $\underline{13}$, 1547 (1974)
8) T. Hino, and R. Yamashita, Trans. Inst. Electr. Eng. Jpn. A95, 79 (1975)
9) T. Miyamoto, and K. Shibayama, Kobunshi Kagaku. $\underline{29}$, 301 (1972)
10) R.W. Warfield, and M.C. Petree, Macromol. Chem. $\underline{58}$, 139 (1962).
11) J.C. Maxwell, in Electricity and Magnetism (Oxford University Press, London, 1892), Vol. 1, p.452.

CHARGE CARRIER BEHAVIOR IN MULTILAYER DIELECTRICS WITH LIQUID CRYSTAL

Kyoichi Shibayama, Hiroshi Koezuka, Hirozoh Kanegae and Hiroshi Ono
Central Research Laboratory, Mitsubishi Electric Corporation

Minamishimizu, Amagasaki, Hyogo, Japan 661

Introduction. Liquid-crystal displays (LCD) have been widely used at present. No resistive current is necessary for the operation of field-effect LCD, but the resistive current flows under DC or AC excitation because of the existence of ionic impurities contained in commercially available liquid crystal materials. So that the consumption of electric power increases and ionic impurities might cause the misalignment of liquid-crystal molecules during the operation to shorten the life of LCD. On the other hand, some of commercial LCD cells have insulating films such as silicon dioxide and polyimide on the electrodes in order to prevent unfavorable electrode reactions under operation. It is, therefore, valuable to investigate the behaviors of charge carriers and the effects of the insulating films used as blocking electrodes in terms of the reliability of LCD. This study will deal with the behaviors of charge carriers in liquid crystal materials mainly by means of thermal depolarization current (TDC) (1).

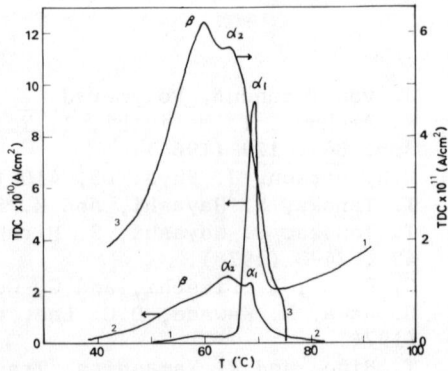

Fig. 1 TDC spectra of various samples. Tp: 79 °C; tp: 1 min;
(1) Al-LC-Al, (2) Al-SiO$_2$-LC-SiO$_2$-Al, Ep: 8 x 10^2 V/cm;
(3) Al-PI-LC-PI-Al, Ep: 1.6 x 10^3 V/cm.

Experimental. Three systems, (I) Al-LC-Al, (II) Al-SiO$_2$-LC-SiO$_2$-Al, and (III) Al-PI-LC-PI-Al, were prepared for the measurements. A sample (IV), InO$_x$-SiO$_2$-LC-SiO$_2$-InO$_x$, was used in order to investigate the effects of the electrode materials. p-Cyanophenyl-p'-caproyloxybenzoate, p-C$_5$H$_{11}$C(O)OC$_6$H$_4$C(O)O-C$_6$H$_4$CN-p, was used as LC, whose mesorage has been determined using the polarizing microscopy with a heating plate to be 64.5 to 101°C. PI stands for polymide (de Beers Laboratories, Incorporated DE910 103C). The thickness of the coated polyimide and the evaporated silicon dioxide is 1200 and 1000 A, respectively. The liquid crystal is injected by capillary action into the 50-μm thick cell, which is kept by a Teflon spacer. The area of the electrode is 5.3 cm^2. The sample was polarized with electric field, E_p, at a temperature, T_p, for a poling time, t_p. The measurements of TDC were carried out at the rate of 0.6 to 0.75 °C/min.

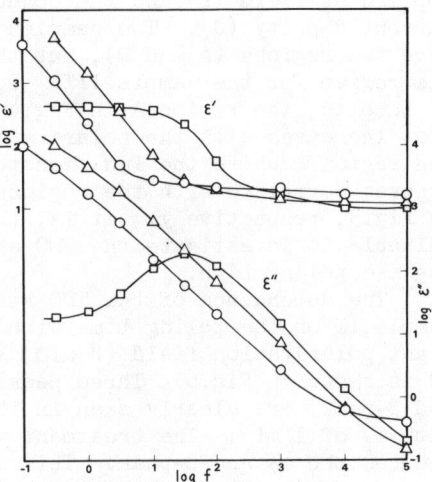

Fig. 2 Frequency dependence of ε' and ε''.
○: Al-LC-Al at 84 °C; △: Al-SiO₃-LC-SiO₃-Al at 80 °C;
□: Al-PI-LC-PI-Al at 81 °C.

Results and Discussion. The TDC spectra of samples I, II, and III are shown in Fig.1. The samples I and II were polarized with E_p of 8×10^2 V/cm at T_p of 79 °C for t_p of 1 min. The polarization of the sample III was carried out with E_p of 1.6×10^3 V/cm. Two peaks have been observed for the sample I (α_1 and β -peak) and III (α_2 and β -peak) near the transition temperature between the solid state and the mesophase. The sample II has three peaks (α_1, α_2 and β -peak). The polarized charge (10^7 c/cm^2) calculated from the observed current for the α_1 -peak of the sample I is far larger than that (10^{-10} c/cm^2) calculated by Debye's dipole model. It is, therefore, considered that the α_1 -peak found is due to the space charges near the electrodes.

The data of ε' and ε'' at various frequencies for three samples (I, II and III) are shown in Fig.2. The slopes of log ε' versus log f and log ε'' versus log f curve are -3/2 and -1 for both samples I and II, respectively (2). This supports the existence of the electrode polarization corresponding to the α_1-peak. On the other hand, a dielectric dispersion corresponding to the α_2 -peak different from the electrode polarization has been observed for the sample III (Fig.2). The ε'' versus ε' curve fits the Cole-Cole plot (Fig.3), which supports that the α_2-peak is due to the Wagner-type interfacial polarization between LC and PI. The α_2-peak of the sample II would be due to the LC-SiO$_2$ interfacial polarization although the corresponding dielectric dispersions have not been observed.

Fig.4 shows the polarization field dependence of the maximum current density (Imα_1) of the α_1-peak for the samples I and II polarized with various E_p at 77 °C for 20 min. The Imα_1 linearly increases to take the maximum and then decreases with the polarization field for the samples I and II. This is a unique behavior for the TDC spectra due to the space charge polarization. It is interesting to study this phenomenon in detail. The Imα_2 for the sample III shows the similar behavior to that of the TDC due to the space charges because of mobile ions, as usually seen.

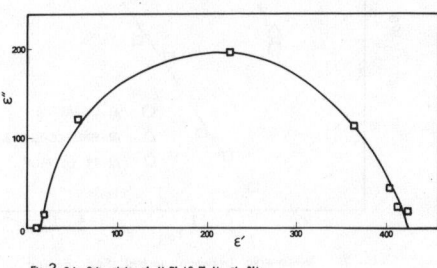

Fig. 3 Cole-Cole plots of Al-PI-LC-PI-Al at 81 °C.

Fig.5 shows the relationship between the

applied DC field (E) and the conduction current density (J). The samples I and II have two regions (A and B), but there is only one region for the sample III. For the I and II samples, the region A in Fig.4 where the $Im\alpha_1$ increases with the polarization field and the region B where the $Im\alpha_1$ monotonically decreases correlate with the regions (A and B) in Fig.5, respectively. It is, therefore, valuable to investigate the TDC spectra varying the poling time.

The dependence of the TDC spectra of the sample II on the poling time with the constant polarization field (8×10^2 V/cm) at 79 °C is shown in Fig.6. Three peaks (α_1, α_2 and β-peak) are clearly seen in the spectrum

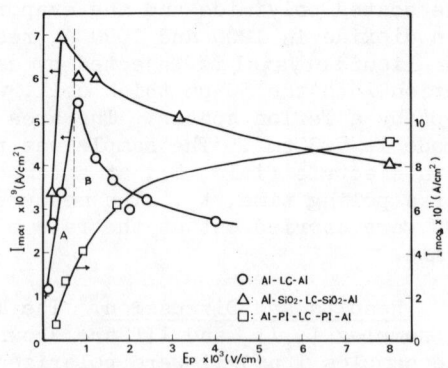

Fig. 4 Dependence of the maximum current density, $Im\alpha$, of α-peak on the polarizing field, Ep. Tp:77°C; tp:20 min.

with t_p of 1 min. The treatment with t_p of 5 min increases the α_1-peak and reduces the α_2 and β-peak. This suggests that three peaks are due to the same species of ions. With the increase of the poling time, the current of the α_1-peak decreases to change the sign. This behavior shows the trapping in or on SiO_2 layer of the negative ions formed near the cathode because the same procedures for the sample I do not cause the appearance of homo-charges (3). These trapped negative ions form the Poisson's field in the sample so that the TDC might emerge in the direction of the polarization (homo). The similar measurements were carried out for the same sample under the poling condition of the higher polarization field (1.6×10^3 V/cm) at 79 °C. The current of the α_1-peak is all in homo region for the poling time of 1 to 20 min. When the sample was polarized with E_p of 4×10^3 V/cm at 79 °C for 20 min, the α_2-peak reappears in the hetero region. These behaviors might be elucidated by that the trapped

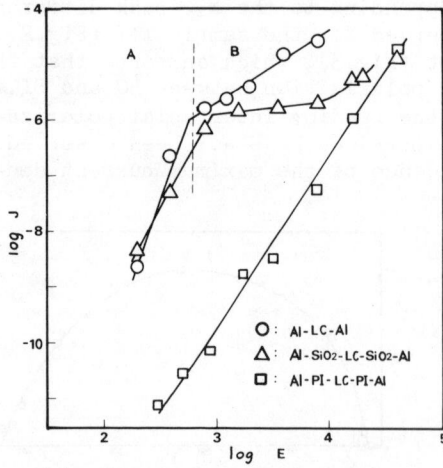

Fig. 5 Relationship between applied field, E(V/cm), and current density, J(A/cm²), at 79 °C.

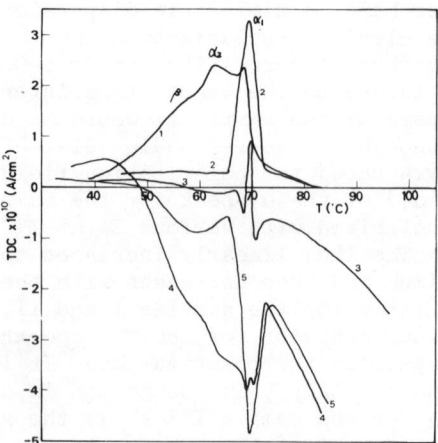

Fig. 6 Dependence of the TDC spectra of Al-SiO₂-LC-SiO₂-Al on the poling time. Tp: 79°C; Ep: 8×10^2 V/cm; tp(min): (1) 1, (2) 5, (3) 10, (4) 20, (5) 30.

ions near the cathode travel to the bulk to reach the counter SiO_2 to form the LC-SiO_2 interfacial polarization under the high electric field.

The measurements of the in-phase (resistive) current (I_r) at 32 Hz were carried out for all samples after applying DC field for 20 min and relaxing at 79 °C (Fig.7). Each measured in-phase current was normalized by that (I_{r_0}) of the virgin sample. The I_r of the sample I increases with DC field. The samples II and III are not changed by DC supply. Comparing the sample II with the sample IV, it is suggested that the Al-SiO_2 system is more stable than the InO_x-SiO_2 for the DC operation. Further, the above idea is also supported from the result that the sample IV has shown no homo-charges in the TDC spectra when the sample

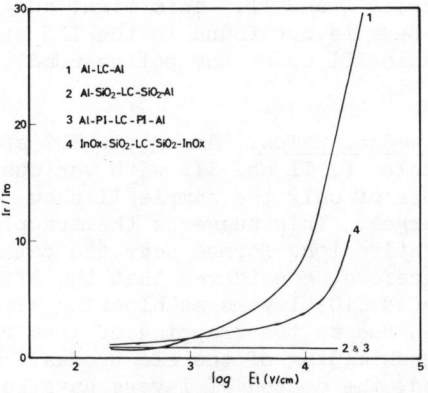

Fig. 7 Influence of the relative in-phase current (I_r/I_{r_0}) at 79°C, 32 Hz by the treatment of applying DC field (E_t) at 79°C for 20 min.

was treated under the same poling conditions as the sample II. Fig.7 shows the existence of the threshold voltage where the sample I system begins to electrolyze. The voltage is near that where the $Im\alpha_1$ takes the maximum (Fig.4). Thus, the decrease of the $Im\alpha_1$ with the polarization field might be attributed to the electrolysis of the sample as one factor. The decreasing tendencies of the $Im\alpha_1$ for the samples I and II are alike and the $Im\alpha_1$ of the sample II reduces more gradually than that of the sample I. It is, therefore, considered that the reducing trend of the $Im\alpha_1$ for the sample II might be attributed to the synergetic effect due to the change of the sample and the trapping of the negative ions formed near the cathode. This is confirmed by the existence of the electrode polarization from the dielectric measurements and the appearance of homo-charges in the TDC spectra.

The TDC spectra of the sample III with various polarization field are shown in Fig.8. Only a α_2-peak as above mentioned is evidently appreciated at low polarization field ($\sim 3.2 \times 10^3$ V/cm). As the polarization field increases, a β-peak clearly emerges. The relationship between the applied DC field and the conduction current density has one region, which is different from the cases of the samples I and II (Fig.5). The result of no existence of the electrode polarization from the dielectric measurements for the sample III coincides with no appearance of the α_1-peak until the polarization field of 1.6×10^4 V/cm. In the TDC spectra of the sample III with E_p of 1.6×10^3 V/cm at 79 °C varying the poling time, a α_1-peak is observed (Fig.9). Three peaks (α_1, α_2 and β-peak) are attributed to the same species of ions. The α_1-peak would be due to the electrode polarization formed by the penetration of small cations such as a proton through PI layer. The poling time at which the α_1-peak appears is longer than

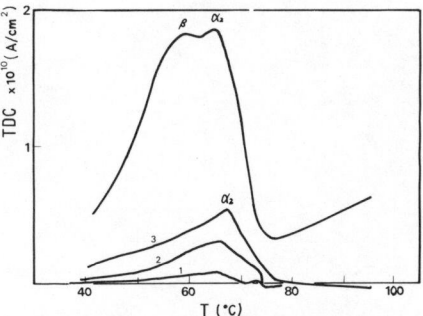

Fig. 8 Dependence of TDC spectra of Al-PI-LC-PI-Al on the polarizing field. T_p: 79°C; t_p: 20 min; E_p(V/cm): (1) 3×10^2, (2) 9×10^2, (3) 3.2×10^3, (4) 1.6×10^4.

samples I and II. This might suggest that the α_1-peak is not found in the TDC spectra of the sample III under the poling conditions as shown in Fig.8.

Conclusion. Among the TDC spectra of the samples I, II and III with various poling time, those of only the sample II show the homo-charges. This suggests the trapping of the negative ions formed near the cathode. It is, therefore, considered that the effect of evaporated SiO_2 layers as blocking electrodes has been due to the trapping of ions rather than the shielding of the electrodes. On the other hand, the coated PI layers have been shown to nearly separate the liquid crystal materials from the electrodes by means of the TDC and the dielectric measurements.

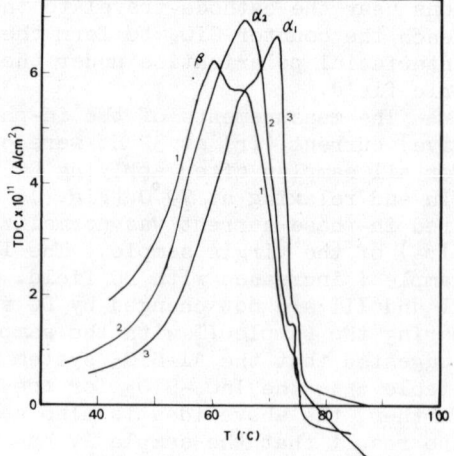

Fig. 9 Dependence of the TDC spectra of Al-PI-LC-PI-Al on the poling time. T_p: 79°C; E_p: 1.6×10^3 V/cm; t_p(min): (1) 1, (2) 10, (3) 20.

References.
(1) T. Tachibana, T. Takamatsu and E. Fukada, Chem. Lett., 907 (1973).
(2) S. Uemura, J. Polymer Sci. A-2, <u>10</u>, 2155 (1972); S. Uemura, ibid, <u>12</u>, 1177 (1974).
(3) A. Sussman, J. Appl. Phys., <u>49</u>, 1131 (1978).

MODEL TSC CALCULATIONS FOR AMORPHOUS MATERIALS

J.Jeszka*, M.Zieliński**, M.Kryszewski**

*Technical University of Łódź, Żwirki 36, 90-448 Łódź, Poland

**Centre of Molecular and Macromolecular Studies
Polish Academy of Sciences, Boczna 5, 90-362 Łódź, Poland

1. INTRODUCTION

In recent years the most widely used method of investigation of the charge carrier transport phenomena in insulators and semiconductors, both organic and ionorganic, is the time-of-flight technique. The Scher Montroll theory [1] or Monte Carlo calculations for exponential trap distribution [2] provide a theoretical background to understand and analyse the strongly dispersive signals that are found for amorphous materials such as e.g. poly(N-vinylcarbazole), molecularly doped polymers or As_2Se_3.

The method of time-of-flight measurements has however the serious limitation, that it cannot be applied in case, of low mobility materials and/or at low temperatures. For long transit times corresponding to low mobility the current value is too small to allow the measurements, and increasing the current by the increase of the injected charge is limited by space charge effects.

Thermally stimulated current (TSC) experiments can help to overcome this difficulty and provide an independent information on transport processes, especially on their temperature dependence. In the case, where the band model can be applied and the transport properties of the material are determined by trapping processes, the TSC spectra give informations on the parameters and distribution of trapping centers.

The simple solution of the TSC equations can be obtained for the cases where either retrapping can be neglected or strong retrapping occurs. Trapping phenomena influence strongly the effective drift mobility of the charge carries and are shown to determine the time-of flight signals. Regardless of the different models describing the two types of processes, results obtained for the case, where hopping transport can be assumed, are similar to that obtained for the case of strong retrapping. Analytical solution of the TSC equations for the dispersive hopping transport or for more complicated trap distributions cannot be obtained. Attempts were made to resolve this problem making some simplifying assumptions valid in special cases, or trying to solve the TSC equations numerically [3]. The first method, however, has serious limitations, while the second leads to hyperbolic partial differential equations, which also make the computations very difficult.

In the present paper we propose an alternative approach to numerical simulation of the TSC spectra. The concept is based on the Scher-Montroll model of but instead of analitical solution for random walk uses numerical computer calculations.

2. THEORY

In this work we make use of the basic assumptions of the Scher-Montroll model but we apply a different calculation procedure to make the numerical calculations possible and to introduce the temperature dependence, which is absent in the Scher-Montroll theory.

We assume a sample having initially introduced (injected) carriers with a certain spatial distribution. The planar sample is divided into cubic cells forming n layers in the direction of applied electric field. Instead of looking for the random walk of the carrier on the cubic array we group the carriers in the cells. The carriers in a cell are divided into m groups of different mobilities and hence of different times τ of staying in the cell (hopping times). The hopping probability distribution $\Psi=f(\tau)$ for a single carrier is replaced by the carrier density distribution $\rho=f(\tau)$ in the cell. The grouping is made so, that the difference of the hopping times between subsequent groups is τ.

After the time $t=\tau$ the fastest carriers are transfered to the neighbour cells and redistributed according to the initial distribution $f(\tau)$. The four cells in the same layer (equidistant from the electrode) absorb 4/6 of jumping carriers, so this part remains in the same layer but is redistributed. The remaining 2/6 is divided between backward and forward directions and the relative numbers of these carriers are made field-dependent. All other carriers that stay in the layer are shifted to the next mobility group, to account for the time flow. The current is calculated from the total charge transfered in one step, divided by the time step Δt.

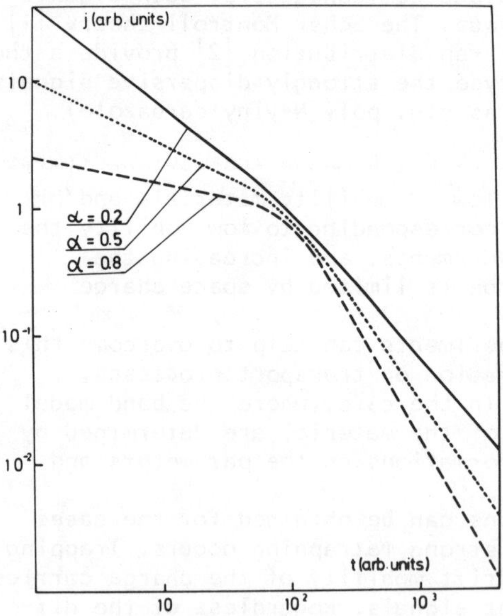

Fig.1. Current vs. time curves (time-of-flight experiment) predicted by Scher and Montroll for three values of α: 0.2, 0.5, 0.8.

In the case of time of flight simulations the time step Δt is made constant equal to the hopping time τ of the fastest carrier group.

In the simulation of the TSC we assume a certain polarization temperature T_o at which the time is zero. The hopping time τ is assumed to depend on the temperature according to the Arrhenius law $\tau=\tau_o \exp(A/kT)$ where τ_o is the pre-exponential factor and A is the activation energy. We can now calculate the time step Δt and the temperature T at which the fastest carrier group jumps

$$T_{i+1}=T_i + \beta\tau_o\exp(A/kT_{i+1})$$

$$\Delta t_{i+1}= \tau_o\exp(A/kT_{i+1})$$

where initially i=0 and β is the linear heating rate.

Thus in the case of TSC the time intervals Δt in which the groups of carriers jump decrease with time and temperature but all the operations on spatial and mobility distributions remain the same.

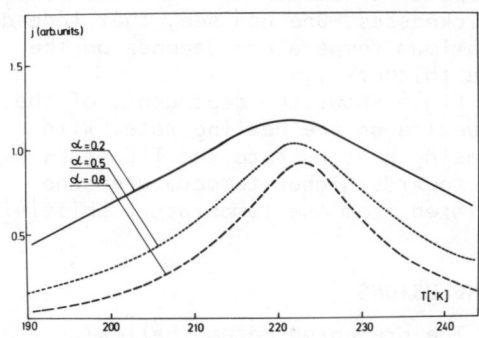

Fig.2. TSC maxima calculated for the Scher-Montroll curves with different α. Activation parameters: $\tau_o=4.0\cdot 10^{-12}$ s, A=0.5 eV.

The feature of such calculation technique for the TSC is that, if the distribution function $f(\tau)$ is assumed only in the pre-exponential factor, the transformation from the time-of-flight result to TSC curve is not dependent on the particular choice of the model parameters, i.e. the number of layers n, groups m or a spatial distribution. Thus every time-of-flight signal, also an experimental one, can be transformed to obtain the corresponding TSC spectrum.

In this work we present the results obtained for the time-of-flight signals predicted by the Scher-Montroll theory.

3. RESULTS AND SISCUSSION

According to the Scher-Montroll stochastic hopping theory the time-of-flight curves plotted in log j - log t coordinates consist of two parts with the slopes $1+\alpha$ and $1-\alpha$ respectively, where α is the parameter of the hopping time probability distribution $f(\tau)\sim\tau^{-(1+\alpha)}$, $\alpha\in(0,1)$. In the calculations of the corresponding TSC maxima we assumed thermal activation parameters corresponding to PVK, $\tau_o=4\cdot 10^{-12}$ s and A=0.5 eV. The heating rate is $\beta=0.1$ °K/s unless otherwise stated.

Fig.1 shows the time-of-flight signals and Fig.2 the corresponding TSC spectra for the fixed transit time and three different distribution parameters α. One can see, that the maximum appears at the same temperature in all cases, and that the TSC spectra broaden with increasing α. The dependence of the peak shape on α implies that the activation energy determined by initial rise method is α dependent. Fig.3 shows the low temperature parts of the TSC maxima replotted in Arrhenius coordinates. One can see from the values listed in the inserted table, that in all cases the slopes of the lines yield values close to the assumed activation energy multiplied by distribution parameter α. The obtained TSC maxima are approximately symmetric in relation to the maximum temperature and have a long tail at high temperatures. It is the feature rather unusual in theoretical results but often found in the experiments.

As the TSC maximum is related to the transit time it is not surprising, that the position of the maximum depends, like the transit time, on the sample thickness. Fig.4 presents the time-of-flight curves and Fig.5 the corresponding TSC maxima for three sample thicknesses. One can see, that indeed the maximum temperature depends on the sample thickness.

Fig.6 shows the dependence of the TSC spectra on the heating rate. With increasing heating rate the TSC peaks shift towards higher temperatures and slightly broaden. The activation energy calculated from the temperature shift [4] is in good agreement with the assumed value.

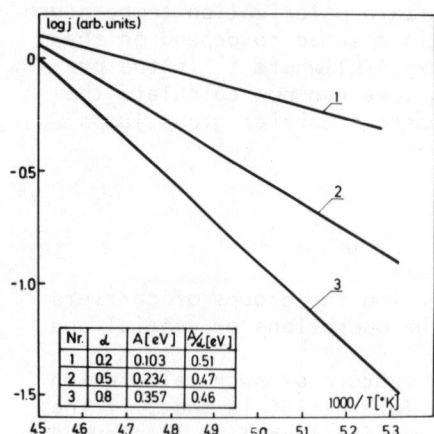

Fig.3. Arrhenius plots for the increasing portions of TSC peaks depicted in Fig.2.

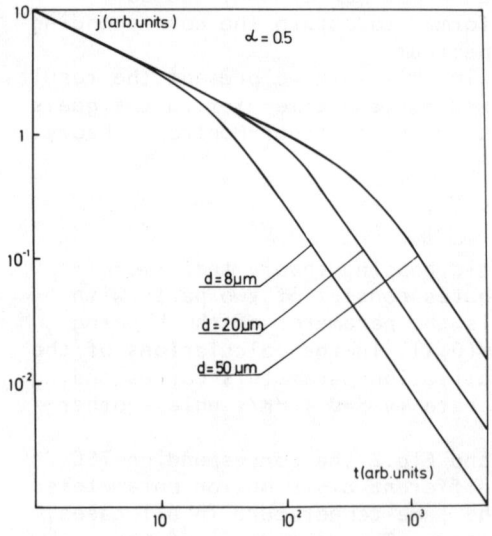

Fig.4. The Scher-Montroll current vs. time curves for three different sample thicknesses: 8µm, 20µm, 50µm.

4. CONCLUSIONS

The presented method, allow to predict theoreticaly the thermally stimulated currents for amorphous systems, for which hopping transport mechanism is applicable. The method is based on the model used by Scher and Montroll in their theory of stochastic hopping, but uses different calculation procedure. It is used to transform the results of time-of-flight measurements into TSC.

The calculations of the TSC spectra, corresponding to the time-of-flight curves predicted by Scher-Montroll model yield broad, almost symmetrical current maxima, the width of which depends on the distribution parameter α. The activation energy calculated from the shift of TSC maxima with heating rate is equal to the assumed one. The "initial rise" activation energy is too low as compared with the assumed one and depends on α while the quotient A/α is approximately constant. The interesting feature of so predicted TSC maxima is that the maximum

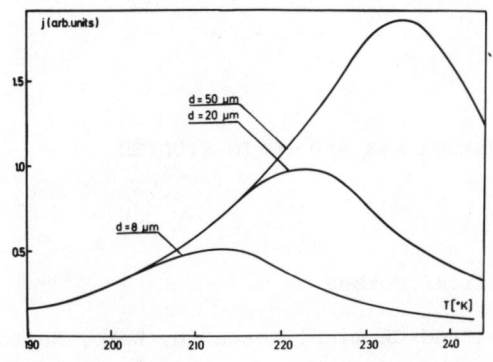

Fig.5. TSC maxima obtained for the Scher-Montroll curves for different sample thicknesses: 8μm, 20μm, 50μm.

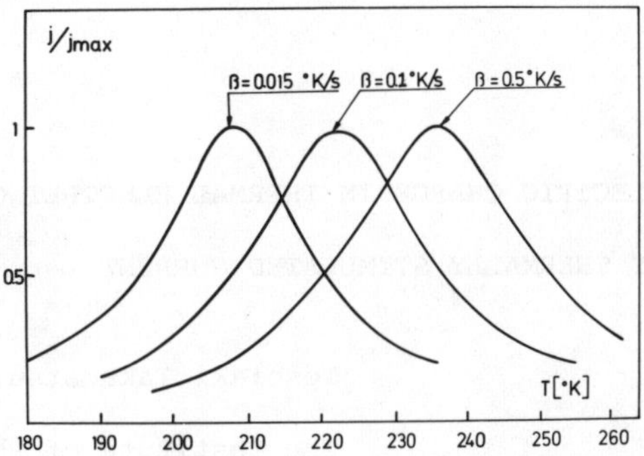

Fig.6. The effect of the heating rate β on the position of the calculated TSC peaks: α=0.5.

position is dependent on the sample thickness.

The model computer calculations permit to investigate the influence of various parameters such as initial spatial distribution, external field, mobility distribution in a straightforward manner, avoiding matematical difficulties.

LITERATURE
1. H.Scher and E.Montroll, Phys.Rev., B12, 2455 (1975).
2. M.Silver and L.Cohen, Phys.Rev., B15, 3276 (1977).
3. I.Chen, J.Appl.Phys., 47, 2988 (1976).
4. R.Chen, J.Mat.Sci., 11, 1521 (1976).

ELECTRIC CHARGE IN THERMAL ELECTRET OF CARNAUBA WAX AND ROSIN STUDIED

BY THERMALLY STIMULATED CURRENT

Toshiaki Takamatsu and Eiichi Fukada

The Institute of Physical and Chemical Research, Wako, Saitama

Eguchi's original electrets are of the equivolume mixture of carnauba wax and rosin. Some of Eguchi's electrets are preserved in National Science Museum in Tokyo. They still exhibits homocharge 50 years after preparations. In order to investigate the mechanism of this excellent stability of charge, we have undertaken the study of the electret properties for the constituents of Eguchi's electret.

Experimentals

Main components of Eguchi's electret are carnauba wax and rosin.[1] In this presentation, we shall use abbreviations: CW for carnauba wax, R for rosin, and CR for the mixture of carnauba wax and rosin. The time change of surface charge and the thermally stimulated current for these electrets have been investigated in detail. Each substances were heated above the melting temperature or the flow temperature and cast on stainless steel plate to form thin plates with approximate dimensions of $3.5 \times 3.5 \times 0.05$ cm^3.

The steel plate at the sample bottom served as a lower electrode. A piece of brass wrapped by aluminum foil was placed on the surface of the samples as an upper electrode. Applying a poling field between 10 and 70 kV/cm at temperatures between 25° and 75°C, the thermal electrets were prepared for each substances.

Corona charged electrets for these substances were also prepared by giving a voltage of 6 kV to a needle electrode and causing corona discharge for 30 seconds.

The surface charge of these electrets were determined by a vibrating reed electrometer placing a measuring electrode 1 mm above the surface of the sample. The thermally stimulated current was also measured with a contactless electrode at a heating rate of 1°C/min.

Results and discussion

Figure 1 shows the initial heterocharge for each substance poled at an electric field of 40 kV/cm and at different poling temperatures. The initial heterocharge for rosin, R, gradually increases with poling temperature. However, those for carnauba wax, CW, and the mixture of carnauba wax and rosin, CR, increase rapidly and gradually level off.

Figure 2 shows the time change of surface charge on CW, CR, and R, all polarized under 40 kV/cm at 60°C. For CW, the transition from heterocharge to homocharge took place at about 3 days after preparation. For CR, the transition took place at about 100 days after preparation. For rosin, no transition was observed. This decay behavior of initial heterocharge clearly distinguishes the properties of carnauba wax, CW, the mixture of carnauba wax and rosin, CR, and rosin, R.

In order to investigate in more detail this decay of heterocharge, we maintained the electrets at different annealing temperatures and determined the time of sign reversal of the surface charge.

Figure 3 shows the relation between ratio of surface potential V to the initial surface potential V_0 and the time of annealing for carnauba wax electrets prepared at 67°C under 70 kV/cm. When the annealing temperature is increased, the decay of heterocharge becomes faster and the time of reversal from hetero to homocharge becomes shorter. For both CW and CR electrets, passing through the hetero-homo transition, the surface charge approaches to the final value of homocharge which is well stabilized at room temperature.

Figure 4 shows the initial heterocharge and the final homocharge for CW and CR electrets which were annealed at different temperatures for a period of 4 hours. The initial heterocharge was almost the same. The final homocharge decreased when the annealing temperature was higher than about 45°C. This decay of homocharge should be due to the higher electric conductivity at the higher temperature.

The classical theory of Gubkin,[2] Perlman,[3] and others for the decay of surface charge in electrets assumes the superposition of homocharge and heterocharge. Homocharge is produced by the discharge in air gap between the poling electrode and the surface of the sample. Homocharge is the space charge consisting of ions and electrons and its relaxation time is assumed to be the product of the resistance R and the capacitance C of the sample as shown in equation (1).

$$\sigma_{homo} = \sigma_o \exp(-t/RC) \sim \sigma_o \quad (RC \gg 1) \quad (1)$$

Since the electrical insulation for wax and rosin is very good, we may assume that the relaxation time RC is much larger than the time of observation. Then the homocharge is approximated to be its initial value σ_o as shown in equation (1).

Heterocharge is produced by either the rotation of dipoles or the separation of ions. The decay constant of heterocharge is given by α in equation (2).

$$\sigma_{hetero} = \sigma_{fo} \exp(-\alpha t) \quad (2)$$

At low temperatures, the decay of heterocharge in wax and rosin appears to be mainly caused by the disorientation of dipoles. The initial value of heterocharge is denoted as σ_{fo} in equation (2).

The total charge is the sum of homocharge and heterocharge as shown in equation (3).

$$\sigma = \sigma_{homo} + \sigma_{hetero} \quad (3)$$

At t=0, when the poling process was finished, the initial value of charge $\bar{\sigma}_o$ is given by equation (4).

$$\bar{\sigma}_o = \sigma_o + \sigma_{fo} \quad (4)$$

At $t=t_{rev}$, when the reversal of the sign of surface charge takes place, we have equation (5).

$$0 = \sigma_o + \sigma_{fo} \exp(-\alpha t_{rev}) \quad (5)$$

By combining equations (4) and (5), we have the expression for the decay constant of heterocharge σ in equation (6).

$$\alpha = \frac{1}{t_{rev}} \ln\left(\frac{\sigma_o - \bar{\sigma}_o}{\sigma_o}\right) \qquad (6)$$

Here α is given by the observed values, that is, $\bar{\sigma}_o$ the initial value of heterocharge, σ_o the final value of homocharge, and t_{rev} the time of reversal from heterocharge to homocharge. We determined using equation (6) as a function of temperature.

The decay constant for CR electrets are more than one order smaller than that for CW electrets. This means that the decay of heterocharge for CR electrets is more than one order slower than for CW electrets at all range of temperature.

The activation energies obtained from the slopes of these lines were about 2 ev both for CR and CW electrets.

In order to study the characteristics of surface charges for these electrets, we have carried out the measurements of the thermally stimulated current, using a contactless electrode. By having a small air gap between the electrode and the surface of sample, the electrical image force between the electrode and the surface charge is reduced. Therefore the sign of the thermally stimulated current may indicate the direction of the current across the thickness of the electret.

For the CW electret, as shown in Figure 5 we obtained a heterocharge current which rised up at 30°C and showed a peak. The current changed the sign at 55°C, above which a large peak of homocharge current was observed. At 90°C near melting temperature, the current decreased to zero.

For the CR electret, the heterocharge current rised up at 35°C, which is higher than the rising temperature for the CW electret, indicating a smaller decay constant of heterocharge for the CR electret. The current changed the sign at about 65°C and then a peak of homocharge current appeared which was smaller compared to the CW electret.

We interprete the heterocharge peaks observed around 50°C for CW and CR should be mainly due to the thermally induced disorientation of dipoles existing in carnauba wax and rosin. As has been suggested by Perlman,[3] the apparent dipoles caused by pairs of trapped ions may also be randomized around this temperature. The homocharge peaks around 70°C should be due to the neutralization of injected ions which was caused by the increased conductivity at high temperature.

For the R electret, a heterocharge peak was observed at 45°C, which should be due to the disorientation of dipoles. To the contrary of CW and CR electrets, at 75°C a peak of heterocharge was observed. The heterocharge peak is considered to show the neutralization of impurity ions which was originally present in rosin and separated during the poling process.

It is interesting to note that at higher temperature range, CW shows the current due to homocharge ions and R shows the current due to heterocharge ions. It seems that the production of homocharge is difficult for rosin but easy for carnauba wax. The reason for this is not completely understood yet. In comparison, rosin is amorphous and soft at the poling temperature of 55°C, while carnauba wax is crystalline and quite hard at the poling temperature of 55°C which is much below the melting temperature. The homocharge is considered to be produced by the discharge in the minute air gaps between the electrode and the surface of sample. The contact between the poling electrode and the surface of the sample may be much better for rosin than for carnauba wax because of the softness of rosin. The closer contact may decreases the possibility of discharge in minite air gaps which will produce the homocharge.

The homocharge is also produced by the injection of charge from the poling electrode to the sample. The conductivity of rosin is nearly three orders smaller than the conductivity of carnauba wax at the poling temperature of 55°C. Such high insulating capacity of rosin may prevent the injection of charges directly from the electrodes.

Another explanation might be that rosin contains relatively large numbers of impurity ions although the conductivity as a whole is small. Consequently the heterocharge due to the displacement of impurity ions overcome the homocharge and thus the sign of the current is that of heterocharge.[4]

Figure 6 shows the TSC curves for corona charged electrets of CW, CR, and R. In this graph, the sign of the ordinate is negative and we have current peaks corresponding only to homocharge. At high temperature, the corona electret of rosin shows a large peak of homocharge in contrast to the heterocharge peak for thermal electret shown in the figure 5.

It is interesting to note that all electrets of CW, CR, and R show the current peaks around 45°C, the sign of which is the same as the homocharge current. At this low temperature, the electric conductivity is still low and the cancellation of homocharge ions can not take place. It is only expected that the motion of dipoles is thermally activated. In corona charged electrets, a large amount of homocharge exist on the surface of electret. This homocharge should give rise to a large electric field inside the sample across its thickness. If the temperature is raised to allow the movement of dipoles, this internal electric field would orient the dipoles in the direction of the field. This polarization current is observed in the external circuit and the direction of the current is the same as that of the neutralizing current of homocharge. Thus the thermally stimulated current around 45°C for corona charged electrets is caused by the orientation of dipoles. It is a polarization current and not a depolarization current. When the corona charged rosin is once warmed to about 60°C, cooled down to room temperature, and immersed into water to remove the surface homocharge, and then the TSC measurement is carried out, a positive current peak is observed around 40°C, which indicates the depolarization current due to the disorientation of dipoles, which were oriented during the previous warming period.[4]

When the corona charged electrets of CW and CR are warmed up above 60°C, the electrical conductivity sharply increases and the homocharge current should be produced like in rosin. However, at such high temperature the orientation of dipoles should be randomized by the thermal motion of molecules, which will produce a heterocharge current. Therefore, the summation of the homocharge current due to corona surface charge and the heterocharge due to the randomization of oriented dipoles should result in small peak or shoulder as seen in the figure.

Reference
(1) M. Eguchi; Phil. Mag., 49, 178 (1925)
(2) A. N. Gubkin; Soviet Phys.- Tech. Phys., 2, 1813 (1957)
(3) M. M. Perlman; J. Appl. Phys., 42, 2645 (1971)
(4) T. Takamatsu and E. Fukada; Proc. Symp. Thermal. Photo Stimulat. Current Insulators p.201, Edit. D. M. Smyth, The Electrochem. Soc. Inc. (1976)

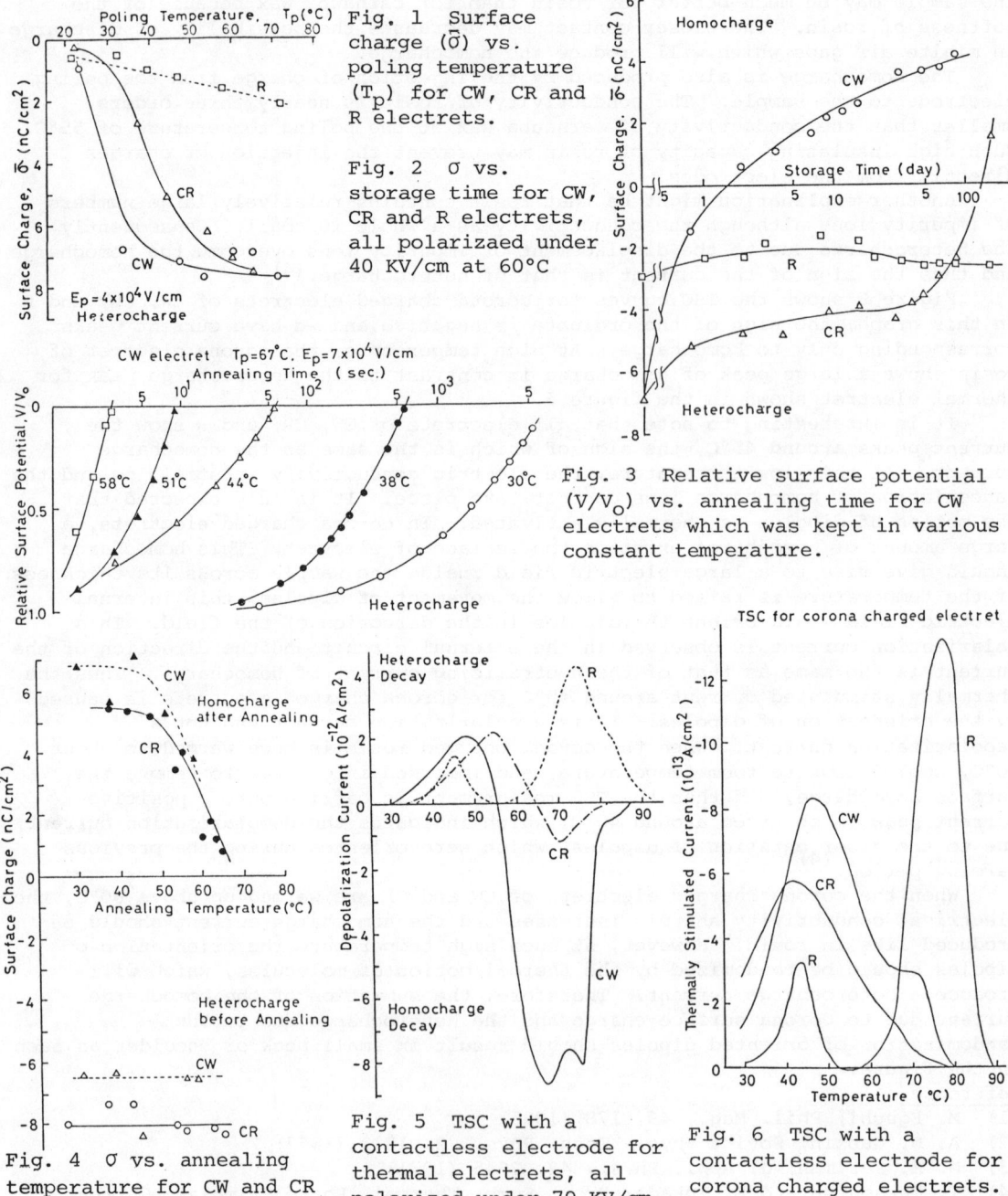

Fig. 1 Surface charge ($\bar{\sigma}$) vs. poling temperature (T_p) for CW, CR and R electrets.

Fig. 2 σ vs. storage time for CW, CR and R electrets, all polarized under 40 KV/cm at 60°C

Fig. 3 Relative surface potential (V/V_o) vs. annealing time for CW electrets which was kept in various constant temperature.

Fig. 4 σ vs. annealing temperature for CW and CR electrets, all polarized under 70KV/cm at 60°C.

Fig. 5 TSC with a contactless electrode for thermal electrets, all polarized under 70 KV/cm at 55°C.

Fig. 6 TSC with a contactless electrode for corona charged electrets.

DIELECTRIC RELAXATION MEASUREMENTS IN Vc:VAc COPOLYMER THIN FILMS

P. C. Mehendru, K. Jain and N. P. Gupta

National Physical Laboratory

New Delhi-110012 (India)

Introduction: Dielectric behaviour of most commonly used organic polymers belonging to polar class such as PVC (Polyvinyl Chloride) (1) and PVAc (Polyvinyl Acetate) (2) in bulk form has been studied in some details and indeed very few investigations on thin film (thickness $\leq 1\,\mu m$) of such materials have been reported. In view of the possible applications of the insulation, isolation and passivation properties of thin polymer films (3,4), a typical copolymer of Vc:VAc (Vinyl Chloride-Vinyl Acetate) containing 3% of Vinyl Acetate (by weight) was chosen. In order to understand the conduction mechanism, polarisation and depolarisation behaviour etc. the depolarisation (TSD) current characteristics have been measured as a function of E_P (polarising field), t_p (polarising time) and T_P (polarising temperature).

To obtain further support of the results so observed, studies of the direct current measurements as a function of temperature have also been carried out. The dielectric loss factor (ϵ') was determined using a conversion relation (5) between the ϵ'' and TSD spectrum and the ϵ'' values calculated by this relation were compared with the ϵ'' values obtained from direct current measurements (6).

Experiment: Thin films of thickness $8000\,\overset{\circ}{A}$ from the solution of the copolymer of Vinyl Chloride-Vinyl Acetate in iso-Butyl methyl ketone were deposited on cleaned glass slides by the isothermal immersion technique in the usual way (3). Silver electrodes were deposited across a bare copper wire, thus giving rise to a film width of 0.056 cm between the two electrodes. Electrical contacts to the silver electrodes were made by an air drying type of silver paint. After polarising the specimen (1) the TSD currents were measured while heating the specimen at a constant rate of $4^\circ K/min$ as described elsewhere (3,4) and whenever overlapping peaks occur a partial heating technique (7) has also been employed.

An application of step d.c. voltage to the specimen, after it has attained the desired temperature, produces a transient charging current and subsequent removal of the d.c. voltage and on short-circuiting the specimen gives rise to a discharging current. Each charge and discharge current was measured from 5 to 10^4 sec at different temperatures with constant applied step d.c. voltage.

Results: Curves A, B, C, D and E in fig. 1 show the TSD spectra of Vc:VAc copolymer films polarised under identical conditions at 373°K for 2 hrs and correspond to E_p = 6.25, 9.37, 12.50, 18.75 and 22.50 kV/cm, respectively. As seen from these curves, three peaks centered around 340 ± 1, 412 ± 2 and $460\pm4^\circ$K are observed in each case and are designated as peaks I, II and III respectively. It is seen from these curves that although with the increase in E_p the shape and positions of the peaks I, II and III remains unchanged, the magnitude of the charge (Q) corresponding to peak I increases linearly with the increase in E_p whereas the charges corresponding to the isolated peaks II & III remain almost the same.

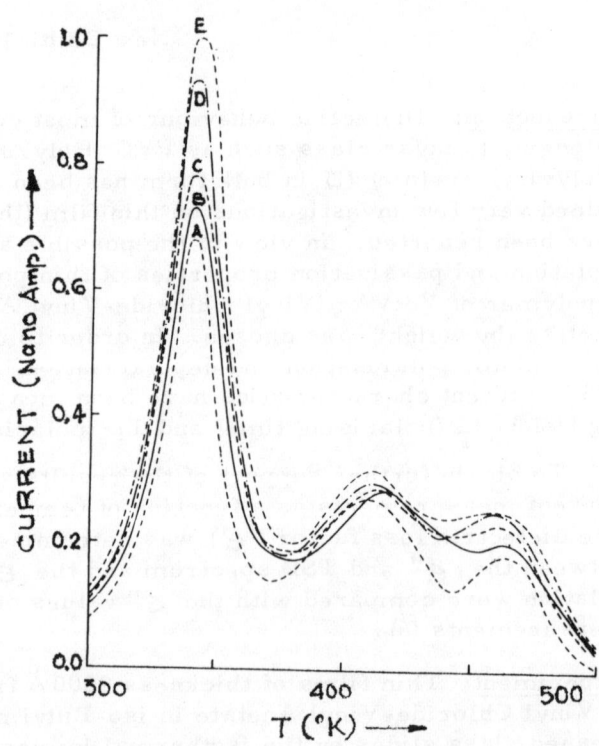

Fig. 1: TSD curves of Vc:VAc copolymer films of thickness 8000 Å, polarised for 2 hrs at 373°K with different E_p. Curves A, B, C, D and E correspond to E_p as 6.25, 9.37, 12.50, 18.75 and 22.50 kV/cm, respectively.

The dependence of depolarising currents on the time of polarisation (t_p) was also investigated in Vc:VAc copolymer films, polarised at 373°K with E_p = 9.37 kV/cm for different time durations. Curves A, B, C, D and E in fig. 2 corespond to t_p as 2, 2.5, 3.5, 7 and 14 hrs, respectively. It is seen from fig. 2 that each curve exhibits the same three peaks I, II & III and with gradual increase in t_p, there is a continuous shift in the positions of peaks I and II towards higher temperatures whereas the temperature of maximum current (T_M) corresponding to peak III (which appears as a shoulder on the peak II) remains unaltered. However, the magnitude of maximum current (I_M) of all the three peaks and also the charges associated with them are found to increase with the increase of t_p. Similar results were observed when the specimen were polarised under identical conditions with E_p = 9.37 kV/cm for 2 hrs at different T_p (305-425°K).

The values of activation energies (U) calculated for the isolated processes under the peaks I and II are also found to increase with the increase of T_p and t_p whereas the activation energy of peak III remains almost unaffected. Since T_M and U increases with T_p and t_p therefore the distribution of relaxation times may be postulated.

The isothermal absorption current from the Vc:VAc films have also been measured at different temperatures ranging from room temperature to ∼373°K (results not shown). It is observed that at temperatures ∠290°K the charging current did not deviate from the discharging current curve, however for temperatures >303°K the form of the charging current is found to be different from the discharging current which is probably due to the onset of steady state conduction current.

The calculated values of ϵ'' from TSD spectra were compared with the ϵ'' values determined by direct current measurements in fig. 3. Since these methods are valid for the charge carriers having a single relaxation time and activation energy, the slight discrepancy in the ϵ'' values determined by two different techniques may be due to the distribution of relaxation times and activation energies in addition to the

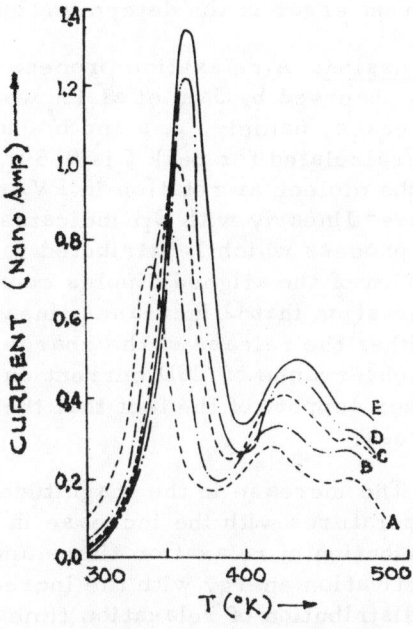

Fig.2: TSD curves of Vc:VAc copolymer films of thickness 8000 Å, polarised at 373°K with E_p as 9.37 kV/cm for different t_p. Curves A, B, C, D and E correspond to t_p as 2, 2.5, 3.5, 7 and 14 hrs, respectively.

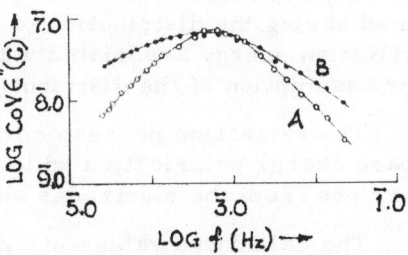

Fig.3: Log log plot of $\omega\epsilon''$ vs frequency, A - calculated from TSD measurements, B - calculated from direct current measurements.

inherent error in the determination of τ_0

Discussion: A relaxation process, namely, α-process in PVC thin film has been observed by Jain et al (3) whereas in PVAc thin films three relaxation processes, namely, β, α and α'-have been reported (4). Since average value of U calculated for peak I is $0.59\pm.02$ eV and it is very close to the value of U for the molecular rotation in PVAc (2,4) and PVC (3) and the charge (Q) increases linearly with E_P indicates that peak I is associated with the α-relaxation process which is attributed to the molecular motion involving the depolarisation of the aligned dipoles connected to the main polymer chain. The observation that Q increases linearly with E_P shows that peak I cannot be due to either the release of the charges from the traps or to a phase change. The non-observance of TSD current on heating the unpolarised specimen is in further support of the fact that the peak I is associated with the α-relaxation process.

The increase in the magnitude of I_M and the shift in T_M towards higher temperatures with the increase in T_P and t_p can be explained by assuming the distribution of relaxation times and activation energies. The linear increase of activation energy with the increase in T_P is in support of our assumption of the distribution of relaxation times (8,9). Since the distribution of dipoles in the amorphous phase is most likely randum it is concluded, therefore, that α-process is a complex process having a distribution of relaxation times and activation energies.

The area under isolated peak II is not a linear function of E_P and its activation energy is as calculated always $\geqslant 1$ eV which indicates that the mechanism associated with this peak is ionic in nature rather than dipolar or some molecular rotation. This may further be speculated on the basis of the value of U ($\geqslant 1$ eV) which are usually associated with ionic traps in amorphous materials (10,11). Since most of the dipolar amorphous polymers exhibit multiple dielectric relaxation process, it is expected that the traps are distributed having the distribution of activation energies. The linear dependence of activation energy associated with peak II on the T_P and t_p is again in support of our assumption of the distribution of traps.

The relaxation process corresponding to peak III is associated with the space charge polarisation which is a resultant charge of the injected charge carriers from the electrodes and already present carriers.

The calculated values of ϵ'' from TSD spectrum (5) were found to be in good agreement with those calculated by direct current measurements (6).

Acknowledgements: The authors are grateful to Dr. A.R. Verma, Director, National Physical Laboratory, New Delhi, for his deep interest throughout the progress of this work. One of the authors (NPG) is also thankful to the University Grants Commission, Government of India, for the award of a Teacher Fellowship.

Reference
1) P.K.C.Pillai, K.Jain and V.K.Jain, Phys.Letters, $\underline{39A}$, 216 (1972); Ind.J.Pure Appl.Phys., $\underline{II}$, 597 (1973).
2) P.K.C.Pillai and V.K.Jain, J.Phys.D.Appl.Phys., $\underline{3}$, 829 (1970); Nuovo Cimento, $\underline{28B}$, 152 (1975).
3) K.Jain, A.C.Rastogi and K.L.Chopra, Phys.Stat.Sol., $\underline{20a}$, 167 (1973)
4) P.C.Mehendru, K.Jain, V.K.Chopra and P.Mehendru, J.Phys.D.Appl.Phys., $\underline{8}$, 305 (1975).
5) T.Hino, Jap.J.Appl.Phys., $\underline{12}$, 651 (1973).
6) B.V.Hamon, Proc.Inst.Elect.Eng.(London), $\underline{99}$, 151 (1952).
7) M.M.Perlman, J.Appl.Phys., $\underline{42}$, 2645 (1971).
8) P.Fischer and P.Rohl, J.Polym.Sc.Polym.Phys., 14, 531 (1976).
9) N.P.Gupta, K.Jain and P.C.Mehendru, J.Chem.Phys.June (1978).
10) R.Chen, J.Appl.Phys., $\underline{40}$, 570 (1969)
11) A.K.Johnscher Thin Solid Films, $\underline{1}$, 213 (1967).

THERMALLY STIMULATED CURRENT OF POLYETHYLENE TEREPHTHALATE

Kimio Shindo

Department of Physics, Shiga University

2-5-1, Hiratsu, Otsu-shi, Shiga-ken, 520 Japan

We have tried to make use of thermally stimulated polarization current as well as thermally stimulated depolarization currrent. These two kinds of current give us some clues to distinguish between the sorts of polarization in a dielectric substance.

TSDC and TSPC - - - - When a polarized dielectric is warmed, it produces a current which is called a thermally stimulated depolarization current (TSDC for short). A dielectric which has not been polarized does not produce any current. But if we apply an electric field to it, a current comes out of it by warming. Let's call this current a thermally stimulated polarization current (TSPC for short). It is evident that TSDC is generated in a destruction of polarization, while TSPC is generated in a construction of polarization.

Main causes of polarization - - - - We assume the following three are the most important:
    (1) Orientation of dipoles
    (2) Transference of space charge
    (3) Injected surface charge from electrode
First we consider the orientation of dipoles - - - - Dipoles in a dielectric substance have structural hindrance and can hardly rotate at a low temperature. They begin to rotate at a higher temperature because the energy of thermal motion makes them rotate over a potential barrier. In the case of TSDC dipoles will go on rotating until they are randomly oriented. The polarization reduces less and less. The displacement current caused by this reduction of polarization gives a peak to a thermogram of TSDC.

In the case of TSPC dipoles will go on rotating until they point to a direction which is parallel to an applied field as nearly as possible. The polarization increases and arrives at some equilibrium value. The displacement current caused by this formation of polarization gives a peak to a thermogram of TSPC.

Secondly we consider the transference of space charge - - - - We adopt hereafter the general name ion instead of electron or hole or ion. An ion in a dielectric can transfer from its equilibrium position to another. Evidently this transference makes a polarization. If ions happen to be trapped, some of them can stay on place slightly apart from their equilibrium positions. The polarization made in this manner will hold on after removal of the field at a low temperature. At a high temperature ions begin to get out of the traps and return to their original position, and the polarization due to them will reduce. The displacement current caused by this reduction of polarization gives a peak to a TSDC.

If all the ions stay in their original equilibrium position at a low temperature, they do not transfer to another position even under an external field because they are in positions where potential energy is a minimum. To get out from these troughs they need some amout of energy which will be acquired at a higher temperature. This polarization will grow up gradually with a temperature rise. A displacement current caused by this increasing polarization gives a peak in a thermogram of a TSPC because the transfer can't go endlessly.

Thirdly we consider the injected surface charge - - - - When a test piece of a dielectric is put between two electrodes and a high field is applied on it, some ions jump to the surface of test piece and are trapped on them. These ions make a kind of polarization. An amount of eneegy is necessary for an ion to get out of a trap. This situation is very similar to the case of space charge. Therefore these injected ions produce a similar peak in a thermogram of a TSDC. On the contrary the injection doesn't contribute to a TSPC because jumps of ions between electrodes and surfaces of a test piece arise and finish immediately after a field applied. Thus this kind of polarization contributes only to a TSDC.

Analytical expression of a peak - - - - Firstly we consider the polarization by orientation of dipoles. Dipoles in a dielectric solid have some structural hindrance. In other words, they must overcome some potential barrier to turn from a stable direction to another. This means that their motion is damped in macroscopic sense and the relaxation time will be as follows:

$$\tau = \tau_0 \exp(\Delta E/k T) \tag{1}$$

Here τ_0 is a constant and ΔE is the height of potential barrier, i.e. the activation energy.

We assume that dipoles are distributed in direction in axial symmetry around a specified direction, i.e. the direction of the polarization in the case of TSDC and that of the applied field in the case of TSPC. Some calculus on statistical mechanics gives us a distribution function of these dipoles.

In the case of TSDC a test piece which has been polarized under a field F at a high temperature T_P is cooled down to an adequate low temperature T_0 and warmed again. Then the distribution will be as follows:

$$f(\theta,t) = \frac{1}{4\pi}\left(1 + \frac{F}{kT}\cos\theta \exp\left(-\int_0^t \frac{dt'}{\tau}\right)\right)$$

where μ is the moment of a dipole. We obtain an average component of dipole moment to the direction of the field, which is now removed, as follows: $(\mu^2 F/3k\, T_p)\exp\left(-\int_0^t dt'/\tau\right)$. Multiplying this average component with the density of dipoles N we get a polarization. The displacement current J can be derived from the polarization by differentiation with respect to time. Ordinarily temperature is controlled to rise at a definite rate, i.e. $T = T_0 + bt$ which makes the formula simpler:

$$J = A\exp\left(-\frac{\Delta E}{kT} - B\int_{T_0}^{T}\exp\left(-\frac{\Delta E}{kT'}\right)dT'\right) \tag{2}$$

where A is $(-N\mu^2 F/3k\, T_p \tau_0)$ and B is $1/b\tau_0$. This formula shows a peak at T_m which has relation with B as follows:

$$B = (\Delta E/k\, T_m^2)\exp(\Delta E/k\, T_m)$$

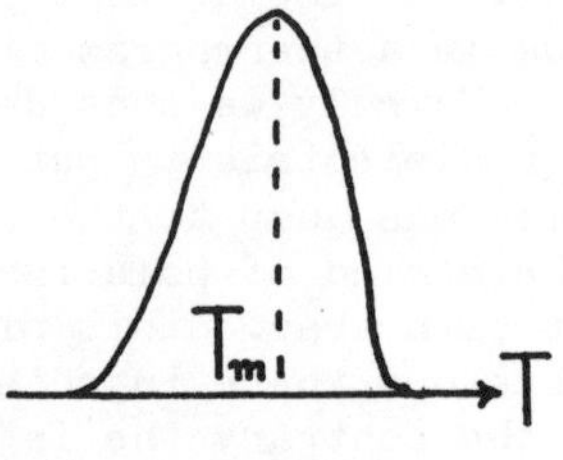

In the case of TSPC a test piece with no polarization is cooled down to an adequate low temperature T_0 and warmed again under a field. The distribution function will be derived as follows:

Fig. 1

$$f(\theta,t) = \frac{1}{4\pi}\left(1 + \frac{\mu F}{k}\cos\theta \exp\left(-\int_0^t \frac{dt'}{\tau}\right)\int_0^t \frac{1}{T\tau}\exp\left(\int_0^{t'}\frac{dt''}{\tau}\right)dt'\right)$$

Multiplying the average component of moment with a density of dipoles, we get an expression of polarization. The displacement current is obtained by differentition of polarization with respect to time:

$$J = \frac{N\mu^2 F}{3k\, T_c \tau_0}\exp\left(-\frac{\Delta E}{kT} - \frac{1}{b\tau_0}\int_{T_0}^{T}\exp\left(-\frac{\Delta E}{kT'}\right)dT'\right)$$

$$\times \left(1 - T_0\int_{T_0}^{T}\exp\left(\frac{1}{b\tau_0}\int_{T_0}^{T'}\exp\left(-\frac{\Delta E}{kT''}\right)dT''\right)\frac{dT'}{T'^2}\right) \tag{3}$$

This equation is similar to eq.(2) except for the last factor which is supposed to be almost equal to unity for a pretty long time.

Secondly we consider the transference of space charge. Some ions are supposed to be fairly free in comparison with other ions which are rigidly bound to their definite position. Space charge is composed of these free ions. Free as they are called, they are not perfectly free. They seem to be trapped in some sort of potential minimum. Some ions may get out of the trap, stray around and get into another vacant trap.

If an ion changes position, a virtual charge of opposite sign appears at the place where the ion was located originally. This pair of charge produces a dipole. If there is a field they will tend to be paralled to the field. They make a kind of polarization which will disappear with the field. But if we cool the specimen sufficiently before the removal of the field, the polarization will survive. We use a simple model. Fig.1 shows three potential minima, the central one is original equilibrium position. An ion will get over the potential barrier E_α with a transition probability which is proportional to $\exp(-E_\alpha/kT)$. The inverse of this probability is a time constant τ_α which has the same character as the one shown in eq.(1). We notice the probabilities of ions staying at A, B and C. They will be derived as functions depending on time. If an ion stay at B it will produce a dipole moment qa in the direction from A to B, where q is the charge of an ion and a is the interval between A and B. That is the same about C. If a field is applied in the direction from C to B the heights of potential barriers will be modi-

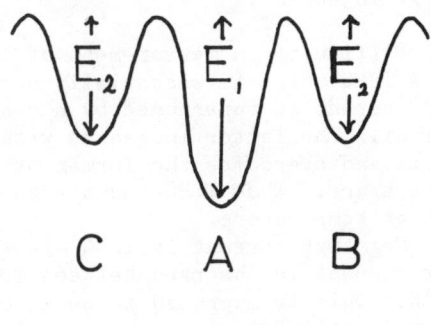

Fig. 2

fied slightly, and transition probability from A to B becomes greater than that from A to C. The ultimate value of the total moment will be 2 μ F/(+ 2)kT. If the specimen is cooled to a low temperature under the same field, this polarization will be maintained. After removal of the field the umbalance of ions will begin to decrease and the polarization will be reduced. This reduction will proceed more and more with temperature rise. After some tedious calculation we obtain an expression for a peak in TSDC as follows:

$$J = \frac{4N\mu^2 F \exp(E_2/kT_p)}{k T_p (\tau_{1p} + 2\tau_{2p})} \exp(-\frac{E_2}{kT} - \frac{1}{b\tau_c}\int_{T_0}^{T} \exp(-\frac{E_2}{kT'})dT') \qquad (4)$$

This is formally similar to that of dipolar deorientation.

We investigated TSPC where the ions transfer gradually under a field with a temperature rise and they bring up a polarization. In this case, we need a somewhat more complicated calculation to find the displacement current similar to the one shown in eq.(2).

Thirdly we consider the injected surface charge - - - - The behavior of trapped ions is very like to that of space charge and the result is similar too, i.e. the depolarization current can be expressed by an equation similar to eq.(4).

Comparison of TSDC with TSPC - - - - According to the investigation stated above, comparison of TSDC with TSPC will enables us to discriminate between polarizations caused by injected surface charge.

A fairly clear result has been obtained with polyethylene-terephthalate. Fig.3 shows thermograms of TSDC and TSPC. They are separated into several peaks.

The most characteristic features, maximum temperature T_m and activation energy E are shown in the Table. Peaks are numbered in order of increasing temperature.

The peaks 2,3,4 and 5 in the TSDC have their correspondents in the TSPC, hence they are considered to be caused by dipole orientation or space charge transfer. While the peak 1 has no correspondent and it is supposed to have its origin in surface charge injection.

Difficulty in measurement of TSPC - - - - A TSPC which is essentially a displacement current is superposed by a conduction current. The latter increases with temperature and overcomes the former at high temperature. Thus TSPC can not be measured at high temperature.

Negative current in TSPC - - - - Negative current is observed between 200°K and 300°K. This is supposed to be an effect of thermal motion of dipole. A dipole may vibrate around a stable direction. This thermal motion may diminish the component of moment in the direction of the field. We have some evidence for this explanation but not sufficient.

REFERENCES

1) For comprehensive description about electret
   M.M.Perlman: ELECTRETS, The Electro-Chemical Soc., 1973

2) P.Debye: Polar Molecules, Reinhold Publishing Co., 1929

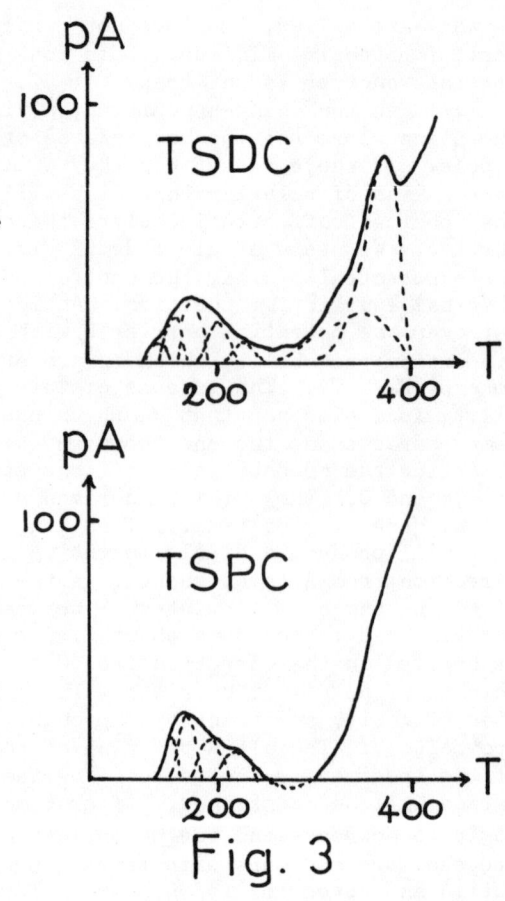

Fig. 3

Table of T_m and E

	TSDC		TSPC	
	T_m(°K)	E(eV)	T_m(°K)	E(eV)
1	105	0.0955	---	----
2	137	0.188	125	0.145
3	161	0.216	163	0.159
4	181	0.352	188	0.396
5	197	0.276	205	0.335
6	351	----	---	----
7	377	0.976	---	----

THERMALLY STIMULATED CURRENT IN POLYETHYLENE

Noriaki Fukushi

Central Research Laboratory, Showa Denko Co., Ltd.

No.2-24-60, Tamagawa, Ota-ku, Tokyo 146, Japan

Today, PVF_2, FEP-Teflon and the other polymer electrets are widely used as acoustic elements and the others. But, the mechanisms of polarization is not completely understood. In this paper, in order to elucidate the nature of charge traps in polymer electrets, polyethylene whose structure and properties have been studied most of all polymer materials was used as samples. And, the effects of branching, thermal treatment, molecular weight, additives, and moreover, charging methods on polyethylene electrets were investigated by using the thermally stimulated current (TSC), and the mechanism, the energy depth, the life time, and the density of charge traps in polyethylene were examined. Consequently, the nature of charge traps in polyethylene changes according to molecular motions, trap centers and trapped charges.

The first problem is concerning the surface charge density (SCD) of polyethylene electrets. In this case, samples used, about 100 μm thick, are high density polyethylene films without additives molded by a hot press method, and the density is 0.951 g/cm^3. This polyethylene has no branching, and the one end of a polymer chain is double-bonded, $=CH_2$, and the another is $-CH_3$.

The experimental procedure is as follows. SCD was measured by an electrostatic induction method using a vibrating reed electrometer, and then, TSC was measured from room temperature to the melting point as a rate of 2°C/min. in the conditions that the measuring electrode is not contact with the surface of electrets.

Corona discharging gave the largest SCD, several hundreds cgs.esu./cm^2, but the charge decayed very rapidly. The TSC spectrum of a corona charged film is shown on Fig.1 (broken line). The TSC peaks appear at about 70°C and 135°C.

The next charging method is "peeling" of a polyethylene film laminated on an aluminum foil. By this procedure, a polyethylene film was negatively charged, and the charge was relatively stable. On Fig.1, the solid line shows the TSC spectrum of such an electret. The TSC peaks appear at about $25^\circ C$, $55^\circ C$, $85^\circ C$ and $135^\circ C$. In this case, SCD of the film was about 40 cgs.esu./cm^2.

The SCD of a thermoelectret with a heterocharge was the smallest, about 10 cgs.esu./cm^2. Fig.2 shows the TSC spectrum of a thermoelectret made by applying the electric field (Ep) of 50 KV/cm at $120^\circ C$. The TSC peaks appear at about $55^\circ C$, $70^\circ C$, $85^\circ C$, $120^\circ C$ and $135^\circ C$.

The next problem is concerning the behavior of electrons injected from the electrode. We found a new method for giving very stable homo-charge electrets. It is the method such as applying a high electric field above 100 KV/cm at room temperature for a short time using sandwitched-type planar electrodes. In this case, SCD of the electrets increased in proportion to Ep, and saturated at about 800 KV/cm. By using this charging method, the following were known. The decay of SCD of polyethylene electrets was decreased with the increase of degree of orientation, and with additives. And, in general, polyethylene is easily negatively charged, but when positively charged by this method, the positive charge decayed rather slowly than the negative charge.

In the following experiments, in order to clarify the nature of charge traps in polyethylene, this charging method was used. In this case, only trapped electrons injected from the evaporated metal-electrode could be discussed. On Table 1, the density and crystallinity of samples used are shown. TSC was measured with no applied field. The direction of TSC was the same as that of current during Ep was applied.

Fig.3 shows the effect of degree of branching on TSC. The TSC peaks appeared at about $70^\circ C$, $105^\circ C$ and $135^\circ C$. The height and temperature of the $70^\circ C$ peak increased with the decrease of degree of branching, in other words, with the increase of degree of crystallinity. On the other hand, the height of the $105^\circ C$ peak increased with the increase of degree of branching.

Fig.4 shows the effect of annealing on TSC. The magnitude of TSC was decreased by annealing, and largely decreased with elevating the annealing temperature.

The effect of an additive on TSC was examined. The TSC peak in samples with an additive appeared at about $105^\circ C$, and the peak height increased in proportion to Ep. The $70^\circ C$ peak which appeared in samples without additives disappeared. In these cases, the energy depth of charge traps at about $70^\circ C$ and $105^\circ C$ obtained from the initial slope of TSC against reciprocal temperature were about 1.5 eV and 0.8 eV, respectively.

Next, the effect of melt-index as the measure of molecular weight on TSC was examined. The samples used, in this case, were made by an inflation method, and contain the same additives. On Table 2, the density and melt-index of samples used are shown. The difference between melt-index affects greatly the morphology of samples. By scanning electron microscopy, spherulites of about micron-meters in size were observed in the samples of which melt-index is larger, but not observed in the samples of which it is smaller. Fig.5 shows the TSC spectra of these samples. The TSC peak of samples of which melt-index is larger appeared at about 70°C, on the other hand, in the samples of which it is smaller, the TSC peak appeared at about 55°C.

In brief, the TSC spectra showed peaks at about 25°C, 40°C, 55°C, 70°C, 85°C, 105°C, 120°C and 135°C, although the height and temperature of these peaks varied with the samples and charging methods. And moreover, the height of the 25°C peak increased by oxidation or chlorination, and those of the 40°C and 55°C peaks increased with additives. Those of the 70°C and 85°C peaks increased with the increase of degree of crystallinity and orientation, and those of the 105°C and 120°C peaks increased with the increase of degree of branching, and with additives. That of the 135°C peak increased with the increase of degree of orientation.

In order to investigate the correlation between molecular motions and the mechanism of charge trapping, the measurement of the temperature dispersion of dynamic elastic modulus and loss modulus, and differential thermal analysis (DTA) were carried out. The height and temperature of loss modulus peak at about 60°C increase with the increase of degree of crystallinity, and therfore, this molecular motions are called "crystalline dispersion". The endothermic behavior of DTA curves above about 105°C is known as the thermal expansion of crystal lattice and the thickening of crystal lamellae.

From the comparison among these facts and the TSC data, charge traps in polyethylene should lie, with each individual energy depth, in the surface region of crystals. Trap centers should be branching points, additives, double-bonds, and catalyst residues, and in oxidized or chlorinated polyethylene, these polar groups, also.

The author wishes to thank Dr. E. Fukada and Dr. T. Takamatsu for very useful assistance.

References

(1) T.Takamatsu, E.Fukada, Kobunshi Kagaku, 29 505 (1972) (in Japanese), Polymer J., 1 101 (1970)
(2) M.M.Perlman, J. Appl. Phys., 41 2365 (1970), J. Elecrochem. Soc., 119 892 (1972)
(3) M.Ieda, et. al., Japan. J. Appl. P., 13 1547 (1974), J. Phys. D., 10 1985 (1977), J. Appl. Phys., 48 2408 (1977)
(4) C.Lacabanne, et. al., Macromol. Chem., 177 1583 (1976)
(5) J.van Turnhout, Thermally Stimulated Discharge of Polymer Electrets, Elsevier, Amsterdam 1973

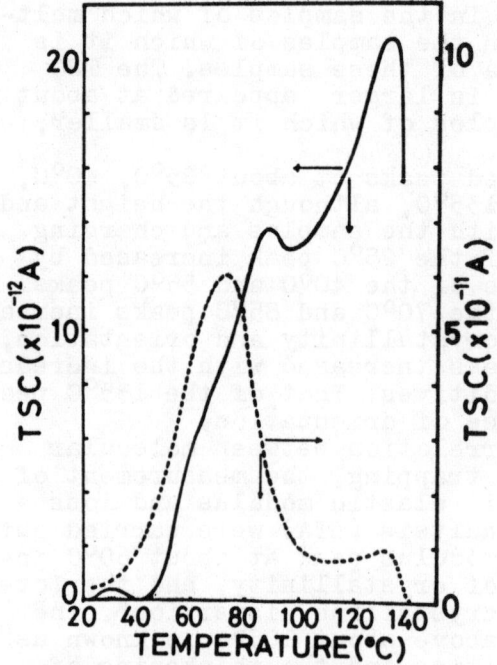

Fig.1. TSC Spectra of HDPE Films Charged by Corona Discharging and Peeling. (broken line: Corona Discharging, solid line: Peeling)

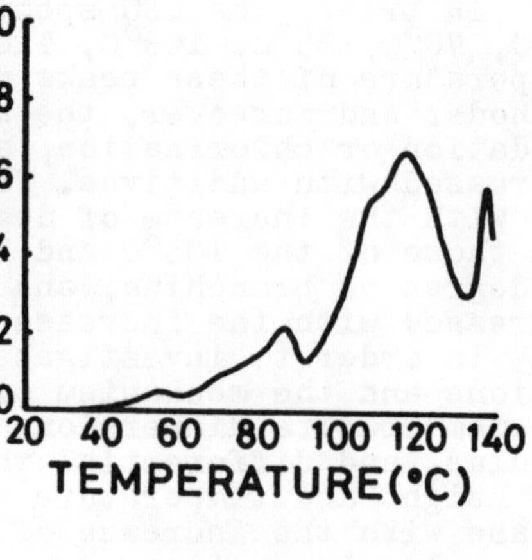

Fig.2. TSC Spectrum of a Thermoelectret.

Table 1. Density and Crystallinity of Samples.

Sample No.	I	II	III	IV	V	VI	VII	VIII	IX	X
Density(g/cm³)	0.951	0.942	0.936	0.932	0.920	0.949	0.971	0.971	0.974	>0.974
Crystallinity(%)	70	65	61	57	50	69	83	83	85	>85

(1) No.I-V: Degrees of Branching are various.
(2) No.VI: an Additive was added to No.I.
(3) Except No.VI, without Additives.
(4) No.I-VI: quenched.
(5) No.VII: slowly cooled.
(6) No.VIII-X: annealed, Annealing Temp., 120°C (No.VIII) 126°C (No.IX), 130°C (No.X)

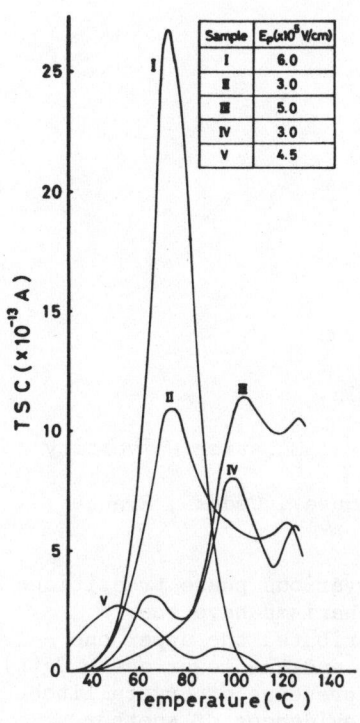

Fig.3. Effect of Degree of Branching on TSC in PE.

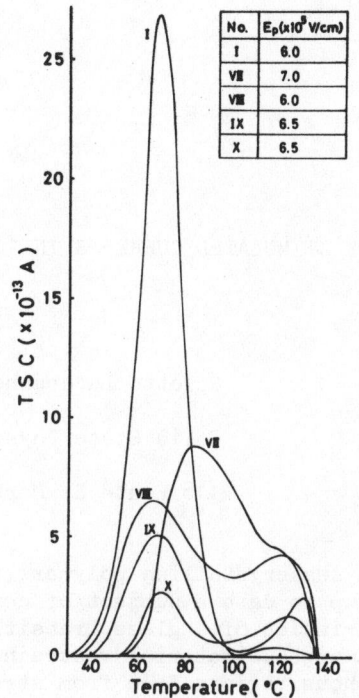

Fig.4. Effect of Thermal Treatment on TSC in HDPE.

Table 2. Density and Melt-Index of Samples.

Sample No.	Density	Melt-Index
A	0.955	5.0
B	0.949	1.3
C	0.954	0.3
D	0.934	0.2

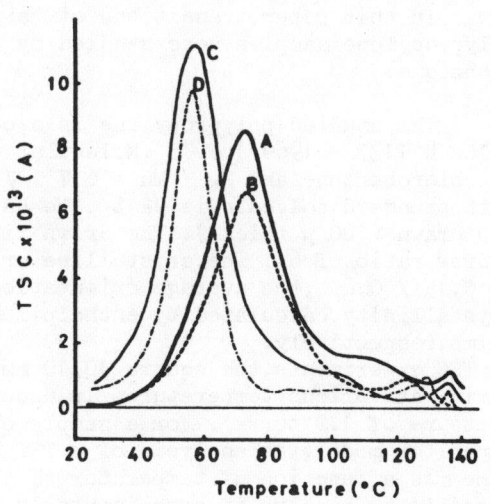

Fig.5. Effect of Melt-Index on TSC in PE.

THERMALLY STIMULATED CURRENTS IN POLYPROPYLENE

Colette Lacabanne and Daniel Chatain

Solid State Physics Laboratory , Paul Sabatier University

118 Route de Narbonne 31077 , Toulouse , Cedex , France

In semicrystalline polymers,the origin of the various phase transitions remains up to date a subject of controversy. Two mechanisms have the characteristics of a glass transition. Boyer (1) attributed the upper one - $T_g(U)$ - to the amorphous material restrained by crystallites and the lower one - $T_g(L)$ - to amorphous regions free from stress caused by the presence of crystallites. Above the glass transition temperature,experimental evidences of another transition are given in the literature (2). Since the temperature region above the glass transition temperature - T_g - is generally considered to represent the liquid state of a polymer,this transition was called the liquid-liquid transition.

In this paper,transitions of varying tacticity and crystallinity polypropylene samples were studied by the Thermally Stimulated Currents - TSC - technique.

The studied polypropylene is a commercial material of Melt - Index (ISO R 1133 - 1969) 5.75 . Molecular weight is $\overline{M}n$ = 60 000 at 140 °C in trichlorobenzene and $\overline{M}w / \overline{M}n$ = 6.7 . Index of isotacticity,obtained by extraction with standard solvents,is 92 % . We have used blown films,undrawn (37 µ thick) and drawn (20 µ thick). The drawn films were prepared in an oven at 130 °C at a draw ratio of 6 . The crystalline orientation function of undrawn / drawn films is 0.15 / 0.80 ; the average orientation function is 0.11 / 0.91 . The degree of crystallinity,calculated by enthalpimetry,is 55 % and 60 % for undrawn and drawn films,respectively.

In TSC experiments,the square 10/10 mm^2 samples were polarized with 500 V for 2 min. at various temperatures T_p,under an atmosphere of dried nitrogen at a pressure of 1.2 torrs . For a sample obeying a series coupled resistance capacitor model,the analysis of a TSC spectrum gives the dielectric relaxation time τ as a function of temperature T . When the TSC spectrum is complex,it is possible to resolve it experimentally,by using the technique of fractionnal polarizations (3).

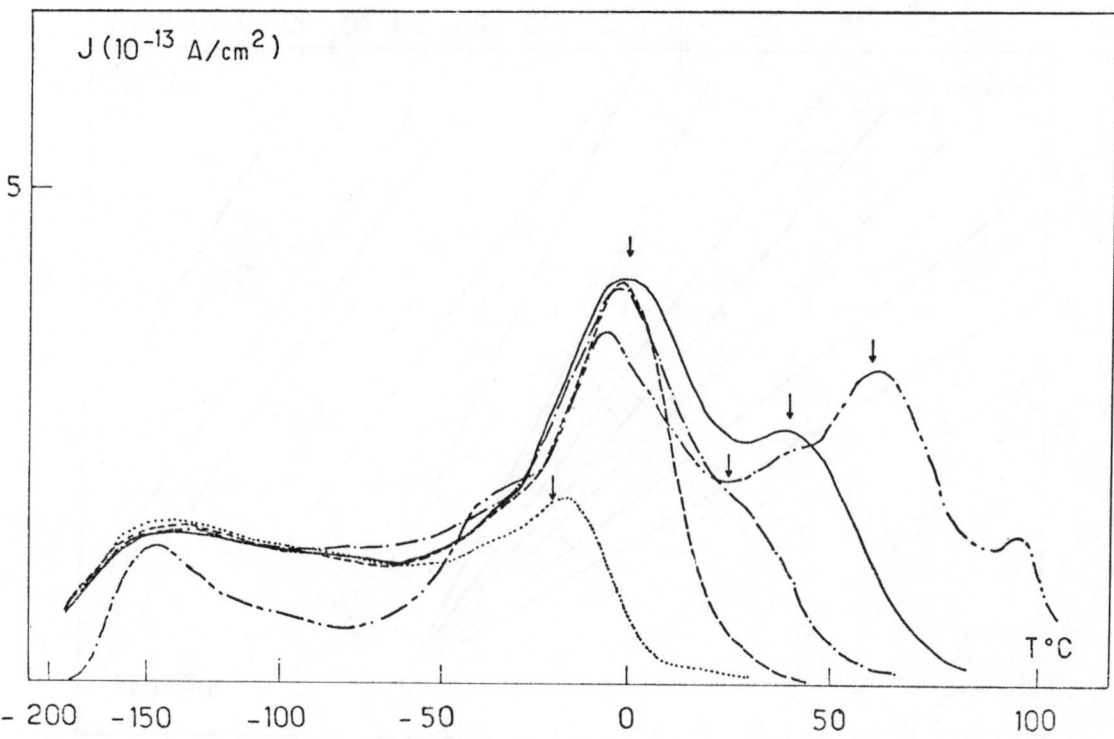

Figure 1 - TSC Spectra of Undrawn Polypropylene.

Figure 1 shows the TSC spectra of undrawn polypropylene for different polarization temperatures. The small arrows indicate the temperature of polarization for each spectra. The chief difference between TSC spectra of undrawn and drawn polypropylene comes from the decreasing in the intensity - J - of the peak observed at - 10 °C with drawing. These complex spectra have been resolved experimentally into elementary processes characterized by a relaxation time. We reported on the same Arrhenius plot the relaxation time versus temperature for the various processes. Figure 2 corresponds to undrawn polypropylene. For each peak, the temperature of the maximum of TSC is indicated by a small arrow. Except for two peaks, the relaxation times follow an Arrhenius equation

$$\tau = \tau_o \exp (U/kT) \tag{1}$$

where τ_o is the pre-exponential factor
and U is the activation energy.

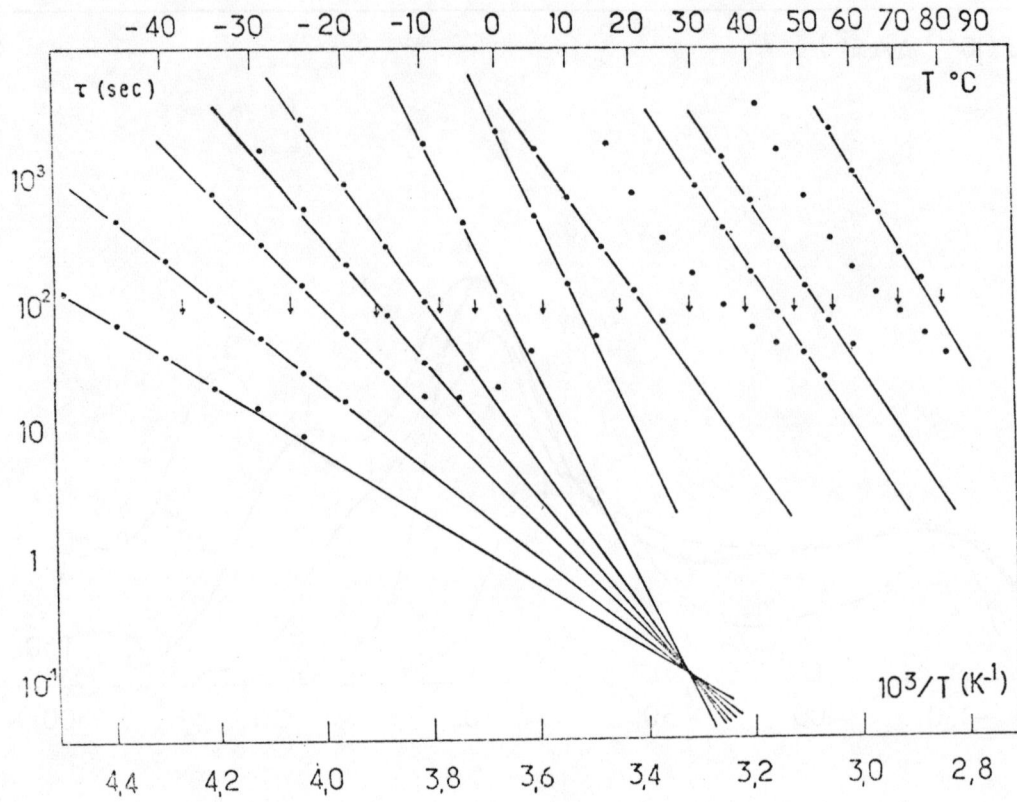

Figure 2 - Arrhenius Plot of Undrawn Polypropylene.

Figure 2 shows an interesting property for the six processes observed at the lowest temperatures, in undrawn polypropylene: The relaxation time of the various mechanisms takes the same value - τc = 0.12 sec. - at a "compensation" temperature - Tc = 26 °C -. Then, the various relaxation times can be written as:

$$\tau = \tau_c \exp\left(\frac{U}{k}\left(T^{-1} - Tc^{-1}\right)\right) \tag{2}$$

For these processes, the pre-exponential factors τ_o and activation energies U defined by Equation 1 are related by:

$$\ln \tau_o = \ln \tau_c - U/kTc \tag{3}$$

The same property is observed in drawn polypropylene but $\tau c = 10^{-4}$ sec. and Tc = 90 °C : τc decreases strongly with drawing while Tc increases.

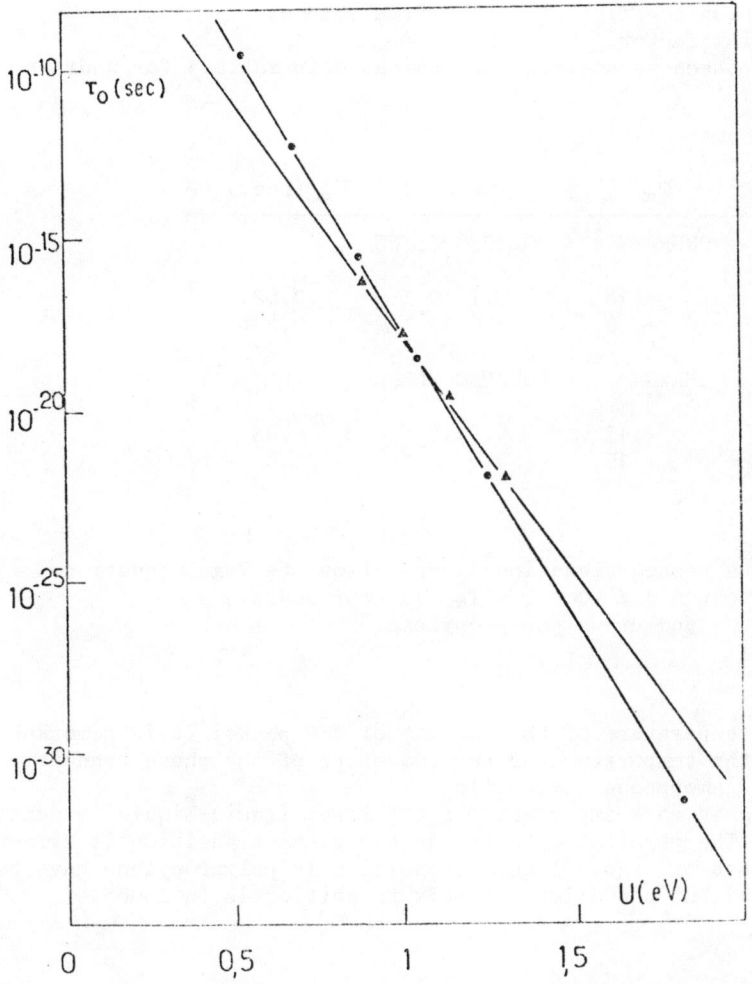

Figure 3 -
τ_o as a function of U for undrawn polypropylene (●) for drawn polypropylene (▲)

On Figure 3, the pre-exponential factor τ_o of the relaxation time is plotted as a function of the activation energy U. Dots are for undrawn polypropylene; triangles are for drawn polypropylene. The lines correspond to Equation 3.

In both undrawn and drawn polypropylene, two TSC peaks have a relaxation time that do not obeys an Arrhenius equation. Then, it exists a "critical" temperature T_∞ which linearizes the variation of $\ln \tau$ versus $1/(T - T_\infty)$. τ is given by the Vogel Equation:

$$\tau = \tau'_o \exp(1/(\alpha(T - T_\infty))) \qquad (4)$$

α is the thermal expansion coefficient of the free volume;
τ'_o is the pre-exponential factor.
The numerical values of these parameters are indicated in Table 1 for undrawn and drawn polypropylene.

T_m (°C)	T_∞ (°C)	α (°C^{-1})	τ'_o (sec.)
UNDRAWN	POLYPROPYLENE		
+ 39	− 18	3.7 10^{-3}	0.62
+ 68	+ 2	2.8 10^{-3}	0.40
DRAWN	POLYPROPYLENE		
+ 28	− 13	5.9 10^{-3}	0.93
+ 55	+ 15	9.9 10^{-3}	8.6

TABLE 1 − TSC peaks whose relaxation times follow the Vogel Equation
$\tau = \tau'_o \exp(1/(\alpha(T - T_\infty)))$ for undrawn
and drawn polypropylene.

T_m indicates the temperature of the maximum of TSC peaks. It is remarkable that T_m coincides with the temperature of the two steps of the phase transition smectic − amorphous and amorphous − monoclinic.
The influence of drawing is more important for the lower liquid-liquid transition than for the upper one: The parallel with the double glass transition is evident (2). Experimental evidence of liquid-liquid transition in polypropylene have been also given by Gillham (4) but the origin of this transition is in doubt.

REFERENCES

(1) − R.F. Boyer,"Encyclopedia of Polymer Science and Technology",Suppl. Vol.II, Boyer and Boyer,Midland,1977,p. 745.
(2) − R.F. Boyer,J. Macromol. Sci.-Phys.B8(3-4),503(1973)
(3) − C. Lacabanne,D. Chatain,J.Guillet,G. Seytre and J.F. May,J. Polym. Sci.-Phys. Ed.13,445(1975)
(4) − J.K. Gillham,AIChE J.20,1066(1974)

THERMALLY STIMULATED CURRENT IN ELONGATED POLYETHYLENE BY GAMMA-IRRADIATION

Shigeru Kobayashi and Kichinosuke Yahagi

Waseda University

3-4-1 Ohkubo, Shinjuku-ku, Tokyo 160, Japan.

I. Introduction

Some change in characteristics of polymer sample has been often introduced to examine the nature of electron trapping sites by means of gamma-ray irradiation, introducing some additives or oxidation.[*1,2] Detrapping process of trapped electrons in polymers is greatly governed by the onset of some molecular motion as first suggested by Nikolski and Buben.[*3] Drawing of polymeric films is expected to be a useful experimental technique to obtain informations about the correlation between the molecular motion and the electric properties because drawing of polymeric films does change the morphology considerably.[*4]

The enhancement of a thermally stimulated current (TSC) peak at 45°C in low density polyethylene film elongated at room temperature was previously reported.[*5] Further experiments on more highly elongated samples have revealed that the TSC peak is not only increased but also decreased for the draw ratios over 200%. It may be considered that the result is explained by the elongation mechanism of polyethylene film.

II. Experimental

Samples used were of inflation film of low density polyethylene (Lexron: density = 0.918 g/cc) and were free from any additives. They were drawn paralell to the machine direction of the inflation at room temperature with a drawing rate of 7 mm/min. Draw ratio was measured around the center of the sample film, and defined as

$$\text{DRAW RATIO} = (1 - l_o)/l_o \times 100 \ [\%]$$

where l_o is the length before the elongation and l is the length after the elongation. Though the sample thickness decreases with drawing, the thickness of the elongated samples was kept constant at 30 μm throughout the experiments by choosing the thickness before the elongation. Silver painted electrodes were adopted and were 12 mm in diameter. The elongated sample was fixed on the electrode assembly as it was maintained to be elongated.   The schematic diagram of the measuring circuit is shown in Fig. 1. All experiments were carried out in dry nitrogen gas at atmospheric pressure.

The elongated sample was irradiated by gamma-rays in air with a dose rate of 500Gy/hr for 2 hours at the excitation temperature of 17°C.   Then the sample was short-circuited under an application of the collecting bias field of 30kV/cm for 2 hours so that the charging current became negligibly small. The collecting bias field was set to be much smaller than the threshold field capable of detecting the TSC peak.*5   The sample was warmed up to 75°C with a warming rate of 1°C/min to measure the TSC.

III.   Results and Discussion

TSC curves obtained from the elongated samples were shown in Fig. 2.   Curves (a) to (e) are for the draw ratios of 0%, 100%, 200%, 300% and 360%, respectively.   It can be seen that three TSC peaks exist at 35°C, 45°C and 60°C in every draw ratio.   These three TSC peaks are labelled P1, P2 and P3, consecutively from low to high temperatures.   None of the peak temperatures is affected by drawing.   Since the P1 and the P2 peaks are in competition with each other for their hights and the peak temperatures especially in the case of 0% elongated condition, they were unable to be well resolved into two by suitable experimental techniques.   The P3 peak

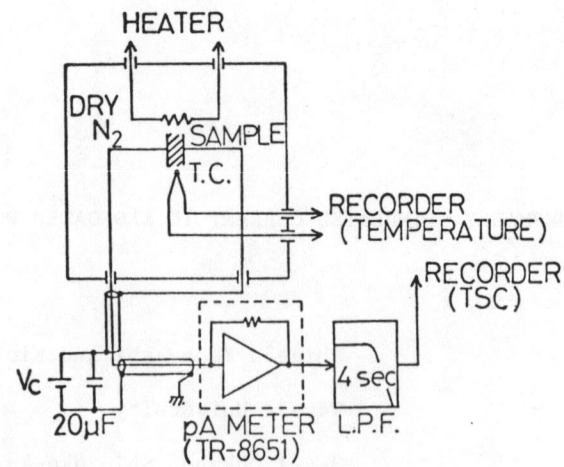

Fig. 1   Schematic diagram of the measuring circuit.

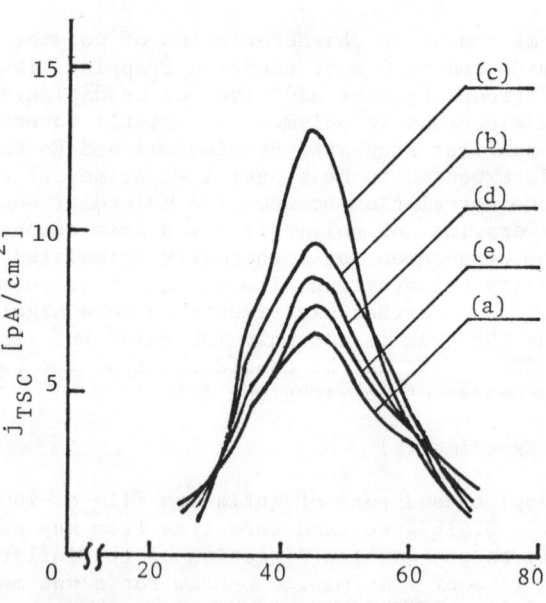

Fig. 2   TSC curves obtained from elongated samples, (a) 0%, (b) 100% (c) 200%, (d) 300%, (e) 360%.

is relatively small, so it is often masked by increasing conduction current and/ or the tail of the lower temperature peak. Thus the nature of the P3 peak was not examanined in detail.

The P1 and the P2 peaks are seemed to increase together with increasing the draw ratio, and they reach a maximum at the draw ratio of 200%. The P1 peak in 200% elongated sample was very often observed as a mere shoulder or disappeared into the lower temperature tail of the P2 peak. It may be considered that it is the P2 peak that is mainly affected by drawing, and that the P1 and the P2 peaks are remaining at their original heights.

It has been raported in the measurement of X-ray long spacing in high density polyethylene elongated at room temperature that the long spacing is increased for the draw ratio up to 40%.*6 It rapidly decreases for the draw ratio from 40% to 120%, and gradually decreases for the draw ratio over 120%. The initial rise in the long spacing has been considered to be due to the elongation of the amorphous parts between the crystal blocks by drawing. It has been thought that the rapid drop in the long spacing is caused by the disruption of the crystalline lamellae, and that the gradual decrease corresponds to the melt-recrystallization due to drawing. The melt-recrystallization is considered to occur as a result of local necking in the bulk of polymer due to the large deformation. The melt-recrystallization is considered to start at the draw ratio of 100 - 200%.

It may be expected that the concentration of the physical deformity which is presumed to be located on surfaces of the crystallites is increased with the disruption of the crystallites, and that it will be decreased from the draw ratio where the melt-recrystallization starts to occur predominantly. Physical deformity has been proposed for effective traps in polyethylene as "cavity trapping". Such cavity traps have been considered to be formed by some suitable molecular configuration.*7 If above process of the elongation is assumed to be applicable to the elongation in low density polyethylene sample, and such physical deformity as is expected to be located on surfaces of the crystallites forms suitable trapping sites responsible for the P2 peak, the increase and the decrease in the height of the P2 peak against the draw ratio may be explained provided that the height of a peak is proportional to the concentration of the responsible trapping sites.

Blake et al. obtained a similar glow curve of thermoluminescence in low density polyethylene to that of Fig. 2 curve (a) with two peaks at 52°C and 71°C.*8 Much higher warming rate in their experiments should shift the peaks towards higher temperatures than those obtained with 1°C/min.*9 The higher warming rate causes lower resolution among the peaks in temperature, and the dual structured-peak such as the P1 and P2 peaks is sometimes observed as a single-structured peak.*10 It is possible to deduce that their peak at 52°C corresponds to the P1 and P2 peaks, and that their peak at 71°C corresponds to the P3 peak.

Blake et al. tentatively assigned the peaks at 52°C and 71°C to be originated in the non-crystalline and the crystalline regions, respectively. If the tentative assignmet is adopted for the location of the responsible trapping sites for the P1 and P2 peaks, the traps may be presumed to be lying in the interfacial regions between the amorphous and the crystalline parts.

Keller and Priest have reported that 90% of terminal vinyl group is located on surfaces of the crystallites in polyethylene.*11 Zurkov et al. have reported on the mechanism of submicrocrack generation in polyethylene that vinyl group will be generated at new chain ends by scissions of molecular chains when the sample is elongated.*12 However, Kobayashi and Yahagi have reported that the increase of vinyl group by drawing cannot be detected in 100% elongated low density polyethylene film in which the TSC peak is considerably enhanced.*5 Thus the trapping sites associated with vinyl group is not likely accepted. Even if the terminal vinyl group is assumed to be the origin of the responsible trapping sites, the reduction in higher draw ratio cannot be clearly explained.

A TSC curve obtained from a non-elongated sample with a slower warming rate of 0.2 °C/min is shown in Fig. 3. This sample was excited by applying a high electric field of 0.67MV/cm at 17°C for 1 hour. The P1 and the P2 peaks are shifted to lower temperatures as expected. However, the peak separation between them became poorer in opposition to the analysis by Randall and Wilkins.*9 They reported that a decrease in warming rate shifts the peak maximum towards lower temperatures and this shift is samller for a peak with higher activation energy. This experimental result suggests the existence of some interaction between the corresponding trapping sites of the P1 and the P2 peaks.

The charge distribution in the bulk and the relaxation of the strain during the measurement of TSC are not taken into consideration, and the discussion about these effects is remained in the future.

IV. Conclusion

Three TSC peaks were found at 35°C, 45°C and 60°C (labelled P1, P2 and P3, respectively) in low density polyethylene film excited by gamma-ray irradiation at 17°C. The P2 peak was considered to be increased in elongated samples for draw ratio up to 200%, and decreased for draw ratio over 200%. While the P1 and the P3 peaks were remaining unchanged. The peak temperatures of these three TSC peaks were not affected by the elongation.

Since the increase of vinyl group due to drawing was not detected 100% elongated sample in which the P2 peak was considerably increased, the trapping sites associated with vinyl group were ruled out for the P2 peak.

The increase and the decrease of the P2 peak with drawing are possibly explained by the increase and the decrease in the concentration of physical deformity on surfaces of the crystallites by the disruption of the crystalline lame-

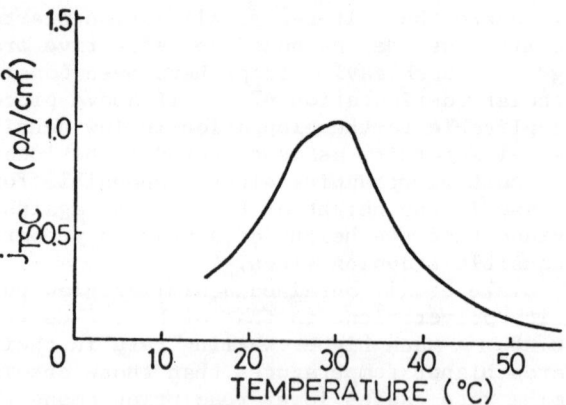

Fig. 3 TSC curve obtained with the slower warming rate of 0.2°C/min.

llae and by the melt-recrystallization, respectively, with the assumption that the surfaces of the lamellae are the location of suitable cavity trapping sites responsible for the P2 peak. It can be deduced from available data on thermoluminescence in low density polyethylene that the P1 and P3 peaks, and the P2 peak are originated in the non-crystalline and the crystalline regions, respectively.

The result of an experiment with a slower warming rate suggests the existence of some interaction between the corresponding trapping sites of the P1 and the P2 peaks.

Acknowledgement

The authors thank Mr. T. Inabu, Mr. H. Ushiyama, Mr. Y. Sada and Mr. M. Suzuki for their help in making the experiments. They also thank Dr. Y. Ohki for help in preparing the paper and Mr. H. Sakamoto for suggestive discussion and encouragement.

References

1) I. Boustead: Proc. Roy. Soc. A-318 (1970) 459.
2) K. Araki, M. Endo and K. Yahagi: Japan. J. appl. Phys. 13 (1974) 1787.
3) V. G. Nikolski & N. Ya. Buben: Proc. Acad. Sci., USSR Phys. Chem. 147 (1960) 896.
4) T. Oda & H. Kawai: Kobunshi (polymer) 18 (1969) 851.
5) S. Kobayashi & K. Yahagi: Japan. J. appl. Phys. 16 (1977) 2053.
6) T. Kawai, Y. Matsumoto, M. Sano and H. Maeda: Zairyo (Materials) 20 (1971) 601.
7) R. H. Partridge: The Radiation Chemistry of Macromolecules (Academic Press, New York and London, 1972), vol. 1, Chap. 10, p.196.
8) A. E. Blake, A. Charlesby and J. T. Randle: J. Polym. Sci., Polym. Lett. Edi. 11 (1973) 156.
9) J. T. Randall & M. H. F. Wilkins: Proc. Roy. Soc. A-184 (1945) 366.
10) I. Boustead & A. Charlesby: Proc. Roy. Soc. A-316 (1970) 291.
11) A. Keller & D. J. Priest: J. Macromol. Sci.-Phys. B2 (1968) 479.
12) S. N. Zurkov, V. A. Zakrevski, V. E. Korsukov and V. S. Kuksento: J. Polym. Sci., Part A-2, 10 (1972) 1509.

CARRIER TRAPPING IN POLYMERS

Masayuki Ieda, Teruyoshi Mizutani, Yasuo Suzuoki
and Masao Shimozato

Department of Electrical Engineering, Nagoya University

Chikusa, Nagoya, Japan

1. Introduction

Carrier trapping phenomena in polymers greatly influence not only their electric conduction process but, through the introduction of space charge, also influence the dielectric breakdown phenomena. Recently, thermally stimulated currents (TSC) and thermoluminescence (TL) from polymers have been widely investigated to clarify the nature of carrier traps. However, their nature has not yet been fully explained even in polyethylene (PE), the most fundamental and practically the most important polymer.

This paper deals with the investigation of carrier traps in PE from x-ray induced TSC.

2. Experimental

The specimens were films of high-density polyethylene (HD-PE, Yukalon PX-40, 25 μm) and low-density polyethylene (LD-PE, Yukalon LK-30, 30 μm) prepared by Mitsubishi Petrochemical Co., Ltd. They are nominally free from additives such as antioxidants. The oxidized specimen was prepared at room temperature by exposure to an ozone atmosphere produced by the gas discharge in oxygen. To prevent the further oxidation, it was degassed for about an hour at around 0.1 Pa. The extent of oxidation was estimated from the infrared absorption of the main carbonyl stretching at about 1720 cm^{-1}. The γ-ray irradiated specimen was irradiated with γ-rays from ^{60}Co at room temperature in an ampoule evacuated to 0.1 Pa. The absorption dose was 17 - 20 Mrad. Under these irradiation conditions, specimens were hardly oxidized (infrared absorption coefficient of the irradiated specimens at about 1720 $cm^{-1} \simeq 0$).

Gold was evaporated to form electrodes of 30 mm diameter on both sides of each specimen.

The experimental procedures were as follows. After the specimen was cooled in a vacuum (∼ 0.1 Pa) to 90 K, it was irradiated with x-rays (35 kV, 20 mA) for 20 minutes under a short circuit condition in order to fill the traps with carriers. After the bias voltage of 270 V was applied to the specimen, it was heated up at a constant rate of 6 K/min. The TSC was detected with a vibrating reed electrometer (Takeda TR-84M). The TSC without preceding x-ray irradiation was also obtained from the same specimen under the same condition but without irradiation. The difference between these two TSC spectra was ascribed to the release of trapped carriers due to irradiation.

3. Results and discussion
3.1. Virgin polyethylene

Figure 1 shows the typical TSC spectra in HD-PE. A large difference between TSC with and without x-ray irradiation suggests an abundance of carrier traps in PE.[1,2] Most of TSC peaks correspond to the onset of molecular motions in PE.[1,2] The P_2 peak corresponds to γ-relaxation, the P_3 peak to β-relaxation in the amorphous region and the P_5 and P_5' peaks to the α-relaxations in the crystalline regions. The dependence of TSC peak temperatures on the amount of vinyl acetate copolymerized with ethylene is just the same as that of the mechanical loss peak temperatures.[1] Furthermore, photo-stimulated detrapping current (PSDC) analysis shows that at 90 K where molecular motions are frozen up traps have much larger depths than around temperatures where molecular motions exist.[3] These suggest that carrier detrapping is a dual process arising from thermal erosion of the trap plus the thermal excitation of carriers from the traps.

Carrier trapping sites are considered[1,2] to be located in the regions where the corresponding molecular motion occurs.[1,2] The traps responsible for the P_2 peak are in lamellar surfaces in HD-PE and in the amorphous region in LD-PE. Those for the P_3 peak lie in the amorphous region and those for the P_5 and P_5' in crystalline regions. Some trapping sites have been reported to be in the amorphous-crystalline boundaries of PE.[4,5] Therefore, the P_4 peak which appears in the temperature range between the α- and β-relaxations may be associated with boundary regions.

The TSC strongly depends on the morphology of the specimen and does not strongly depend on the amount of vinyl acetate. The TSC spectra are essentially the same for different specimens, i.e. HD-PE, LD-PE and ethylene-vinyl acetate copolymers which are considered to contain different kinds of impurities and additives. These suggest that the origin of carrier traps in virgin PE may be physical defects such as cavities formed by local arrangement of molecualr chains.[1] Such traps would be very likely to be influenced

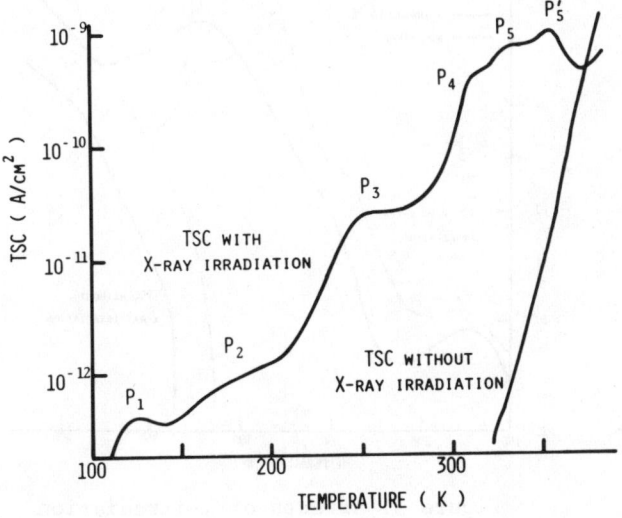

Figure 1. TSC from HD-PE.

by molecular motions.

3.2. Effects of γ-ray irradiation

Figures 2 and 3 show the resultant TSC spectra in γ-irradiated HD-PE and LD-PE, respectively. The TSC with x-ray irradiation has a new TSC peak P_7 in the high temperature region and in the low temperature region it is much smaller than that of PE without γ-ray irradiation. The TSC without x-ray irradiation in γ-irradiated PE is also much smaller than that of virgin PE.

This enhancement of P_7 peak suggests that γ-irradiation in a vacuum introduce deep trapping sites in PE.[7] The decrease of the $P_1 - P_5'$ peaks can be ascribed to the retrapping of detrapped carriers by the deep traps and/or to the decrease of trapped carriers due to the competition with deep traps. The decrease of TSC without x-ray irradiation can also be ascribed to the decrease of carrier mobility and/or the decrease of mobile carrier density due to the increase of deep traps.

Takai et al reported on the basis of photo-stimulated detrapping current analysis that the crosslink due to γ-irradiation in a vacuum acts as a deep trapping site at energies above 3 eV.[6] Consequently, it is considered that such deep traps due to radiation defects both introduce a new TSC peak in the high temperature region and reduce the low temperature TSC and TSC without x-ray irradiation.

3.3. Effects of oxidation

Figures 4 and 5 show the effects of oxidation on TSC spectra in HD-PE and LD-PE. The oxidizing periods were 450 minutes for HD-PE and 720 minutes for

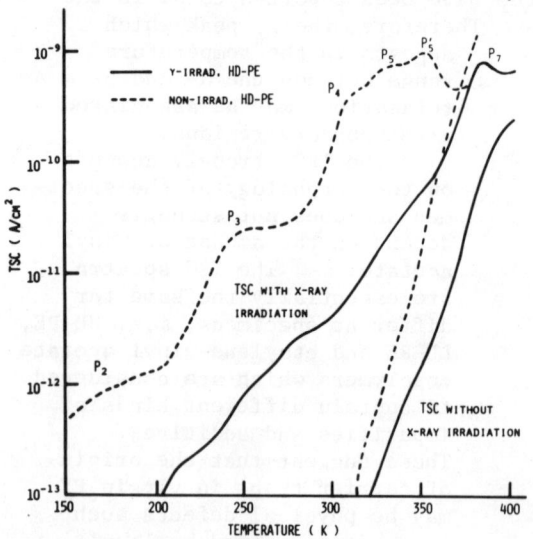

Figure 2. Effetts of γ-irradiation on TSC from HD-PE. (absorption dose = 17 Mrad)

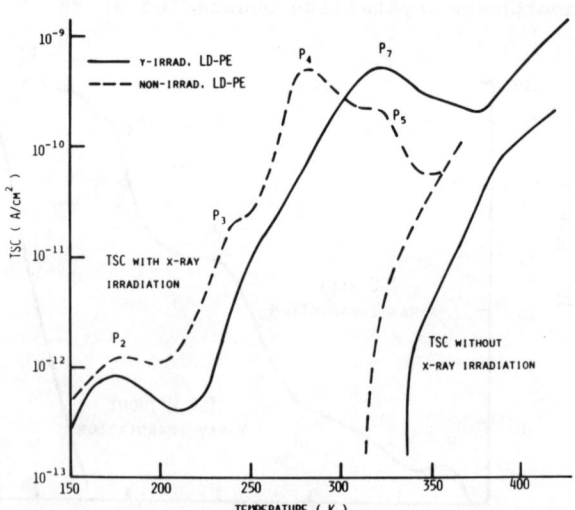

Figure 3. Effects of γ-irradiation on TSC from LD-PE. (absorption dose = 20 Mrad)

LD-PE.

In slightly oxidized HD-PE, a new peak P_3'' appears just below room temperature and is removed by further oxidation.[7] It is considered to be due to unstable intermediates in the oxidation process. It is not refered to any further, as it is only an intermediate effect.

With increasing oxidation TSC decreases in the low temperature region. The TSC peaks around 360 K in oxidized HD-PE and 310 K in oxidized LD-PE, designated P_5'', increase with oxidation. The TSC without x-ray irradiation in oxidized HD-PE is smaller than that in unoxidized HD-PE. These phenomena are very similar to those in γ-ray irradiated PE. Therefore, oxidation seems to introduce deep trapping sites in PE.

However, in the case of LD-PE, oxidation enhances TSC without x-ray irradiation. This phenomenon cannot be explained only by the introduction of deep traps by oxidation. The increase of TSC without x-ray irradiation can also be observed even in HD-PE, when the conductivity is once reduced to a small value by γ-ray irradiation in a vacuum. Therefore, oxidation of PE is considered to have generally two effects, i.e. the introduction of deep traps and the generation of carriers. In HD-PE, the effects of deep traps predominate over those of carrier generation. Contrarily the latter effect dominates the TSC without x-ray irradiation in LD-PE.

Deep trapping sites responsible for the P_5'' peak induced by oxidation are considered to be related to stable oxidation products such as carbonyl groups, as the TSC spectrum of the greatly oxidized specimen is not changed by annealing.[7]

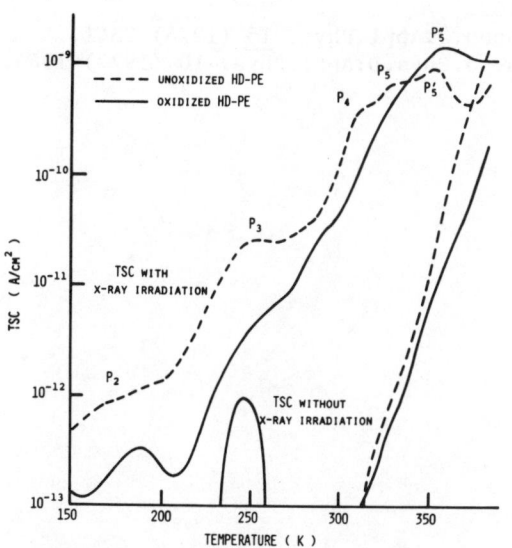

Figure 4. Effects of oxidation on TSC from HD-PE.

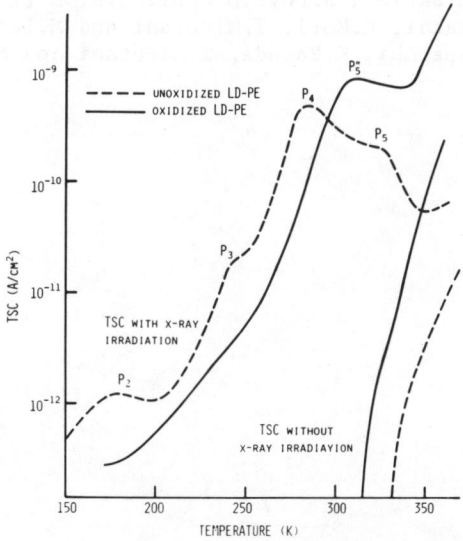

Figure 5. Effects of oxidation on TSC from LD-PE.

4. Conclusions

Carrier traps in PE were investigated by observing x-ray induced TSC. Virgin PE has 5 - 6 kinds of carrier traps. Detrapping of carriers from the traps is closely related with the release of molecular motions and the trapping sites are considered to be located in the regions where the molecular motions occur. The traps responsible for the P_2 peak in HD-PE lie in lamellar surfaces and those for the P_2 peak in LD-PE in amorphous regions. Those for the P_3 peak are in amorphous regions and those for the P_5 and P_5' are associated with the crystalline region. The traps for the P_4 peak may lie in the amorphous-crystalline boundaries.

The detrapping of carriers may be dual process arising from thermal erosion of trapping sites and thermal excitation of carriers.

The origin of these traps may be physical defects such as cavities.

Both γ-irradiation in a vacuum and oxidation introduce deep trapping sites. The origin of these deep traps may be radiation defects such as crosslinks and stable oxidation products such as carbonyl groups.

References

1) T.Mizutani, Y.Suzuoki and M.Ieda: J.Appl.Phys. <u>48</u> (1977) 2408.
2) Y.Suzuoki, K.Yasuda, T.Mizutani and M.Ieda: Japan.J.Appl.Phys. <u>16</u> (1977) 1339.
3) Y.Suzuoki, Y.Takai, T.Mizutani and M.Ieda: Japan.J.Appl.Phys. <u>16</u> (1977) 1929.
4) M.Ieda, M.Kosaki, Y.Yamada and U.Shinohara: 1964 Annual Report:Conference on Electrical Insulation (National Academy of Sciences-National Research Council, Washington, D.C.) p.29.
5) D.K.Davies: J.Phys.D:Appl.Phys. <u>5</u> (1972) 162.
6) Y.Takai, K.Mori, T.Mizutani and M.Ieda: Japan.J.Appl.Phys. <u>15</u> (1976) 2341.
7) Y.Suzuoki, K.Yasuda, T.Mizutani and M.Ieda: J.Phys.D:Appl.Phys. <u>10</u> (1977) 1985.

ELECTRICAL CONDUCTION AND SPACE CHARGE ACCUMULATION

IN POLYETHYLENE TEREPHTHALATE

Keiichi Miyairi	Masayuki Ieda*
Shinshu University	*Nagoya University
Nagano 380, Japan	*Nagoya 464, Japan

1. Introduction

The study of electrical conduction and electrical breakdown in insulating materials continues to attract considerably interest. This interest arises from an increasing application of insulators in solid state electronic devices, power apparatus and power cables, and also because the underlying physical mechanisms responsible for the current have not been uniquely determined.

Recently epoxy resin and other thermally stable high polymers attract the attention to the use for insulating materials at high temperatures. Generally, polymers contain many impurities, additives and imperfections. These impurities and additives are dissociated and contribute to electrical conduction especially at high temperatures, and it is said that ionic conduction is to be dominant at high temperatures whereas electronic one is important at low temperatures. Ionic species accumulate in front of the blocking electrode and around the crystalline region which might result in space charges and affect transient and stationary currents. The transient current decay after the voltage application to the insulators is caused by various ways:(1)dipolar relaxation (2)charge injection leading to trapped space charge effects (3)electrode polarization and (4) tunneling to empty traps. In the case of dipolar relaxation the inherent conductivity may be determined after the decay of polarization current. The second concept states that after the voltage application the carriers injected from the metal electrode accumulate near the injecting electrode and prevent the following injection. The time dependent current corresponds to the dynamics of the charge accumulation. The third concept states that the actual conductivity corresponds to the current flowing immediately after the voltage application. Further, the charge carriers are accumulated in front of the electrode, so a polarization voltage arises with direction opposite to the applied voltage and the current decreases inside the bulk. The electrode polarization occurs in ionic materials. The tunneling model assumes the presence of trap levels in the dielectrics, and electrons are assumed to tunnel into the distributed trapping levels from the metal electrode under dc bias.

We tried to solve the influence of ionic space charge on the electrical conduction by obtaining as much of reliable information as possible by dc conduction and TSC measurement.

2. Experimental method

Currents were measured with polyethylene terephthalate(PET) films by applying constant electric fields and charging currents were monitored for about 15 hours in order to investigate the hetero space charge accumulation in front of the electrode due to the migration of ionic species in the bulk. The procedure for obtaining the TSC curve is shown in Fig.1. Firstly, the sample was set between electrodes in a vessel at atomospheric pressure and thermally treated by raising the temperature up to 250°C to reduce the initial suprious stored charges and to inhibit the change of crystalline state of polymer during the measurement. After this treatment, the temperature was lowered and kept at a polarization temperature T_p. Samples polarized at this temperature by dc stress(E_p) for t_p minutes were cooled, then TSC was observed as the depolarization current in heating of constant rate(β) after removing the dc stress.

3. Results and discussion

The charging current decreased gradually with time at first, and then it almost reached a steady state value after few hours. Stationary currents which were measured at about 15 hours after the application of electric field are presented in Fig.2. The current shows a sublinear characteristics($J \propto V^n$; $0 < n < 1$) and inclines to be saturated in the magnitude of the current at 160°C, while the current at 100°C increases linearly and then increases superlinearly with increase in the electric field.[2] The same nature is demonstrated for polyvinyl chloride by Dr.I.M.Talwar.[3]

The current shows the anomalous time dependence when the polarity of an applied voltage is reversed at a certain time after the voltage application at the temparature higher than 120°C.[2] Figure 3 shows the present experimental result of anomalous time dependent current. A dc charging voltage of 30v was applied for 20 minutes and then the polarity was reversed. Many different values of dc voltage were used for the polarity reversal. The time to the current maximum (t_{max}) from polarity reversal decreases and the magnitude of the maximum current increases with increase in the reversal voltage. These results suggest the possibility of ionic space charge accumulation in front of the electrode and their liberation with polarity reversal. This speculation is also supported by the following discussion. The glass transition temperature of PET is around 80°C and above this temperature segmental motion occurs which results in an increase in the free volume in the bulk. The increase in the free volume would facilitate the motion of an ionic charge carrier. Besides the above fact, a dc conductivity of PET is affected considerably by the use of catalysts: residual catalysts themselves are ionized in the polymer and contribute to dc conduction.[4] From above description, electrical conduction in PET at high temperatures is to be dominated by ionic process and the result shown in Fig.3 can be considered as follows. One of the mechanisms to explain this phenomena is an effective decrease in the density of mobile ions within the sample due to the arrival at the electrode. The second is an effective drop of electric field in the bulk due to the space charge effect of drifting ions to the electrode. From the sake of simplicity, electric field in the bulk assumed to be equal to the mean applied field, and the change in current with respect to time may be subjected to the carrier density. Thus only as a first approximation disregarding the space charge effect, carrier transit time might be correlated to t_{max} in anomalous currents, and we get the relation between mobility μ and t_{max} as expressed in eq.(1)

$$t_{max} = L^2/[\mu \times V] \qquad (1)$$

Hence, ion mobilities are estimated to be $3 \times 10^{-12} cm^2/V \cdot sec$ and $3 \times 10^{-11} cm^2/V \cdot sec$ at 140°C and 160°C respectively.

We examined the TSC at high temperatures in order to obtain the information of electrode polarization. Four peaks which are defined P_1, P_2, P_3 and P_4 are present in Fig.4.[5] TSC P_1 peak which is observed around the glass transition temperature shows no thickness dependence of the sample. This P_1 peak has been already explained to be due to the depolarization current of dipoles.[6] TSC P_2 and P_3 peaks show somewhat complex features. A large P_4 peak observed at the temperature higher than 160°C shows the sharp thickness dependence, the result of which excludes the possibility of depolarization of dipoles, since the depolarization of dipoles takes place throughout the bulk and the depolarization current is considered to be independent of the sample thickness. This sharp thickness dependence suggests that P_4 peak might be associated with macroscopic space charge accumulation and it can be expected that P_4 peak is likely to be due to the electrode polarization and has a close relation with the anomalous current and the sublinear characteristics in the electrical conduction. The electrode polarization is achieved by carrier transport through the bulk to the electrode, therefore, the inherent conductivity of the bulk should affect the characteristics of the electrode polarization. Then we examined the TSC of different samples of the same density and maybe the same crystallinity 54% according to Cobbs.[7] The peak current of PET-A is considerably higher than that of PET-B, while in other peaks P_1 and P_2, the difference is not observed. The difference of the magnitude of the peak current is probably due to the difference of the impurities, such as the residual catalysts or additives. Unfortunately, the impurities could not be found by X-ray micro analyser. Therefore the influence of the impurities on TSC P_4 peak remains uncertain.

In general, four transitions may take place in saturated linear polyesters: glass transition, cold crystallization, premelt crystallization and melting. These transition temperatures are reported to be around 80°C, 120°C, 220°C and 265°C respectively in PET.[6] The temperatures of TSC P_1, P_2 and P_4 peaks are likely to be consistent with those of glass transition, cold crystallization and premelt crystallization respectively. This fact suggests that the molecular chain motion is closely related to the depolarization of dipoles and the liberation of the trapped carriers.

The P_4 peak shifts toward the higher temperatures with increase in the polarization voltage or with increase in the polarization time. The activation energy obtained by initial rise method is shown in Fig.6. It increases with the polarization field E_p at a given polarization time, and also increases with t_p at a given polarization field E_p. Activation energies obtained under different conditions of E_p and t_p fall on a curve and they seem to be expressed as a function of $E_p \times t_p$. As $E_p \times t_p$ is closely related to macroscopic displacement of charge carriers in the bulk, the increase in the activation energy, the detailes of this mechanism should be further investigated, might be attributed to the increase in the accumulation of ionic space charges. Then a possible mechanism responsible for TSC P_4 peak could be due to the liberation of ionic space charges near the blocking electrode.

4. Conclusions

It has been shown that nonlinear characteristics in electrical conduction at high temperatures would have a close relation with the accumulation of ionic space charges near the electrode. The details of the mechanism are now being investigated and will be reported in the future.

5. Acknowledgement

This work was partly supported by the scientific expenditure from the Ministry of Education, Japan.

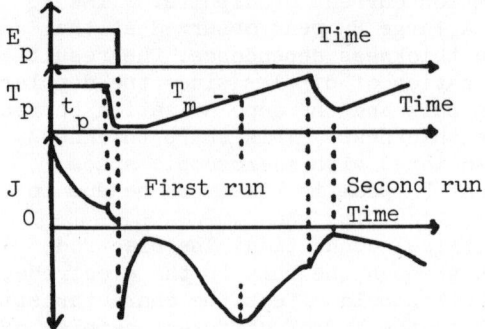

Fig.1 The procedure for obtaining TSC

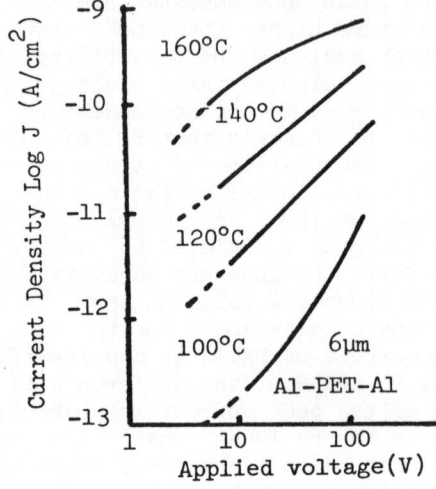

Fig.2 Leakage current in PET at high temperatures. Currents were measured at ca. 15 hours after the voltage application.

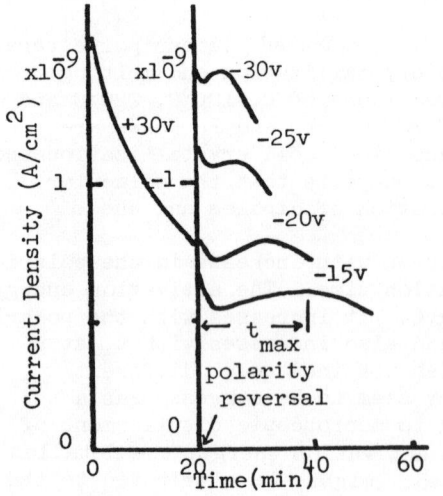

Fig.3 Anomalous currents after the polarity reversal.

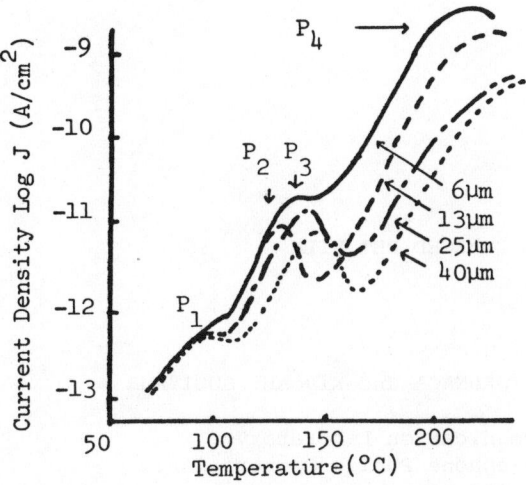

Fig.4 The general features of TSC observed above room temperatures.

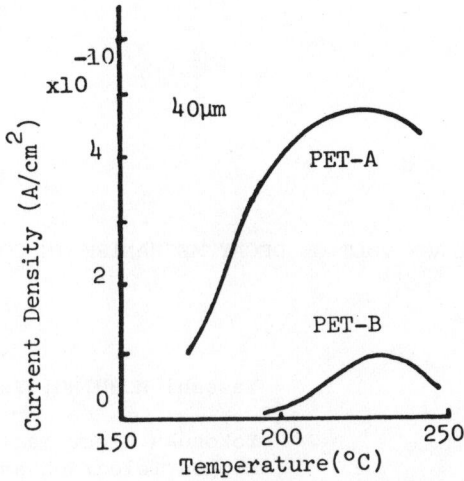

Fig.5 TSC of different samples. These curves show thermally cleaned P_4 peaks.

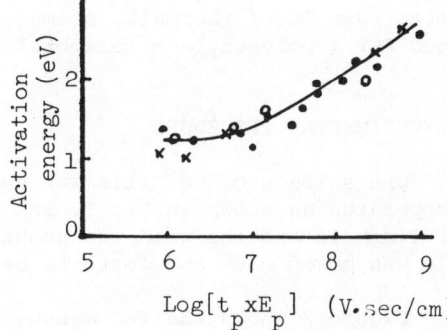

Fig.6 The variation of activation energy of P_4 peak with $t_p \times E_p$.
- ○ : $E_p = 3.85 \times 10^3$ v/cm
- ● : $E_p = 1.54 \times 10^4$ v/cm
- × : $E_p = 7.69 \times 10^4$ v/cm

References

(1) H.J.Wintel: J.Non-Cryst.Solids 15 (1974)471
(2) K.Miyairi and M.Ieda: J.Inst.Elect. Engrs. 96A (1976)25[in Japanese]
(3) I.M.Talwar and D.R.Bhawalkar: Ind. J.Pure & Appl.Phys. 7 (1969)681
(4) H.Sasabe, K.Sawamura, S.Saito and K.Yoda: Polym.J. 2 (1971)518
(5) K.Miyairi and I.Yanagisawa: Japan J.appl.Phys. 17 (1978)593
(6) T.Hino: Japan J.appl.Phys. 12 (1973)611
(7) W.H.Cobbs and R.L.Burton: J.Polym. Sci. 10 (1952)275

SURFACE VOLTAGE DECAY MECHANISM IN CORONA CHARGED PET FILM

Yasushi HOSHINO, Yukio TOKUNAGA and Kiyoshi SUGIYAMA

Yokosuka Electrical Communication Laboratory,
Nippon Telegraph and Telephone Public Corporation,
Yokosuka-shi, Kanagawa, 238, Japan.

I. INTRODUCTION

Recently, there has been a lot of interest in charge storage phenomena in insulating films. Charge storage phenomena have practical importance in electrets, electrostatic recording, and electrophotography. To investigate these phenomena, a contactless TSC (thermally stimulated current) technique is applied on corona charged PET (polyethylene telephtalate) film.

II. EXPERIMENTAL PROCEDURE

One surface of PET film was charged by the corona discharge method using the apparatus as shown in Fig.1. Its grid had a 0.4 mm pitch mesh. The gap between the PET sample and the grid was about 1 mm. During applying the voltage, the PET sample was moved back and forth to be charged uniformly.

Figure 2 shows the TSC measuring apparatus. The temperature, which was programmed by a motor driven potentiometer and regulated by an electronic PID controller, was raised at the constant rate of 5 °C/min. from 30 to 160 °C. The TSC measurement involving an air gap between the sample and the upper electrode, so-called contactless TSC, is a good method to measure surface charge decay because the PET sample charge cannot move to the upper electrode.

III. RESULTS AND DISCUSSION

Figure 3 shows the dependence of the TSC curves on the PET states, i.e. drawn state and crystalline state. This figure shows that the peak magnitude in the lower temperature region is greatly affected by the PET crystallinity.

The PET film TSC under various experimental conditions were measured. The experimental conditions varied were the contact condition between the PET film and the lower electrode, the surface voltage polarity, and the film thickness. Table 1 shows the TSC peak temperature under the different experimental conditions. In samples No.1 through 5, surface voltages and sample thicknesses are the same, but the lower electrode contact condition and the surface voltage polarity are different. They are arranged in order of second peak temperature.

Firstly, the Table shows that the position of the lower temperature peak is unchanged by the sample contact condition. Thus it can be concluded that it originates from the bulk properties of the PET film. Moreover, this peak does not appear in the crystalline PET as shown in Fig.3, so it is considered that it is due to the orientation of the dipoles in the PET amorphous regions. The position of the lower temperature peak corresponds to that of the PET thermoelectret lower temperature peak[1], indicating that it corresponds to heterocharges in the PET thermoelectret.

Secondly, it shows that the position of the high temperature peak is affected by the sample contact condition. This means that the peak is not dependent on the PET bulk properties, i.e. conductance or charge carrier mobility. It is also found that the position of the peak is almost entirely independent of the contact condition, when the sample is negatively charged. This result suggests that this peak is controlled by the charge injection mechanism from the negatively corona charged surface state.[2],[3]

Thirdly, the results shown in the Table 1 also suggest that the main carriers in the PET are electrons. When the sample is positively charged, the peak temperature depends largely on the lower electrode contact condition. This is interpreted as electrons being easily injected from the silver evaporated electrode. This interpretation is supported by the reported photoconductivity data.[4]

To further investigate the charge injection mechanism from the negatively charged PET surface, the surface voltage dependance of the TSC was measured. Fig.4 shows the dependence of the TSC on initial surface voltage. It is found that the second peak shifts to a lower temperature as the surface voltage is increased. This means that charge injection from the corona charged surface has electric field dependence. Usually, electric field dependence is explained by the Poole-Frenkel theory. So, it can be considered that the libration of the charge trapped by the surface state is enhanced by a Poole-Frenkel effect. The transit time that it takes for librated charge to move through the PET film can be neglected, because compared to the surface state trapping time, the transit time is very small, as shown in the lack of difference between the 25 and 50 μm samples in the Table 1.

Using the above assumptions, these phenomena are analyzed as follows. The model is shown in Fig.5. Charge is only distrbuted near the PET surface. X-axis is taken in the direction to the lower electrode with the surface level as zero. Using these conditions, the electric field E at X can be written with a good approximation.

$$E(X) = \frac{1}{\varepsilon_p} \int_0^x e \cdot n(X) \, dx , \qquad (1)$$

where ε_p is the PET dielectric constant, e is the electron charge, and $n(X)$ is the trapped charge density at X. On the assumption of a Poole-Frenkel effect, the decaying rate equation of the trapped charge can be expressed by

$$\frac{dn(X)}{dt} = \frac{1}{\tau_0} \exp\left(-\frac{\varepsilon_{acti} - \alpha E^{\frac{1}{2}}}{kT}\right) n(X) , \qquad (2)$$

where τ_0 and α are constant, ε_{acti} is the activation energy, k is Boltzmann's constant and T is the absolute temperature. Integrating Eqs.(1) and (2), the differential equation about surface charge σ is expressed as follows,

$$\frac{d\sigma}{dt} = -\frac{\varepsilon_p}{\tau_0} \exp\left(-\frac{\varepsilon_{acti}}{kT}\right) \left\{ \frac{2}{\beta^2} \left(\beta \left(\frac{\sigma}{\varepsilon_p}\right)^{\frac{1}{2}} \exp\left(\beta\left(\frac{\sigma}{\varepsilon_p}\right)^{\frac{1}{2}}\right) - \exp\left(\beta\left(\frac{\sigma}{\varepsilon_p}\right)^{\frac{1}{2}}\right) + 1 \right) \right\} , \qquad (3)$$

where $\beta = \frac{\alpha}{kT}$. (4)

The TSC I can be calculated to be

$$I = -\frac{\varepsilon_{air}}{\varepsilon_p} \cdot \frac{\ell}{d} \cdot \frac{d\sigma}{dt} . \qquad (5)$$

Numerical calculations on Eqs.(4) and (5) were carried out using the Runge-Kutta method. Figure 6 shows the dependence of the peak temperature on surface voltage. The points in Fig. 6 are experimentally obtained data and the solid line is the numerically calculated data. The values used in these numerical calculations are ε_{acti} =1.2eV, τ_0 =2.0 × 10^{-13} sec, and α =8.3 × 10^{-6} eV m$^{\frac{1}{2}}$ V$^{-\frac{1}{2}}$. The value α obtained by this fitting is smaller than the value predicted by the Poole-Frenkel theory. Reasons are considered as follows.
(1) The charge is assumed to be continuously distributed at the PET surface.
(2) Activation energy distribution is ignored.
(3) The trapping site may be rather a neutral site[5] than the charged site as assumed in the normal Poole-Frenkel theory.

IV. SUMMARY

Contactless TSC was measured on corona charged PET film. Two peaks were obtained in this TSC experiment. The lower temperature peak is due to dipole orientation. The origin of the higher temperature peak is investigated. When the sample is negatively charged, it is found that negative charge is injected from the charged surface state, and that the injection has electric field dependence. The electric field dependence is explained by Poole-Frenkel type injection efficiency enhancement.

Acknowledgement

The authors wish to express their thanks to Dr. Ichizo Nakano for his encouragement and support on the work.

References

(1) K.Sasaki and S.Fujino: Kobunshi Ronbunshu [in Japanese] 32(1975)349.

(2) A.C.Lilly,Jr., R.M.Henderson and P.S.Sharp: J.appl.Phys. 41(1970)2001.

(3) E.A.Baum, T.J.Lewis and R.Toomer: J.Phys.D 10(1977)487.

(4) Y.Takai, T.Osawa, T.Mizutani and M.Ieda: J.Polymer Sience (Polymer Phys. Ed.) 15(1977)945.

(5) C.G.Garton and N.Parkman: Proc. IEE 123(1976)271.

Table 1. EXPERIMENTAL CONDITIONS AND PEAK TEMPERATURES

Sample No.	Experimental Condition			Peak Temperature	
	Counter Electrode Condition	Charging Voltage V	Sample Thickness μm	Lower °C	Higher °C
1	blocking contact	+600	25	90.4	133.2
2	blocking contact	-600	25	90.9	127.1
3	silver evaporated condition	-600	25	91.7	126.1
4	positively charged condition	-600	25	90.5	124.6
5	silver evaporated condition	+600	25	90.9	120.3
6	silver evaporated condition	-1200	50	90.4	126.1

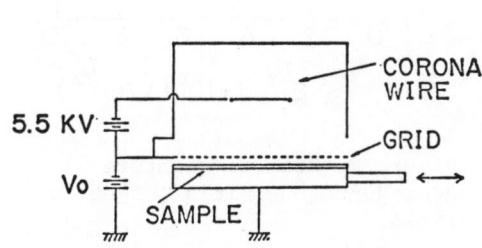

Figure 1. Corona discharge apparatus

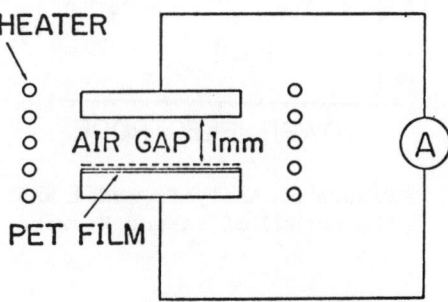

Figure 2. TSC measuring apparatus

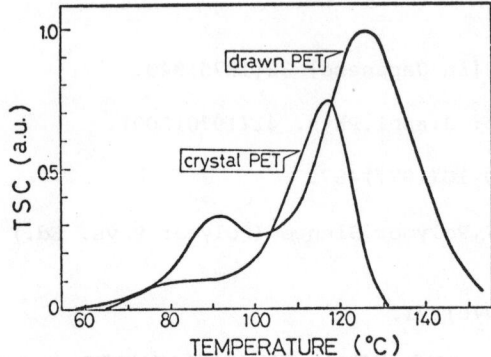

Figure 3. TSC dependence on PET state.

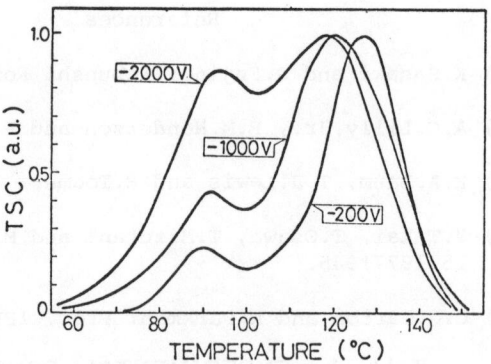

Figure 4. TSC dependence on initial surface voltage. Sample thickness is 25 μm.

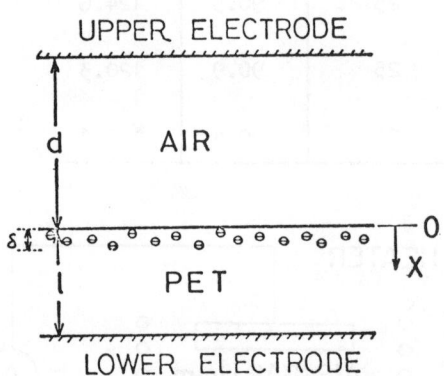

Figure 5. Analysis model and the method of taking X-axis.

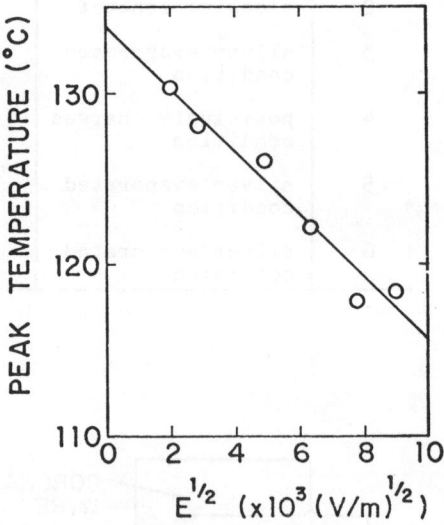

Figure 6. Peak temperature dependence on initial electric field.

THERMALLY STIMULATED SURFACE POTENTIAL OF HIGH DENSITY

POLYETHYLENE FILM CHARGED BY CORONA IRRADIATION

Mitsuru Matsui and Norio Murasaki

Tokyo University of Agriculture and Technology

Nakamachi, Koganei, Tokyo 184, Japan

1. Introduction

Recently, the measurement of thermally stimulated surface potential (TSSP) has been proposed as a good approach to investigate behaviors of space charge in dielectric materials [1,2]. In such TSSP measurements, corona irradiation is, sometimes, used to charge the sample. The sample charged by such a method has some surface charges deposited on the surface as well as some space charges injected into the bulk, and furthermore there is no quantitative co-relation between both the charges. Therefore, the TSSP of such sample is affected by the behavior of both the charges, and it gives a complicated result.

This paper describes the TSSP of high density polyethylene (HDPE) film charged by corona irradiation, whose surface was wiped with a methanol-wetted tissue paper in order to remove the surface charge, prior to the measurement. This method will be applicable to observe properties of electret having no electrode.

2. Experimental Procedure

The sample used in the experiment was a HDPE film of 50 μm thickness. Test pieces of 100 x 100 mm^2 were cut out from it and aluminum metal was evaporated on the bottom surface of them as an electrode of 90 mm diameter disk.

As shown in Fig.1-(a) sample S is put on a hot plate with silicone grease for good heat conduction, and the bare surface of the sample is exposed to the corona discharge from the needle point N_p, at the charging temperature of 363 K for the charging period of 2 minutes. The surface potential of the sample exposed to the corona discharge is controled to be at -4.2 KV, with a biased grid electrode G covering the sample film. The needle point was connected to the high voltage source of -8 KV. After the charging period, the sample-holding hot plate was quickly cooled by a stream of water. When the temperature of sample fell to 298 K, the irradiation was stopped, and furthermore it was cooled down to room temperature, then the surface of the sample was wiped with a methanol-wetted tissue paper. Hereinafter we will call the methanol wiping treatment as the wiping treatment.

After stored for some minutes t_r, the sample was heated up with a constant temperature-rising rate of 3.3 deg/min for the TSSP measurement. The TSSP was measured with a field mill set at 10 mm over the surface of the sample, as shown in Fig.1-(b).

While, the samples which had the same electrodes on both the surfaces were prepared, and they were charged by directly applying a dc bias voltage to the electrodes, under the same charging conditions of temperature, period and voltage that they were subjected to the corona irradiation. The TSSP and also thermally stimulated current (TSC) of them were measured. The surface of these sample was not wiped with methanol after the charging. In the following, the former sample is called as the corona charged sample and the latter as the field injected sample.

3. Results and Discussion

A typical curve of the relation between conduction current and applied voltage obtained from the sample used for the field injection, at the charging temperature of 363 K, is shown in Fig.2. The conduction current increased significantly around the charging voltage. It is suggested that the increase is due to the charge injection from the electrode into the bulk. Furthermore, it is suggested that the charge injected at the high temperature has been frozen within the bulk of the sample as the sample is cooled down to room temperature.

A model of sample charged with unipolar charge is shown in Fig.3. As seen in Fig.3-(a), the sample charged by corona irradiation has some space charge Q injected into the bulk as well as some surface charge Q_s deposited on the surface. Furthermore, there is no relation between the density of both the charges. In this case, there is no way to know the density of both the charges through a simple method. However, since the wiping treatment makes a short circuit connection between the both surfaces of the sample through methanol, the surface induces only the charge density corresponding to that of the space charge distributed in the bulk. The surface charge density Q_{s1} is determined by the spatial distribution of the space charge and the position x_0 of the zero-field-plane in the bulk, as written in Fig.3. Consequently, the same situation of charge distribution as on the field injected sample connected between both the electrodes with a short circuit connection is established by the wiping treatment on the corona charged sample.

The TSSP curves of the corona charged sample obtained after such wiping treatment are shown with solid lines in Fig.4. Dotted lines show the TSSP curves obtained from the field injected sample. These TSSP curves rose with increasing temperature and the each showed a peak. The profile of TSSP curve obtained from two kinds of the samples agreed, in spite of the difference of the charging method. From this agreement, it seems that the species of charge contributing to the TSSP in both the samples are the same kind, it is likely electrons.

While, the TSSP curves obtained from the sample in which no space charge was built up, are shown in Fig.5. These curves were obtained after the sample was charged up to various values of the surface potential by irradiating the corona at room temperature, and they showed no peak.

Figure 6 shows the TSC curves obtained from the field injected sample. Curve (a) was obtained without the collecting bias voltage; $V_c = 0$. Curves (b) and (c) were obtained with $V_c = -450$ and $+450$ V, respectively. In this figure, positive sign of TSC means the direction of the charging current, ie. the current flows in the direction to the counter electrode from the injecting electrode through the bulk of the sample, and negative sign means the direction of the discharging current. On curve (a), two peaks appeared at the temperature of 324 and 350 K, and the each temperature agreed with the temperature region in which the TSSP rose and fell, respectively. The negative TSC of curve (a) indicates that the charges released from the traps drift to the injecting electrode, ie. the backward TSC. The TSC of curve (b) accelerated to the back side with the collecting bias voltage showed two peaks, a small increment and a large peak, as well as curve (a). However, on curve (c), no peak appeared in the lower temperature region since the backward current giving the small increment would be masked by the large forward current accelerated with the collecting bias voltage.

The energy depth of the trap corresponding to the each peak on the backward TSC was estimated from an Arrhenius plot of the TSC measured by using thermal-cleaning method, as 0.9 eV for the shallow trap giving the small increment and 1.5 eV for the deep trap giving the large peak, respectively.

According to Hino and Kaneko [2], the TSSP can be observed only in the fast-retrapping case, as far as the unipolar charge is drifted along to Poisson's field formed by the space charge in the bulk. In the present case, the contribution of the dipolar orientation and the diffusion of charge are negligible to raise the TSSP, since the sample used, HDPE, is a non-polar polymer, and since the density of free carrier is very small in such good insulator. Thus, it is suggested that the TSSP well describes the behavior of the charges released from the traps with increasing temperature of the sample.

From the result mentioned above, we may propose a model to explain the TSSP obtained from HDPE film charged by corona irradiation, as follows. It is reasonable to presume that the density of space charge trapped in the deep trap decreases gradually from the surface into the bulk, as seen on a dotted line in Fig.7, and the main part of the space charge trapped in the shallow trap is distributed near the surface, as seen on a dushed line. After the wiping treatment, the charges released from the traps, with increasing temperature of the sample, are drifted to the surface and/or to the bottom along to Poisson's field in the bulk. At first, the shallow traps release the charges in the lower temperature region. They are drifted to the surface due to the field formed in the left hand side of the zero-field-plane, and they raise the surface potential of the same sign with that of the injected charge. As a result, the field strength in the left hand side of the zero-field-plane decreases. As the temperature rises in the higher temperature region, the surface potential falls down due to the drift of the charges released from the deep traps to the bottom of film.

4. Summary

The results and the discussion are summarized as follows.
(1) The TSSP of the corona charged film can be observed as same as that of the film having the electrodes on the both surfaces, by using the wiping treatment.
(2) The corona charged sample of HDPE showed the same TSSP profile as that of the field injected sample. The agreement in the profile suggests that the same species of charge contributes to the TSSP in both the sample, in spite of the difference of the charging method.
(3) The surface potential during the TSSP of HDPE rises due to the drift of charges released from the shallow traps and falls down due to the drift of charges released from the deep traps.

References

[1] M.M. Perlman; J. Electrochem. Soc., **119**, 892 (1972)
[2] T. Hino and F. Kaneko; Trans. I.E.E. Japan, **96-A**, 309 (1976)

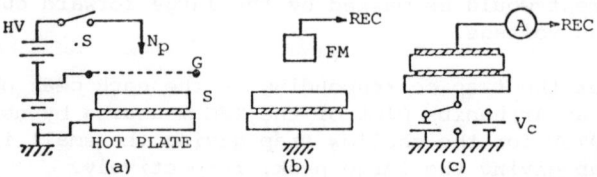

Fig.1 Schematic diagram of experiment; corona irradiation (a), TSSP (b) and TSC (c) measurement

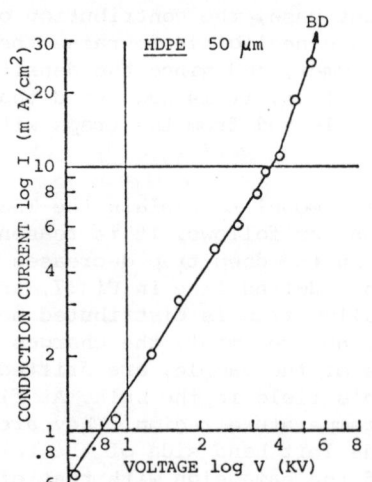

Fig.2 V-I characteristic curve of the sample at 363 K

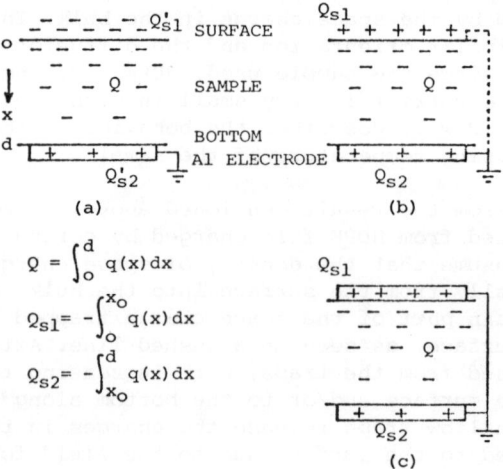

Fig.3 A model of sample charged with unipolar charge; before the wiping treatment (a), after the wiping treatment (b) and the field injected sample (c)

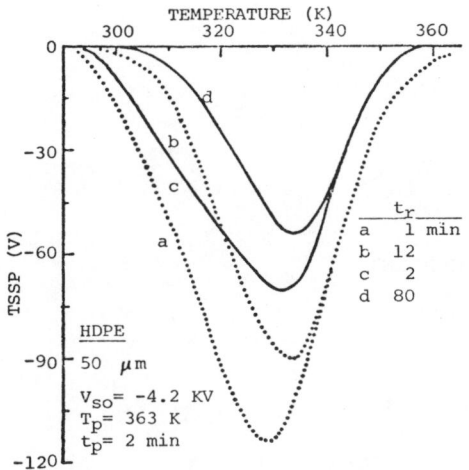

Fig.4 TSSP curves of corona charged sample

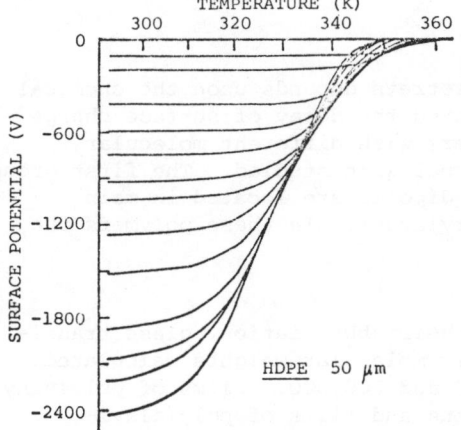

Fig.5 TSSP due to only surface charge

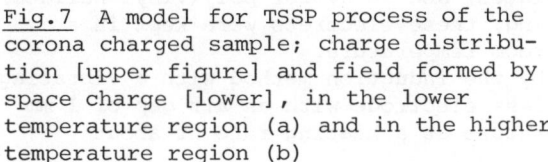

Fig.7 A model for TSSP process of the corona charged sample; charge distribution [upper figure] and field formed by space charge [lower], in the lower temperature region (a) and in the higher temperature region (b)

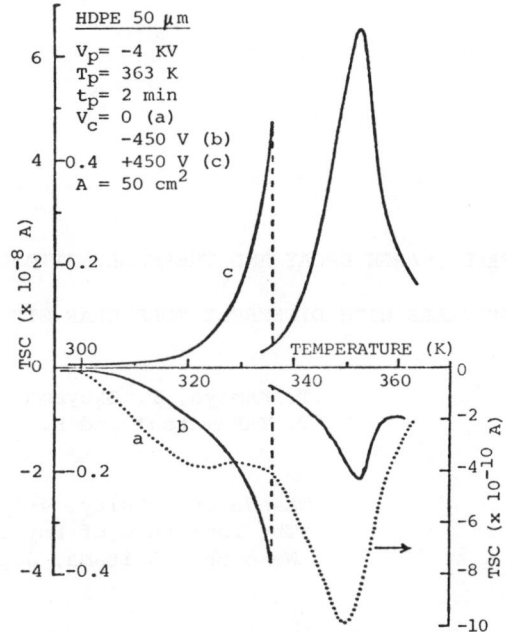

Fig.6 TSC of field injected sample; $V_c = 0$ (a), $V_c = -450$ V (b) and $+450$ V (c)

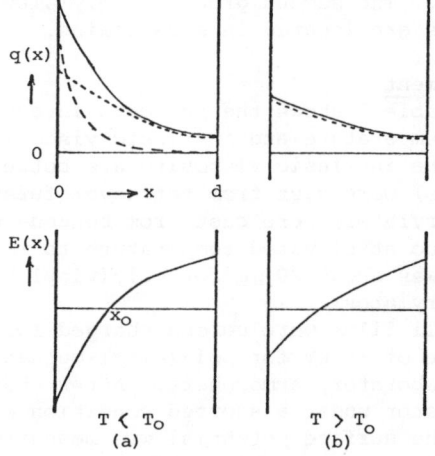

SURFACE CHARGE DECAY AND THERMALLY STIMULATED CURRENT

IN POLYMERS WITH DIFFERENT MOLECULAR STRUCTURES

T. Kamiya, T. Okuyama, I. Shinohara,
T. Takamatsu* and E. Fukada*

Waseda University, 4-170 Nishi-okubo, Shinjuku-ku, Tokyo, Japan
*The Institute of Physical and Chemical Research, 2-1, Hirosawa
Wako-shi, Saitama, Japan

Introduction

The decay of surface charge of polymer electrets depends upon the chemical structure of the polymer.[1-3] We have investigated the decay of surface charge and the thermally stimulated current for polymers with different molecular structures. Two different series of polar polymer were studied. The first group is poly(vinyl halides). In these polymers the dipoles are located in main chains. The second group is poly(alkyl-methacrylates). In these polymers the dipoles are located in side chains.

Experiment

Table 1 shows the polymers investigated, their abbreviation, glass transition temperature and intrinsic viscosity. Their molecular weights calculated from the intrinsic viscosity are between 50,000 and 200,000. Films of poly(vinyl halides) were cast from tetrahydrofuran solutions and films of poly(alkyl-methacrylates) were cast from benzene or toluene solutions. All films were kept in vacuo at elevated temperature to remove the solvent. The thickness of the films was about 20 μm for poly(vinyl halides) and about 100 μm for poly(alkyl-methacrylates).

All films were corona charged for 30 sec. by a needle electrode with a voltage of +6 kV for poly(alkyl-methacrylates) and -6 kV for poly(vinyl halides) in a laboratory atmosphere. After charging, these electrets were kept in a desiccator under a shorted condition at room temperature.

The surface potential was measured by the induction method with a vibrating reed electrometer. The TSC by a contactless electrode with an air gap of 1 mm was measured from 0°C to softening temperature at a heating rate of about 2°C/min.

Table 1 Glass transition temperature and intrinsic viscosity of polymers

Polymer	Abb.	T_g (°C)	[η]
poly(vinyl bromide)	PVBr	80	0.19 (in THF at 25°C)
poly(vinyl chloride)	PVC	55	1.37 (in THF at 25°C)
poly(vinylidene fluoride)	PVDF	-35	1.09 (in DMAA at 25°C)
poly(methyl methacrylate)	PMMA	95	0.53 (in benzen at 30°C)
poly(ethyl methacrylate)	PEMA	60	0.32 (in benzen at 30°C)
poly(n-butyl methacrylate)	PnBMA	22	0.46 (in benzen at 30°C)
poly(iso-propyl methacrylate)	PiPMA	90	0.73 (in benzen at 30°C)
poly(tert-butyl methacrylate)	PtBMA	120	0.64 (in n-butylacetate at 25°C)

Result and Discussion

Figure 1 shows the decay of the surface potential for the corona charged electrets of poly(vinyl halides). The ordinate is the logarithm of the ratio of the surface potential E to the initial potential E_0 and the abscissa is the storage time in a unit of day. As shown in Fig. 1, the decay is fairly fast. In 20 days, the surface potential for PVBr decreased about one order of magnitude. The glass transition temperature is 80°C for PVBr, 55°C for PVC and -35°C for PVDF. The decay is faster for the polymer with the lower glass transition temperature, which shows the higher conductivity at room temperature.

Figure 2 shows the temperature dependence of the dielectric loss ε" measured at 10 Hz for poly(vinyl halides). The peaks due to the glass transition are seen at 75°C for PVBr, 60°C for PVC, and -35°C for PVDF. The magnitude of ε" is largest for PVDF and smallest for PVBr at room temperature.

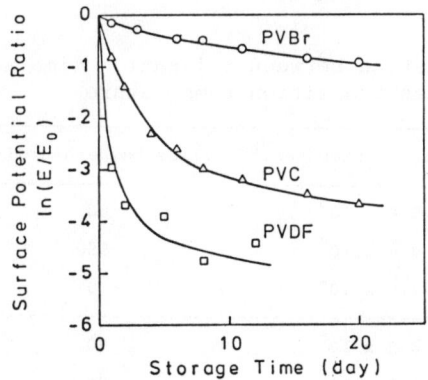

Fig. 1 Relation between surface potential ratio and storage time for corona charged poly(vinyl halides) electrets.

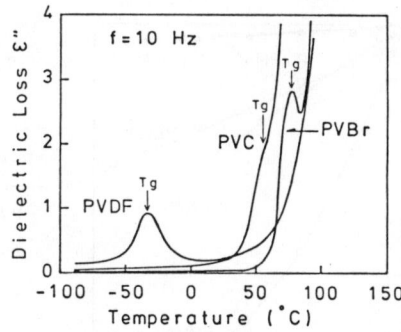

Fig. 2 Temperature dependence of dielectric loss ε" measured at 10 Hz for poly(vinyl halides).

Figure 3 shows the TSC curve with a contactless electrode for poly(vinyl halides). The sign of the current corresponds to the homocharge. The peak temperatures are below glass transition temperatures shown in Fig. 2. By these reasons, the TSC peaks in this figure should be caused by the neutralization of surface homocharge due to the increased conductivity.

Figure 4 shows the decay of the surface potential for corona charged electrets of poly(alkyl-methacrylate). The ordinate is the logarithm of the ratio of the potential E to the initial potential E_o. The abscissa is the storage time in a unit of day. Compared to poly(vinyl halide), the decay of surface potential is very slow. Particularly for poly(isopropyl-methacrylate) PiPMA and poly(tertiary-butyl methacrylate) PtBMA, the decay of the surface potential is only about 10 % in 20 days after corona charging. From the initial part of the decay curves, the relaxation time of charge, that is, the reciprocal of the decay constant, was calculated. The results are shown in Table 2.

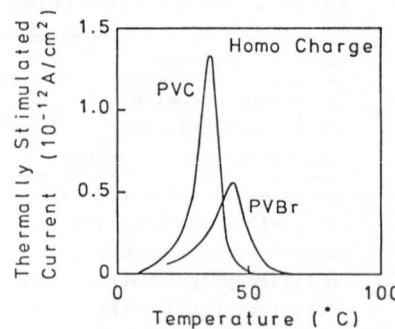

Fig. 3 TSC spectra of corona charged poly(vinyl halides) electrets, with a contactless electrode method. $\beta \approx 2°C/min$.

Table 2 shows the relaxation time for the decay of charge for polymers investigated in this study. Compared to the poly(vinyl halides), the relaxation time for poly(alkyl-methacrylates) is larger by one or two orders of magnitude. Among poly(alkyl-methacrylates), when the end group of side chain becomes more bulky, the relaxation time becomes more prolonged.

The glass transition temperatures determined by dielectric measurement are also shown in the table. With the introduction of longer hydrocarbons, the glass transition temperature decreased and then increased. This indicates the change of molecular flexibility and interactions between polymers.

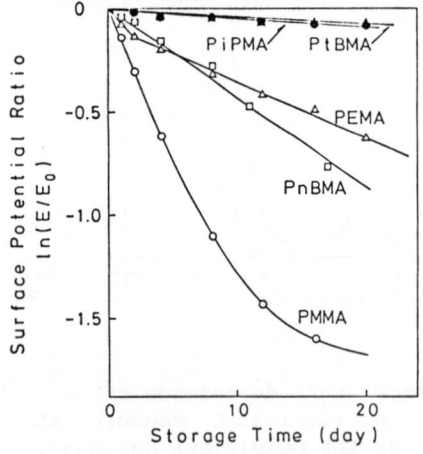

Fig. 4 Relation between surface potential ratio and storage time for corona charged poly(alkyl-methacrylates) electrets

Table 2 Relation between relaxation time and glass transition temperature

polymer	relaxation time (sec.)	glass temperature (°C)
PVBr	5.6×10^5	80
PVC	9.4×10^4	55
PVDF	2.9×10^4	-35
PMMA	6.2×10^5	90
PEMA	1.2×10^6	70
PnBMA	2.4×10^6	40
PiPMA	4.6×10^6	90
PtBMA	6.4×10^6	120

Figure 5 shows the temperature dependence of dielectric loss ε" measured at 10 Hz for poly(alkyl-methacrylates). In general three peaks of ε" were observed against temperature. Taking PMMA for example, the peak at about 90°C shows the glass temperature relaxation, the peak at about 20°C shows the side chain relaxation, and the peak at about -100°C shows the local mode relaxation.

The peaks for the highest temperatures should indicate the glass transition temperature. These temperatures were listed in Table 2. With the substitution of methyl group in side chains PMMA by longer hydrocarbons, the glass temperature first decreased and then increased. While the peak temperature for side chain relaxation only increased gradually.

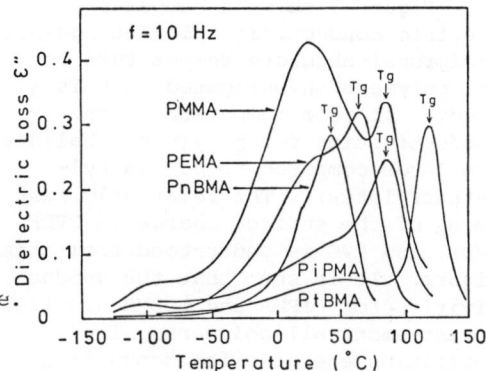

Fig. 5 Temperature dependence of dielectric loss measured at 10 Hz for poly(alkyl-methacrylates).

Figure 6 shows the TSC curves for some poly(alkyl-methacrylates). For each polymer two peaks are observed both with the sign of homocharge. First we shall consider the origin of the peaks at the lower temperature, at 55°C for PMMA and PiPMA and at 75°C for PtBMA. The glass temperatures determined by dielectric measurement at 10 Hz as shown in Fig. 5 are 90°C for PMMA and PiPMA and 120°C for PtBMA. Since the equivalent measuring frequency for TSC is approximately 10^{-3} Hz, the glass relaxation temperature could shift to the lower temperature.[4]
Therefore the TSC peaks at 55°C and 75°C should be related to the thermal motion of dipoles of molecules. However, the sign of current shows that of homocharge, which is opposite to the depolarization of oriented dipoles. We assume that the TSC peaks at this temperature range should be caused by the polarization current of dipoles. The large amount of homocharge on the surfaces which was injected by corona charging is persistent in this temperature range and gives an internal field across the film, which causes the orientation of thermally defrozen dipoles. The TSC peaks at the lower temperature are polarization currents due to orientation of dipoles at glass transition temperature.

The TSC peaks at higher temperatures should be the neutralizing current of trapped homocharge due to an increased conductivity at higher temperature. The magnitude of peaks should be reduced by the depolarization current with an opposite sign due to the disorientation of dipoles which were aligned at the lower temperature.

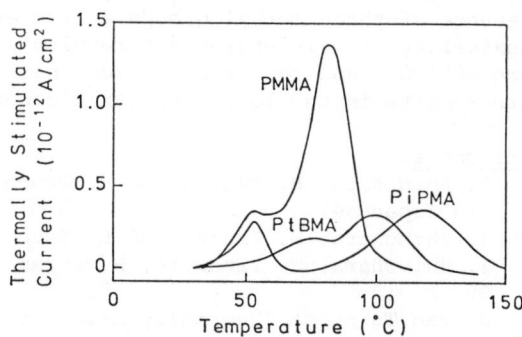

Fig. 6 TSC spectra of poly(alkyl-methacrylates electrets with a contactless electrode method. $\beta \approx 2°C/min$.

Figure 7 shows a comparison of electric conductivity plotted against reciprocal absolute temperature for all polymers investigated in this study. At room temperature, the conductivities for poly(vinyl halides) are large compared to poly(alkyl-methacrylates). The relatively fast decay of the surface charge in PVDF, PVBr, and PVC is understood from this figure. It is seen that the conductivities for PiPMA and PtBMA are the lowest among all polymers. The activation energies for conduction are about 38 kcal/mol for these two polymers which are higher than about 30 kcal/mol or less for other polymers. This lowest conductivity for PiPMA and PtBMA will account for the smallest decay of surface charge on these polymers.

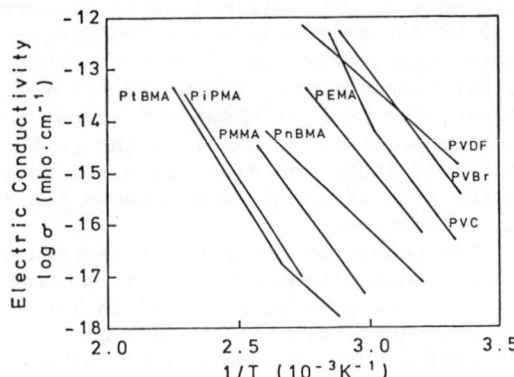

Fig. 7 Relation between electric conductivity and reciprocal absolute temperature (1/T).

The correlation between the chemical structure of polymer and the electric conductivity is not well understood. However, the substitution of the methyl group of side chain in PMMA by longer hydrocarbons gave an interesting effect on the electrical property of polymer molecules. Taking PMMA as a reference substance, substitution of methyl group by ethyl group (C_2H_5) and butyl group (C_4H_9) decreased the glass transition temperature and increased the conductivity. Substitution of methyl group by iso-propyl group ($CH(CH_3)_2$) or tertiary-butyl group ($C(CH_3)_3$) increased the glass transition temperature and decreased the conductivity. In contrast to the glass transition temperature, the temperature for side chain relaxation only increased with the substitution of the methyl group by longer hydrocarbons.

PtBMA has the best stability of surface charge. It is observed that the presence of three methyl groups at the end of polar side chains decreases the flexibility of main chains and results in the highest glass transition temperature of 120°C and the highest side chain relaxation temperature of 50°C, which in turn results in the lowest electric conductivity.

References
1) I. Shinohara, H. Tsuiki and E. Tsuchida, Kogyo Kagaku Zasshi 73, (1970) 1460 (in Japanese)
2) I. Shinohara, H. Tsuiki and E. Tsuchida, ibid., 73, (1970) 1436 (in Japanese)
3) I. Shinohara, F. Yamamoto, H. Anzai and S. Endo, J. Electrostatics 2, (1976) 99
4) J. van Turnhout,"Thermally Stimulated Discharge of Polymer Electrets" Thesis Leiden (1972) T.N.O.

Thermally Stimulated Current of Cellulose Electrets

Atsushi SAWATARI

Faculty of Agriculture, Tokyo University of Agriculture and Technology
Fuchu-shi, Tokyo, Japan 183

1 Introduction

Electret thermal analysis technique has been shown to be excellently sensitive as a method for determining the temperature where one of various transitions or initiation of molecular motion in either crystalline or amorphous region of polymers takes place. Sharp peaks of depolarization current appear at temperatures that correspond to transition temperatures determined by viscoelastic, dielectric and other methods. This method can detect with a high sensitivity the temperature at which certain changes in polymer properties occur and even transition temperature to which the ordinary viscoelastic methods are insensitive[1].

By using the ordinary methods, it is relatively difficult to determine temperatures at which certain transition in cellulose occur, because, in many cases, peaks for celluloses obtained by the ordinary methods appear to be broad and not clear[2,3]. Baum has applied this technique to regenerated cellulose film in the temperature range from -88°C to 47°C, and found peaks of depolarization current at -58°C and 29°C[4]. But, attempts to assign the peaks observed were not made.

In this study, thermally stimulated current (TSC) spectra were examined for celluloses, and furthermore, in order to elucidate the molecular motion of cellulose, some attempts to assign each peak observed in TSC spectra were undertaken.

2 Experimental

For obtaining TSC spectra for dry celluloses, experiments were carried out under the following procedure. Regenerated cellulose (saponified cellulose diacetate), Whatman cellulose, Avi-cel, triphenylmethyl (trityl) cellulose, corn-cob xylan and pine glucomannan were prepared in powdered state. Each of powders prepared was pressed into disc. In addition, film specimens were also prepared for regenerated cellulose (containing no glycols), cellulose triacetate and trityl cellulose. In a word, cellulose derivatives and other materials were used as model compounds. Silver electrodes were vapor deposited directly onto the specimens.

Electret thermal analyser (made by TOYOSEIKI SEISAKU-SHO LTD.) was used to obtain TSC spectra for the discoidal specimen from each prepared powder and film specimen.   Samples were dried for two hours at 120°C, 10^{-2} Torr. in the measuring cell and then dry nitrogen gas was introduced into the evacuated cell.

In that case, the nitrogen gas was severely dried by passing through the concentrated sulfuric acid, over phosphorous pentoxide and potassium hydroxide.

Unless otherwise indicated, polarizing temperature was fixed at 130°C, polarizing electric field was 30 kV/cm, polarizing time was 30 min. and heating rate was 2 (°C/min.).

3 Results and Discussion

TSC spectra obtained are shown in Figs.1——5.   For cellulose powder, a broad peak at about -130°C, a peak at about -70°C and a peak at about 60°C were observed (Fig.1).   Peaks at similar temperatures were also observed for regenerated cellulose film (Fig.2).   In these cases, the peaks observed were seen more clearly than those by ordinary methods.

From Fig.1 and Fig.2, it is shown that the current value of the peak at about -70°C was reduced by drying under a severe condition as compared with that under a mild condition.   Further, as shown in Fig.2, when some moisture was given to the regenerated cellulose film at 5.5 %RH, the current value at about -70°C was increased.   Therefore, appearance of the peak at about -70°C is considered to be related to small quantity of adsorbed moisture.

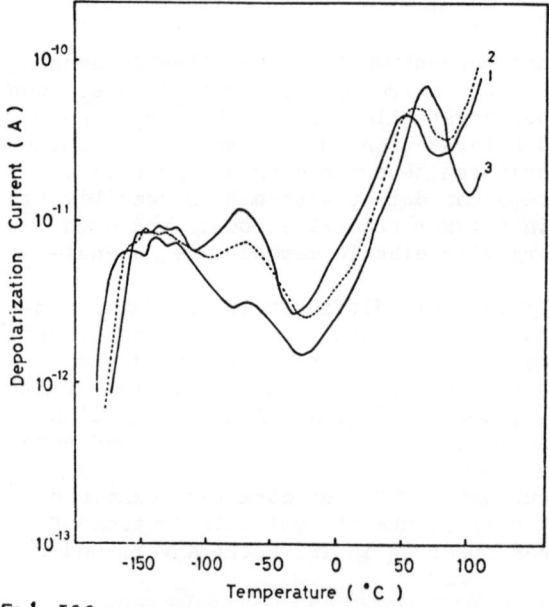

Fig.1 TSC spectra for regenerated cellulose powder (Saponified acetate),
Ep (Polarizing electric field) = 3 x 10 KV/cm , Tp (Polarizing temperature) = 130°C , tp (Polarizing time) = 30min, β(Heating rate)= 2°C/min.

1 : Sample was dried at 10^{-2} Torr and 20°C for 20 min. in the measuring cell, and then dried air was introduced into the evacuated cell.
2 : Sample was dried at 10^{-2} Torr and 120°C for 2hrs. in the measuring cell, and then dried air was introduced into the evacuated cell.
3 : Sample was dried at 10^{-2} Torr and 120°C for 2 hrs. in the measuring cell, and then severely dried nitrogen was introduced into the evacuated cell.

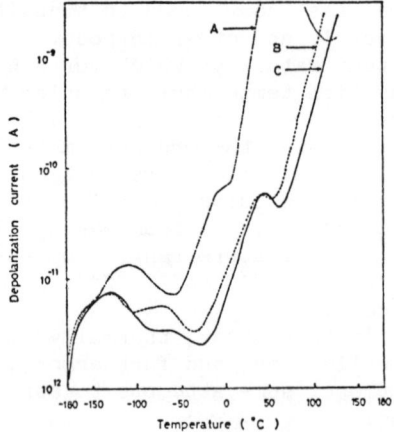

Fig.2 TSC spectra for regenerated cellulose film under the various drying condition

A : Sample was conditioned at 5.5%RH and evacuated at 20°C for 3min. and then dried air was introduced into the evacuated cell. (Tp = 20°C)
B : Sample was dried at 120°C for 2hrs. in the measuring cell, and then dried air was introduced into the evacuated cell.(Tp=130°C)
C : Sample was dried at 120°C for 2hrs. in the measuring cell, and then severely dried nitrogen was into the evacuated cell.(Tp=130°C)

The current value of the peak in the vicinity of -130°C was reduced by tritylation and acetylation of cellulose (Fig.3). Xylan powder had no peaks at about -130°C except fine structures (Fig.4). From Fig.4, the appearance of the peak at about -130°C is considered to be related to primary hydroxyl group in glucose residue of cellulose. For, the tritylation of cellulose proceeds preferentially at primary hydroxyl groups and all of the hydroxyl group in cellulose triacetate are blocked. Besides, xylan has no primary hydroxyl group. It is shown in Fig.5 that the peak at about -130°C appeared for glucomannan, which has primary hydroxyl groups. Furthermore, the current value at about -130°C was shown to decrease with an increase in trityl D.S.[5]

By dry milling treatment of regenerated cellulose powder, the current value of the peak at about -130°C was increased (Fig.6). On the other hand, in the case of crystalline Whatman cellulose and Avi-cel powder, the current value of the peak at about -130°C decreased with increasing crystallinity index of celluloses.

To sum up, the appearance of the peak at about -130°C is considered to be due to the primary hydroxyl group of glucose residue in the amorphous region of cellulose.

As a systematic change was not found out for the peak at about 60°C, the origin of the peak could not be explained at present.

For each peak, the depolarization current was almost directly proportional to the polarizing electric field (Fig.7). Consequently, each peak is considered to be related mainly to the dipole orientation described by Bucci and co-worker's theory[6].

By the way, dielectric relaxation has been observed for dry cellulose at about -60°C in the frequency range from 30 Hz to 1 MHz[3,7-9]. In Fig.8, the dielectric relaxation is shown as the frequency dependence of dielectric loss factor.

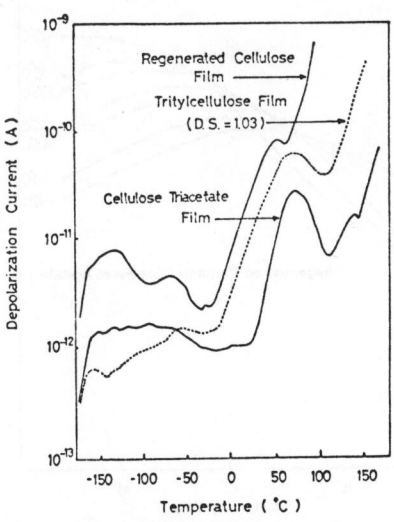

Fig.3 TSC spectra for regenerated cellulose film, tritylcellulose film and cellulose triacetate film.

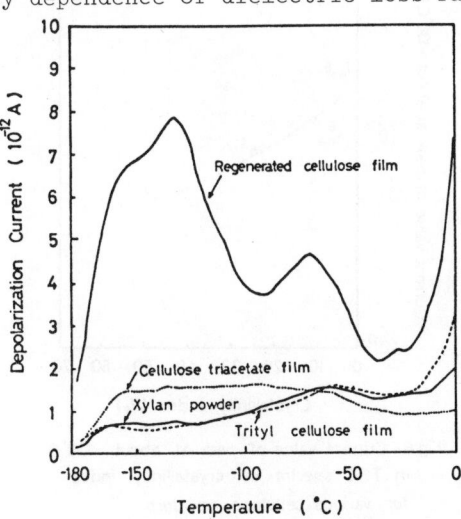

Fig.4 TSC spectra for regenerated cellulose film, cellulose triacetate film, trityl cellulose film and xylan powder

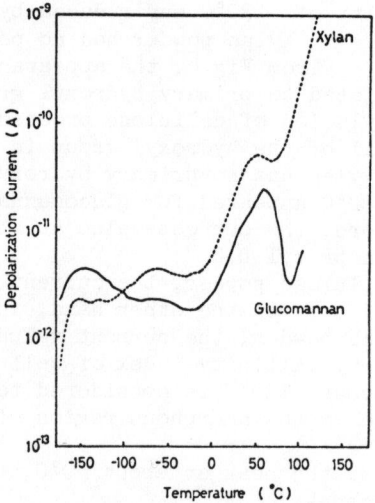

Fig. 5 TSC spectra for glucomannan powder and xylan powder

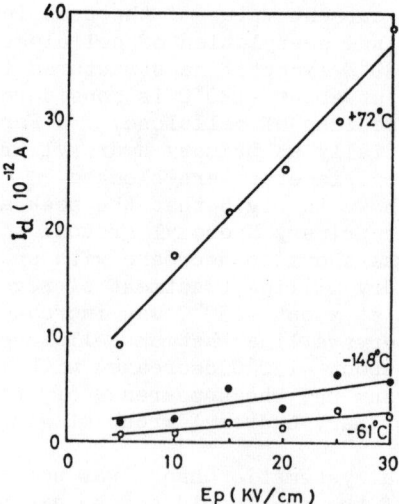

Fig. 7 Relationship between depolarization current at respective temperatures (I_d) and polarizing electric field (E_p) for regenerated cellulose powder (saponified acetate)

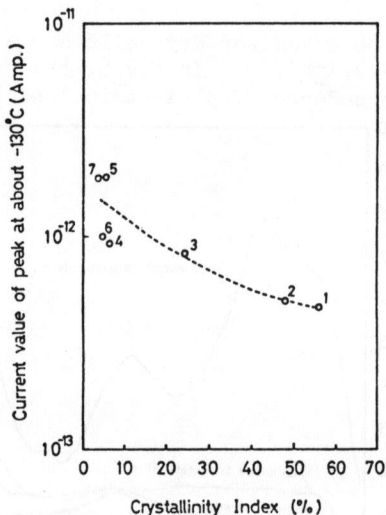

Fig. 6 Current value of peak at about -130°C in TSC spectra vs. crystallinity index for various cellulose powders.
1. Whatman cellulose powder, 2. Avicel powder
3. Regenerated cellulose (saponified acetate) powder
4. Regenerated cellulose powder (ground for 1 hr.)
5. Regenerated cellulose powder (ground for 2 hrs.)
6. Regenerated cellulose powder (ground for 4 hrs.)
7. Regenerated cellulose powder (ground for 6 hrs.)

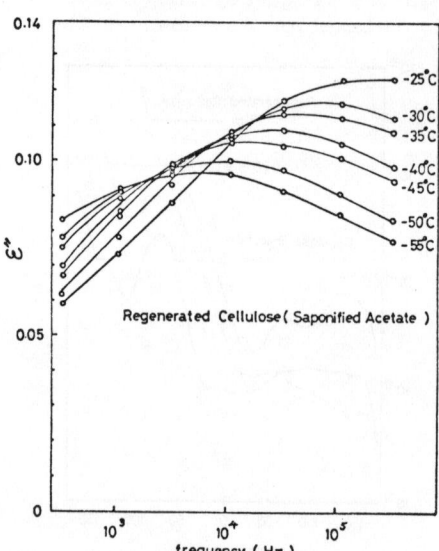

Fig. 8 Frequency dependence of dielectric loss factor (ε'') for saponified cellulose acetate powder (ground for 5 min.)

This relaxation is confirmed to be due to the orientation polarization of primary hydroxyl group of glucose residue in amorphous region of cellulose[8-12]. Norimoto et al. assigned this relaxation to rotation of primary hydroxyl groups around C_5—C_6 axis by electric field[10]). As mentioned above, TSC peak is mainly attributed to the dipole orientation. In this connection, the TSC peak is considered to have the same mechanism as dielectric relaxation. In general, equivalent frequency is proposed to connect TSC with dielectric relaxation[13,14]). As shown in Fig.9, solid line is calculated from the results of the dielectric relaxation in Fig.8. Equivalent frequency calculated from TSC and T^{-1} for Avi-cel is plotted in the dispersion map (Fig.9). Then, the point calculated from TSC peak falled on the point near the broken straight line which is extended from the solid line. This result is considered to show that the TSC peak at about -130°C has the same relaxation mechanism as the above-described dielectric relaxation. Accordingly, the validity of the interpretation concerning the TSC peak at about -130°C is considered to be proved.

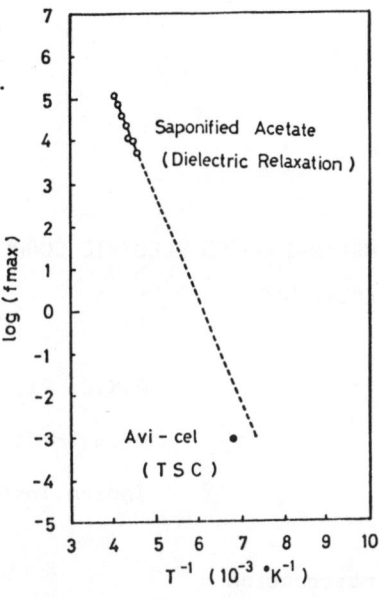

Fig.9 Plots of $\log(f_{max})$ against T^{-1}

4 Conclusion

TSC spectra were examined for dry celluloses. A broad peak at about -130°C, a peak at about -70°C and a peak at about 60°C were observed with an excellently high sensitivity. The peak at about -130°C was attributable to the orientation polarization of the primary hydroxyl group of glucose residue in the amorphous region of cellulose. The peak observed at about -70°C was found to be related to the small quantity of adsorbed moisture. At present, the origin of the peak in the vicinity of 60°C is not found out.

References

1) T.Takamatsu and E.Fukada, Polymer J.,1, 101 (1970)
2) S.Nakamura, J.K.Gillham and A.V.Tobolski, Repts. Prog. Polym. Phys. Japan, 12, 373 (1969)
3) R.Seidman and S.G.Mason, Can. J. Chem., 32, 744 (1954)
4) G.A.Baum, J. Appl. Polym. Sci., 17, 2855 (1973)
5) A.Sawatari,H.Kurihara and T.Takashima, J.Japan Wood Res. Soc.,84,224(1978)
6) C.Bucci, R.Fieschi and G.Guidi, Phys. Rev., 148, 816 (1966)
7) Y.Ishida, M.Yoshino and M.Takayanagi, J. Appl. Polym. Sci., 1, 227 (1959)
8) G.P.Mikhailov, A.I.Arthyukov and T.I.Borisova, Visokomolek. Soed., B 8, 138 (1966)
9) G.P.Mikhailov, A.I.Arthyukhov and V.A.Shevelev, Polym. Sci. U.S.S.R., 11, 628 (1969)
10) M.Norimoto and T.Yamada, Wood Res., No.59/60, 106 (1976)
11) A.Sawatari, T.Yamada and T.Takashima, Japan Tappi, 29, 84 (1975)
12) H.Kurihara, A.Sawatari and T.Takashima, J.Japan Wood Res. Soc.,24,60 (1978)
13) J. Van Turnhout, "Thermally Stimulated Discharge of Polymer Electret", p.34, Elsevier Pub. Co., Amsterdam (1972)
14) T.Takamatsu, Proc. Inst. Electrostatics Japan, 1, 166 (1977)

THERMO-INDUCED ELECTRIC CURRENT FROM PAPER CELLULOSE SANDWICHED IN BETWEEN METAL ELECTRODES

P.K.C. Pillai and M. Mollah[*]

Department of Physics

Indian Institute of Technology Delhi, New Delhi 110 029, India

Introduction

In the absence of any field, previously unpoled samples of paper cellulose sandwiched in between metal electrodes have been found to generate electric current. Three types of paper, viz., filter paper, tissue paper and bond paper were studied. Other types of paper are also expected to show the same property. Recently (1-3), it has been observed that some polymeric materials (viz., flax, cotton and viscose rayon), having cellulose as the major constituent, sandwiched in between metal electrodes generate electric current with heating. Thus it is inferred that cellulose, the major constituent of paper, is responsible for the production of current. There are some reports (4-6) about the generation of electric current with heating from some polymers (viz. PE, PET, PS, PC, PVC, EVA and nylon 66) sandwiched in between metal electrodes. The origin of current, however, is yet to be explained properly. The results presented here may help in explaining the phenomenon of spontaneous current emission from polymers. Since thermally stimulated current is being observed from even unpolarized polymers and other dielectrics, much care should be taken in analysing the thermally stimulated current (TSC) data from polarized polymers. Since the phenomenon observed in this investigation is, to some extent, similar to pyroelectric effect, some practical applications (as in the case of pyroelectric materials) of the phenomenon can be thought of.

Experimental

The samples used were filter paper (Whatman No.1), tissue paper and bond paper having thicknesses 118 μ, 115 μ and 70 μ respectively. Temperature dependence of short-circuit current or open-circuit voltage was measured by inserting

---

[*] On leave from Jahangirnagar University, Savar, Dacca, Bangladesh.

the sandwich cell in a specially prepared oven which was heated at a constant rate by means of a program controller and an electric heater.

Results and Discussion

When the sample is placed in between two electrodes an open-circuit voltage difference ($V_o \simeq 0.4$ Volt) is observed between the electrodes. On short-circuiting, small current ($I_s \simeq 10^{-9}$A) is seen to flow. The magnitude and direction of current are seen to be dependent on electrode materials. Apart from Volta Potential between the electrodes, electrochemical reaction between metal and water (or ions in water) seems to be the origin of current. The shape of the I_s vs Temperature plot (discussed below) also proves the involvement of water which is responsible for the production of current.

Short-circuit current (I_s) as a function of temperature (T) from Cu-paper-Al systems using three kinds of paper are shown in Fig. 1. The magnitude of current is seen to be highest in the case of tissue paper probably because of the fact that % of water content in tissue paper is highest because of its high amorphous crystalline ratio. For the same reason, the magnitude of current in the case of filter paper is expected to be higher than that in the case of bond paper. But due to larger thickness, current is less in the case of filter paper. Figure 2 shows the plots of I_s vs T for four repeated heatings using the same sample. It is seen that the magnitude of current is highest during first heating. This happens due to the fact that the amount of water in the sample at the start of first heating is higher than that at the start of repeated (2nd - 4th) heatings.

In an experimental study of pyrolysis (thermal degradation) of cellulose, Tang and Bacon (7) found that, upto 150°C, physical desorption of sorbed water occurs. The rise in current with heating is attributed to this desorbed water. With the start of coming out of this water, conductance (1/R) of the sample increases. Figure 3 shows the plot of (1/R) vs T and V_o vs T. It is seen that V_o is almost constant. Plot of (1/R) vs T is seen to be similar to I_s vs T. Thus, variation in current (with heating) is related to (or proportional to) variation of conductance which in turn is related to accumulation (freeing) of bound water. Near about 100°C, evaporation of this accumulated water begins to occur which lowers the conductance (and hence current). Thus, initial 'increase' and ultimate 'decrease' in current causes a peak to occur in the I_s vs T plot. Possibility of relation of the peak with glass-transition temperature of cellulose (8) can not also ruled out.

Rise in (1/R) or I_s above 150°C is ascribed to dehydration (7) of water from the equatorial hydroxyl groups in the cellulosic units. The magnitude of current in this region ($>$150°C) is almost same for all types of paper (Fig.1) and for all repeated heatings (Fig. 2); whereas magnitude varies for the region 'from room temperature to 150°C'. This indicates that two types of water are there in the two different regions. Figure 4 shows that current can not be maintained constant at low constant temperature (40°C); but it is almost constant at high constant temperature (220°C) possibly due to the fact that two types of water are involved in these two different regions (viz. $<$150°C and $>$150°C).

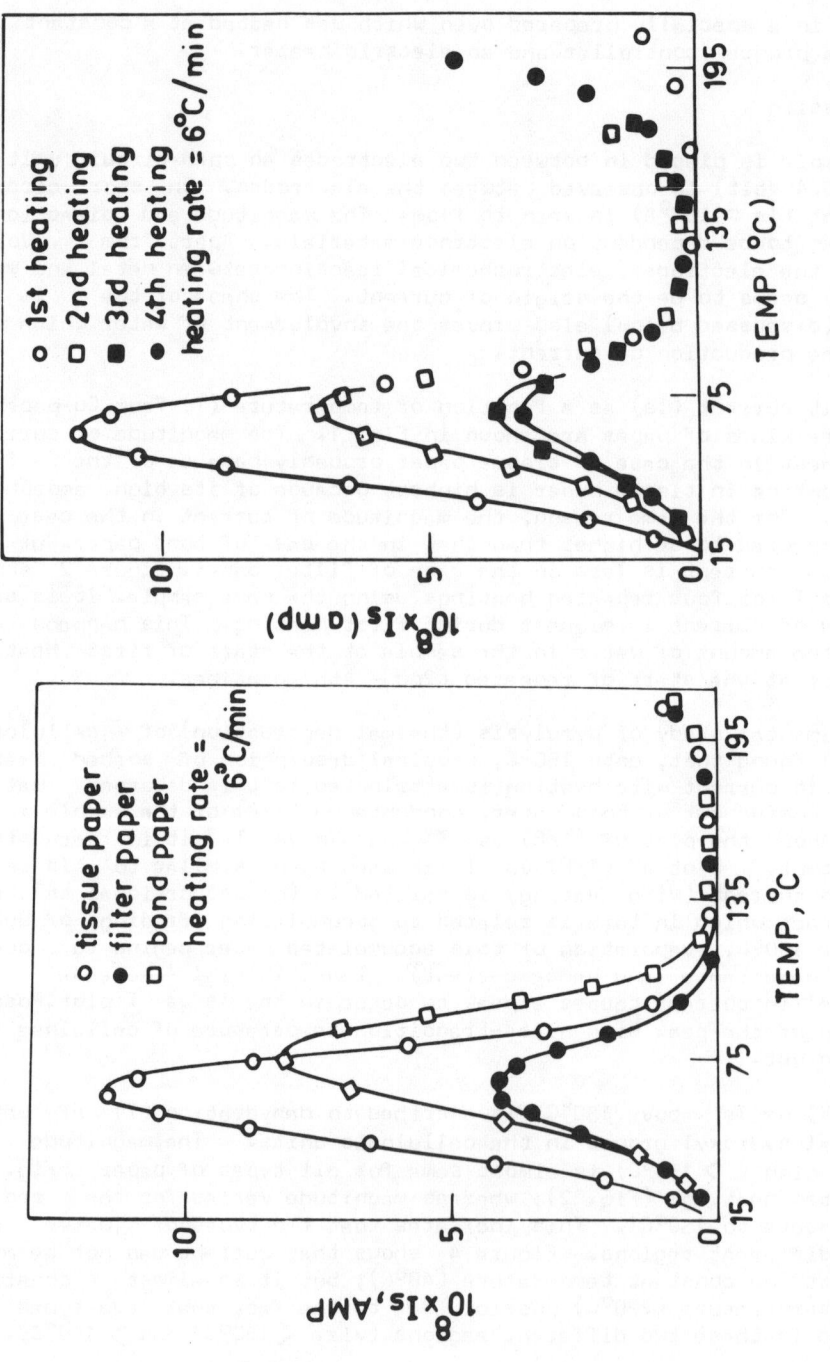

Fig. 1. Is vs T ; Cu-paper-Al system

Fig.2. Is vs T; Cu-tissue paper-Al system. Time intervals for 2nd, 3rd and 4th heatings are 22½, 3¼ and 46½ hours respectively. Currents (magnitude) for 2nd - 4th heatings are shown ten times magnified.

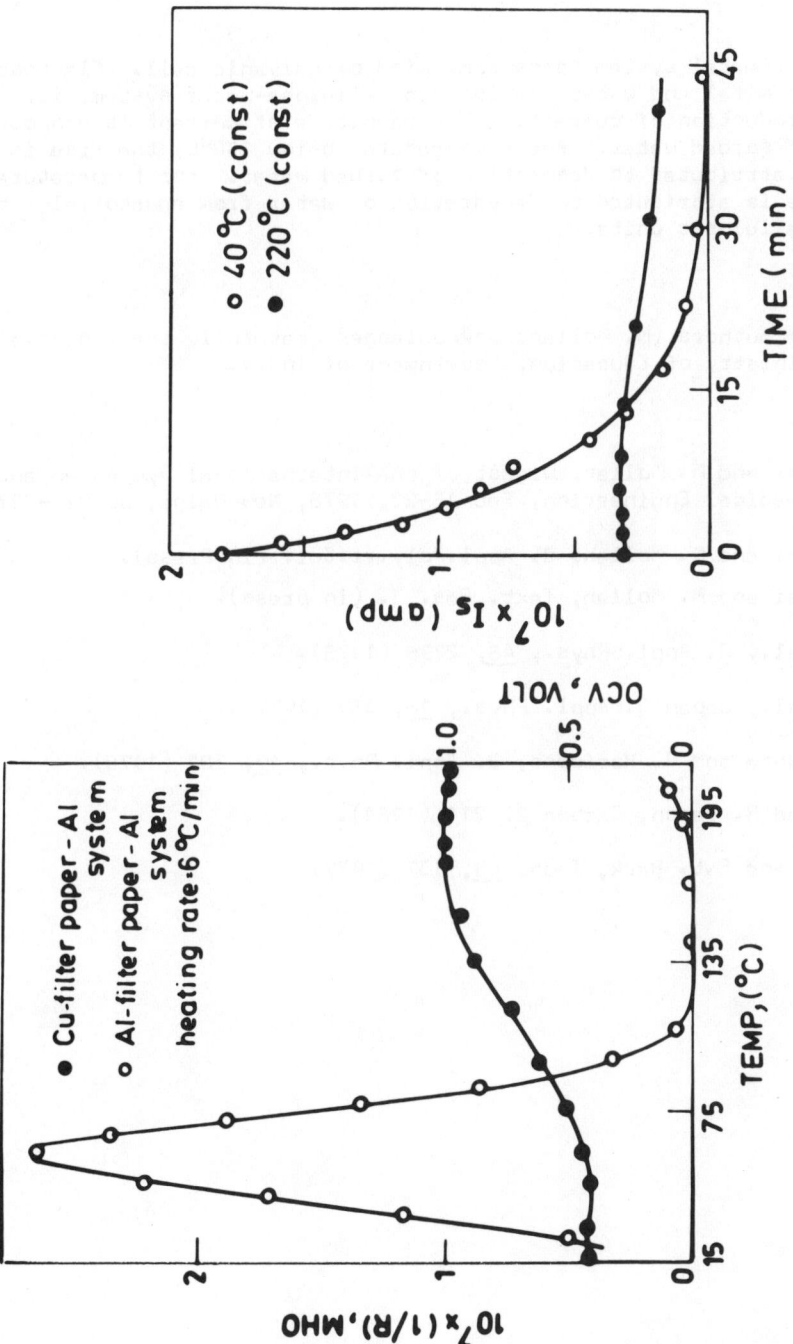

Fig. 3. ● (OCV) vs T ; ○ (1/R) vs T.   Fig. 4. Is vs time; Cu-tissue paper-Al system

Conclusions

Metal-paper-metal system forms some kind of galvanic cell. Electrochemical reaction between metal and water (or ions in cellulose-water system) is responsible for the production of current. The magnitude of current is proportional to the amount of sorbed water. For temperatures below 150°C, the rise in current with heating is attributed to desorption of sorbed water. For temperatures above 150°C, this rise is attributed to dehydration of water from equatorial hydroxyl groups in the cellulosic units.

Acknowledgement

One of the authors (M. Mollah) acknowledges gratefully the financial support from the Ministry of Education, Government of India.

References

1. P.K.C. Pillai and M. Mollah, Digest of the International Symposium and Workshop on Biomedical Engineering, Feb 15-22, 1978, New Delhi, pp 226-227.

2. P.K.C. Pillai and M. Mollah, J. Appl.Polymer.Sci. (in Press).

3. P.K.C. Pillai and M. Mollah, Text. Res. J. (in press).

4. M. Ieda et al., J. Appl. Phys., **46**, 2796 (1975).

5. G. Sawa et al., Japan J. Appl. Phys., **16**, 359 (1977).

6. S. Radhakrishna and S. Horidoss, J. Appl. Phys., **49**, 301 (1978).

7. M.M. Tang and R. Bacon, Carbon **2**, 211 (1964).

8. N.L. Salman and E.L. Back, Tappi **60**, 137 (1977).

VI
ELECTRICAL CONDUCTION AND BREAKDOWN

A CLASSICAL MODEL FOR ELECTRONIC CONDUCTIVITY IN POLYMERS

J. H. Calderwood

Department of Electrical Engineering

University of Salford, Salford M5 4WT, UK

1. **Introduction** It is well known that electronic conductivity in polymers generally does not change very much with field at low field strengths, but increases, usually quite steeply, with increasing field. Attempts to explain this behaviour are usually either based on the postulate that current is controlled by electron emission from the cathode or by electron liberation from sites in the bulk of the material. In an attempt to produce good predictions of experimentally observed behaviour, these postulates have been subject to various elaborations and refinements, and some quite complicated models have resulted. Even so, a given model may be applicable only over a specified field range, outside of which another model must be substituted. However, V Adamec and the author have jointly re-examined the possibility of prediction based upon the concept of a bulk controlled conductivity mechanism, and have found that a simple classical model can give satisfactory results for a wide range of materials, particularly polymers. A brief account follows.

2. **Frenkel's theories of conduction in insulators** In 1938, in a letter to the Physical Review (1), Frenkel explained his idea concerning the mechanism of electronic conductors in insulators. The basic concept, based upon an earlier postulate of Poole (2), was that the neutral atoms of the material could become thermally ionized by losing an electron, which contributes to conduction when an electric field is applied. Later developments have assumed that electrons cannot escape from every atom, but only from certain sites in the material referred to as ionizable centres.

If we have a number density N of ionizable centres, Frenkel showed that in the absence of an applied field the number density n_o of centres which would be ionized at any time as a consequence of thermal ionization is given by

$$n_o = N \exp\left(-\frac{W_i}{2kT}\right) \tag{1}$$

where W_i is the ionization energy, k is Boltzmann's constant, and T is the absolute temperature. This is the number density of electrons available to give the material a temperature dependent zero field conductivity, σ_o. When a field E is applied, it interacts with the field of the positive ion of an ionizable centre so that the energy required for an electron to escape in a direction opposite to the field direction (i.e. in the field assisted or 'forward' direction) is reduced by an amount ΔW_i, where

$$\Delta W_i = \beta_F E^{\frac{1}{2}} \qquad (2)$$

The term β_F is called the Poole-Frenkel coefficient, which a simple calculation gives as

$$\beta_F = \left(\frac{e^3}{\pi \varepsilon_o \varepsilon}\right)^{\frac{1}{2}} \qquad (3)$$

where e is the electronic charge, ε_o is the permittivity of free space, and ε is the relative permittivity of the material. An elementary calculation also shows that the potential energy maximum which occurs in the forward direction is situated at a distance x_o from the ion, where

$$x_o = \left(\frac{e}{4\pi \varepsilon_o \varepsilon}\right)^{\frac{1}{2}} E^{-\frac{1}{2}} \qquad (4)$$

For all but near-breakdown fields, this distance is rather long on the atomic scale. To give some idea of the orders of magnitude involved, for polythylene (PE) at 50°C,

if $E = 10^6$ V/m, then $\Delta W_i \simeq 0.05$ eV and $x_o \simeq 250$ Å;

if $E = 10^8$ V/m, then $\Delta W_i \simeq 0.5$ eV and $x_o \simeq 25$ Å.

The flatness of the maximum of the potential energy curve helps to justify the classical calculation for electron escape and the ignoring of the possibility of tunnelling.

Because of the reduction in escape energy ΔW_i, Frenkel[1] pointed out that the number density n of free electrons would be increased to

$$N \exp\left(-\frac{W_i - \Delta W_i}{2kT}\right) \qquad (5)$$

so that the conductivity σ is given by

$$\frac{\sigma}{\sigma_o} = \exp\left(\frac{\Delta W_i}{2kT}\right) \qquad (6)$$

or

$$\frac{\sigma}{\sigma_o} = \exp\left(\frac{\beta_F E^{\frac{1}{2}}}{2kT}\right) = \exp A \qquad (7)$$

This expression is derived on the assumption that the conductivity is determined entirely by the number density of free electrons, and that their mobility µ does not vary significantly as field and temperature are varied. Although this formula gives reasonably good predictions for a number of materials over a limited range of field strengths, it tends to predict values of conductivity which are too high at higher field strengths, the error increasing to perhaps an order of magnitude as the field strength is raised to near the breakdown value.

In his book published in 1948, Frenkel(3) considered how free electrons move through a structure of neutral potential wells which would govern their mobility. These wells were supposed to have a distance δ between their centres, and to be of depth W_μ. Frenkel considered a one-dimensional model, in which electrons could jump from one well to another in a direction either with or against the field, the attempt-to-escape or vibrational frequency being ν. This led to the development of an expression for mobility μ, namely

$$\mu = \frac{\nu\delta}{E} \exp\left(-\frac{W_\mu}{kT}\right) \sinh \frac{eE\delta}{2kT} \qquad (8)$$

From this we see that the mobility μ tends towards a constant value μ_0 at very low fields, given by

$$\mu_0 = \frac{\nu e\delta^2}{2kT} \exp\left(-\frac{W_\mu}{2kT}\right) \qquad (9)$$

Hence

$$\frac{\mu}{\mu_0} = \frac{2kT}{eE\delta} \sinh \frac{eE\delta}{2kT} \quad \text{or} \quad \frac{\mu}{\mu_0} = \frac{1}{B} \sinh B \qquad (10)$$

If we now assume that the number density n of free electrons does not vary significantly as the field and temperature are raised, and the conductivity is determined entirely by electron mobility, which is the opposite assumption to that made earlier for the derivation of the first Frenkel conduction equation, we arrive at Frenkel's second conduction equation, namely

$$\frac{\sigma}{\sigma_0} = \frac{2kT}{eE\delta} \sinh \frac{eE\delta}{2kT} \quad \text{or} \quad \frac{\sigma}{\sigma_0} = \frac{1}{B} \sinh B \qquad (11)$$

In order to obtain the best possible fit with experimental results, usually it is necessary to choose a value for δ rather higher than that which might be expected from physical reasoning, say about 50Å. Even so, the fit obtained is usually not good.

3. <u>Modifications to Frenkel's theories</u> Several proposals have been put forward based on Frenkel's idea, modified in ways which it was hoped would bring them more into accordance with physical processes in real insulating materials. One evident point is that Frenkel considered a one-dimensional model, but the material is three-dimensional. An electron is in principle capable of escaping from an ionizable centre in any direction which has a component in the forward direction. For escape in such a direction making an angle θ with the applied field, a cosine factor can be introduced into the amount by which the potential barrier is lowered. A treatment of escape in the reverse direction is more

troublesome. Jonscher(4) assumed that escape in any direction which had a component in the reverse direction was impossible, while Hartke(5) assumed that such escape took place at the same rate as would be the case if no field were applied. A more sophisticated treatment was provided by Ieda et al(6) who proposed that when an electron attempting escape in the reverse direction had become separated from its parent atom by a certain critical distance it should be considered as being free, on the grounds that the possession of an energy of the order of kT made recapture unlikely. They were able to obtain good predictions of conductivity as a function of applied field and temperature for SiO films. Other models with considerable degrees of sophistication have been proposed from time to time: see for example the work of Hill(7,8), and Lewis(9). However, it still seemed worthwhile to revert to the original ideas of Frenkel, and to examine whether good predictions would result if both the supply of carriers from ionizable centres, and their subsequent mobility, were taken into account in a three dimensional model.

4. **Two-stage three-dimensional model** This model was first proposed by Adamec and Calderwood(10) in a simple form, in which electron escape from ionizable centres, and their subsequent jumps between the potential wells controlling the mobility, only took place in six principal directions along three mutually perpendicular axes, one of which was in the field direction. Later, the authors(11) refined the calculation so that both escapes and jump could take place in any direction. It is only the results of the refined calculation which are given below.

(i) Calculation of electron concentration. Electron escape in directions in the 'forward' hemisphere can be dealt with by the introduction of a cosine factor into the amount by which the field lowers the potential barrier. The situation concerning escape in directions in the 'reverse' hemisphere is more obscure. The two extreme positions to take up are to imagine with Jonscher that escape in the 'reverse' hemisphere is impossible, or to imagine with Hartke that escape in the 'reverse' hemisphere is completely unaffected by the presence of the applied field. We therefore adopted an empirical approach and increased the magnitude of the barrier for escape in a given direction in the 'reverse' hemisphere by the same amount as it was reduced for escape in the opposite direction. This seemed a more reasonable procedure than adopting either of the extreme positions. Carrying out the appropriate integration gives:

$$\frac{n}{n_0} = \frac{2}{A^2}(1 + A \sinh A - \cosh A) \qquad (12)$$

where

$$n_0 = N \exp\left(-\frac{W_i}{2kT}\right) \qquad (13)$$

(ii) Calculation of electron mobility. Again we consider electron escape in any direction, but now from the neutral potential wells controlling mobility. We must take account not only of the variation with direction of the potential barrier to be overcome, but also remember that the contribution to mobility of an electron moving in a direction at an angle α to the applied field is proportional to $\cos \alpha$. Appropriate integration gives:

$$\frac{\mu}{\mu_0} = \frac{3}{B^3} (B \cosh B - \sinh B) \tag{14}$$

where
$$\mu_0 = \frac{\nu e \delta^2}{6kT} \exp\left(-\frac{W_\mu}{kT}\right) \tag{15}$$

(iii) Expression for electronic conductivity. From the above expressions for n and μ the conductivity σ follows:-

$$\frac{\sigma}{\sigma_0} = \frac{6}{A^2 B^3} (1 + A \sinh A - \cosh A)(B \cosh B - \sinh B) \tag{16}$$

where
$$\sigma_0 = \frac{Ne^2 \nu \delta^2}{6kT} \exp\left(-\frac{W_\sigma}{kT}\right) \tag{17}$$

and
$$W_\sigma = \tfrac{1}{2} W_i + W_\mu \tag{18}$$

The conductivity equation results from the product of terms representing the field dependence of the charge carrier concentration and the field dependence of their mobility. Both terms approach unity with a slope tending to zero as the field is reduced towards zero, thus expressing the tendency towards ohmic behaviour of the material at low field strengths.

It is interesting to examine the relative importance of each of these terms. At a field strength of 1MVcm^{-1}, the charge concentration (for ε = 3) would be about one thousand times higher than that at low fields, while in order to obtain the same increase in the mobility term we should have to assume a value for δ of about 50Å. However, our assumption that a realistic estimate of the jumping distance in solid insulating materials is only a few Ångström units (say 5Å) would result in a rise in mobility very much smaller than the rise in charge concentration. If we therefore neglect the field dependence of mobility, we obtain the simplified expression for conductivity:

$$\frac{\sigma}{\sigma_0} \simeq \frac{2}{A^2} (1 + A \sinh A - \cosh A) \tag{19}$$

5. **Predictions of the model compared with experiment** The validity of the proposed model has been checked by comparing the observed with the calculated field dependence of conductivity for several polymeric insulating materials. As well as our own results on polyethylene terephthalate, published data of other investigators on polyethylene(12), polyvinyl chloride(13) polyvinyl formal(14) and polyimide(15) were used. Very good agreement was shown between theory and experiment, as already reported(10).

In the case of polyethylene terephthalate and polymide, for which the field dependence of conductivity over a comparatively wide range of field strengths is available, a comparison between the applicability of some published formulæ has been carried out. The field dependence of the relative conductivity σ/σ_o was calculated according to the formulae published by Frenkel(1), (3), Mead(16), Jonscher(4), Hartke(5) and Adamec and Calderwood(10). The prediction of the Schottky formula(17) was also included, although this is based on electrode emission, rather than ionization in the bulk of the material. It was shown(10) that the best agreement between the experimental and calculated variation of σ/σ_o with field is provided by our equation.

In order to eliminate the possibility of electron injection, an experiment using contactless electrodes was carried out in which the current was measured in a polyethylene terephthalate film sandwiched between two polystyrene films. There is not space to describe the details, which have already been published (11), but the results are as predicted by our equation, and give strong support to the suggestions that, even in the non-ohmic region when fields of about 1MVcm^{-1} are applied to the specimen, the current is bulk controlled.

Of course, even the excellent fit of experimental results with our equation does not mean that the model yielding the equation is the only one possible. Nevertheless more sophisticated models can only be justified if they have greater predictive powers. For models with the same scope, simplicity is always to be preferred to complication.

6. **Criterion for thermal breakdown** It has been long established that the temperature dependence of electric strength is of the same general form for all polymeric insulating materials. It is almost constant at low temperatures and falls off rapidly above some critical temperature. There have been several explanations of this drop in breakdown strength, one of which postulates that it is caused by thermal instability. We have used our expression for conductivity to calculate conduction losses in polyethylene and polymethyl methacrylate, employing experimentally determined values for σ_o and ε over a range of temperatures, the latter being necessary for the evaluation of β_F.

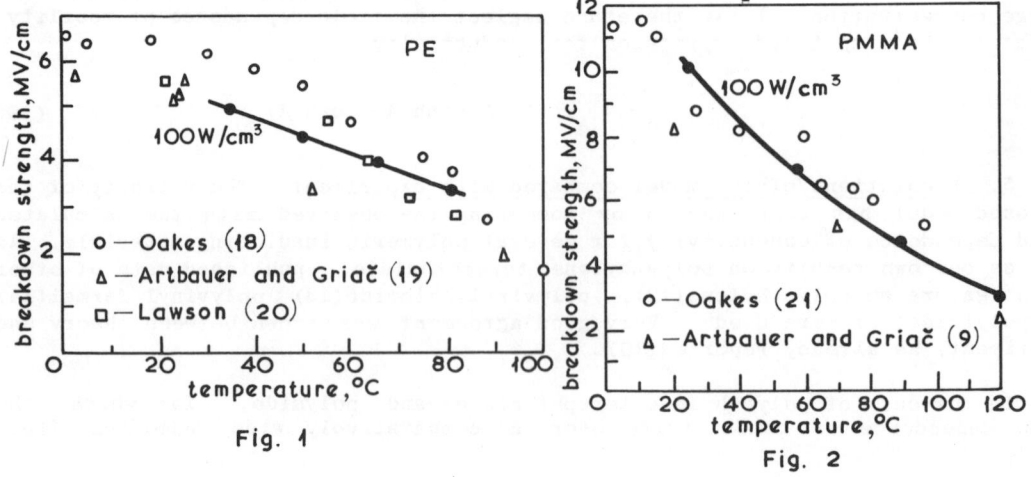

Fig. 1

Fig. 2

It is then possible to determine the field strength at which a certain level of conduction power loss is attained, for a particular specimen temperature. Figures 1 and 2 show curves indicating field strengths causing an arbitrary chosen level of 100 Wcm^{-3} to result for polyethylene and polymethyl methacrylate respectively. These curves fit surprisingly well the values of electric strength found by some investigators, shown by variously marked points in the figures. This agreement suggest the possibility that the observed fall in breakdown strength above a critical temperature might be explained by thermal failure.

7. **Evaluation of the constants of the conductivity equation** Equation(17) for low field conductivity, when re-arranged, gives

$$N = \sigma_o \frac{6kT}{e^2 \nu \delta^2} \exp\left(\frac{W_\sigma}{kT}\right) \quad (20)$$

Substitution of appropriate values in this equation allows an estimate to be made for N. As an example, we shall consider polyethylene-terephthalate (PET) at a temperature of $80^\circ C$. Experimental results at that temperature give the values $\sigma_o = 1.5 \times 10^{-15}$ S/m and $W_\sigma = 2.16 \times 10^{-19}$ J(= 1.35 eV). We can make the reasonable assumptions that $\nu = 10^{13}$ Hz and $\delta = 5 \times 10^{-10}$ m (= 5Å). Also, we take established constants $e = 1.5 \times 10^{-19}$ C and $k = 1.38 \times 10^{-23}$ J/$^\circ$K. Substitution of these values in equation(20) gives $N = 1.3 \times 10^{28} m^{-3}$. For comparison, the concentration N_m of monomer units in PET is $N_m = 2.7 \times 10^{27} m^{-3}$ and the concentration N_a of atoms is $N_a = 5.9 \times 10^{28} m^{-3}$.

It is not possible to determine n_o and μ_o from conductivity experiments alone, but a rough estimate of these quantities can be made on the assumption that the activation energy for conductivity found from irradiation experiments is equal to the activation energy for mobility, i.e. that charge generation occurs entirely at the cost of radiation energy. A typical value for W_μ obtained in this way is 0.3eV. Taking $W_\mu = 4.8 \times 10^{-20}$ J (= 0.3eV), we can then substitute in equation(15) to find μ_o. This gives a value of $\mu_o = 7.1 \times 10^{-10} m^2/Vs$, which is not unreasonable.

Now substituting the values above for W_σ and W_μ gives us $W_i = 3.4 \times 10^{-19}$ J (= 2.1 eV). Finally substitution in equation(13) gives $n_o = 1.3 \times 10^{13} m^{-3}$. A summary of these basic constants for PET at $80^\circ C$ is given in Table 1.

$N_a (m^{-3})$	$N_m (m^{-3})$	$N (m^{-3})$	$n_o (m^{-3})$	W_i (eV)	W_μ (eV)	W_σ (eV)	$\mu_o (m^2/Vs)$
5.9×10^{28}	2.7×10^{27}	1.3×10^{28}	1.3×10^{13}	2.1	0.3	1.35	7.1×10^{-10}

Table 1

From Table 1 we see that in PET at $80^\circ C$, in the absence of an applied field, only about 1 out of every 10^{15} ionizable centres is actually ionized at any instant. If, for the purpose of a rough approximation, we assume uniform distributions in a cubic array, we find that while the average nearest neighbour distance between ionizable centres is 4.3Å, the corresponding distance for ionized centres is 10^5 times greater. A calculation for n using equation(12) indicates that this factor will only be reduced by about an order of magnitude even when fields of the order of 1MV/cm are applied.

It therefore appears unlikely that the question of overlapping potentials from neighbouring ionized centres is likely to be of significance for our calculations.

It is interesting to consider how the distance of 4.3Å between ionizable centres might be related to the molecular structure of the material. Making the same assumption regarding distribution as before, we find that the mean monomer spacing is about 7.2Å while the mean atomic spacing is 2.6Å. It might be possible to identify ionizable centres with molecular features of the polymer chain. It is also possible that the potential wells controlling mobility, assumed to be spaced at about 5Å, are simply ionizable centres which are not ionized. But these suggestions must remain for the present in the realm of speculation.

8. <u>Conclusions</u> It is evident that the three-dimensional two-stage model based on the Poole-Frenkel effect is capable of making satisfacory predictions of electronic conductivity for a number of materials over a wide range of field strengths. The inference is that in these materials, over the range of conditions examined experimentally, the electron current is bulk controlled, rather than controlled by electrode injection. Experiments using a specimen sandwiched between layers of a different dielectric material give results which support this view. It is possible to apply the conductivity equation so that the general behaviour of electric strength as a function of temperature is predicted.

It will be noted that although the model is two-stage, the second stage, embodying the change of mobility with field strength, was not really used. However, in some cases it might not be possible to neglect it, and it would be desirable to carry out experiments where the variation of the mobility term would be expected to be significant, so that this aspect of the theory would be properly tested.

Finally, it will be noted that we have fixed values for W_i and W_μ, rather than a distribution. This of course does not mean that such distributions do not exist, but rather that their effect, if they do exist, is the same for the purpose of conductivity calculations, as if appropriate fixed values are taken for W_i and W_μ. It would obviously be possible to postulate various distributions which, if used in our calculations, would give the same result as our fixed values. If experimental evidence should appear which allows such distributions be specified, our model can accommodate such a development. Meanwhile, we have not introduced the additional complication of distributions, since the fixed value model makes satisfactory prediction.

9. <u>Acknowledgement</u> This work was carried out jointly with Dr V Adamec, to whom thanks are due for many stimulating discussions over a long period of collaboration.

(1) Frenkel J, 1938, Phys. Rev. $\underline{54}$, 647-648.
(2) Poole H H, 1916, Phil. Mag. $\underline{32}$, 112-129.
(3) Frenkel J, 1946, Kinetic Theory of Liquids (Oxford).
(4) Jonscher A K, 1967, Thin Solid Films, $\underline{1}$, 213-234.
(5) Hartke J L, 1968, J. Appl. Phys. $\underline{39}$, 4871-4873.
(6) Ieda M, Sawa G and Kato S, 1971, J. Appl. Phys. $\underline{42}$, 3737-3740.
(7) Hill R M, 1971, Phil. Mag. $\underline{23}$, 59-86.
(8) Hill R M, 1971, Phil. Mag. $\underline{24}$, 1307-1325.
(9) Lewis T J, 1974, Proc. Conf. Electrical Properties of Organic Solids, Scient. Papers of the Inst. of Org. & Phys. Chem. Wrocław Tech. Univ. $\underline{7}$, 146-161.
(10) Adamec V and Calderwood J H, 1975, J. Phys. D: Appl. Phys. $\underline{8}$, 551-560.
(11) Adamec V and Calderwood J H, 1975, I.E.E. Conference on Dielectric Materials, Measurements and Applications, Cambridge, Publ. No. 129, 265-268.
(12) Taylor D M and Lewis T J, 1971, J. Phys. D: Appl. Phys. $\underline{4}$, 1346-1357.
(13) Kosaki M, Sugiyama K and Ieda M, 1971, J. Appl. Phys. $\underline{42}$, 3388-3392.
(14) Lengyel G, 1966, J. Appl. Phys. $\underline{37}$, 807-810.
(15) Hanscomb J R and Calderwood J H, 1973, J. Phys. D: Appl. Phys. $\underline{6}$, 1093-1104.
(16) Mead C A, 1962, Phys. Rev. $\underline{128}$, 2088-2093.
(17) Schottky W, 1914, Z Phys. $\underline{15}$, 872-878.
(18) Oakes W G, 1948, Proc. I.E.E. $\underline{95}$ Pt 1, 36-44.
(19) Artbauer J and Griač J, 1965, Proc. I.E.E. $\underline{112}$, 818.
(20) Lawson W G, 1965, Brit. J. Appl. Phys. $\underline{16}$, 1805-1812.
(21) Oakes W G, 1949, Proc. I.E.E. $\underline{96}$ Pt 1, 37-43.

A.C. IMPEDANCE MEASUREMENTS ON CdS:Cu BINDER LAYERS

P.K.C. PILLAI, C.K. PILLAI AND R.G. MENDIRATTA

Department of Physics

Indian Institute of Technology Delhi, New Delhi 110 029, India

Introduction

The capacitance and the dielectric loss of a photoconductor are known to change when illuminated with radiations of appropriate wavelength. Such an effect, termed as the photodielectric effect (PDE), was first noticed by Gudden and Pohl (1). Many phenomenological theories (1-4) have been proposed to explain this effect. Kallmann et al. (2) have proposed that the PDE is due to the photoconductivity of the grains and the whole sample can be considered as an insulator interspread with conducting regions. Garlick and Gibson (3) suggested that the polarisation of traps containing loosely bound electrons may be the origin of the PDE. Krispin and Ludwig (4) interpreted the effect on the basis of space charge polarisation. Kneppo and Cervenak (5) derived expressions for the fractional changes in capacitance and conductance on the basis of the electrical inhomogeneity of the medium. Later, Pillai et al.(6) modified the above model by considering a single level of traps. According to this revised model, the variation in capacitance of the sample will be less for materials containing more number of traps.

In the present investigations, CdS:Cu photoactive material was used with polystyrene as binder. Variations in the capacitance and dielectric loss of the sample with frequency of the applied a.c. electric field under different intensities of illumination were studied. The observed results are explained on the basis of space charge polarisation.

Experimental Details

Fine powder of CdS:Cu was mixed with 40% by weight of polystyrene dissolved in benzene. The uniformly mixed solution was poured over a conducting glass plate to form a uniform layer and was allowed to undergo slow evaporation in benzene atmosphere. The sample was kept in between two conducting glasses which formed the electrodes. An L.C.R. bridge was employed for the capacitance and the loss measurements in the frequency range of 50–10^4 Hz of the applied field. The sample was illuminated with radiations from a mercury vapour lamp of intensity ranging from 15 to 1000 μWcm^{-2}.

Results and Discussion

The dark capacitance and the resistance of the sample were 64 pf and 6×10^9 Ω respectively. The respective thickness and the area of the layer were 230 μm and 3.25 sq.cm. The experimental results are presented in Figs. 1–3.

Figure 1 shows the observed variation of the fractional change in capacitance with frequency of the applied field when the sample was under illumination. In dark also the capacitance was seen to decrease with frequency. These results can be explained on the basis of space charge formation near the electrodes. The fractional changes in capacitance, $\Delta C/C$, were observed to decrease with frequency and increase with intensity of illumination. At low frequencies it is possible for the space charge formation to be in step with the variation of the applied electric field (7). At higher frequencies, however, this is not possible and as such polarisation never attains its maximum value. This explains the higher values of $\Delta C/C$ at lower frequencies. Since in the illuminated sample, the photocarriers are also present in addition to the equilibrium carriers, one expects the values of $\Delta C/C$ to be higher at all frequencies. A possible reason for this behaviour may be the presence of more number of trapped electrons in the binder layer. The carriers being bound in traps do not contribute much to polarisation and hence result in a slower variation of $\Delta C/C$ with frequency.

The variations of loss factor, $\tan \delta$, with frequency of the applied field for different intensities of illumination are shown in Fig. 2. The loss factor decreases faster at lower frequencies and attains a minimum value at a particular frequency. It is observed that $(\tan \delta)_{min}$ shifts to higher frequency side when the intensity of illumination is increased. This behaviour can also be explained on the basis of space charge formation. Space charge is an electrical inhomogeneity appearing in the material (8). Any such inhomogeneity causes dielectric loss At lower frequencies of the field the space charge formation being more, the dielectric losses are also more and the opposite is true at higher frequencies. The shifting of $(\tan \delta)_{min}$ shows an increase in d.c. conductivity of the medium. Larger dielectric losses with increasing intensity of illumination are due to the addition of photogenerated charge carriers resulting in an increase in the space charge.

Fig. 3 gives the variation of a.c. conductance, G, as calculated from the observed values of $\tan \delta$ at different frequencies. The conductance is observed to increase with the frequency of the applied field and with the intensity of illumination. This observed variation of a.c. conductance also supports the space charge polarisation concept.

Fig. 1 Variation of fractional change in capacitance with frequency of the applied electric field.
(a) 100 μWcm^{-2} (b) 200 μWcm^{-2}
(c) 500 μWcm^{-2}

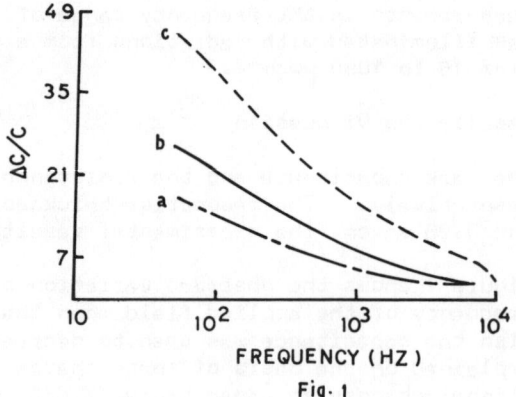

Fig. 1

Fig. 2 Variation of tan δ with frequency of the applied electric field.
(a) dark (b) 200 μWcm^{-2}
(c) 500 μWcm^{-2}

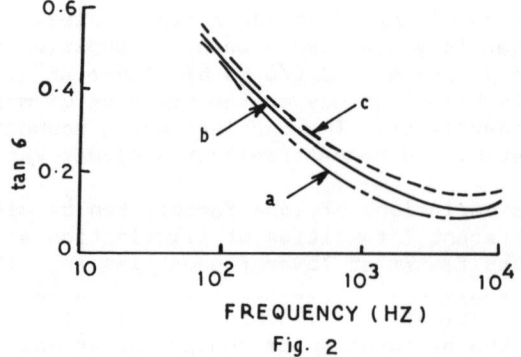

Fig. 2

Fig. 3 Variation of a.c. conductance with frequency of the applied electric field.
(a) dark (b) 100 μWcm^{-2}
(c) 500 μWcm^{-2}

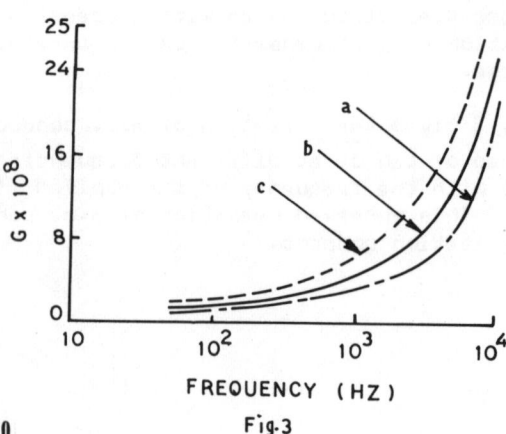

Fig. 3

References

1. B. Gudden and R. Pohl : Z Physik **1**, 365 (1920)

2. H. Kallmann, B. Krammer and P. Mark : Phys. Rev. **99**, 1328 (1955)

3. G.F.J. Garlick and A.F. Gibson : Proc. Phys. Soc. **62A**, 731 (1949)

4. P. Krispin and W. Ludwig : Phys. Stat. Sol. **12**, 595 (1965)

5. I. Kneppo and J. Cervenak : Solid State Electronics, **15**, 587-593 (1972)

6. P.K.C. Pillai and R. Nath : Phys. Stat. Sol. **37**, 491 (1976)

7. J.R. Macdonald : Jr. of Chem. Phys. **23**(2), 276 (1955); Phys. Rev. **92**(1) 11 (1953).

8. B. Tareev : Physics of Dielectric Materials (Moscow: Mir), 147 (1975)

Space charge effects in rutile ceramics

L. Badian, S. M. Gubański

Inst.Podst.Elektr. Technical University of Wrocław

Wrocław ul. Wyb. Wyspiańskiego 27, Poland

1. Introduction.

Anomalous conductivity effect [1,2] is based on an increase of a conduction current with time when the constant electric field E = const is applied (fig.1). In this paper the attempt was made to explain the anomalous conduction effect using the space charge model [3]. The experimental investigations were made for rutile ceramics. The theoretical analysis [4] has shown that for adequately chosen boundary conditions for space charge limited currents, the increasing with time solutions for j(t) can be obtained. In accordance to the space charge model, the time changes of current j(t) would have to be connected to the different distributions $q_v(x,t)$.

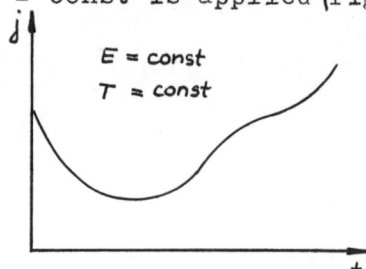

Fig. 1. A typical anomalous current effect.

2. Experimental technique.

Investigations of $q_v(x)$ distribution were carried out by use the potential probe method [5,6]. The potential distributions V(x) along the sample were measured (fig.2), and than in accordance to the Poisson equation the distributions of the field E(x) and the charge $q_v(x)$ were obtained by application the graphic differentiation method. This type of measurements are very difficult becouse of the influence of the following factors:
- influence of the input impedance of the electrostatic voltmeter

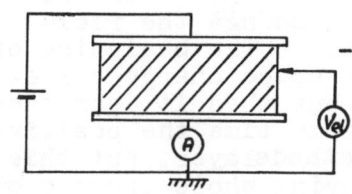

Fig. 2. Schematic view of measuring system.

on the time constant of the system and a measured value of $V(x)$,
- influence of the bulk conduction distribution $\sigma(x)$ inside of the dielectric on the resultant potential distribution $V(x)$ at the edge of the sample.

The influence of the voltmeter impedance was minimalized empirically. The influence of the $\sigma(x)$ distribution on $V(x)$ and $q_v(x)$ can be estimated theoretically. The influence of the distortion of the field at the edge of the sample has however not been fully studied yet. The correctness of the applied method was estimated only experimentally by simultaneous measurements of the individual elements of the stacks of samples, using the probe method and the Faraday cage method. The positive (in the range of a half order of magnitude) results of these investigations have encouraged to assent the method as an admissible one, and to apply it at **least** in a comparative sence for the analysis of the anomalous conduction effect and some ageing effects in dielectrics.

3. Results and their discussion.

Simultaneous investigations of the potential distribution $U(x)$ along the sample and time dependence of the current $j(t)$ (fig.3) have shown the strong correlation between them.

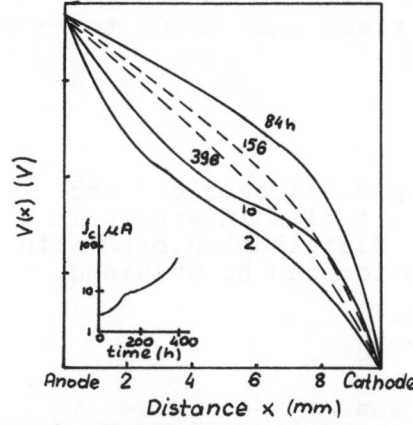

Fig. 3. Potential distributions $V(x)$ at different times. The corresponding time dependence of current is also shown.

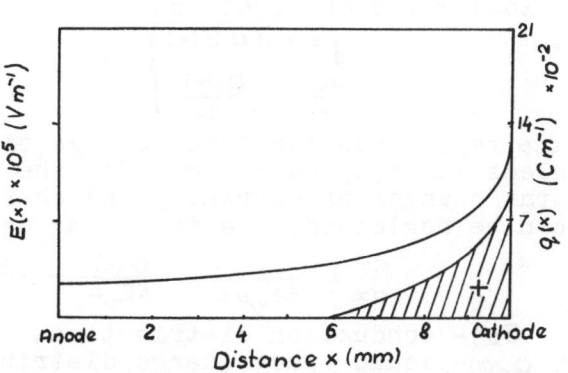

Fig. 4. Steady-state distributions of field $E(x)$ and space charge $q_v(x)$ derived from fig.3 (156 h)

When the conduction current $j(t)$ flows through the dielectric

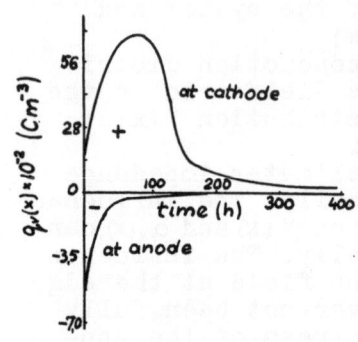

Fig. 5. Space charge concentrations at anode and cathode as a function of time.

the space charge distribution $q_v(x,t)$ will be created in it, and it will change the field distribution $E(x)$ as well. At the beginning of the process the negative space charge collects at the anode layer, but then it decreases rapidly (fig. 5). At the same time the positive charge collects at the cathode layer, but this time it remains there during whole process of the current increase. The analysis of the experimental results and literature data allows to presume the ionic nature of the positive space charge. This is probably coused by positively charged anionic vacancies in rutile lattice. The concentration of the positive charge at the cathode (fig. 5) is strictly correlated to the time dependence of the current (fig. 3). When the current remains constant at the beginning of the process the space charge concentration at the cathode increases and the electric field strength increases in this region as well. It causes an increase in the injection level of electrons from the cathode to dielectric. The injecting character of the anomalous current was confirmed by J-V characteristics measurements. Injection of electrons into dielectric leads to a decrease of the charge density at the cathode (fig. 5). The fact that the current remains at the same level or even increases when the field strength decreases is difficult to explain. Nevertheless such a result can be obtained in theory.

The theoretical analysis [7] of the described case needs to solve the following set of equations

$$\left. \begin{array}{l} j = e n \mu E(x) \\ \dfrac{\partial E}{\partial x} = \dfrac{q_v(x)}{\varepsilon \varepsilon_o} \end{array} \right\}$$

where $q_v(x)$ represents the total charge of injected carriers and motionless ionic space charge. For the short time intervals in which the changes of current j and charge distribution $q_v(x)$ with time can be neglected, the following equation can be obtained

$$\frac{dE}{dx} = \frac{j}{\varepsilon \varepsilon_o \mu E} - \frac{\sigma(x)}{\varepsilon \varepsilon_o \mu} - \frac{q_{vi}}{\varepsilon \varepsilon_o}$$

where:
$\sigma(x)$ - conduction distribution,
$q_{vi}(x)$ - ionic space charge distribution.

Each parts of the equation are correlated with the influence of different factors on the electric field distribution in a dielectric, and the $\frac{j}{\varepsilon \varepsilon_o \mu E}$ element refers to the charge of injected electrons, whereas two following elements refer to conduction and ionic space charge distributions respectively. Putting into the equation experimental data for j, $E(0), \sigma(x)$ and $q_{vi}(x)$ obtained from the probe method measurements and using the normalisation condition $\int_0^L E dx = U$, the distributions of the field were obtained which are in good agreement to those ones obtained experimentally (fig.6). It has been found as well, that this agreement can be obtained only when the value of electric field strength at the cathode E(0) decreases, while the value of current j increases.

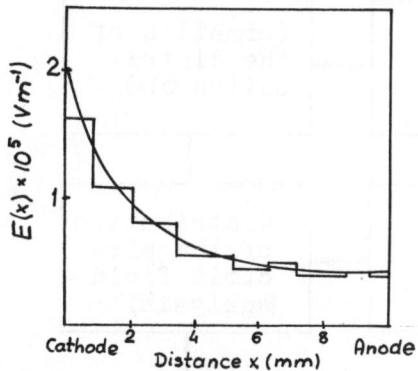

Fig. 6. Theoretically (solid line) and experimentally (step line) obtained field distributions

This result is consistent with the experimental data, however the physical mechanism of the effect remains obscure. There may exist some explanations of it. The neutralisation of positive space charge very near to the cathode may not occur so readily becouse there will be a finite mean free path for electron trapping and becouse the electron energy can be large. More likely reason however for explanation seems to be decreasing of the work function for the electrons at the cathode-dielectric border due to the ionic processes. It also would be possible that one is dealing with a double injection effect, especially becouse of the great importance of the conditions at the anode-dielectric border on the rate of anomalous current increase. From the theoretical analysis it has been also found that for obtention of the theoretical distribution E(x) similar to those obtained experimentally, the value of mobility μ put into the equation can vary in the relatively narrow range. This gives the possibility of application of the described above analysis together with the potential probe measurements for estimation of the mobility of carriers in dielectrics. The value of mobility for studied rutile ceramics was estimated to lie in the range $10^{-4} - 10^{-5}$ m^2/Vs.

On the basis of the carried out experiments and theoretical analysis the model of anomalous conduction can be proposed. It is shown on the following draft.

R e f e r e n c e s
1. L. Badian, S.M. Gubański, T.J. Lewis, 1977, J. Phys. D., 10, (2513)
2. S. N. Koikov, G.B. Sjomuskin, V.G. Landau, 1977, Proc. of 22 Intern. Wiss. Koll. TH Illmenau (71)
3. L. Badian, 1977 Bull. Acad. Pol. Sci. XXV, No 1, (13)
4. Cz. Stec, to be published
5. L. Badian, S.M. Gubański, 1977, Sci. Pap. of IEEF of Wrocław Technical University No 1 (227)
6. K. Lehovec, G. A. Shirn, 1962, J. Appl. Phys., 33, (2036)
7. S. M. Gubański, V. G. Landau, G. B. Sjomuskin - to be published

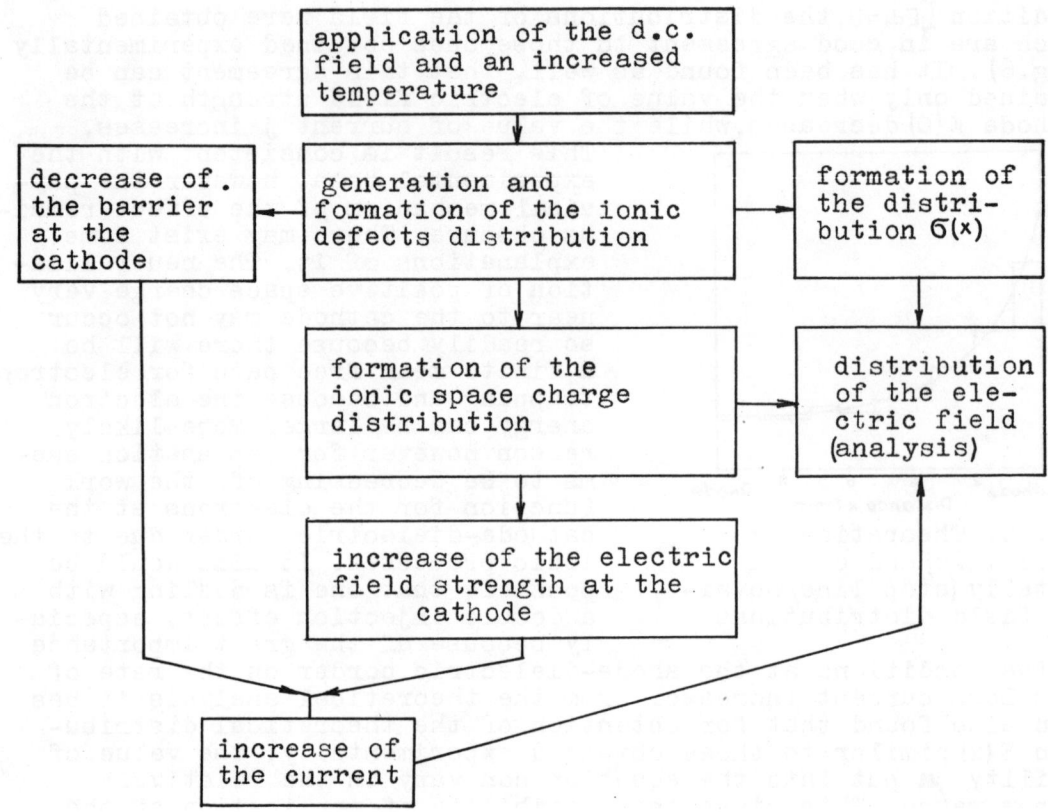

SPACE CHARGE AS AN INSTRUMENT OF INVESTIGATION OF ELECTRIC

PHENOMENA IN DIELECTRICS

L. Badian, E. Smycz

Inst.Podst.Elektr. Technical University of Wrocław

Wrocław ul. Wyb. Wyspiańskiego 27, Poland

Forming and decaying of space charge in dielectric may be of great importance [1,2] for interpretation of some non entirely resolved problems as: the non-exponential absorption and resorption currents, the non-Debye dispersion relations, the anomalous conductivity $\sigma'(t)$ and others. To explain them, it would be sufficient in most cases to clear up the problem of either absorption or resorption current-time function under constant electric field E. It may be shown [3,4] that at least the slow relaxation phenomena of these currents can be easy explained by means of the space charge model /SCM/ and its temporary effects. For the simplest case of plan-parallel specimen the space-charge distribution $q_v(x,t)$ the current functions $j(t)$ and the boundary conditions are studied. Distinction was made between the so called direct problem aimed toward finding the current from the initial condition $q_v(x,0)$ and the inverse problem which consists on indentyfying $q_v(x,0)$, some boundary conditions and on calculating some material constants from currents and other magnitudes being measured.

Theoretical considerations are based on the following system of equations:

$$j(t) = \frac{\partial D(x,t)}{\partial t} + \mu q_v(x,t) E(x,t) + \sigma E(x,t) - \lambda_d \frac{\partial q_v(x,t)}{\partial x} \quad (1)$$

$$\frac{\partial D(x,t)}{\partial x} = q_v(x,t) \quad (2)$$

$$D(x,t) = \varepsilon_0 \varepsilon_\infty E(x,t) + \varepsilon_0 \int_{-\infty}^{t} \alpha(t-t_0) E(x,t_0)\, dt_0 \approx \varepsilon_0 \varepsilon E(x,t) \quad (3)$$

$$\int_0^l E(x,t)\, dx = U \quad (4)$$

Appropriate boundary condition have to be included for definite problems. First of all they are determined by electrode processes.

The following equation is obtained for the simplified case of ideal dielectric /$\sigma=0$/ with instantaneus polarisation /$D = \varepsilon_0 \varepsilon_\infty E$/ and $\mathcal{J}_d = 0$ /diffusion is omited/:

$$j(t) = \frac{\partial D(x,t)}{\partial t} + \frac{\mu}{\varepsilon_0 \varepsilon} D(x,t) \frac{\partial D(x,t)}{\partial x} \quad (5)$$

This equation was later generalized for more complicated situations.

The direct problems of the resorption current $j_r(t)$. (ref. Fig 1)

The initial distribution of the charge density $q_v(x,0)$ is assumed. Application of the well known method of characteristics gives the following results:

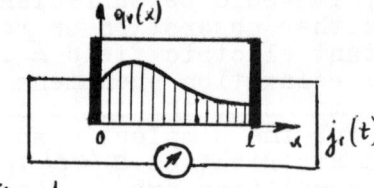

Fig. 1

- equations of trajectories

$$\frac{dx(t)}{dt} = \frac{\mu}{\varepsilon_0 \varepsilon} D[x(t),t] \quad (6)$$

$$\frac{dD(x,t)}{dt} = j_r(t) \quad (7)$$

- for monotonous function all trajectories begin in a segment $[0,l]$ (fig. 2) and hence the knowledge of $D(x,0)$ from $q_v(x,0)$ determines all the function $D(x,t)$. The initial conditions $q_v(x,0)$ is in this case good enough as a boundary condition to solve the problem $q_v(x,t)$ and even to find $j_r(t)$

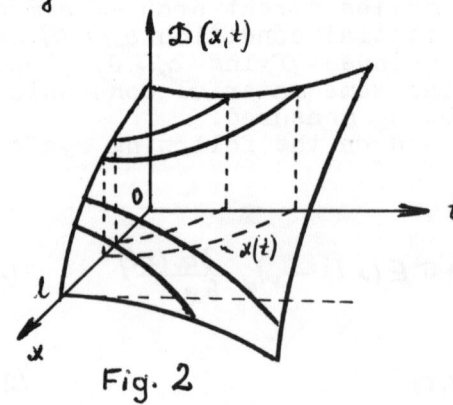

Fig. 2

The following equations are then valid.

$$x(t) = \frac{M}{\varepsilon_0 \varepsilon}\left[t\,D(\lambda) + \int_0^t\int_t^\tau j_r(\tau)\,d\tau\,d\tau\right] \tag{8}$$

$$D[x(t),t] = D(\lambda) + \int_0^t j_r(\tau)\,d\tau \tag{9}$$

$$D(\lambda) = D(x,0) \quad \text{and} \quad x \in [0,l] \tag{10}$$

From here after having reduced the parameter λ

$$D(x,t) = F\left[x,\,t,\,\int_0^t j_r(\tau)\,d\tau,\,\int_0^t\int_0^\tau j_r(\tau)\,d\tau\,d\tau\right] \tag{11}$$

and then the involved equation for current $j_r(t)$ can be obtained

$$\int_0^l F\left[x,\,t,\,\int_0^t j_r(\tau)\,d\tau,\,\int_0^t\int_0^\tau j_r(\tau)\,d\tau\,d\tau\right] = 0 \tag{12}$$

It is relatively simple [4] to find $j_r(t)$ when $q_v(x,0)$ is aproximated with a step-function. Examples are shown on Fig. 3.

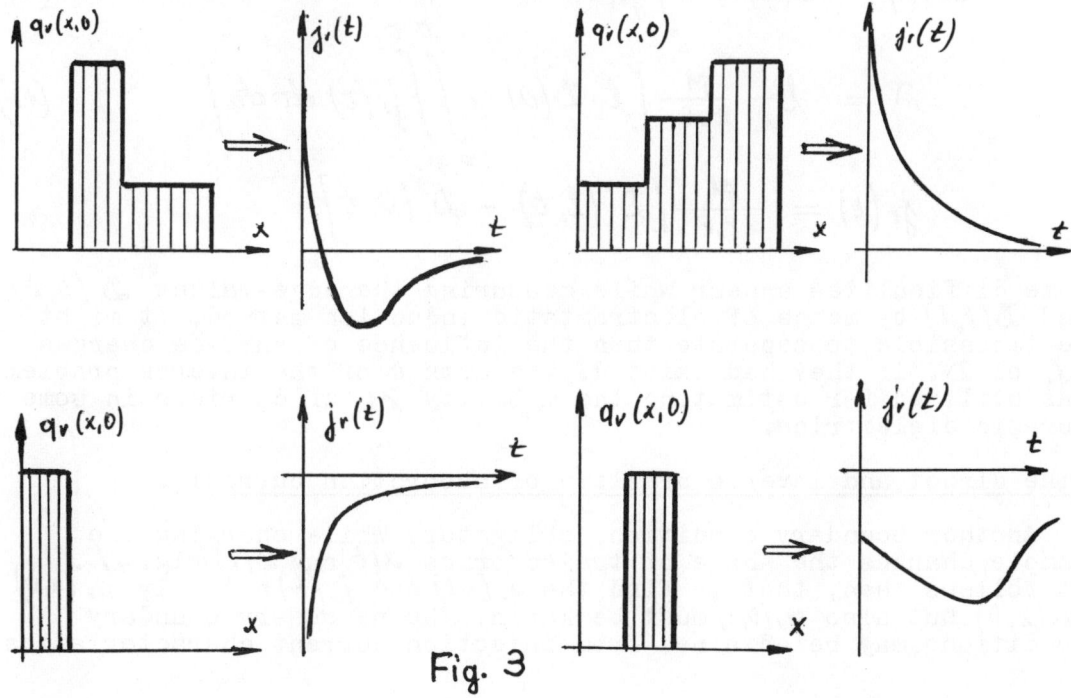

Fig. 3

Two important conclusion may be drawn out:
- It is always possible to find an appropriate distribution of $q_v(x,0)$ leading to non-Debye time function of $j(t)$ for example to power time dependence $j = At^{-n}$
- Even the changes of direction of resorption current can be explained by means of $q_v(x,0)$, without taking into account of polarisation phenomena.

The inverse problem of the resorption current

In this case $q_v(x,0)$ should be found with the knowledge of $j_r(t)$ function from the experiment. It is however necessary to know then also the values $D(0,t)$ and/or $D(l,t)$. Procedure follows immediately from the equations:

$$D(\lambda) = D(0,t) - \int j_r(\tau) d\tau \qquad (13)$$

$$\lambda = \frac{\mu}{\varepsilon_0 \varepsilon}\left[t \, D(\lambda) + \int_0^t\int_0^\tau j(\tau) d\tau d\tau\right] \qquad (14)$$

$$D(\lambda) = D(l,t) - \int_0^t j_r(\tau) d\tau \qquad (15)$$

$$\lambda = l - \frac{\mu}{\varepsilon_0 \varepsilon}\left[l \cdot D(\lambda) + \int_0^t\int_0^\tau j_r(\tau) d\tau d\tau\right] \qquad (17)$$

$$j_r(t) = \frac{\mu}{2\varepsilon_0 \varepsilon}\left[D^2(l,t) - D^2(0,t)\right]$$

Some difficulties appear while measuring the edge-values $D(0,t)$ and $D(l,t)$ by means of electrostatic induction method. It might be impossible to seperate then the influence of surface charges q_s on D, if they had existed. The method of the inverse problem was utilised for estimating the mobility μ of carriers in some ceramic dielectrics.

The direct and inverse problems of absorption currents.

Another boundary condition, obligatory while charging the sample, changes the shape of trajectories $x(t)$ and $D(t)$ /Fig. 4/. It follows then, that to find the $q_v(x,t)$ and $j_a(t)$ not only $q_v(x,0)$ $q_v(x,0)$ but also $q_v(0,t)$ must be known. The necessary boundary conditions may be obtained from injection current characteristics.

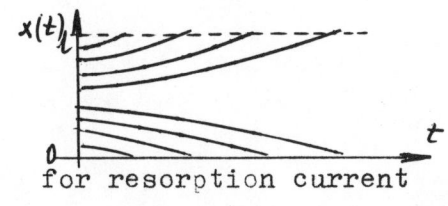

for resorption current

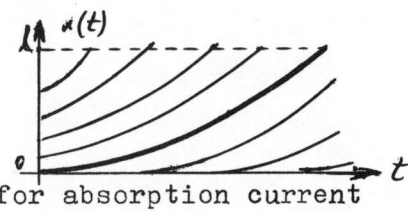

for absorption current

Fig. 4

If in the most simple case the cathode injection current assumed in the form

$$j_k(t) = a + b\, E_k$$

then the $j_a(t)$ should have the shape presented on Fig. 5.

Fig. 5.

Solving the problem under discussion leads to asymptotic distribution $q_v(x,\infty)$ which may be then the starting point for the direct problem of resorption current. In the inverse problem and $j(t)$ at one of the electrodes are assumed to be known from experiment. It is possible then to estimate the characteristics of electrode processes.

It is last of all worth to emphasize that the different demands in relation to the boundary conditions for resorption and absorption currents create the possibility of not-fullfiling the superposition law for the both polarisation current and the conductivity currents.

Reference
1) L. Badian, E. Smycz, Ageing phenomena in dielectrics. Prace IPEiE Politechn. Wrocł. ser. K. No.1, pp.32-34 (1976).
2) L. Badian, Bull. Acad. Sci. Pol. Vol.XXV, No.1, p.13 (1977).
3) Space charge in dielectrics. Coll. Text Book IPEiE Politechn. Wrocł. (1977).
4) IPEiE Politechn. Wrocł. Reports on space charge in dielectrics 1976—1978

ELECTRICAL CONDUCTION IN POLYMERS

K. Kamisako, H. Sasabe and K. Shinohara*

Department of Electronic Engineering, Faculty of Technology,
Tokyo University of Agriculture & Technology,
Koganei, Tokyo 184, Japan,
*Science and Engineering Research Laboratory, Waseda University,
Shinjuku, Tokyo 162, Japan.

§1. INTRODUCTION
So far the question whether the carrier species in polymers is ions or electrons (holes) is an unsolved problem. However, in most polymers both ionic and electronic conduction mechanisms are possible, depending upon the experimental conditions such as temperature, pressure and applied field. Under high applied fields the electronic conduction is dominant and its mechanisms have been discussed in terms of a Schottky emission[1] and/or a Poole-Frenkel effect.[2] The details of the mechanism around the metal electrode - polymer interface have not been clarified as yet. Authors proposed the other model somewhat different with Schottky emission, taking a diffusional effect of injected carriers (electrons) and a space charge effect into consideration.[3] As for polystyrene[3] and polyethyleneterephthalate[4] it was confirmed that the changes of a steady state current through polymers with an applied field and with temperature were well fitted to this model. In this paper the applicability of the model to polyethylenenaphthalate will be checked with experimental results.

§2. EXPERIMENTAL
A polyethylene-2,6-naphthalate (PEN) sample was provided by Dr. I. Ouchi of Teijin Co. in the form of biaxially stretched films. The film thickness was $ca.$ 25 μm on the average. Electrodes for electrical measurements were formed on both side of the film by means of a vacuum deposition technique of gold or aluminum. In order to measure the current only through the bulk of a polymer, a surface conduction current should be eliminated, hence a three terminal method was adopted in our experiments. Before the measurement samples were heat treated at the highest measuring temperature for 6 hours under the vacuum of 10^{-3} Torr. The current measurement was carried out $in\ vacuo$ by using a vibrating reed electrometer (TAKEDA TR-84M). The dielectric constnats of PEN were measured in the frequency range of 60 Hz -100 kHz by using a capacitance bridge (GENERAL RADIO, GR-1620A).

§3. MODEL

In a steady state the current density J can be expressed as follows:

$$J = -qD\frac{\partial n(x)}{\partial x} + n(x)\mu q E(x), \tag{1}$$

where q is the charge of an electron, D the diffusion constant, n(x) the density of electrons at a distance x from the cathode, μ the mobility of electrons, and E(x) the field strength. Putting $qE(x)=-dU(x)/dx$, where U(x) is the potential energy of electrons, and $\mu/D=q/kT$ (the Einstein's equation), the next equation is given from Eq. (1):

$$-\frac{J}{\mu} = kT\frac{dn(x)}{dx} + n(x)\frac{dU(x)}{dx}. \tag{2}$$

In a previous paper[3], we solved Eq.(2) under the proper boundary conditions,

$$J \simeq \frac{N_0 \mu kT}{\int_0^L exp[U(x)/kT]\,dx}, \tag{3}$$

where N_0 is a constant and L is the sample thickness. In this case U(x) was expressed as

$$U(x) = \phi - \frac{q^2}{4\varepsilon x} - q\int_0^x \left[\frac{V}{L} - \frac{4\pi q}{\varepsilon L}\int_0^L\left(\int_0^x \rho(x)dx\right)dx\right]dx \tag{4}$$

$$= \phi - \frac{q^2}{4\varepsilon x} - q\left(\frac{V}{L}\right)_{eff} x, \tag{4'}$$

where ϕ is the work function to bring an electron from the electrode into the polymer, ε the dielectric constant, V the applied voltage, $\rho(x)$ the density of space charges and $(V/L)_{eff}$ the effective field characterized with space charges. The second term in the right hand side of Eq.(4) is the potential energy due to the image force caused by the electrode. $(V/L)_{eff}$ is given as

$$\left(\frac{V}{L}\right)_{eff} = \frac{V}{L} - \frac{4\pi q}{\varepsilon L}\int_0^L\left(\int_0^x \rho(x)dx\right)dx \tag{5}$$

and the second term in the right hand side of Eq.(5) means that the electric field near the cathode is weakened by space charges accumulated in the polyemr.

In practice the denominator in Eq.(3) is roughly approximated in the form of[5]

$$J \simeq \frac{N_0\mu q\left(\frac{V}{L}\right)_{eff} exp\left[-(\phi - \sqrt{\frac{q^3}{\varepsilon}\left(\frac{V}{L}\right)_{eff}})/kT\right]}{ln\,2\sqrt{1 + \frac{2}{kTln2}\sqrt{\frac{q^3}{\varepsilon}\left(\frac{V}{L}\right)_{eff}}}} \tag{6}$$

Experimental results for polystyrene and polyethyleneterephthalate were compared with Eq.(6) as discussed in the previous papers.[3,4] On the other hand, we will calculate Eq.(3) directly by using a programmable calculator with the assumption that $U(x)=-q^2/4\varepsilon x - q(V/L)x$, that is, no space charges exist. In this case the current density J is given by the following equation:

$$J = \frac{N_0\mu kT exp(-\phi/kT)}{\int_0^L exp\left[-(\frac{q^2}{4\varepsilon x} + q\left(\frac{V}{L}\right)x)/kT\right]dx} \tag{7}$$

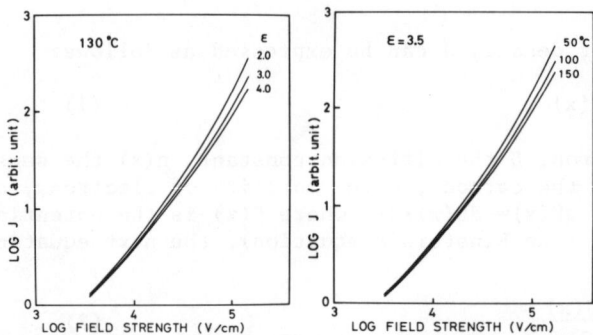

Fig. 1. Calculated curves of Eq.(7) with (a) T=130°C as a function of ε and (b) ε=3.5 as a function of T.

Fig. 1 represents the changes of curvature of $\log J$ vs. $\log V$ plots with ε and T. The vertical position of a curve has some ambiguity because $N_0\mu$ and ϕ are unknown parameters. Therefore the curve is shifted along the longitudinal axis to find a best fit position with experimental data points.

§4. RESULTS AND DISCUSSION

4.1. Field Dependence of Current: Fig. 2 shows the time evolution of current at (a) 105°C and (b) 125°C as a function of applied field. It is clearly seen in this figure that when the lowest field of $7.2\times10^{+3}$ V/cm is applied for the sample the current decreases gradually for more than 1000 min and never approaches a steady value, but that under the higher fields a steady state is accomplished in about 500 min or less. The higher is the applied field, the shorter becomes the time in which the current approaches a steady state value. The temperature gives the similar tendency to the applied field. In general, a current through dielectrics consists of a time dependent current and a steady state current. The time dependent current may be attributable to the re-orientation of a certain kind of dipoles attached to polymer chains, whereas the steady state current may be attributable to the migration of free carriers in the polymer samples. The current discussed

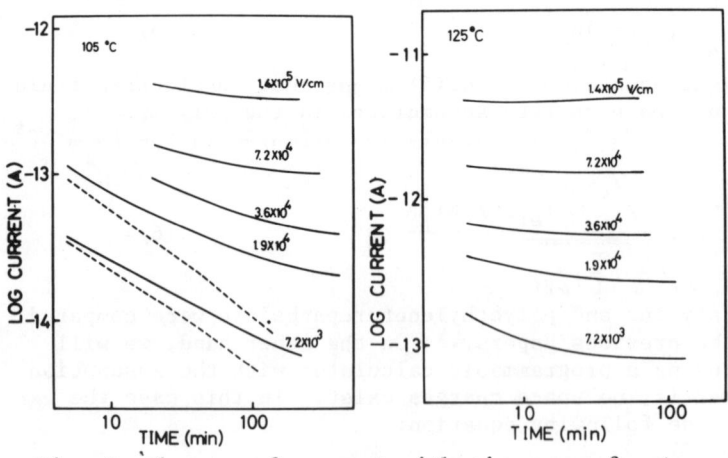

Fig. 2. Changes of current with time as a function of applied field: (a) at 105°C and (b) at 125°C.

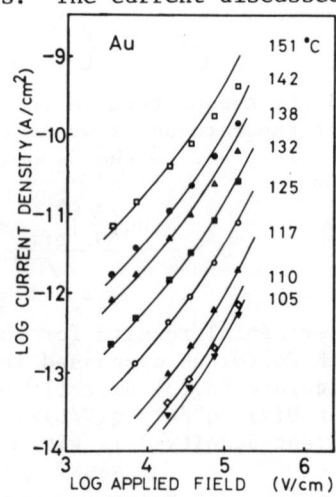

Fig. 3. J-V characteristics in PEN. Solid lines are calculated with Eq.(7).

hereafter is a steady state current defined in this manner.

Fig. 3 represents the field dependence of the steady state current at various fixed temperatures with Au electrodes. The plots in Fig. 3 are experimental results, while the solid lines are the calculated curves of Eq.(3) with the assumption of no space charges. As for the dielectric constant ε, was used the value of 3.5 obtained by the capacitance measurement at 60 Hz. As discussed in the preceding section, the vertical position of a curve cannot be fixed uniquely. That is, the calculated curve was shifted along the log J axis to find a best fit position with experimental points. It is found that the experimental results and the calculated curves agree considerably well with each other at lower temperatures. Some discrepancy at higher voltages and temperatures may indicate that currents were suppressed due to space charges accumulated in polymers, as is in the polyethylene-terephthalate.

4.2. Temperature Dependence of Current: In order to check the temperature dependence of the current density J experimental results given in Fig. 3 were replotted in the Arrhenius form (Fig. 4). In the temperature range of 110°C – 140°C, log J at a fixed applied voltage changes with 1/T in the linear fashion. Below this temperature range the deviation from the linear line is remarkable, which might be caused by the change of the molecular motion of polymer chains at T_g (=113°C). Above 140°C, on the other hand, the deviation might be caused by the change of the crystalline state of a biaxially stretched PEN, in spite of a thermal treatment before the measurement. Note that a dielectric loss peak observed at about 150°C and 1 kHz decreases in its height with the thermal treatment above 150°C. Therefore we will analyze the results only in the range of 110°C – 140°C hereafter. Eq. (7) is rewritten as

$$J/C \equiv \frac{J}{kT} \int_0^L exp[-(\frac{q^2}{4\varepsilon x} + q(\frac{V}{L})x)/kT] dx = N_0 \mu exp(-\phi/kT) \qquad (7')$$

The values of the left hand side of Eq.(7') can be calculated by using the experi-

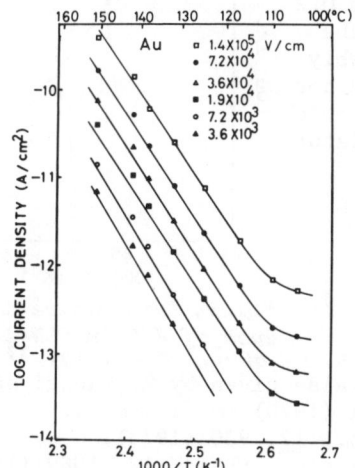

Fig. 4. Arrhenius plots of J in PEN with Au electrodes at various fixed applied fields.

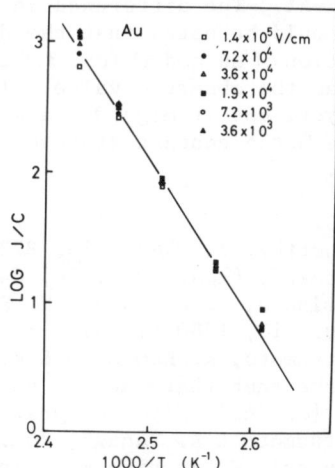

Fig. 5. Temperature dependence of J/C in PEN with Au electrodes.

mental data shown in Fig. 4. *Log* J/C is plotted against the reciprocal temperature at a various fixed applied voltages in Fig. 5. The slope gives an apparent activation energy ΔE_{obs}. As is clear from Eq. (7'), ΔE_{obs} depends scarcely on the applied voltage, then the average value of ΔE_{obs} is estimated $ca.$ 2.4 eV. Assuming that the mobility of electrons μ is expressed as $\mu = \mu_0 exp(-\Delta E/kT)$, where ΔE is an activation energy for the mobility and μ_0 the constant, one can derive the relation $\Delta E_{obs} = \phi + \Delta E$. Since the exact value of ΔE in PEN has been unknown so far, the literature values of other polymers[6][7], $0 \sim 0.3$ eV, are applied for ΔE. Then φ becomes $2.1 \sim 2.4$ eV. As a metal-insulator interface the energy barrier for the electron injection from a metal to an insulator would be lowered by the amount of the electron affinity χ of the insulator. Taking that χ in naphthalene[8] is $ca.$ 2.6 eV into account, the estimated value of φ for the Au-PEN interface may be $ca.$ 2.0 (=4.6-2.6) eV. This agrees fairly well with the observed value (2.1~2.4 eV). Here the work function of Au is estimated as 4.6 eV.

4.3. Electrode Effect:

Results obtained above were on the sample with Au electrodes. In the similar way the experiments with Al electrodes were carried out. Fig. 6 shows the temperature dependence of current J on this sample, and Fig. 7 shows the *log* J/C vs. 1/T plots. It is found that there are some differences in the absolute value of the current and in the activation energy ΔE_{obs} between the samples PEN-Au and PEN-Al. The difference in ΔE_{obs} is about 0.2 eV. It should be noted that the difference in the work function of Au and Al ($ca.$ 3.8 eV) is considerably larger than the observed value. The reasons are not clear as yet, but it might be caused by a microscopic difference in the contact state at the metal-insulator interface.

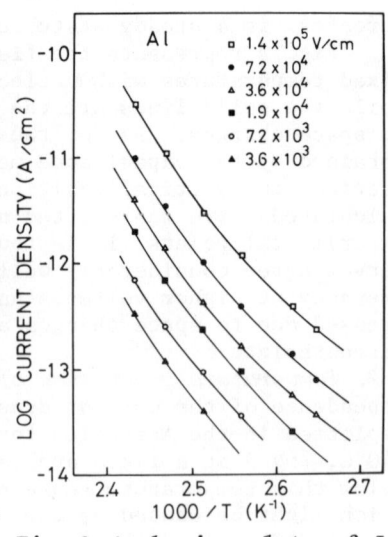

Fig. 6. *Arrhenius plots of J in PEN with Al electrodes.*

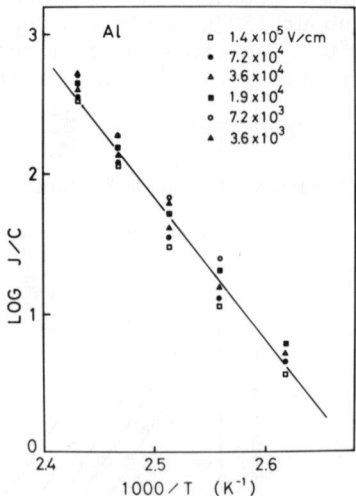

Fig. 7. *Temperature dependence of J/C in PEN-Al.*

REFERENCES:
/1/ W. Schottky, *Z. Phys.*, **15**, 872 (1914).
/2/ J. Frenkel, *Phys. Rev.*, **54**, 647 (1938).
/3/ K. Kamisako, S. Akiyama & K. Shinohara, *Jpn. J. Appl. Phys.*, **13**, 1780 (1974).
/4/ S. Matsumoto, K. Kamisako & K. Shinohara, *Jpn. J. Appl. Phys.*, **15**, 37 (1976).
/5/ The case that there are no space charges is also given by K. Yahagi and K. Shinohara [*J. Inst. Elect. Engrs. Jpn.*, **90**, 741 (1970) *in Japanese*].
/6/ S. Matsumoto & K. Yahagi, *Jpn. J. Appl. Phys.*, **12**, 930 (1973).
/7/ K. Hayashi, K. Yoshino & Y. Inuishi, *Jpn. J. Appl. Phys.*, **12**, 1089 (1973).
/8/ F. A. Matsen, *J. Chem. Phys.*, **24**, 602 (1956).

ELECTRICAL CONDUCTION IN DRAWN POLYMER

Masayoshi MASUI, Hideo NAGASAKA and Kichinosuke YAHAGI*

Ibaraki University, Hitachi 316, Japan

*Waseda University, Tokyo 160, Japan

1. Introduction

It is known that the orientation of polymer chains due to elongation, changes the electrical properties of polymers[1],[2]. Authors reported previously that uniaxial drawing of high density polyethylene increased electrical conductivity and decreased apparent activation energy[3]. This paper deals with the effect of uniaxial elongation on electrical conduction and the apparent activation energy, as well as changes of morphology under the condition of elongation in low density polyethylene pre-irradiated with gamma-rays.

2. Specimen preparation

Additive-free inflated low density polyethylene films were pre-irradiated by 2×10^5 Gy in a vacuum to produce a cross-link with cobalt-60 gamma-rays, mainly for the purpose of drawing the samples at 120°C without danger of melting. Films were 50 μm in thickness. The Melt index was 1.0 and the degree of crystallinity was ca. 48% derived from the estimation of density before irradiation. Degree of cross-linking was estimated to be 56% by the gel fractions. The specimens were drawn uniaxially in a device similar to an Instron stress-strain measuring system at temperature of 20 and 120°C in an air oven. The rate of drawing was 7 mm/min. Sample size was 60 mm wide and sample length varied from 50 to 110 mm. The draw ratio is expressed by the extent of drawing $(1-l_0)/l_0 \times 100$ %, where l_0 and l represent the lengths before and after elongation, respectively. The films were fixed with metallic frame jigs after drawing in order to maintain a constant draw ratio.

3. Measurements of electrical conduction and activation energy

Electrical conductivity was measured at room temperature maintaining a constant draw ratio at ca. 2 hours. Aluminium foil electrodes were attacted by means of silicone grease. After drawing at a certain draw ratio, conductivity decreased with time and became nearly constant in less than 2 hours. An electric field was applied perpendicular to the axis of the stretching direction. Current was measured by an electrometer with a time response of less than 5 seconds. As the electrical con-

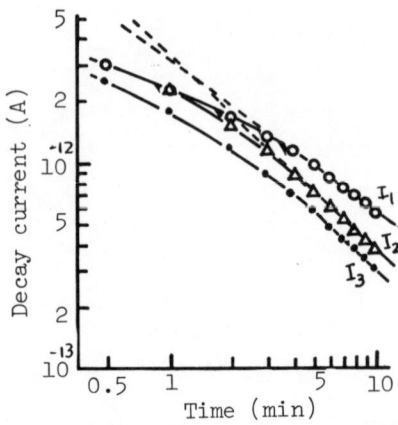

Fig.1. Decay currents as a function of time at room temperature. $I_1 = 3.30 \times 10^{-12} t^{-0.76}$ no drawing; $I_2 = 3.25 \times 10^{-12} t^{-0.94}$ draw ratio d=30%, $I_3 = 2.85 \times 10^{-12} t^{-0.92}$ d=60% at 20°C.

Table I. Values of n in the decay current.

Draw ratio	Drawing temperature 20°C	120°C
0%	0.76	0.76
30%	0.94	....
60%	0.92	1.19
120%	....	1.07

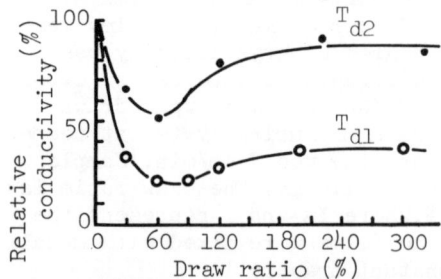

Fig.2. Effect of uniaxial elongation on electrical conductivity. Relative conductivity is expressed in values relative to the standard value of the undrawn sample $2.5 \times 10^{-15} (\Omega m)^{-1} = 100\%$. Drawing temperature $T_{d1} = 120°C$, $T_{d2} = 20°C$.

duction below 7×10^6 V/m was in the Ohmic region, electrical conduction was conveniently measured by applying the electric field of 1.6×10^6 V/m for 10 minutes. A 10 minutes value of current was adopted for estimating electrical conductivity. Samples, which were once applied to voltage, were short-circuited over 1 hour in order to discharge accumulated charges in the polymer.

In order to get activation energy, electrical conduction was measured in the air oven with a temperature controller, whose temperature was changed 20 to 60°C. The relationship between electrical conductivity σ and activation energy E may be expressed as follows:

$$\sigma = \sigma_0 \exp(-E/kT) \quad (1)$$

where K is Boltzmann's constant, T absolute temperature, σ_0 a constant. Activation energy is obtained from the slope of $\ln\sigma$ vs. $1/T$ curve.

4. Results and Discussion

4.1. Electrical conduction   Figure 1 shows the decay of electrical currents with time, where electric field of 1.6×10^6 V/m was applied to specimens with draw ratios of 0, 30 and 60%. Generally, the time dependence of the transient current corresponded to the expression.[4)]

$$I(t) = At^{-n} \quad (2)$$

where I represents the current, t the time after application of the external voltage, A the temperature dependence factor and n the constant. A magnitude of n=0.76 was obtained for undrawn samples; a large value of n was obtained for drawn samples, as shown in Table I. The relationshop between electric conductivity in the Ohmic region and the draw ratio is shown in Fig.2. There is a minimum conductivity at the draw ratio of 60 %.

In the electric field according to Ohm's law, a small number of carriers existing in a bulk of polyethylene may be thermally activated and flow to trap at the interfaces between amorphous and crystalline parts through the bulk. If the current carriers are ions, the increase in macroscopic density of polyethylene may impede ionic conduction. If this is the case, then, ionic conduction may explain results of figure 2 which showed the decrease of conductivity in case of 60% draw ratio. The explanation of the increase in conductivity in case of a draw ratio from 120 to 300% at 20°C, however, is

ELECTRICAL CONDUCTION AND BREAKDOWN

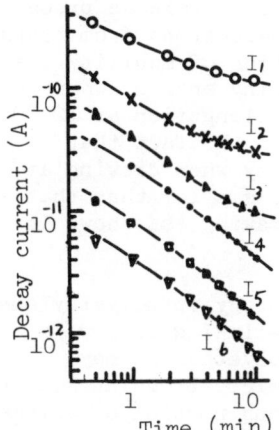

Fig.3. Decay currents for undrawn samples in various electric field strength(V/m). I_1=2.8 x 10^7, I_2=1.9 x 10^7, I_3=1.3 x 10^7, I_4= 9.4 x 10^6, I_5=3.6 x 10^6, I_6=1.6 x 10^6

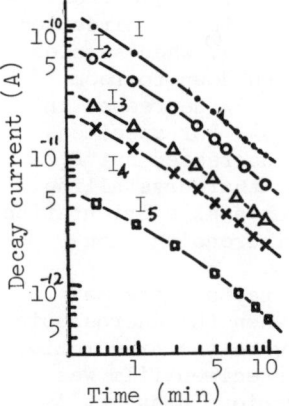

Fig.4. Decay currents for 20°C 300% drawn samples in various electric field strength (V/m). I_1=2.3 x 10^7, I_2=1.5 x 10^7, I_3=7.7 x 10^6, I_4=5.5 x 10^6, I_5= 1.6 x 10^6 (V/m).

quite difficult. On the contrary, the decrease of conductivity in case of a draw ratio up to 60 % may be also explained by assuming electronic conduction,i.e., elongation may generate deep traps for electrons,which then play an role in increasing the rate of current decay with time as shown in Table I. In the case of a high draw ratio at 20°C, increasing macroscopic density in amorphous parts of polyethylene may increase the conductivity due to the decreasing of distance between hopping sites.

In the range of high electric field , in which a relation between current and voltage becomes non-Ohmic, conduction current decayed very slowly in the case of undrawing but decreased rapidly with time in the case of drawing as shown in Figs.3 and 4. The conduction currents,which were measured 10 minutes after voltage supplied,were plotted in Figs.5 and 6 as a function of applied electric field strength. The current was increased super-linearly in the high electric field range. The magnitudes of current of 300% drawn samples as shown in Fig.4 was smaller than those of undrawn samples as shown in Fig.3. It is well-known that electronic charges injected from cathode by the effect of high electric field , immigrate by means of a hopping process. Figure 5 may be explained with the assumption of a particular case of space-charge-limited current injected into polyethylene in the high electric field range. The experimental results may be due to an increase of trap sites by drawing. Electric field strength of the transitional region from Ohm's law to space-charge-limited current was dependent on the draw ratio for samples drawn at 20°C, while at 120°C the electric field strength was almost independent of the draw ratio, as shown in Fig.5.

The difference in high electric field behavior in currents between 20 and 120°C drawn samples, means that elongation at room temperature results in various changes of polymer morphology, e.g.,c-axis orientation, splitting of crystalline lamellae and packing up of amorphous parts. High temperature elongation results almost the same change of morphology irrespective of the draw ratio, because when effecting elongation at 120°C, polyethylene is nearly in an amorphous state. In other words, the cooling down process after high temperature elongation may generate almost constant new electron traps, the number, trap level and the carrier hopping distance of which are independent of the draw ratio. The increase in density of polyethylene and simultaneous decrease in entropy due to the decreasing current when effecting elongation results in

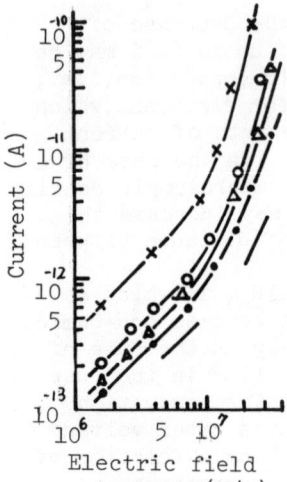

Fig.5. 10 minutes value of currents against the electric field strength for samples drawn at 120°C. Draw ratio d_1=0%, d_2=60%, d_3=120%, d_4=200%.

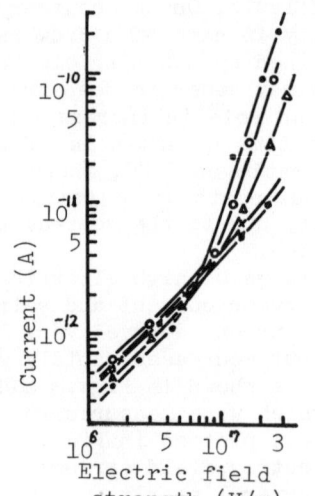

Fig.6. 10 minutes value of currents against the electric field strength for samples drawn at 20°C. Draw ratio -o- 0%, -•- 30%, -△- 120%, -x- 200%, -□- 300%.

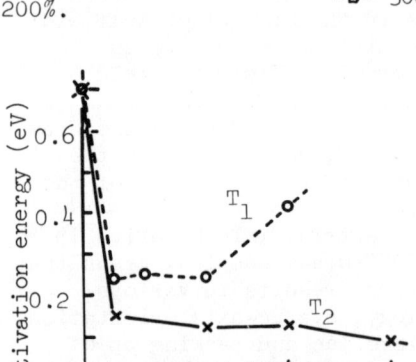

Fig.7. Apparent activation energy vs. draw ratio. Drawing temperature T_1=120°C, T_2=20°C.

the diminished release probability of electrons from traps. The reduction of mobility due to decreasing entropy when effecting elongation at 120°C may explain the fact that conductivity when drawing at 120°C, was smaller than that at 20°C drawing.(as shown in Fig.2)

4.2 Morphology in polyethylene

Cross-linking in low density polyethylene may suppress micro-Brownian motion[5] in amorphous portion of polymer, thus resulting in an increase in density.

It was observed by a method of infra-red dichroism that orientation function F_ε of the c-axis in the crystalline lamella increased and orientation function, F_α of the a-axis decreased when drawing at 20°C, i.e., the c-axis was oriented in the direction of the drawing[6]. when samples were drawn at 120°C and cooled down to room temperature, F_α increased and F_ε decreased in the draw ratio from 30 to 90%. For specimens that were further drawn, F_α decreased but F_ε increased slowly. This means that crystalline lamellae are transformed from a-axis orientation to c-axis orientation with increasing elongation over a 90% draw ratio.

The density of amorphous parts increases with increasing draw ratio when the macroscopic density increases together with the draw ratio. Macroscopic density in the specimen film was measured after drawing at various draw ratios by using a density gradient column method and was found to increase with the increasing draw ratio from the experimental results.[6]

4.3 Activation energy for electrical coduction

A relationship between apparent activation energy and the draw ratio in the Ohmic region is shown fig.7. As described above, uniaxial elongation produces not only many traps, which are due to the increase of interfacial regions between amorphous and crystalline parts, but also decrease of entropy (subsequent to the strong inter-chain attraction and closer packing of molecules). These new deep traps may cause a decrease in conductivity. Increasing temperature to ascertain the level of activation energy may relax internal stress produced by elongation and it may be assumed that these traps may change from deep to shallow states. The change of morphology due to stress relaxation caused by increasing temperature may associated with Brownian motion of amorphous parts.

5. Conculsion

It was found that uniaxial drawing of irradiated low density polyethylene resulted in a rapid rate of conduction current decay and a decrease of electrical conductivity in the Ohmic region. These results are attributed to the increase of deep trapping sites and decrease of entropy due to elongation (assuming electrons fuctioning as carriers). A little increase of electrical conductivity in the draw ratio from 60 to 300%, is due to an increase of carrier mobility in terms of decreasing of distance between shallow hopping sites. A transition electric field strength from Ohm's law to a space-charge-limited current was found to be dependent on the draw ratio for samples drawn at 20°C but it was almost independent on the draw ratio at 120°C. This difference is due to the fact that elongation at 20°C was carried out with polyethylene in an amorphous state.

Acknowledgments

The authors wish to thanks Mr. Yukio Sukegawa of Tokyo Shibaura Electric Co., Ltd. and Mr. Yoshio Maeda of Nippon Telegraph and Telephone Public Corp. for helping with this experiment as well as Nihon Petrochemicals Co.,Ltd. for preparing specimens.

References
1) L.E.Amborski: J.Polymer Science 62 (1962) 331.
2) D.A.Seanor : J.Polymer Science C14 (1967) 195.
3) M.Masui and K.Yahagi : Trans.Inst.Elect.Engrs.Japan 96-A(1976)165 (in Japanese).
4) B.V.Hamon :Proc.Inst.Elect.Engrs.Pt.4.99(1952)151.
5) W.P.Slichter,E.R.Mandell : J.Phys.Chem.62(1958)334.
6) Y.Maeda and K.Yahagi : Japan.J.Appl.Phys.16 (1977) 179

NON-DESTRUCTIVE CHARGE RELEASE FROM

HIGH VOLUME RESISTIVITY POLYMERS

J. E. West

Bell Telephone Laboratories

600 Mountain Avenue, Murray Hill, N.J., 07974 USA

I. INTRODUCTION

Teflon FEP and PTFE, used in electrets,[1] are frequently manufactured with residual non-uniform surface and volume charges[2] with densities greater than 10 nC/cm^2. Undesirable charging is caused by for example: friction,[3] static electrification and radiation. Such residual charges vary with sample area and and are in both shallow and deep traps that previously could not be emptied without damage to the polymer. Annealing and the use of non-penetrating radiation at best slightly reduce or compensate the residual charges. The lifetime of such charges is unpredictable, thus both uniform charge densities and stable electrets are difficult to make.

A new method to "clean" residual charges from Teflon is suggested which uses small doses pf penetrating gamma radiation to temporarily raise the conductivity of the polymer.[4] The radiation induced conductivity[5] (RIC) induced by the gamma radiation can be annealed[6] out restoring the polymer to nearly its virgin low conductivity.

Teflon samples 'cleaned' of residual charges were charged in a controlled manner using a mono-energetic non-penetrating electron beam.[7] The lifetime of the charge decay under various conditions was studied and compared with virgin (non-treated prior to charging) samples.

II. CHARGE RELEASE BY PENETRATING RADIATION

Fowler[4] has shown that penetrating X-ray radiation of polymers with low conductivity generates a large concentration of secondary carriers. These secondary carriers have no preferred orientation in the absence of an applied field. In the present experiment a ^{60}Co cell with a dose rate of 140 rad/sec or 8.5×10^{15} eV/g sec was used to generate secondary carriers. Since 10^2 eV/g sec is necessary to generate one carrier pair, this source has a production rate of about 10^{14} carriers/g sec. With the above dose rate, an induced conductivity in Teflon of about 10^{15} (ohm/cm)$^{-1}$ is obtained after 10^2 sec, or a dose of 24 krad. The induced conductivity in insulators due to radiation has been separated into three parts. The prompt conductivity,[8] having a time constant of a few msec., is essentially the free mobility of the trapped carriers. The radiation induced conductivity (RIC) and the delayed radiation induced conductivity[5] (DRIC) have time constants of seconds and days respectively. The induced conductivity allows the trapped residual charges to recombine with the generated carriers or drift under their own field through the polymer.

III. ANNEALING

A delayed radiation induced conductivity (DRIC) with a time constant of the order of years has been found in the irradiated volume of electron beam charged Teflon, that can be explained[5] by secondary carriers generated by the beam. These carriers are not in deep traps and can recombine by annealing[6] at temperatures between 100°C and 180°C.

Secondary carriers due to penetrating and non-penetrating irradiation have similar characteristics.[10] The carriers due to γ irradiating Teflon (see Sect. II) can be annealed out of the material by heating to 100°C for a period of 10 hours. Shorter annealing times at higher temperatures are also possible.

IV. EXPERIMENTAL DETAILS

Teflon type PTFE and FEP 25 μm thick with a 1000 Å aluminum electrode on one side were cut into circular samples 8.7 cm in diameter from stock rolls 10 cm width and over 500 m in length and framed in 50 μm thick cardboard rings. Initial residual surface charge densities $-10^{-9} < \sigma_I < -10^{-8}$ C/cm^2 were measured with variations at the extreme of σ_I within a few centimeters along the width or length of the Teflon rolls.

These samples were arbitrarily divided equally into three groups as shown in the upper part of Fig. 1. The samples in group (a) were isotropically irradiated in a ^{60}Co cell (Step II) and then annealed in a shielded oven for 10 hours at 100°C (Step III). Group (b) was annealed only (Step III) and the samples in group (c) were untreated.

The residual charge density of samples in group (a), treated by the γ irradiation was reduced to $\pm 5 \times 10^{-11} < \sigma_{II} < \pm 10^{-10}$ C/cm^2 and did not change during annealing (Step III).

The samples were then charged to between 1.6 and 3×10^{-8} C/cm^2 using a monoenergetic non-penetrating 20 keV electron beam[7] (Step IV Fig. 1). All samples were again annealed in open circuit (polymer surface free) at 150°C for 2 hours to eliminate the (DRIC) in the irradiated volume due to the charging electron beam (Step V).[6] The annealing process results in a slight reduction in the measured charge density due to charge movement deeper into the material as a result of the trap modulated mobility.[11]

After electron beam charging (Step IV) and annealing (Step V) the samples were studied under three environmental conditions: laboratory, elevated temperature isothermal conditions, and in humid atmosphere. The samples were stored either in open circuit (polymer surface free) or in short circuit (polymer surface contacted at some points) using pressed on polished aluminum electrodes.

The decay of the charge in polymer electrets is different for storage in open and short circuit.[12] In open circuit where $s_1 \gg s$ (see Fig. 2) the field[9]

$$E_{(a)} = -(\bar{\sigma}/\varepsilon), \tag{1}$$

where $\bar{\sigma}$ is the average surface charge density in the above and $\varepsilon = \varepsilon_1 \varepsilon_0$ is the normalized dielectric constant. The field $E_{(a)} = -10^5$ V/cm for open circuit storage with $E_{(b)} = 0$. Any field dependent decay mechanism, for example electron mobility,[11] would thus cause the charge to move deeper into the material reducing the effective surface charge.

Under short circuit conditions where $s_1 \leq s$ with pressed on shorting electrodes,

$$E_{(a)} = -(\bar{\sigma}/\varepsilon)\bar{r}/s, \tag{2}$$

$$E_{(b)} = -(\bar{\sigma}/\varepsilon)(1 - \bar{r}/s), \tag{3}$$

where s is the thickness of the foil and $\bar{r} = 5$ μm, the depth of the charge centroid at 20 keV.[13] For the same surface charge under short-circuit the compensating charges on the electrodes rearranges: $E_{(a)}$ and $E_{(b)} = -20$ and -80 KV/cm respectively. Since the field $E_{(b)}$ is much higher in the short circuit cases, field dependent decay is expected to be larger. Thus charge can migrate with the field to the surface and recombine with surface iron.

V. RESULTS

The results presented in this section represent data averaged over a least three samples for each data point. Room temperature and humidity cycled samples from groups a, b and c open circuit case show similar results. In short circuit, the observed differences are within experimental error since these samples depend on careful handling to avoid contact charging.

The results for both FEP and PTFE heated in a shielded oven at 100°C in open and short-circuit storage are shown in Fig. 3. Teflon FEP open circuit storage both irradiated-annealed (group a) and annealed, (group b) show a stepwise decay corresponding to previous results obtained at 150°C.[14] The time constant of the step in the present case is of the order of 20 days as compared to 0.2 days at 150°C. Such step decay characteristics further support the likelihood of multiple trap levels.[11] Other measurements on FEP at 100°C[15] indicate a corresponding decay of charge but without a step. It should be noted that such a step is absent from PTFE at 100°C, pointing to some basic differences in trap activation energies between the two Teflons. Under short circuit conditions FEP and PTFE samples behave about the same within experimental error over the measured period. For PTFE, the short circuit decay is faster than the open circuit case due to the considerably larger fields present in the free surface direction and the absence of the stepwise decay at 100°C.

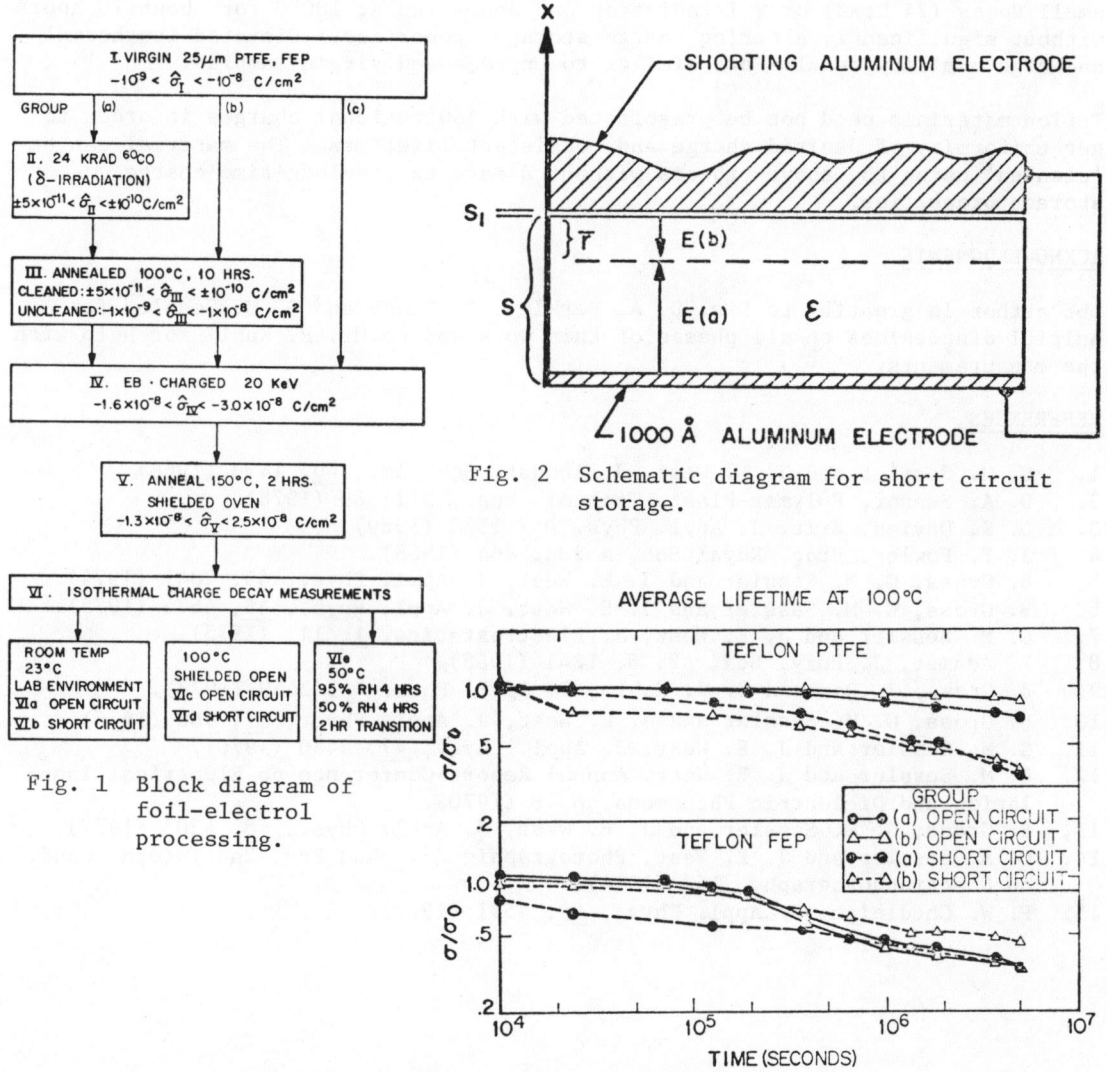

Fig. 1 Block diagram of foil-electrol processing.

Fig. 2 Schematic diagram for short circuit storage.

Fig. 3 Isothermal charge decay of 25 μm FEP, PTFE at 100°C.

VI. CONCLUSIONS

Results of this study clearly show that the residual charge normally found on most Teflon FEP and PTFE due to the manufacturing process can be 'cleaned' by small doses (24 krad) or γ irradiation and annealing at 100°C for about 10 hours without significantly altering charge storage properties. Elevated isothermal and high humidity results are similar to unprocessed virgin samples.

Teflon materials need not be preselected with low residual charges in order to get uniformity of desired charge and consistent lifetimes. The material can be 'cleaned' using the above process without damage to its long-time charge storage properties.

ACKNOWLEDGMENTS

The author is greatful to Drs. D. A. Berkley, B. Gross and G. M. Sessler for helpful discussions on all phases of this work and to Mr. R. Kubli for help with the measurements.

REFERENCES

1. G. M. Sessler and J. E. West, J. Acoust. Soc. Am., 40, 1433 (1966).
2. D. A. Seanor, Polymer-Plast. Technol. Eng., 3(1) 69 (1978).
3. D. K. Davies, Brit. J. Appl. Phys. D 2 1533 (1969).
4. J. F. Fowler, Proc. Royal Soc. A 236, 464 (1968).
5. B. Gross, G. M. Sessler and J. E. West, J. Appl. Phys., 45, 2841 (1974).
6. B. Gross, G. M. Sessler and J. E. West, J. Appl. Phys., 46, 4674 (1975).
7. G. M. Sessler and J. E. West, J. Electrostatics, 1, 111 (1975).
8. V. Adamec, J. Poly. Sci. A2, 6, 1241 (1968).
9. B. Gross, J. Dow and S. V. Nablo, J. Appl. Phys. 44, 2459 (1973).
10. B. Gross, G. M. Sessler and J. E. West, J. Appl. Phys., 47, 968 (1976).
11. G. M. Sessler and J. E. West, J. Appl. Phys., 47, 3480 (1976).
12. G. M. Sessler and J. E. West, Annual Report-Conference on Electrical Insulation and Dielectric Phenomena, p. 8 (1970).
13. B. Gross, G. M. Sessler and J. E. West, J. Appl. Phys., 48, 4303 (1977).
14. G. M. Sessler and J. E. West, Photographic Sci. and Eng. 2nd Intern. Conf. on Electrophotography, 162 (1974).
15. P. W. Chudleigh, J. Appl. Phys., 49, 4591 (1977).

ELECTRICAL CONDUCTION AND FREE RADICALS IN PLASMA POLYMERIZED STYRENE

Shinzo Morita, Meijo University, Tenpaku, Nagoya 468 Japan

Goro Sawa, Mie University, Tsu 514 Japan

Masayuki Ieda, Nagoya University, Chikusa, Nagoya 464 Japan

Mitchel Shen, University of California, Berkeley, Ca., U. S. A.

INTRODUCTION

Plasma polymerized films have potential applications as dielectrics and coatings for several purposes[1,2]. These films have several features distinct from conventional polymers. The film is usually highly crosslinked[3] and there are a lot of residual free radicals in the films[4]. Especially, it is supposed that the role of free radical may be important in the electrical properties of the film because it is active electrically and chemically, it lives in the film for a long time and the amount of residual free radical is large. However, relatively little attention has been directed to a role of free radical in the electrical properties. In this paper, the effects of free radicals on the electrical properties of plasma polymerized styrene(PPS) are discussed.

EXPERIMENTAL

Metal-PPS-metal sandwich specimens were prepared in a manner similar to that described previously[5,6]. The films were polymerized at a frequency of 5KHz, a pressure of 0.5torr and a discharge current of 1mA/cm^2 in styrene vapor. Photo- and darkcurrent and capacitance of the film were measured in vacuo(about 5×10^{-5} torr)[Stage I], in air(1atm, 25°C, 60%RH)[Stage II] and in re-evacuated condition [Stage III]. For this experiment, an apparatus was so designed that the electrical measurement could be made without opening the vacuum system to air after the fabrication of the sample. Three types of specimen were formed such as Au-PPS-Au (A type), Au-PPS-Al(B type) and Al-PPS-Al(C type). The front electrode metal of the specimen for the photocurrent measurement was always gold. The thickness of the metal film of gold was controlled by its conductivity, which was measured by using a resistance meter situated outside of the vacuum chamber. The effective area of electrode for capacitor was 4mm^2. The thickness of PPS film was calculated from the capacitance by assuming $\varepsilon'=3$. A xenon lamp with a power of 1KW was used as the light source. Photons of 200-700mμ were obtained through a monochromater. A positive current is defined as one in which a positive charge moves away from the front surface of the sample into the bulk of the film as shown with an arrow in Figure 1. Only dark current and capacitance were measured for C type specimen. ESR and transmission spectra for PPS were also measured.

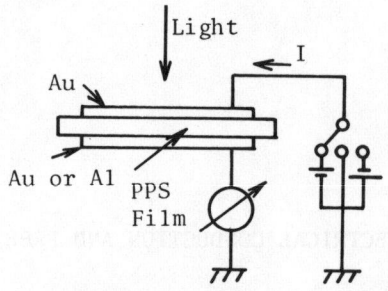

Fig.1. Experimental arrangement for photocurrent measurement. Positive photocurrent was defined as shown with an arrow.

RESULTS AND DISCUSSION

Photocurrents in short-circuit in vacuo are shown in Figure 2 from virgin A type and B type specimen. A polarity of short circuit photocurrent from the both samples was always positive. Photocurrent of A type specimen was larger than that of B type. Negative photocurrent was observed with a large negative applied voltage for the both types of the specimen. By introducing air in the apparatus, the photocurrent decreased invariably.

Square root of the photocurrent per incident photon is plotted against the photon energy for A type specimen in three stages as shown in Figure 3. The threshold energy of photocurrent of virgin film was 1.6eV for both polarities of photocurrent. When the film was exposed to air, the threshold increased and it became 2.1eV for both polarities of photocurrent in re-evacuated condition. In the case of A type specimen, a possible candidate for carrier generation is hole injection from the front electrode or hole generation in bulk. The energy barrier Φ_h for hole injection was estimated at 2.4eV or more by assuming the values of 6-8.8eV as an energy band gap E_g of polystyrene[7,8], 4.76eV as a work function W of gold metal and 1.2eV as an electron affinity χ according to the following equation,

$$\Phi_h = E_g + \chi - W.$$

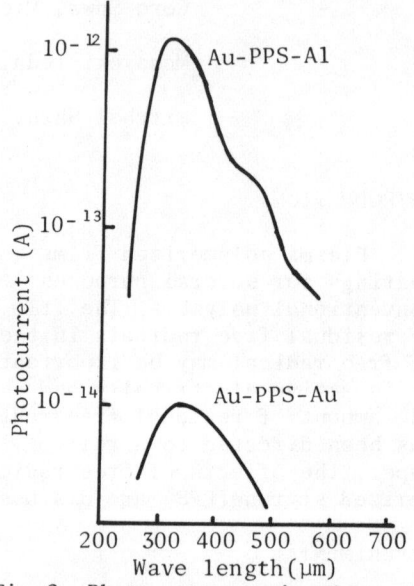

Fig.2. Photocurrents in short circuit in vacuo vs. wave length

The calculated value of energy barrier for hole injection is larger than the experimental value of 1.6eV. Then, hole generation in bulk is the most possible mechanism for Au-PPS-Au specimen. The small value of energy gap for hole generation suggests that there is some acceptor levels because the energy band gap of polystyrene is likely 6eV or more.

The square root of the photocurrent per incident photon for B type specimen is plotted against photon energy, as shown in Figure 4. The threshold energies of virgin film in vacuo are 2.6 and 1.9eV for positive and negative photocurrent, respectively. However, the threshold in re-evacuated condition is 2.0eV for the both polarities. Photocurrent of B type specimen is larger than that of A type specimen and the positive photocurrent flows from the gold electrode to aluminum electrode. The transmission spectrum for PPS film is shown in Figure 5 and an considerable amount of photon can be transmitted throughout PPS film at wave length for 250-600μm in contrast of conventional polystyrene. This fact suggest that an appreciable amount of photon reaches the bottom aluminum electrode. Guastavino et.al.[9] used 1.2eV for the electron affinity of conventional polymer. An electron affinity of PPS is expected larger than this value because the dielectric constant of PPS is usually larger than that of conventional polymer. Therefore, the barrier height for the electron injection may be smaller than 3.2eV. For virgin Au-PPS-Al type specimen, it may be concluded that most positive photocurrent is caused on the electron injection from the bottom aluminum electrode.

For the negative photocurrent with a negative large applied voltage on A type and B type specimen, the electron injection from the front gold

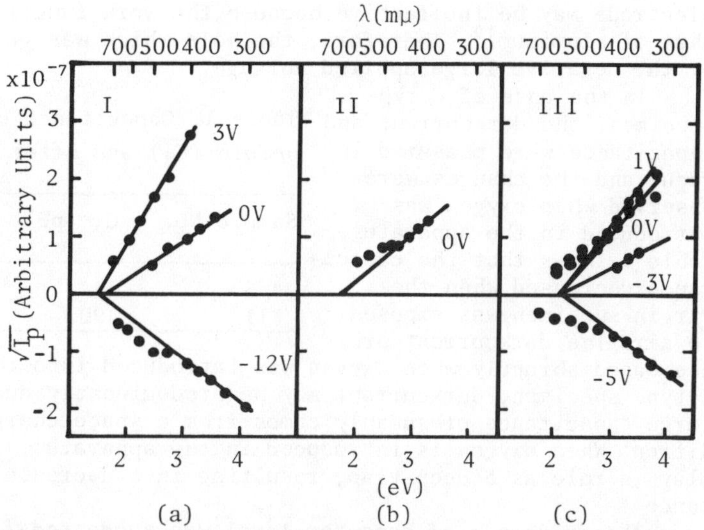

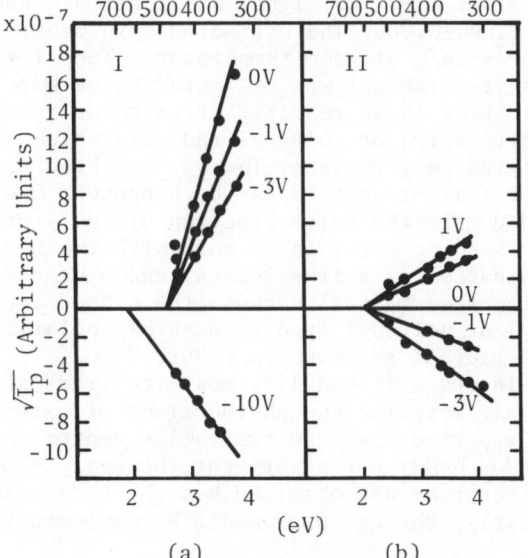

Fig.3. Square root of the photocurrent per incident photon vs. photoenergy for Au-PPS-Au type specimen

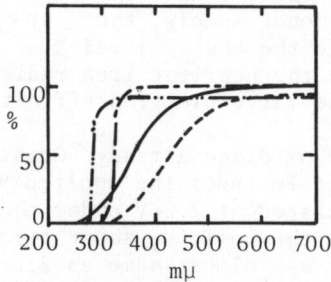

Fig.5. Transmission spectra
——;PPS, thinner than 1μ
----;PPS, thicker than 1μ
—·—;Styrene monomer
—··—;Conventional polystyrene

Fig.4. Square root of the photocurrent per incident photon vs. photoenergy for Au-PPS-Al type specimen.

electrode may be ineffective because the work function of gold is larger than that of aluminum[10]. Therefore, the hole which was generated in bulk may be drifted by the negative large applied voltage.

In the case of C type specimen, the darkcurrent and capacitance were measured in vacuo and the changes were observed when oxygen was introduced in the apparatus. Table I shows that the capacitance decreased when the virgin specimen was exposed to air. The darkcurrent are decreased abruptly when oxygen was introduced into the apparatus[5]. In the case of C type specimen, darkcurrent may be predominantly due to injected electron and large capacitance presumably comes from a space charge in inefficacious trap condition. When oxygen is introduced in the apparatus, free radical is oxidized and plays a role as a deep trap, resulting in a decrease in darkcurrent and capacitance.

Table 1. Capacitance of plasma polymerized styrene before (C_1) and after (C_2) exposer to air.

Sample No.	C_1 (pF)	C_2 (pF)	$\frac{C_1-C_2}{C_1} \times 100$ (%)
# 5	144	122	15.3
#11	190	163	14.2

The existence of acceptor level was suggested from a photocurrent of A type specimen. The acceptor may be caused by the residual free radical, but the free radical was also suggested to act as a shallow trap, from separated experiments. This inconsistence may be explained in following.

From several experimental results, it is supposed that the molecular structure of PPS is quite different from conventional polystyrene. Figure 6 shows the ESR signal of PPS. The signal was isotropic and g value was 2.000±0.005. The estimated spin density was about 5×10^{17} spin/g at room temperature. The life time of this free radical was estimated to be more than a few week. From these results, it is thought that PPS film involve allyl or polyenyl radicals which are observed in highly irradiated polyethylene. The various kind of roles of free radicals in electrical properties may be supposed. One of these is an electron trap due to the large electron affinity of free radical[11]. Another one is an acceptor in co-work with the nearest π-bond. Namely, the combination of a free radical and π-bond may change the energy level of the free radical because the π-bond accompaning the nearest free radical of allyl or polyenyl type is possibly polarized by the large electron affinity of free radical as shown in Figure 7.

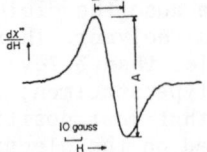

Fig.6. Typical ESR signal

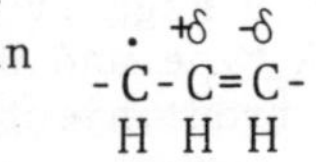

$\cdot$ $+\delta$ $-\delta$
$-C-C=C-$
$\;\;$H H H

Fig.7.

In stage II and III, most free radicals may be oxidized already. Guastavino et. al. measured the photocurrent of oxidized PPS film under the applied voltage and reported that an excitation center in bulk existed at 2.2eV under the conduction band. Our experiment, however, showed that the positive short circuit photocurrent was observed but the threshold energy was almost same as 2.2eV. Probably, the acceptor would be supposed to be exist

CONCLUSION

Photocurrent and darkcurrent of PPS film were measured in vacuo, in air and in re-evacuated condition. The roles of free radicals in the electrical properties were discussed. It was supposed that the free radical may act as either a shallow trap or an acceptor in co-work with the nearest π-bond in virgin film in vacuo.

REFERENCES

1) M. Shen ed.,"Plasma Chemistry of Polymer" Dekker, New York(1976)
2) J. R. Hollahan and A. T. Bell ed.,"Techniques and Applications of Plasma Chemistry" J. Wiely and Sons, New York(1974)
3) S. Morita, T. Mizutani and M. Ieda, Japan J. Appl. Phys., 10(1971)1275
4) S. Morita, M. Shen, G. Sawa and M. Ieda, J. Polymer Sci. Polymer Phys., 14 (1976)1917
5) S. Morita, G. Sawa and M. Ieda, J. Appl. Phys., 44(1973)2435
6) S. Morita, M. Shen and M. Ieda, J. Polymer Sci. Polymer Phys., 15(1977)981
7) Y. A. Cherkasov, L. N. Vinokurova, O. M. Sorokin and V. A. Blanck, Sov. Phys. Solid State, 11(1970)1959
8) M. Ofran, M. Oron and A. Weinreb, J. Chem. Phys., 48(1968)4805
9) J. Guastavino, H. Carchano and A. Bui, Thin Solid Films, 27(1975)225
10) T. Mizutani, Y. Takai and M. Ieda, Japan J. Appl. Phys., 12(1973)1553
11) J. H. Richardson, L. M. Stephenson and J. T. Brauman, J. Amer. Chem. Soc., 97 (1975)2967

RELATION BETWEEN DIELECTRIC BREAKDOWN AND MORPHOLOGICAL STRUCTURE IN
POLYETHYLENE

Kichinosuke Yahagi and Keiji Ishiki

Waseda University

1-4-3, Ohokubo, Shinjuku-ku, Tokyo, Japan

Impulse breakdown strength was obtained in polyethylene drawn with various drawing speeds and draw ratios at various temperatures. Samples of polyethylene were films of ca. 50 μm thick and the density was ca. 0.918 g/c.c.
They were free from any additives. A stainless steel spherical electrode 20 mm in diameter was placed in position along a center of the film drawn by a method similar to an Instron stress-strain measuring apparatus.

The breakdown strength was measured in silicone oil which was held at room temperature. The voltage of 70 % of the breakdown strength was at first applied to the sample and then increased 1 kV by 1 kV up to breakdown. The dichroic ratio was obtained on the drawn film fixed by a metallic frame. Dichroic ratio D is defined as $D = D_\perp / D_\parallel$, where $D_\perp$ and $D_\parallel$ are optical density for polarized lights with electrical vectors perpendicular and parallel to the drawn direction, respectively. An orientation function Fa for a - axis of a lamella of polyethylene crystal is determined from D for 730 cm^{-1} band and Fb for b - axis from D for 720 cm^{-1}. Fc is an orientation function for c - axis and can be calculated from Fc = - Fa - Fb.

Impulse breakdown strength was also increased with draw ratio $(1 - l_o)/l_o$ %, where l or l_o was the effective length of the film with or without drawing. Orientation function of c - axis Fc was monotonously increased but those of a - axis Fa and b - axis Fb decreased with drawing at room temperature. Figure 1 shows the relation between orientation functions and draw ratios. Figure 2 and 3 show the relation between impulse breakdown strength and orientation functions Fc and Fa, respectively. Breakdown strength is increased with increasing Fc and decreasing Fa in both cases of films drawn at room temperature and of films cooled down to room temperature under holding of constant draw ratio after drawing at 120°C.

Breakdown strength is considered to be associated with not only above orientation of axes of crystalline lamellae but also with increasing density of amorphous regions due to packing up of tie chains by drawing. Macroscopic density was increased with increasing draw ratio in films drawn at room temperature and 120°C by means of a density gradient column method.[1] Decrease of rare electron density regions with increasing draw ratio was also confirmed with a measurement of small angle X-ray scattering.[2]

Effect of drawing on breakdown strength may be decrease of free voids with drawing in amorphous regions and increase of main chain orientations perpendicular to the applied electric field.

If it is assumed that I is the energy to ionize polyethylene, especially polarized sites e.g. carbonyl groups, or end group of main chains solvated with small amount of water, and an electron can gain I in a free void by the electric field E_B as follows:

$$E_B = I / \lambda_f e \qquad (1).$$

An applied voltage to induce E_B in the free volume is considered to be a criterion of the breakdown, where λ_f is the mean size of free void which is assumed to be smaller than a mean free path (ca.600Å) in air and expressed by a following equation,

$$\lambda_f = \lambda_{go}(1 + \Delta\alpha_o(T - T_{go}) f_{go}^{-1})^{1/3} \qquad (2).$$

The dependence of a free void on temperature is taken to be the difference $\Delta\alpha_o$ between the thermal expansion coefficient α and α above and below the glass transition temperature, where λ_{go} is λ_f at T_{go} (ca.223K) which is a transition temperature for an impulse breakdown strength to decrease with temperature. T_{go} is considered to be related with the temperature beginning of microbrownian motion in amorphous regions. f_{go} is f at T_{go} and f is v_f/v. v_f and v are free volume and specific volume, respectively.

$$f = f_{go} + \Delta\alpha_o(T - T_{go}) \qquad (3).$$

λ_f and λ_{go} are proportional to $v_f^{1/3}$ and $v_{fgo}^{1/3}$, respectively. E_B can be expressed by

$$E_B = I e^{-1} \lambda_{go}^{-1} (1 + \Delta\alpha_o(T - T_{go}) f_{go}^{-1})^{-1/3} \qquad (4).$$

Elongation increases density of amorphous regions and suppresses microbrownian motion which may raise T_{go}. There is a report of increase T_{go} with drawing in some elastomers.[3]

Electron do not have to gain the full energy I in the free void in the case of increasing of a-axis orientation of crystalline regions. I should be changed with draw ratio because the breakdown was also induced by acceleration of electrons along main-chains in crystalline regions.

The effect of drawing speed on impulse breakdown strength was also investigated. Impulse breakdown strength was measured at room temperature in films which were drawn to a draw ratio 200% at several temperatures with various drawing speeds and then cooled down to room temperature.

Figure 4 shows the relation between an average breakdown strength on ca. 10 measurements and the drawing speed at various temperatures. We find the drawing speed v_m on which the breakdown strength shows min. at various environmental temperatures.

The relation between $\log v_m$ and $(f_{go} + \Delta \alpha_o (T - T_{go}))^{-1}$ has a straight line as shown in Fig. 5, where f_{go} is 0.025 and $\Delta \alpha_o$ is 4×10^{-4}/deg.. We obtain a following expression from Fig. 5,

$$v_m \propto \exp(-a/(f_{go} + \Delta \alpha_o (T - T_{go}))) \quad (5),$$

which is an similar expression to viscosity of amorphous polymers expressed as

$$\eta^{-1} \propto \exp(-bv_o^{-1}/(f_{go} + \Delta \alpha_o (T - T_{go}))) \quad (6),$$

where η is the viscosity, v_o is the specific volume and b is the constant.

Mechanical relaxation process by drawing has a chance to create a free void in amorphous regions which is able to become a trigger site for electrical breakdown. The relaxation time to create the free void is related with the viscous properties. If the time t for drawing to some draw ratio is equal to the relaxation time τ, a probability for creation of the free void may be max. or the largest free void may be created in comparison with the case that t is differnet from τ. If the relaxation time τ is assumed to be proportional to η, drawing speed v_m which has a time equal to τ for drawing to some draw ratio is proportional to η^{-1}. Therefore, v_m is expressed by Eq. (5).

REFERENCES:

1) K.Yahagi and Y.Maeda, Memoirs of the School of Science and Engineering, Waseda University. No.41, (1977) 31
2) K.Yahagi and S.Mita, IEE Conference Pub. No.129 (The Institution of Electrical Engineers, London, 1975)
3) G.Gee, P.N.Hertley, L.B.M.Herbert and H.A. Lancellay, Polymer, 1 (1960)365

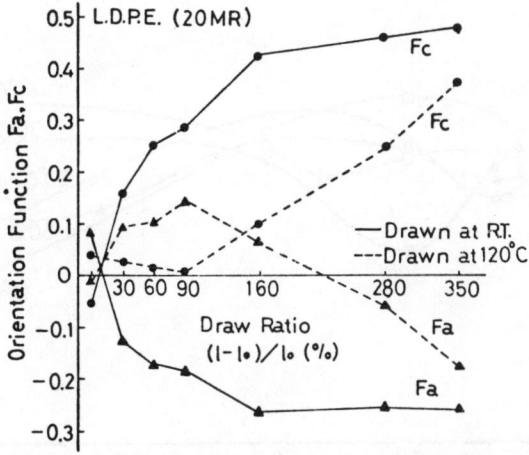

Fig. 1 Relation between orientation functions of c-axis and a-axis, and draw ratio $(l - l_0)/l_0 \times 100\%$, where l or l_0 is the effective length of the film with or without elongation. Solid line: elongated at room temperature. Broken line: elongated at 120°C followed by cooling down to room temperature with keeping constant draw ratio.

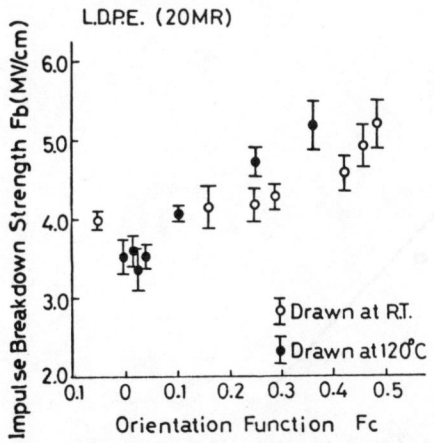

Fig. 2 Relation between orientation function F_c and impulse breakdown strength.

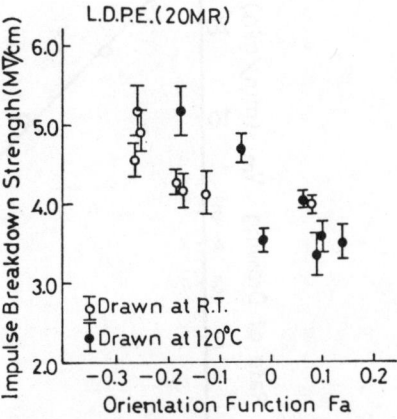

Fig. 3 Relation between orientation function F_a and impulse breakdown strength.

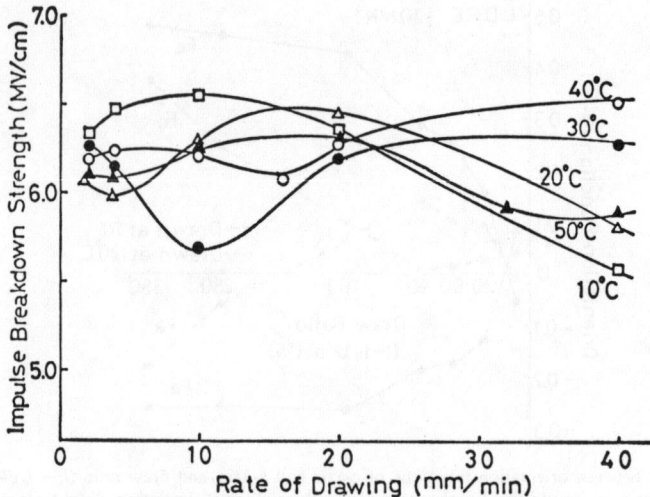

Fig.4 Relation between impulse breakdown strength and drawing rate.

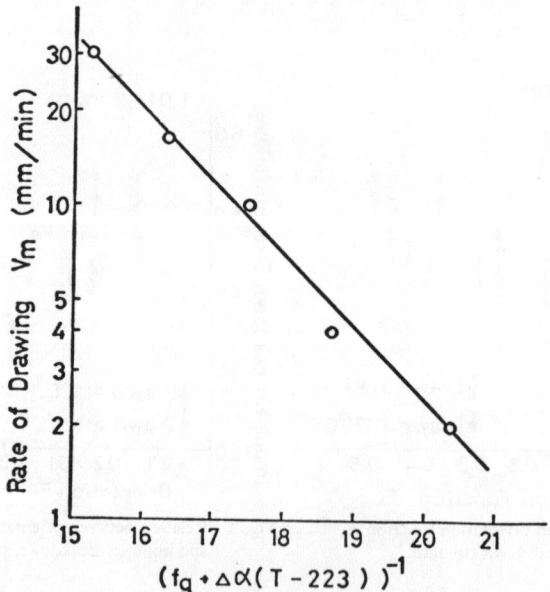

Fig.5 Relation between drawing rate corresponding to min. breakdown strength and $(f_g + \Delta\alpha(T-223))^{-1}$ in LDPE films drawn to 200%.

ON NONLINEAR DIELECTRIC RELAXATION II*

Akio Morita**

Engineering School, Trinity College, Dublin 2, Ireland
and
Department of Chemistry, Akita University, Akita-shi, 010,Japan

In I of this series, the Laplace transform of the polarisation following the sudden application of a unidirectional electric field is obtained exactly in terms of an infinite continued fraction. However, in this paper we shall investigate the theoretical implications of I in detail.

In order to calculate the polarisation, we use the following Smoluchowski equation for the rotational Brownian motion of a symmetrical top:

$$\frac{\partial \rho}{\partial t} = \frac{D}{\sin\theta} \frac{\partial}{\partial \theta} \left[\sin\theta \left(\frac{\partial \rho}{\partial \theta} + e(t) \sin\theta \rho \right) \right] \quad (1)$$

where t is time, ρ is the distribution function, D is the rotational diffusion constant, θ is the angle between the axis of symmetry along which a dipole μ lies and the direction of an applied electric field E(t).
Further in equation (1),

$$e(t) = \mu E(t)/k_B T \quad (2)$$

where k_B is the Boltzmann constant and T is the absolute temperature.
By expanding $\rho(\theta,t)$ in terms of the Legendre polynomials, $P_n(\cos\theta)$, namely,

$$\rho(\theta,t) = \sum_{n=0}^{\infty} a_n(t) P_n(\cos\theta), \quad (3)$$

and putting equation (3) in equation (1), we find the following recurrence relations:

$$\frac{da_0}{dt} = 0 \quad (4)$$

---

*An article in J. Phys. D(Appl. Phys.) 11, 1357-1367(1978) is referred to as I.
**Formerly at Trinity College, Dublin and now, at Akita University.

$$\frac{1}{Dn(n+1)} \cdot \frac{da_n(t)}{dt} = -a_n(t) + e(t)\left(\frac{a_{n-1}(t)}{2n-1} - \frac{a_{n+1}(t)}{2n+5}\right), \quad (n=1,2,3,\ldots) \tag{5}$$

In the case where a unidirectional electric field E_0 is applied at time t=0, on taking the Laplace transform of both sides of equation (5), we obtain

$$\{s + Dn(n+1)\}A_n(s) = Dn(n+1)e_0\left(\frac{A_{n-1}(s)}{2n-1} - \frac{A_{n+1}(s)}{2n+3}\right) \tag{6}$$

where
$$e_0 = \mu E_0/k_B T \tag{7}$$
and
$$A_n(s) = \mathcal{L}[a_n(t)] \tag{8}$$
in which
$$\mathcal{L}[f(t)] = \int_0^\infty f(t)\, e^{-st} dt. \tag{9}$$

From equations (3), (6) and (8), it follows that

$$\mathcal{L}[<\cos\theta(t)>] = \frac{A_1(s)}{3a_0} = \frac{2De_0}{3} \cdot \frac{\Lambda(s,De_0)}{s} \tag{10}$$

where angular brackets represent the ensemble average and

$$\Lambda(s,b) = \cfrac{1}{s+2D + \cfrac{\gamma_1 b^2}{s+2(3D) + \cfrac{\gamma_2 b^2}{s+3(4D) + \cfrac{\gamma_3 b^2}{s+4(5D) + \cdots}}}} \tag{11}$$

in which

$$\gamma_n = \frac{n(n+1)^2(n+2)}{(2n+1)(2n+3)}. \tag{12}$$

Equation (10) is the exact expression for $\mathcal{L}[<\cos\theta(t)>]$.

By expanding the continued fraction in equation (11) as a power series of e_0, we find that

$$\mathcal{L}[<\cos\theta(t)>] = \frac{2De_0}{3}\left[\frac{1}{s(s+2D)} - \frac{\gamma_1 D^2 e_0^2}{s(s+2D)^2(s+6D)} + \frac{\gamma_1 \gamma_2 D^4 e_0^4}{s(s+2D)^2(s+6D)^2(s+12D)}\right.$$

$$\left. + \frac{\gamma_1^2 D^4 e_0^4}{s(s+2D)^3(s+6D)^2} - \cdots\right] \tag{13}$$

Equation (13) may be inverted easily, giving

$$\langle\cos\theta(t)\rangle = \frac{e_0}{3}\left[1 - \frac{e_0^2}{15} + \frac{2e_0^4}{315}\right.$$

$$+ \left\{-1 + \frac{e_0^2}{20}(1+4Dt) - \frac{3e_0^4}{1750}\left(\frac{151}{40} + 6Dt + \frac{35}{3}(Dt)^2\right)\right\}e^{-2Dt}$$

$$\left. + e_0^2\left\{\frac{1}{60} + \frac{e_0^2}{420}\left(\frac{1}{12} - Dt\right)\right\}e^{-6Dt} - \frac{e_0^4}{13125}e^{-12Dt} - \cdots\right] \quad (14)$$

This equation is valid for small values of e_0.
On the other hand, for $t \gg (6D)^{-1}$ by assuming

$$s+2(3D) \simeq 2(3D)$$
$$s+3(4D) \simeq 3(4D)$$
$$\cdots\cdots\cdots$$

in equation (11), we find from equation (10) that

$$\langle\cos\theta(t)\rangle = L(e_0)\{1 - \exp(-Z(e_0)Dt)\} \qquad (t \gg (6D)^{-1}) \quad (15)$$

where the Langevin function $L(z)$ is defined by

$$L(z) = \coth z - \frac{1}{z} \quad (16)$$

and

$$Z(e_0) = \frac{2e_0}{3L(e_0)} \cdot \quad (17)$$

It should be noted that the validity of equation (15) is not restricted to the strength of e_0 but depends on the condition of $t \gg (6D)^{-1}$. The effective relaxation time τ_{eff} defined by

$$\tau_{eff} = \frac{1}{2D}\frac{3L(e_0)}{e_0} \quad (18)$$

is plotted agaist e_0 in figure 1.
Now, we shall investigate how sufficiently equation (15) may be used. To this end, we shall compare $\langle\cos\theta(t)\rangle$ from equation (15) with that from equation (14) for small values of e_0. However, in general, equation (15) should be checked experimentally also for large values of e_0. In figure 2, a plot of $\langle\cos\theta\rangle$ as $t\to\infty$ versus e_0 is shown to determined the range of e_0 in using equation (14). In this figure,

curve a represents $\frac{e_0}{3}$

curve b represents $\frac{e_0}{3} - \frac{e_0^3}{45} + \frac{2e_0^5}{945}$

curve c represents $L(e_0)$

and curve d represents $\frac{e_0}{3} - \frac{e_0^3}{45}$.

In figure 3, values of $<\cos\theta(t)>/(e_0/3)$ both from equations (14) and (15) are plotted versus Dt for several values of e_0. For $e_0^2=0.01$, $<\cos\theta(t)>/(e_0/3)$ from equation (14) agrees very well with that from equation (15), leading to a single curve represented by (—●●●●●—). However, values of $<\cos\theta(t)>/(e_0/3)$ from equation (14) for $e_0=1$ (—△△△△—) and for $e_0^2=2$ (—◇◇◇—) deviate around Dt=1 from those obtained from equation (15) for $e_0=1$ (—▼▼▼▼—) and for $e_0^2=2$ (—◆◆◆◆—), respectively. At the same time, it is seen that values of $<\cos\theta(t)>/(e_0/3)$ from equation (14) agree well with those from equation (15) at small and large values of Dt for $e_0=1$. The agreement at small values of Dt follows from the fact that $<\cos\theta(t)>$ hardly depends on e_0, whereas that at large values of Dt aries from the fact that the condition of Dt>>(1/6) is, in fact, satisfied. The difference in values of $<\cos\theta(t)>/(e_0/3)$ for $e_0^2=2$ at large values of Dt is due to the fact that the curve represented by (—◇◇◇◇—) tends to a value described by curve b in figure 2, while the curve represented by (—◆◆◆◆—) approached to a value of curve c in figure 2. Therefore equation (14) is not sufficient to consider $<\cos\theta(t)>$ for $e_0^2=2$ at large values of Dt.

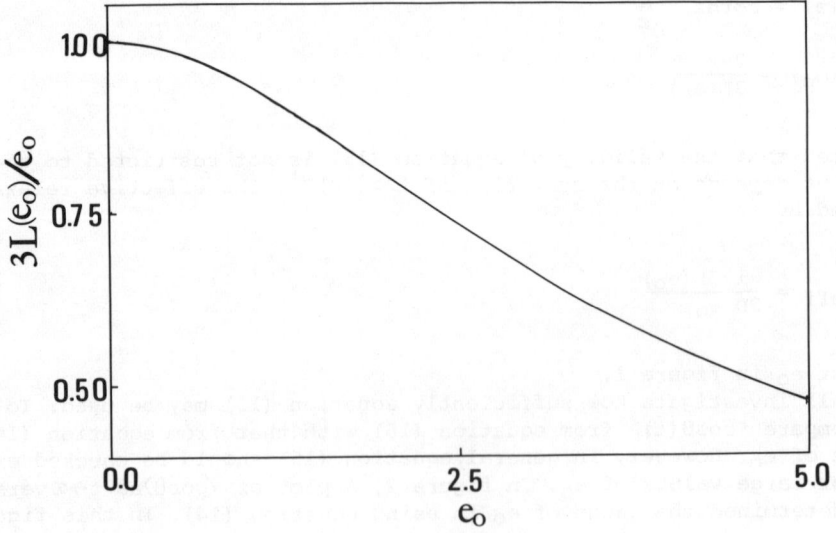

Figure 1

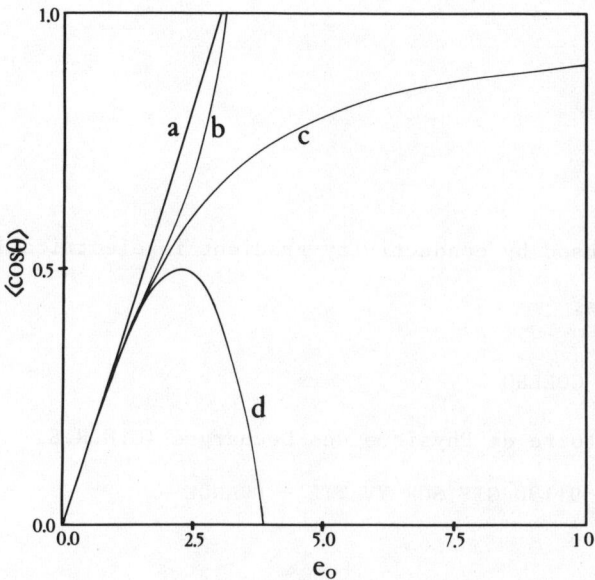

Figure 2

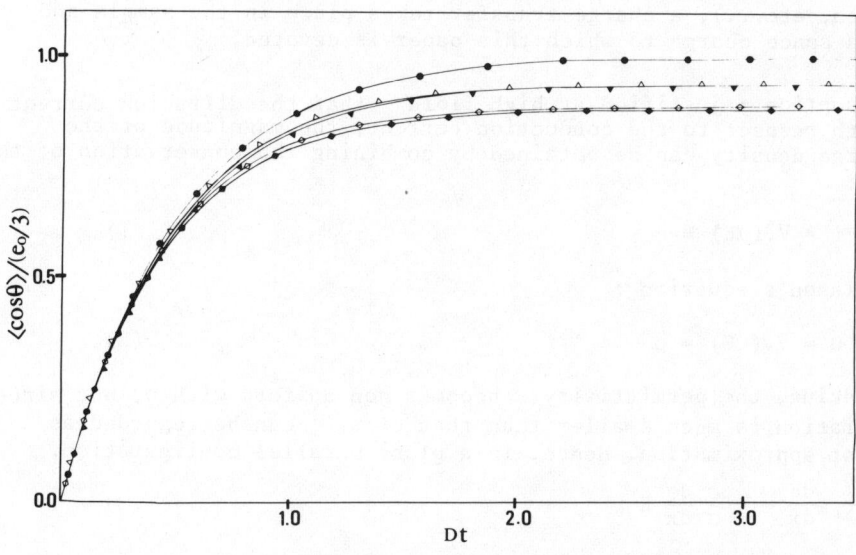

Figure 3

The storage of charge caused by conductivity gradient in electrically stressed semi-insulating materials.

Roland COELHO

Laboratoire de Physique des Décharges (C.N.R.S.)

E.S.E. 91190 GIF SUR YVETTE - FRANCE

Introduction

Let us consider a material of intrinsic conductivity σ submitted to an applied field E_a. If σ is locally distorted by some external means (heat flow, absorbed radiation, etc...), a charge transfer takes place in the sample and builds up in it a space charge to which this paper is devoted.

Under the assuption - justified at high field - that the diffusion current remains small with respect to the conduction current, the magnitude of the steady state charge density can be obtained by combining the conservation of the current density :

$$\nabla . i = \nabla . (\sigma E) = o \qquad (1)$$

together with Poisson's equation :

$$\nabla . D = \nabla . (\varepsilon E) = \rho \qquad (2)$$

Strictly speaking, the permittivity ε becomes non uniform with σ, but since its range of variation is much smaller than that of σ, ε can be regarded as uniform in a first approximation. Hence, in a plane parallel configuration,

$$\rho = \varepsilon \frac{dE}{dx} = - \frac{\varepsilon}{\sigma} \frac{d\sigma}{dx} E \qquad (3)$$

When a temperature difference ΔT is maintained across the sample of thickness L, the heat flux is uniform ; so is the temperature gradient, provided that the thermal conductivity of the sample does not vary appreciably between T and

($T + \Delta T$). Consequently, if σ depends on T exclusively,

$$\frac{d\sigma}{dx} = \frac{d\sigma}{dT} \frac{\Delta T}{L} \tag{4}$$

and, writing $\sigma(T)$ under the form :

$$\sigma(T) = \sigma(T_o) \exp \alpha(T-T_o) \tag{5}$$

a simple relation for ρ is easily obtained :

$$\rho(x)^\bullet = - \varepsilon\alpha \frac{\Delta T}{L} E(x) \tag{6}$$

The total charge $Q = \int_0^L \rho(x) dx$ stored by unit area is, using eqn. (6) for $\rho(x)$:

$$Q = - \varepsilon\alpha \, \Delta T \, E_a \tag{7}$$

where E_a is the applied field

Formulaes for $\rho(x)$, Q, $E(x)$ and the current in the external circuit at time $t = 0+$ (i.e. just after ΔT is applied) and in the steady state ($t = \infty$) are displayed in Table 1 [1], [2]. In this table, I_o stands for the isothermal current $I_o = \sigma(T_o) E_a$, and $Z = \alpha \Delta T$. The table covers ohmic materials in column 1 and "hyperohmic materials" in which $\sigma \propto A(T) E^n$ in column 2. As expected, column 2 reduces to column 1 if $n = 0$.

Table 1

	ohmic materials	hyper-ohmic (E^n)
$E(x, 0+)$	E_a	E_a
$\rho(x, 0+)$	0	0
$I(0^+)/I_o$	$(e^z - 1)/z$	$(e^z - 1)z$
$E(x, \infty)$	$z \exp(xz/L)/(e^z-1)$	$\dfrac{z/(n+1)}{e^{z/(n+1)}-1} \exp\left[\dfrac{xz}{L(n+1)}\right]$
$\rho(x, \infty)$	$-\varepsilon z E(x, \infty)/L$	$-\varepsilon z E(x,\infty)/L(n+1)$
$Q(\infty)$	$-\varepsilon z E_a$	$-\varepsilon z E_a/(n+1)$
$I(\infty)/I_o$	$z/(1 - e^{-z})$	$\left[\dfrac{z/(n+1)}{1 - e^{-z/(n+1)}}\right]^{n+1}$

This table shows in particular that a field dependence (E^n) of σ reduces the stored charge Q (line 6) by the ratio(n + 1). In principle, n can be obtained by comparing the charge actually stored in the sample to value ($-\varepsilon z\, E_a$) for ohmic materials. By checking this value of n against that which results from the s ope of the current-voltage characteristics plotted on log-log graph, it should be possible to assert the validity of the intrinsic conductivity hypothesis.

Experimental

Attemps to measure the charge stored in samples of polyethylene and polystyrene stressed electrically at temperatures up to 70°C in the presence of a thermal gradient (electodes at T and A + ΔT respectively) have been made by three different techniques described below :

a) stress a sample without thermal gradient, cool rapidly under stress (fig. 1), measure its net charge in a total influence capacitor. Repeat with a thermal gradient and compare the two results

b) same as a), but scan both faces of the sample with an electrostatic probe to obtain the net charge. Repeat with a thermal gradient.

c) record - in situ - the transient current following the application of the thermal gradient, and compare the data with the prediction of the theory which has been developed for a step function of ΔT [1].

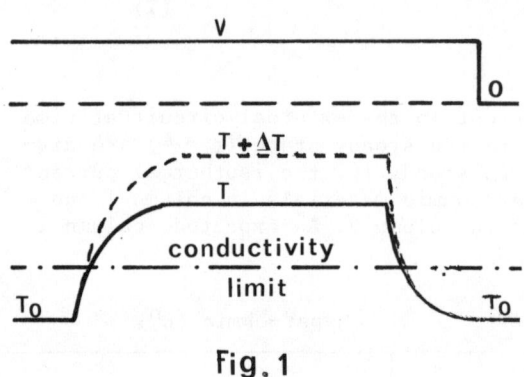

Fig. 1

Techniques a), b) and c) have usually produced data of the expected magnitude ($< \varepsilon z\, E_a$), but some samples dispay a charge of the wrong sign. Furthermore, the agreement between the data given by these techniques was often unsatisfectory. For these reasons a new technique based on thermally stimulated depolarization, specially designed to detect the charge under investigation is being developed, and will be described at the workshop.

Practical importance and possible application of the charges

Although they are not easy to measure, the charges under investigation here seem to be present and to play a significant role in many practical situations. Among these, we shall mention briefly electrothermoconvection in fluids, electroreprography and charge decay in corona irradiated polymers, and divergent field configurations.

Thermoelectroconvection. Application of an electric field to an insulating fluid undergoing natural thermoconvection is expected to enhance the convection, since the lowered permittivity of the heated liquid should cause a dielectrophoretic

repulsion from the high field heated electrode. Experiments carried out with a
wire heated by Joule effect in various insulating liquids have shown that an
electric field may actually increase the temperature of the wire - thus impede
the heat transfert to the liquid - when this liquid contains ionic impurities [3].

According to the model developed above, a space-charge builds up around the
wire in the combined electric field and thermal gradient surrounding it, and the
net electrophoretic force on this space charge has been shown to be oriented
downward - hence to counteract the upward convection. This might account for the
unsuccessful attempts to improve heat transfer between solids and fluids, at least
in laminar convection, by using an electric field.

Electrophotography. In electrophotography, a corona in air from an array of
sharp tips connected to a source of high positive voltage charges the surface
of a photo-conducting plate. The image to be reproduced, formed with photons of
$h\nu$ higher than the energy gap of the photoconductor, is focused onto the plate.

After the photo-excited electrons at the illuminated sites recombine with
the positive ions on the surface, surface carriers have - so to speak - be replaced by bulk carriers. (fig. 2a)

From our electrostatic model, the enhanced conductivity in the absorbing
layer, combined with the field due to the surface charge, build up a positive
space charge under the illuminated surface (fig. 2b)

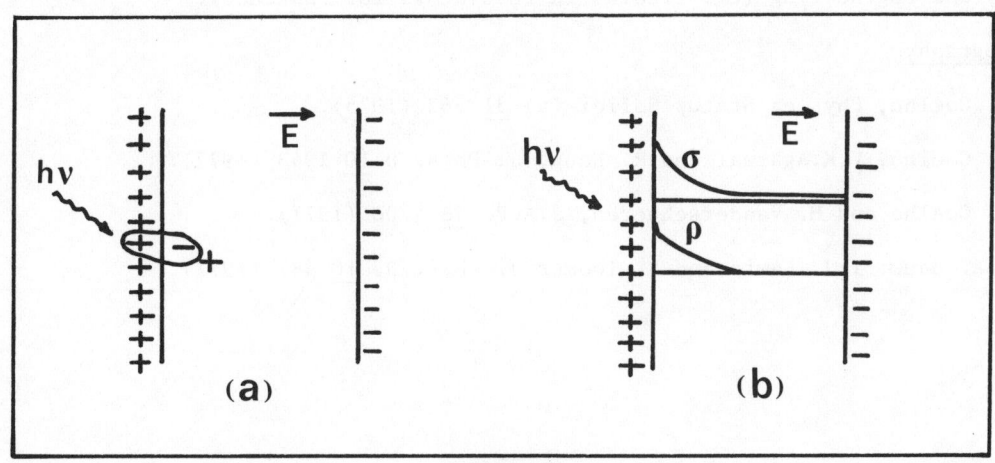

Fig.2

Hence, the model appears as the macroscopic analog of the more usual microscopic model (fig. 2a) of the electrophotographic process, and this approach
might prove of practical interest.

Charge decay in corona irradiated polymer sheets

It has been suggested by Baum et al. [4] that ultraviolet radiation emitted by the glow favors the decay of charges on a corona charged polymer sheet.

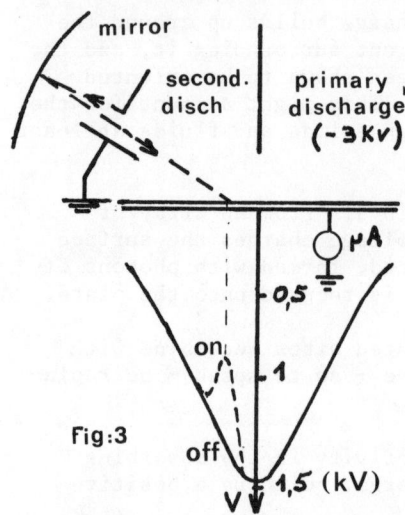

Fig:3

We have confirmed this by scanning, with an electrostatic probe, a polyethylene sample charged by a "primary" discharge, on which the light from an "auxiliary" discharge (of same spectrum but higher intensity) was focused by a spherical aluminum mirror. In fact, recordings of the decay seem to indicate that the light of the "auxiliary" discharge does not just enhance the decay rate, but also contributes to the build up of a space charge, according to the above model.

Divergent field configurations. Due to the hyperohmic nature of most semi-insulating materials, space charges of the type investigated here tend to develop many configuration - even isothermal - in which the Laplace field is not uniform, that is in all electrode configurations except the plane/parallel one. This applies, in particular, to the tip/plane and wire/cylinder systems commonly used in physics, chemistry and engineering (cf. treeing in insulators for instance).

Bibliography

(1) R. Coelho, Physica Status Solidi (a) 31 563 (1975).

(2) R. Coelho, V.K. Agarwal and R. Haug, J. Phys. D 10 1943 (1977).

(3) R. Coelho and H. Vanderschueren, J.A.P. 48 4700 (1977).

(4) E.A. Baum, T.J. Lewis and R. Toomer J. Phys. D. 10 487 (1977).

VOLTAGE-INDUCED CHARGE TRANSFER AND ITS EFFECTS ON TREE INITIATION IN POLYETHYLENE

Toshikatsu Tanaka & Allan Greenwood*
Central Research Institute of Electric Power Industry
Komae-shi, Tokyo, Japan,
and
*Rensselaer Polytechnic Institute
Troy, New York, U.S.A.

Introduction

Treeing is a process of partial electric breakdown in a solid dielectric occurring at very high electric field regions. It requires some time interval to take place, after voltage application. It is called the incubation period of treeing. Nothing appears to happen in this period, after which a tree grows to make visual inspection and electrical measurement possible. However, something which is of accumulative character must take place in the period, resulting in tree initiation. Electric charge carried by either electrons or holes can be injected into a dielectric from a metal electrode when it is biased by d.c. voltage. When it is short-circuited, almost all of the charge is extracted, but some of it remains in the dielectric. When a.c. voltage is applied, charge is considered to be injected and extracted, too, according to voltage polarity. Such a repeated process of accumulative character must be responsible for the treeing breakdown of polymer dielectrics such as polyethylene (1)(2)(3). This paper gives some evidences for the above phenomena and proposes a model for the initiation of the treeing breakdown of polymers.

Injected charge

Injected charge is measured in needle-plane electrode system by a short-circuiting methode as shown in the upper illustration of Fig. 1. Injected charge obtained is shown in Fig.1. as a function of applied voltage with respect to polyethylene and polyethyleneterephthalate. Figure 1 suggests three important aspects of charge injection characteristics of polymers: (i) Injected charge increases with applied voltage. (ii) Charge injection level depends on the kind of materials: polyethylene exhibits the higher injection level than polyethyleneterephthalate in this case. (iii) For the two kinds of polymers described above, the injected charge appears to be larger with positive voltage applied to the needle tip than with negative voltage.

Trapped Charge and Its Liberation by the TSC Method.

Some amount of injected charge remains trapped in the specimen even after it is short-circuited. This can be liberated by the thermally stimulated current (TSC) method. A TSC experimental method in the paper is different from the usual way in the following two aspects. First, it is categorized as non-uniform field charge injection rather than the uniform electric field charge injection. A razor blade to plane electrode system was used instead of the plane parallel electrode system, giving rise to the divergent electric field distribution. Second, the forming (or injecting) temperature is same with the TSC measurement starting temperature in the experiments, while the former is usually higher than the latter. It is to verify that the charge injected at a certain temperature can not be discharged completely in the short-circuit condition at the same temperature.

Figure 2 shows the total charge liberated up to 85°c by the quasi TSC method as a function of forming (or injecting) time at room temperature. This indicates that charge carriers are injected at room temperature and some of them ramain or are trapped in polyethylene even under short-circuit condition. Trapped charge tends to increase with voltage application time. Another outstanding feature of the figure is that trapped charge is more when the razor electrode has been negatively biased than when it has been positively biasesd. Figures 1 and 2 suggest that negative charge carriers are less injected and more trapped than positive charge carriers. Negative and positive carriers are considered to correspond to electrons and positive holes, respectively.

Repeated Process of Carrier Injection and Extraction and Germination of Trees.

The phenomenon described above takes place under d.c. condition. When a.c. voltage is applied to a treeing specimen, electrons and holes will be injected from a needle electrode into the dielectric during the negative and positive rising quarter cycles, respectively. And they will be almost drawn back to the electrode within the followed falling quarter cycles. Some of the injected carriers will remain even at the zero cross points of the voltage and will be extracted compeletely within the next quarter cycles by the external voltage of opposite polarity. In other words, there are two kinds of charge carriers. Carriers of the first kind are free to move in phase to applied a.c. voltage and therefore show quick response to the phase change of the voltage. Carriers of the second kind are, on the other hand, hard to move because of their trapped state, and therefore show out-of-phase or retarded response to applied voltage.

Carriers of the first kind will mitigate the electric field around the needle electrode due to the in-phase formation of a homo-charge layer. Carriers of the second kind will contribute to intensify the electric field around the needle electrode, when the voltage reaches the beginning of the second half cycles. They form homocharge in the injection half-cycles, but turn out to be heterocharge in the followed half-cycles. This phenomenon takes place due to their retardation character.

Behaviors speculated above are schematically illustrated in Fig.3. Both electorons and positive holes are injected. According to the experimental results described in the preceeding sections, electrons are less injected but more trapped in polyethylene than holes. For simplicity, it is considered in the figure that holes are not trapped, while some of the injected electrons are trapped. Figure 3 shows one of those patterns of response of injected charge and electric field near the needle electrode tip that are most probable. Electrons make a retarded move against voltage change and therefore are drawn back after the injection cycles to give rise to peaking of the electric field due to heterocharge against the polarity of applied voltage. Holes, on the other hand, follow the phase change of the voltage.

The process described above——charge injection and quick and retarded extraction——is repeated according to altenating voltage. On each cycle of voltage, electrons or holes have a chance to gain energy sufficient to cause some polymer decomposition to lower molecular products and gas. Repetition of such a process will result in the formation of a hollow channel large enough to accomodate gaseous discharges or in the reduction of breakdown strength due to material degradation to cause a local dielectric breakdown. It means the start of tree formation. As is clear, this process is of accumulative type.

Either trapped or free carriers of injected ones or both will initiate tree germination. Which carriers will start a tree in dielectrics is uncertain. No direct evidence for it is available yet. But it is interesting to note that trapped carriers responsible for the intensification of the electric field near the needle tip are considered to gain higher energy than free carriers. In this case, a tree will be germinated at the time of electric field maximum as shown in Fig.3 (lower trace).

For treeing experiments, a single needle electrode specimen was utilized, having a needle electrode of 1 mmϕ diameter with a 3μm tip radius, and 3 mm electrode spacing. Space for tree germination was found to be up to 20μm apart from the needle tip. Within this range of space, carrier injection and extraction effective for tree initiation are made, depending on the magnitude of voltage. Charge injected to the tree-formation effective region is uncertain. But the injection of charge 20 pC equivalent to 10^8 electrons or holes would produce the space charge electric field of the order of 10^6 v/cm in the sphere of 20μm diameter with carrier density of $3 \times 10^{16} cm^{-3}$.

Voltage vs. Tree Initiation Time Characteristics

Since tree initiation is caused accumulatively by the repetition of the carrier injection and extraction process, the relation between tree initiation voltage and time can be derived if the injection process involved is definite. Effects of work functions of electrode metals and surface states of dielectrics were investigated (2)(3). This suggests that carrier tunneling would take place via the field emission.

A formula is derived for such a V-t characteristic on the basis of the following three assumptions: (i) Charge carriers are injected via the field emission. (ii) Carriers can contribute to tree initiation only if they have energy exceeding a certain critical energy. (iii) They have a cumulative effect on the tree initiation, i.e., the same amount of total energy can cause the same amount of damage to a dielectric. It is indefinite whether forward or backward movement is more significant and whether free or trapped carriers are more important in case of the back ward movement.

The three assumptions stated above will give

$$\log t_I = a\phi^{3/2}/V + \log[b/\{a.\phi^{3/2}(1/V_0 - 1/V)\}] \quad (1)$$

where ϕ: The effective work function, V_0: a critial voltage below which no treeing breakdown would be anticipated, V: applied voltage, a & b: constants. The effective work function ϕ is usually the difference between the work function of a metal and the electron affinity of a dielectric, and furthermore includes the effect of surface states of the dielectric. This can explain experimental results better than the common law ($t_1 \propto V^{-n}$) for lifetime.

If V-t_I characteristics are obtained for different electrode materials with exactly the same dielectric and electrode configuration, the vagueness of V can be eliminated. Then we obtain

$$\log(t_1/t_2) = (a/V)(\phi_1^{3/2} - \phi_2^{3/2}) + \alpha \quad (2)$$

where suffices 1 and 2 denote different electrode materials and α is constant. Figure 4 shows a V-t_I characteristic represented in the way of formula (2) for polyethylene specimens with steel or silver needles as electrodes. It is of interest to note that there seems to be observed a straight line between $\log(t_1/t_2)$ and $(1/V)$. This result supports the model proposed in the paper.

Conclusion

It is proposed that a repeated process of voltage stimulated charge transfer causes tree germination in polyethylene via material degradation. A new formula is derived for V-t_I characteristics. Retarded carrier movement is possible to initiate a tree. Further study is needed to establish a detailed model for tree initiation in polymers.

This research was financially supported by the National Science Foundation which the authors would acknowledge.

References
(1) G. Bahder, T. W. Dakin & J. H. Lawson, CIGRE 15-05(1974)1-16.
(2) T. Tanaka & A. Greenwood, IEEE Trans. EI, in the press.
(3) T. Tanaka, C. W. Reed, J. C, Devins & A Greenwood, Paper to be presented at the 1978 Conf. EI & DP Pocono, Pa., Oct 29 to Now. 2, 1978.

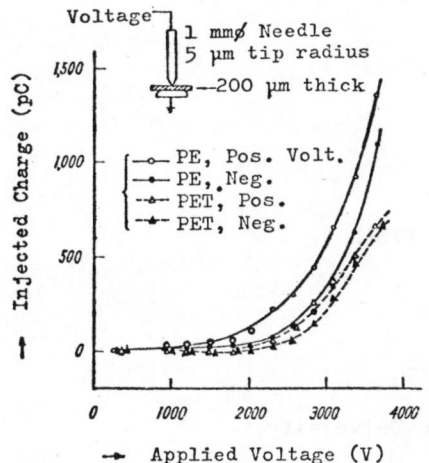

Fig. 1. Voltage dependence of injected charge for PE and PET films in a needle-plane electrode system.

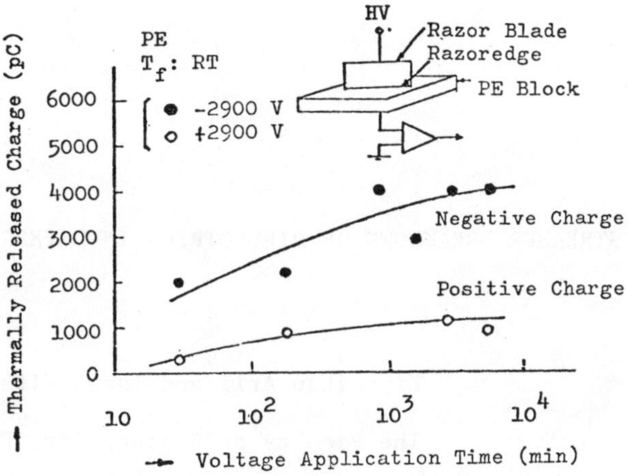

Fig. 2. Charge liberated by the quasi TSC method vs. voltage application time (Charge is injected from a voltage-biased razor blade at room temperature).

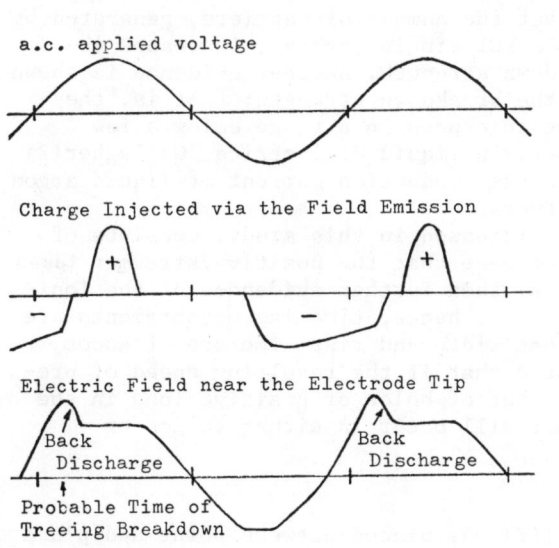

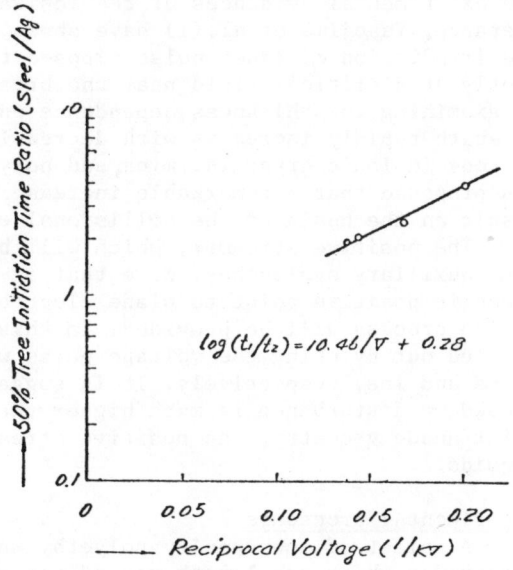

Fig. 3. Speculative response of injected charge and electric field near the electrode tip as regards tree initiation.

Fig. 4. $\log(t_1/t_2)$ vs. $(1/V)$ characteristics.

THE STREAMER BREAKDOWN OF DIELECTRICS IN DIVERGENT FIELD

Kiyomitsu Arii and Isamu Kitani

The Faculty of Engineering, Ehime University

Bunkyo-cho, Matsuyama, Ehime, Japan

Introduction

One of the fundamental processes in the breakdown phenomena is a carrier multiplication that results from the ionization caused by electron collisions. The experimental evidences of the ionization have been shown by many authors; for instance, Yasojima et al.(1) have shown that the number of carriers, generated by the irradiation of laser pulse imposed to a KCl single crystal, increases abruptly at a critical field near the breakdown strength. Another evidence is shown by examining the thickness dependence on the breakdown strength; that is, the strength rapidly increases with decreasing thickness in a range below a few microns in ionic crystals, mica, and polymer. In liquid dielectrics, Gallagher(2) has proposed that a remarkable increase in the conduction current of liquid argon result on the basis of the collisional effects.

The positive streamer, which will be discussed in this study, consists of many auxiliary avalanches. Note that in the case that the positive streamer takes place in positive point to plane electrodes, then further evidences of the ionization process will be provided. In this study, hence, time lag measurements are carried out by using the voltage pulse whose width and rise time are of about 100ns and 1ns, respectively. It is suggested that if the resulting speed of pre-breakdown disturbance is much higher than that of holes or positive ions in the point anode geometry, the positive streamer will occur in either solids or liquids.

Experimental procedure

A pellet of low density polyethylene(PE) was placed between point and plane electrodes whose gap length was adjusted at a suitable value, and then melted in a vacuum of 10^{-3} mmHg to be brought into a good contact with the electrodes. Finally, it was cooled down to the room temperature in the vacuum. Care was taken to avoid the change in gap length during the thermal process; namely, the expansion coefficient of an insulator was chosed to be equal to that of metal electrodes. Polystyrene(PS) and polymethylmethacrylate(PMMA) were synthesized from the respective monomers between the electrodes. Polycarbonate(PC) and polyethyleneterephtalate(PET) films had an evaporated silver electrode on one side, and the point electode was slightly inserted into the other side. These samples were put into a vessel filled with silicone oil to prevent corona discharge. The liquids, used in this experiment, were GR grade n-pentane, n-hexane, and n-octane. Isooctane and benzene were spectroscopic grade. Pentane and hexane were distilled. Benzene was dehydrated with $CaCl_2$ for ten hours and then distilled with Na in the flask. The liquid was then poured into the test cell(in Fig.1).

The point electrode was a sawing needle of steel and etched by electrolytic polishing so as to have a radius less than 1μm. The finite radius might result in some scattering, but it could not affect the time lag. Fig.2 shows the schematic diagram of pulse generator: D.c. high voltage is applied to high pressure gap through a pulse forming coaxial cable, connected with a resistor of 1MΩ. The generated pulse propagates along the transmission line and reflects when it reaches the sample until breakdown occurs. A capacitive divider is mounted at the end of the transmission line in order to pick up the voltage signal, which is sent to an oscilloscope. The voltage reflection had no influence on the measurement since only the first pulse was observed.

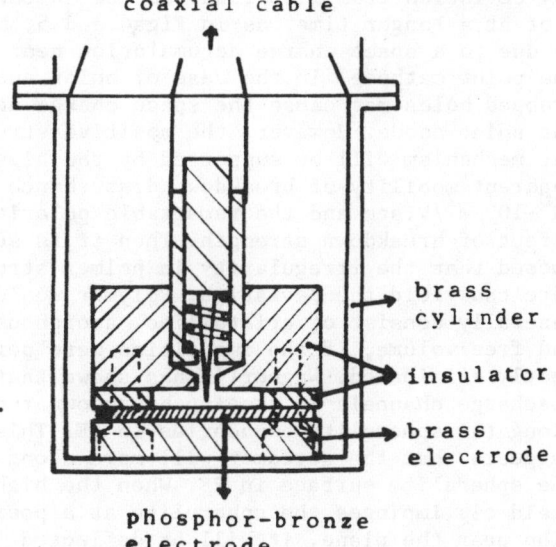

Fig.1 The test cell

Results and Discussions
(1) Solids

In ionic crystals, KCl and NaCl, pre-breakdown luminescence was observed along the crystal axis in the point to plane electrodes in previous paper(3). When a positive voltage was applied to the point electrode(point anode), the lower breakdown strength and the higher speed of breakdown also yielded. The apparent mobility of breakdown disturbance was of the order of $10^2 cm^2/V \cdot sec$ which was much higher than that of holes and of electrons. These confirm that the positive streamer mechanism should occur in the ionic crystals in the point anode system.

In polymers, Fig.3 shows "L" shape of curves in Laue plots of PE in the point anode geometry. The similar properties are also in the time lag of PET,PC, and PMMA. But in PS, time lags are rather widely distributed, as shown in Fig.4, and the mean breakdown strength is the highest

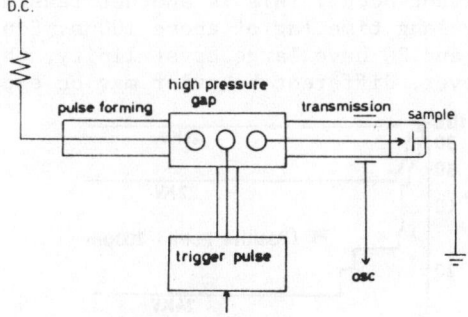

Fig.2 The schematic diagram of pulse generator

among them. Qualitatively similar Laue plots are obtained when a negative voltage is applied to the point electrode. The time at which n_t/n in Laue plot becomes 100% is estimated to be the formative time of breakdown. Figs.5 and 6 show that the formative time depends on applied voltages and on its polarity. We could not obtain the finite formative time of PE for point cathode, since the breakdown often occurred at the wave front. Any deviation from linear dependence in Laue plot at a longer time, as in Figs.4 and 5, may be due to a space charge accumulation near the point cathode. In the case of point anode, trapped holes may cause the space charge near the point anode. However, the positive streamer mechanism will be supported by the high apparent mobility of breakdown disturbance of 10^0-$10^1 cm^2$/V.sec and the remarkable polarity effect of breakdown strength. Then it is supposed that the irregularity in polmer structure can yield the deviation. Polymer would generally consist of cristalline, amorphous, and free volume, if the impurities were perfectly eliminated. Wagner(4) has shown that discharge channels in treeing breakdown run along the spherulite boundaries in PE. This suggests that the streamer will move along the spherulite surface in PE. When the high field tip impinges the spherulite at a position near the plane, it will be deflected along the surface whose direction is perpendicular to the original field. Then the streamer will stop there, and the breakdown will not occur. This is another reason of very long time lag of above 100ns. Since PET and PC have large crystalinity, the streamer will propagate as well as in PE. However, different behavior may be expected in PMMA and PS because of amorphous.

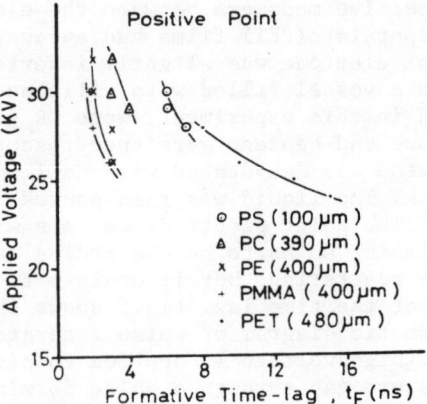

Fig.5 Formative lags of polymers for point anode

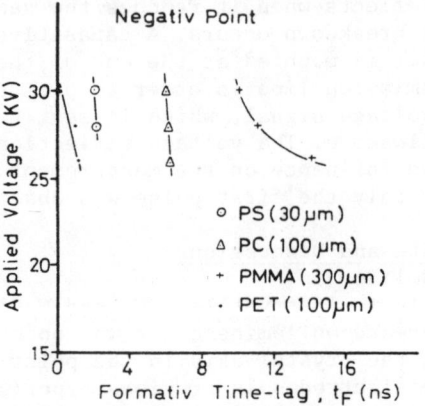

Fig.6 Formative lags of polymers for point cathode

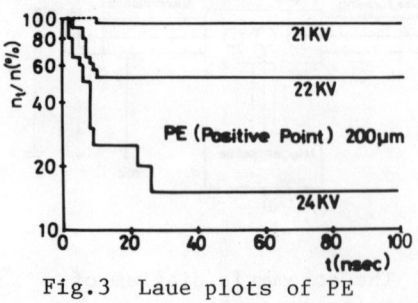

Fig.3 Laue plots of PE

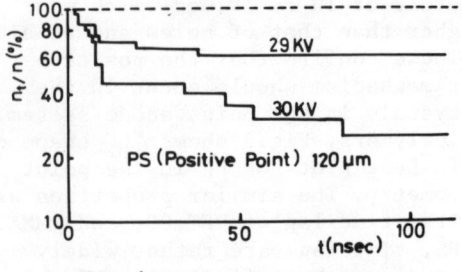

Fig.4 Laue plot of PS

Homogeneous solid, paraffin, shows the exponentially distributed time lag, as in Fig.7. From the obtained shape of curves in Laue plot, it is concluded that the breakdown is very sensitive to the structure of materials, though the space charge cannot be neglected.
(2) Liquids

The liquids are more isotropic than solids, the exponential function will be appeared in time lag. The time lags of silicone oil which has several thousands of molecular weight was distributed exponentially(5). In liquid alkane, isooctane, the normally distributed time lag appears, as shown in Fig.8. In n-pentane, n-hexane, and n-octane, similar time lags are measured. However, in benzene, breakdown does not occur below the field of 1MV/cm.

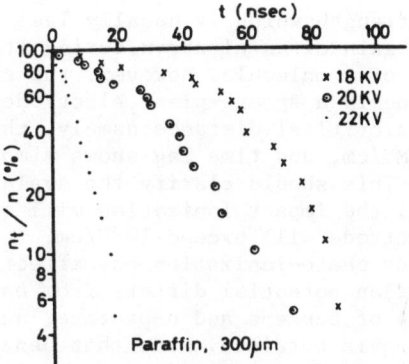

Fig.7 Laue plots of paraffin

It is important whether bubble formation dominates the breakdown process in liquids or not. Then time lag measurement was carried out in the pressures below $10kg/cm^2$. The obtained time lag is independent of hydrostatic pressure, and the bubble mechanism is denied in this experimental condition. Therefore, the positive streamer will be a suitable mechanism for breakdown. The streamer mechanism is also supported by the following facts; the breakdown strength for point anode is lower than that for point cathode, and the apparent mobility of breakdown disturbance is of $3cm^2/V\cdot sec$ in n-hexane, while the electron mobility is only $0.01 cm^2/V\cdot sec$. Only the positive streamer will result such a high mobility.

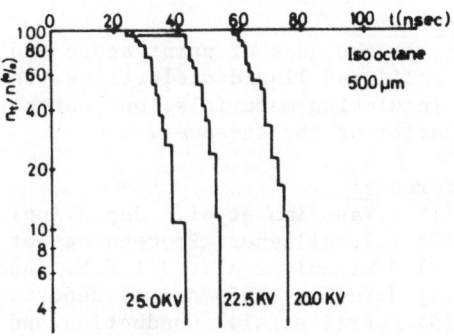

Fig.8 Laue plots of isooctane

The mean time lag depends on the liquid density; namely, Fig.10 shows that the high density causes the long time lag. The similar properties have been obtained in uniform field(6). However, it is interesting that it appears in the point anode geometry. The mean time lag, t_0, is given by

$$t_0 = 1.9 \times 10^{-20} n(CH_2) + 3.4 \times 10^{-20} n(CH_3) - 603,$$

where t_0 is in ns and $n(CH_2)$ and $n(CH_3)$ are the number of CH_2 and CH_3 in unit volume, respectively.

The positive streamer needs photo- and collisional ionization for its development. It has been considered that the impact ionization by electrons hardly occurs at the break-

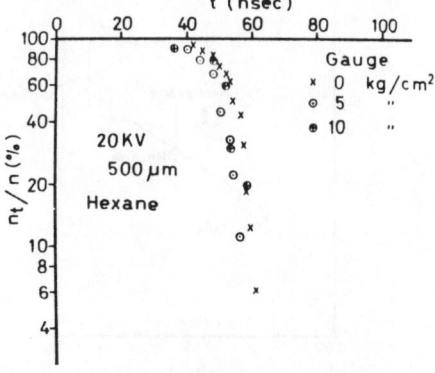

Fig.9 The pressure dependence on the time lag of n-hexane

down strength which is usually less than 1MV/cm in the uniform field, since the energy gain of an electron, during its mean free time, is less than the ionization energy of a molecule. However, the authors(7) have shown that the time lag of n-hexane in a sphere-plane electrodes is independent of gap length in the ranges above a critical distance; namely, the critical distance is between 40μm and 50μm at 3.5MV/cm, and time lag shows almost the same distribution within experimental error. This should clarify the avalanche-streamer transition. Thus in divergent fields, the impact ionization will occur, since the field strength near the tip of electrode will exceed 10MV/cm.

The photo-ionization may affect the time lag of the mixture of liquids whose ionization potential differs from each other. Fig.11 shows the time lag of the mixture of benzene and n-pentane, and at the rate of 50%, the widely distributed time lag is obtained. Note that benzene does not break down at the field strength. The difference in ionization potential, 1.13eV, may enhance the speed of the streamer, and π-electrons in benzene may suppress the streamer development.

Conclusion

In the gaps of point anode and plane cathode, the positive streamer occurs in solid and liquid dielectrics. The path of streamer depends on the structure of insulating materials, but further experiments are needed to elucidate the behavior of the streamer.

References

(1) Y.Yasojima et al.: Jap.J.Appl.Phys.,14,815 (1975)
(2) T.J.Gallagher: Proceedings of the Grenoble Conference,113 (1968)
(3) I.Kitani et al.: J.I.E.E. Japan,95,165 (1975)
(4) H.Wagner: 1974 Annual Report, Conf. Elect.Insul.Dielec.Phenom.,43
(5) K.Arii et al.: Conduction and breakdown in dielectric liquids,(Proc. 5th Int. Conf.) 163, Delft Univ. Press (1975)
(6) T.J.Lewis:Prog.Dielec.1, 97 (1959)
(7) K.Arii et al.: to be published

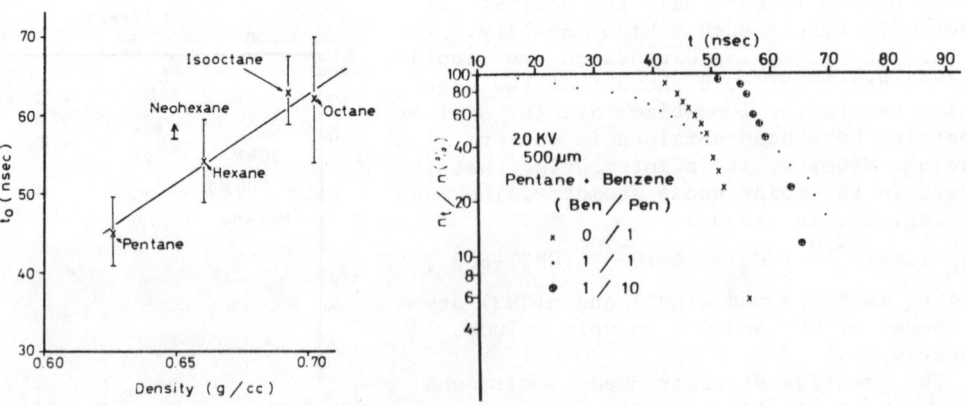

Fig.10 The density dependence on the mean time lag of alkanes

Fig.11 Laue plots of the mixture of n-pentane and benzene

IONIC CONDUCTION CURRENT RELATED TO NEUTRALIZATION

AT INSULATOR-METAL INTERFACE

K. Yamashita, F. Kaneko, D. Y. Kang and T. Hino

Department of Electrical and Electronic Engineering
Tokyo Institute of Technology

2-12-1 O-okayama, Meguro-ku, Tokyo 152 Japan

1. Introduction

Certain kinds of synthetic polymers contain mobile ions and ionic conduction is dominant at elevated temperatures in such polymers. These ionic conduction characteristics have been frequently measured,[1]-[4] but effects of space charge polarization and charge exchange at the polymer-metal electrode interface have not been sufficiently investigated. These phenomena are important for polymeric insulators as electrical breakdown is affected by the space charge field. In the present paper, ionic conduction currents in polyethylene terephthalate (PET)[1],[2] are measured at the temperatures higher than the glass transition temperature. And the properties are discussed by using results of thermally stimulated currents (TSC).

2. Experimental

Samples are commercially used PET (Mylar) films having a thickness of 19 μm. Aluminum was evaporated both sides of the films as electrodes whose area is about 8 cm^2. All currents were measured in a temperature controlled system kept in a vacuum of about 10^{-5} Torr. A measuring process of TSC is as follows. A bias voltage V_b is applied to a sample at an elevated temperature T_b for a constant time t_b. And then the sample is cooled to a low temperature with applying V_b. From this temperature, the sample is heated at a constant rate β with measuring the current, i.e., TSC, in the short circuit condition. Charging currents at constant temperatures are also measured until the currents become constant to obtain the values of conduction currents.

3. Results

Figure 1 shows TSC and TSSP (thermally stimulated surface potential) in PET which were reported by the present authors.[5] It is considered that TSC peaks A and B are due to dipolar polarization[6] and peak C is due to ionic space charge polarization.[7] Peak D is the largest peak but corresponding TSSP is very small.

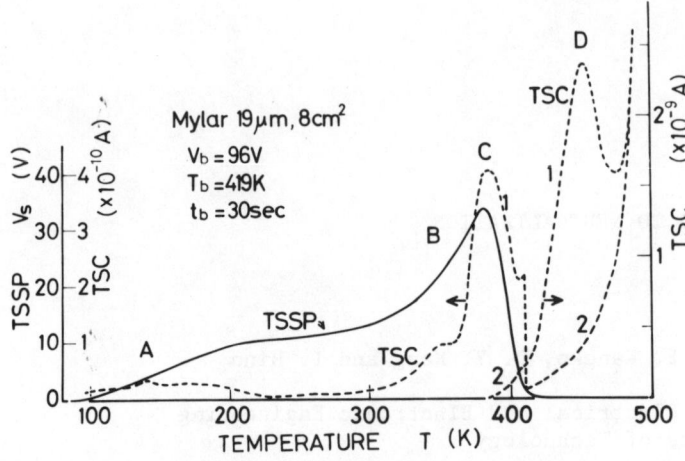

Fig. 1   TSC and TSSP in polyethylene terephthalate.

Fig. 2   Charging currents in PET.

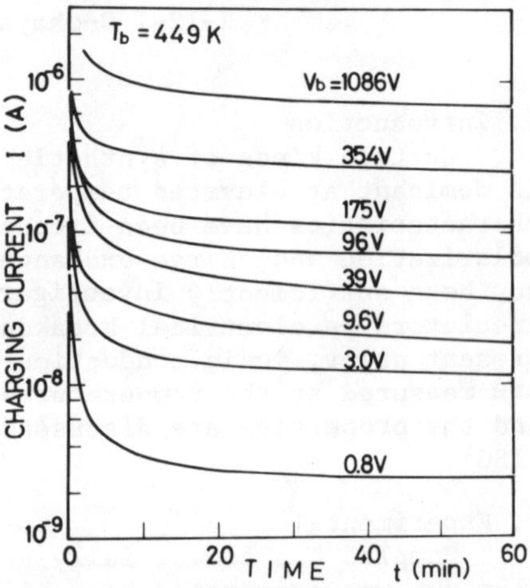

From these facts, this peak is considered to be due to ionization of neutralized ions at PET-metal electrode interface.[5] In the present paper, conduction currents are investigated in the temperature region of peak D.

Figure 2 shows transient charging currents in PET at 449 K as a parameter of applied voltage V_b. The charging currents become constant within considerably short time. Seeing the characteristics, it is difficult to judge whether currents after long elapsed time indicate real steady-state conduction currents or charging currents decaying slowly. But TSC has a possibility to obtain the evidence of the judgement.

Figures 3-5 show TSC peak D for various charging times t_b in (a), and charge Q_{TSC} calculated from the area of TSC peak and a corresponding charging current measured before TSC in (b). In the low field case (V_b = 0.2 V), charge Q_{TSC} saturates for $t_b > 4$ min as shown in Fig. 3(b). The time required to reach the constant charging current corresponds to the charging time required to reach the saturation value of Q_{TSC}. In the intermediate field case (V_b = 9.6 V), TSC peak D is split into D_1 and D_2 as shown in Fig. 4(a). For $t_b \leq 90$ sec, the peak D_1 increases as t_b increases, but peak D_2 does not appear clearly. For $t_b > 90$ sec, on the other hand, D_1 decreases and D_2 increases. However, as shown in Fig. 4(b), Q_{TSC} of total peak D ($=D_1+D_2$) is almost constant for $t_b \geq 10$ min. Charging currents in this case continue to decay slowly for a long time. In the high field case (V_b = 175 V), D_1 decreases and D_2 increases from short time t_b as shown in Fig. 5(a). Both peaks reach saturation at about t_b = 15 min and Q_{TSC} of D is almost constant in Fig. 5(b).

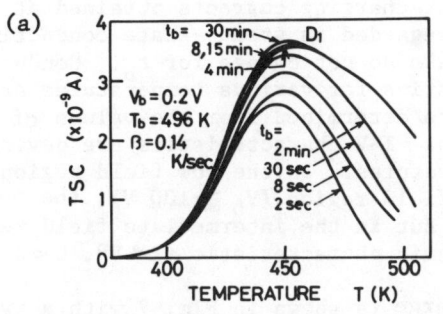
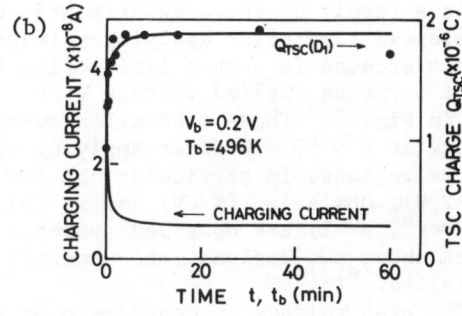

Fig. 3 (a) TSC for various charging times in the low field case.
(b) TSC charge and a charging current.

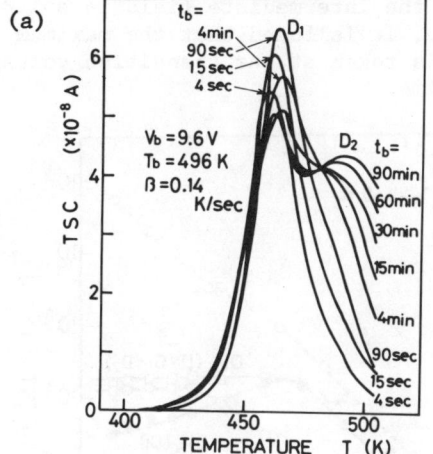
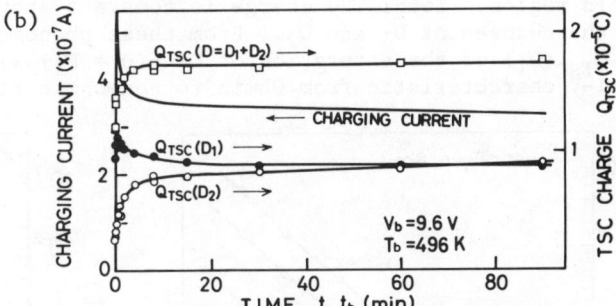

Fig. 4 (a) TSC for various charging times in the intermediate field case.
(b) TSC charge of the peaks D_1, D_2 and D, and a charging current.

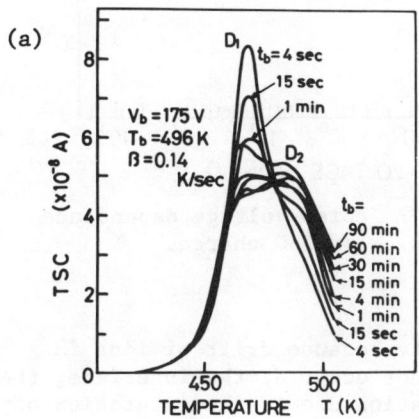
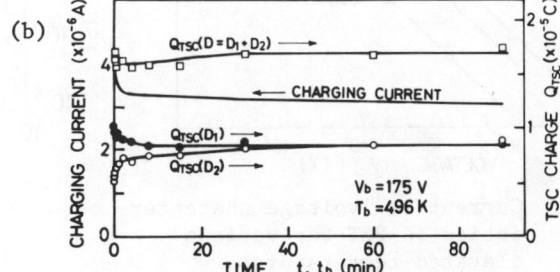

Fig. 5 (a) TSC for various charging times in the high field case.
(b) TSC charge of the peaks D_1, D_2 and D, and a charging current.

As a result of above experiments, constant charging currents obtained at a long elapsed time after applying voltage are regarded as steady-state conduction currents because in such a time region TSC peaks do not change for t_b. Conduction current I versus applied voltage V characteristics for various temperatures are shown in Fig. 6. These conduction currents are determined from the values of currents at t = 90 min after applying voltages. I-V characteristics are devided to three regions, in particular, at low temperatures. In the low field region ($V_b \lesssim 2$ V), Ohm's law ($I \propto V$) and in the high field region ($V_b \gtrsim 100$ V), the power law ($I \propto V^n$, $n>1$) are observed respectively. But in the intermediate field region between above two regions, the abnormal non-Ohmic characteristic ($I \propto V^n$, $0<n<1$) is obtained.[8],[9]

The bias voltage V_b dependence of TSC charge is shown in Fig. 7 with a typical I-V characteristic. In the low field region, peak D_1 is only appear and its charge $Q_{TSC}(D_1)$ is proportional to V_b. Above the intermediate field region $Q_{TSC}(D_1)$ decreases and peak D_2 begins to appear. Charge $Q_{TSC}(D_2)$ saturates in the high field region. Total TSC charge is constant above the intermediate field in spite of the changes of D_1 and D_2. From these phenomena, it is found that the maximum of $Q_{TSC}(D_1)$ or the saturation of $Q_{TSC}(D = D_1+D_2)$ is taken at the transition voltage of I-V characteristic from Ohmic to non-Ohmic regime.

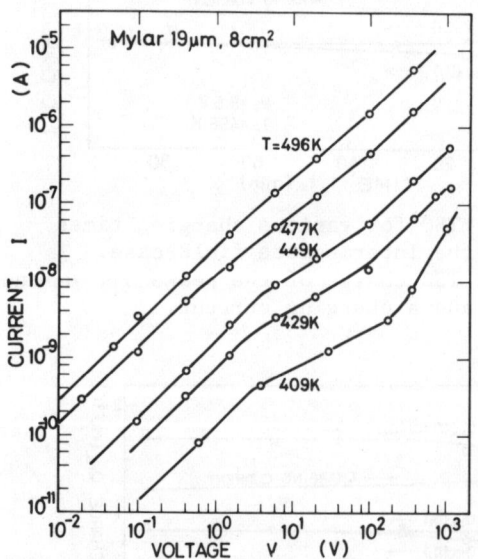

Fig. 6   Current vs. voltage characteristics in PET for various elevated temperatures.

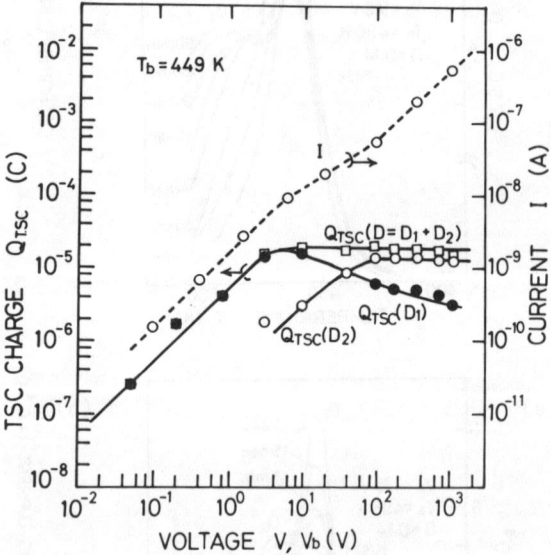

Fig. 7   Bias voltage dependence of TSC charge.

4. Discussion

Ionic conduction is usually in non steady state because drift of ions is blocked by the electrode. If neutralization does not occur at the interface, the charging current continues to decay until its cessation due to the saturation of ionic space charge polarization. And even if neutralization occurs, the current may decay due to accumulation of neutralized particles near the electrode. For a mechanism of steady-state ionic conduction current, it must be considered how ionic carriers are provided continuously. Neutralized particles near the interface are possible to diffuse to the opposite direction to the applied field. If the diffused particles in the PET are ionized again, the ions are drifted to the interface by the applied field and neutralized there. This cyclic process is considered to result steady-state ionic conduction currents. Ohmic conduction is expected if supply of carriers is enough in the low field region.

The abnormal non-Ohmic characteristic in the intermediate field region is related with the carries which generate peak D_1, because $Q_{TSC}(D_1)$ decreases above this field region as shown in Fig. 7. Also the decay of charging current for a long period in the intermediate field region means the decrease of carriers of peak D_1 with elapsed time as shown in Fig. 4. Ions which generate peaks D_1 and D_2 seem to be the same kind as total TSC charge $Q_{TSC}(D)$ is constant for V_b and t_b. Two different conditions of neutralization will result peaks D_1 and D_2. As increasing V_b or t_b, the condition transfers from D_1 to D_2. And carriers in a neutralized condition which results peak D_2 do not contribute the steady state current. Carriers in a neutralization condition which results peak D_1 cause the cyclic mechanism described above. Therefore the decrease of peak D_1 corresponds to occurrence of abnormal I-V characteristics. The mechanisms of these neutralization condition have not been clarified yet.

5. Conclusion

I-V characteristics in PET are measured at elevated temperatures. Ions are dominant in the carriers of the current. The evidence that the current is in steady state is obtained by TSC measurements. A mechanism of steady-state ionic conduction is suggested in relation to neutralization and ionization. The abnormal non-Ohmic characteristic is observed in I-V characteristics and the mechanism is discussed by TSC results.

References
(1) L.E.Amborski; J.Polymer Sci., 62, 331 (1962)
(2) H.Sasabe et al; Polymer J., 2, 518 (1971)
(3) D.A.Seanor; J.Polymer Sci.A-2, 6, 463 (1968)
(4) M.Kosaki et al; J.Appl.Phys., 42, 3388 (1971)
(5) T.Hino & F.Kaneko; 1977 NRC Conf. on Elect. Insu. & Dielect. Phenomena, C-7, Albany, USA (Oct.,1977)
(6) T.Hino; J.Appl.Phys., 46, 1956 (1975)
(7) F.Kaneko & T.Hino; Trans.IEE Japan, 98-A, 101 (1978)
(8) A.A.Shilyaev & L.G.Gindin; Soviet Phys.Solid State, 14, 1094 (1972)
(9) K.Miyairi & M.Ieda; Trans.IEE Japan, 96-A, 25 (1976)

INTERFACIAL PHENOMENA IN POLYMERS AND THEIR APPLICATION FOR DETERMINING CARRIER SPECIES

Shuhei Nakamura, Goro Sawa and Masayuki Ieda[*]

Mie University, Tsu, Japan 514

Nagoya University, Nagoya, Japan 464[*]

In spite of primitive problem it is by no means clear whether carrier species in polymers are ionic or electronic in low field conduction or natural condition. Difficulties for the application of conventional methods arise through low current density based on the good insulating properties of polymers.

In only a few cases, a direct evidence of ionic charge transfer has been obtained from the examination of Faraday's law of electrolysis.[1-4] However, an electrolytic current for polymer is so small that the products are hardly detected. In most investigations only indirect evidence for determining carrier species has been, therefore, employed such as the change in conductivity associated with crystallization, crosslinking, glass transition, applied pressure and plasticization.[5-9]

Previously, we have reported that the M_0-M_1-polymer-M_2-M_0 system forms a galvanic cell, where polymer film was sandwiched in between two different kinds of metals M_1 and M_2 and those metals were connected with copper wires M_0, and presented a new approach to the determination of the kind of carriers predominant in organic polymers by using the galvanic cell system.[10,11] The purpose of this paper is to specify the temperature region where ionic conduction is predominant at a low electric field, for various organic polymers by using that method.

Polymers used were nylon 66, plasticized polyvinyl-chloride(PVC), PVC, polyethylene oxide(PEO), high density polyethylene(HD PE), ethylene-vinyl acetate copolymer(EVA), polyethylene terephthalate(PET), polyvinyl-fluoride(PVF), polyvinylidene fluoride(PVDF) and polypropylene(PP). The chemical structures and thicknesses of these polymers are shown in Table 1. These polymers were in form of film and were sandwiched between evaporated Au and Al electrodes, in a diameter of 3 cm. All measurements of the open-circuit voltage V_0, the short-circuit current I_S for the M_0-Au-polymer-Al-M_0 system and the conduction current I for the M_0-Au-polymer-Au-M_0 system were made in dry nitrogen gas with a vibrating reed electrometer(VRE;TR-84M,Takeda Riken Ind. Co.,Ltd.) during rising temperature at a constant rate of 0.67 °C/min. In the measurements of the

Polymer	Structural Unit	Thickness (μm)	Remarks
nylon 66	$-N-(C)_6-N-C-(C)_4-C-$ with H,H,O,O	25	Toray Ind. Inc.
pl.PVC	-C-C- with Cl	50	Mitsubishi Monsanto Chem. Co. (containing DOP 33 wt%)
PVC	-C-C- with Cl	40	
PEO	-C-C-O-	250-300	Meisei Chem. Ind. Co.
HD PE	-C-C-	25	Mitsubishi Petrochem. Co. (PX-40)
EVA	-C-C- + -C-C- with C=O,C	30	Mitsubishi Petrochem. Co. Ltd. (containing vinyl acetate 22 %)
PET	$-O-C-\bigcirc-C-O-(C)_2-$	25	Toray Ind. Inc. (type T)
PVF	-C-C- with F	60	E.I. duPont de Nemours & Co. (Tedlar)
PVDF	-C-C- with F,F	25	Kureha Chem. Ind. Co.
PP	-C-C- with C	20	Mitsubishi Petrochem. Co.

Table 1. Polymers used in experiments.

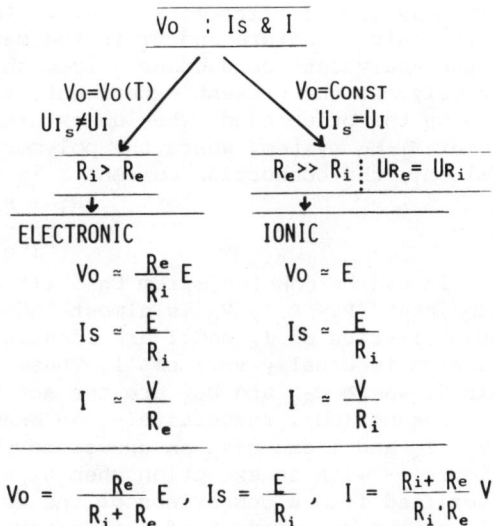

Table 2. Schematic diagram for determining the predominant carrier species using the galvanic cell.

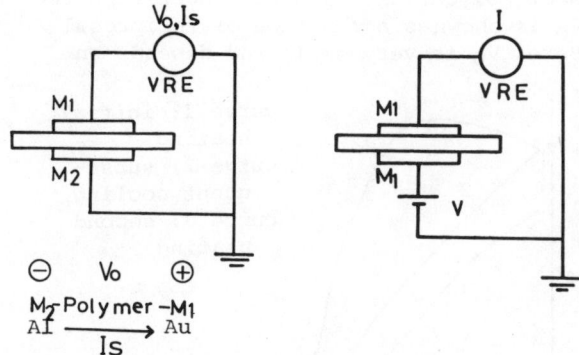

Fig. 1. Schematic diagrams of the measuring circuit (a) for the short-circuit current I_s and open-circuit voltage V_0 and (b) for conduction current I. The positive directions of I_s and V_0 are defined as shown.

conduction current, an external voltage of about 1.6 V was applied to the M_0-Au-polymer-Au-M_0 system. In the open-circuit voltage, the electric potential of Au was always positive against Al. The short-circuit current from the M_0-Au-polymer-Al-M_0 system flowed in such a direction as a positive charge carrier moved from Al to Au through the polymer, within the limit of the present experiment.

If the electrical conduction of polymer is purely ionic, the M_0-M_1-polymer-M_2-M_0 system forms a galvanic cell.[11] According to Wagner's theory the electromotive force E' of a galvanic cell consisting of an electrolyte involving both ionic and electronic conductions[12] is given as

$$E'/E = \bar{t}_i = \frac{1/R_i}{1/R_i + 1/R_e} = \frac{R_e}{R_i + R_e}, \quad (1)$$

where E is the electromotive force of the galvanic cell when the electrolyte is purely ionic in nature and t_i is the mean transport number for ions, $1/R_i$ and $1/R_e$ are the equivalent conductances from the ionic and electronic conduction, respectively. In the present experiment, V_0 corresponds to E'.

On the other hand, when an external voltage V is applied to the M_0-M_1-polymer-M_1-M_0 system, where the polymer is sandwiched between two identical metals M_1, the conduction current I is given as

$$I = \frac{R_i + R_e}{R_i R_e} V . \qquad (2)$$

It can be concluded from Eqs. (1) and (2) that, if the ionic conduction is predominant ($R_i \ll R_e$), V_0 is almost independent of temperature and the temperature characteristics of I_s and I are identical, since the change in E with temperature variation is usually very small. These arguments can be summarized as shown in Table 2, where U_{Re} and U_{Ri} are the activation energies for the electronic and ionic conductions, respectively. An examination of the temperature characteristics of V_0, I_s and I can give an answer to the question whether predominant conduction is ionic,[10] with an exception when $U_{Re} = U_{Ri}$. To be brief, the ionic conduction can be verified from a comparison of the activation energies between I_s and I when V_0 is almost independent of temperature.

As an example, V_0 is shown in Fig.2 for nylon 66 which has been reported to involve ionic conduction above 120 °C from the detection of H_2 gas evolution.[1] The value of V_0 rises with ascending temperature from room temperature to about 70 °C and is nearly constant from 70 °C to 150 °C. Both I and I_s show the identical temperature dependence from room temperature to 150 °C.[10]

Another example is CQ complex salts of polycation which have been reported to be electronic in nature.[13] In Fig.3, V_0 is shown as a function of reciprocal temperature in a semilogarithmic scale. Here, V_0 is very small and depends on

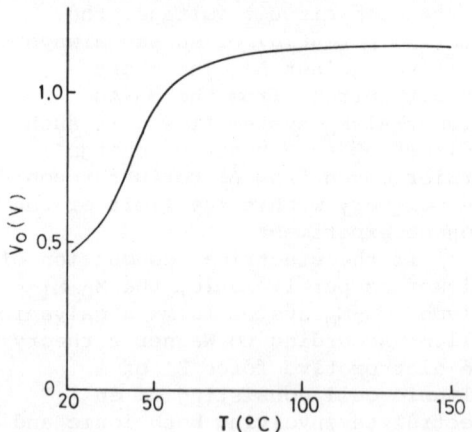

Fig. 2. V_0 for Au-nylon66-Al system. (Source; ref.10. Reproduced by kind permission of AIP.)

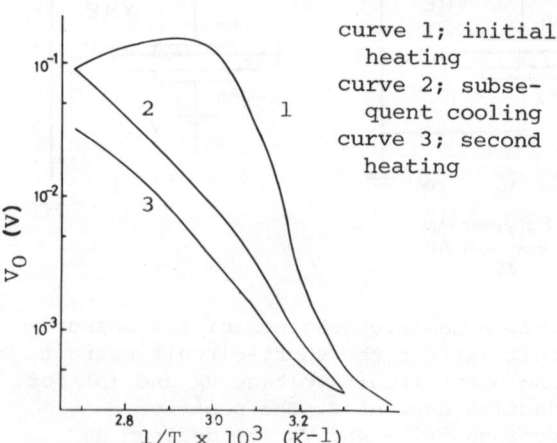

curve 1; initial heating
curve 2; subsequent cooling
curve 3; second heating

Fig. 3. V_0 as a function of reciprocal temperature in the logarithmic scale for the Au-Pix-CQ complex salt-Al system. (Source; ref.10. Reproduced by kind permission of AIP.)

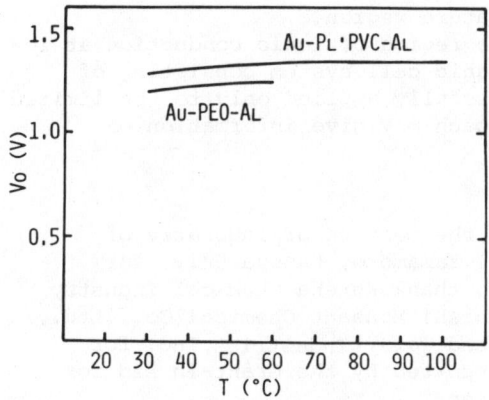

Fig. 4. V_0 for Au-pl.PVC-Al and Au-PEO-Al systems.

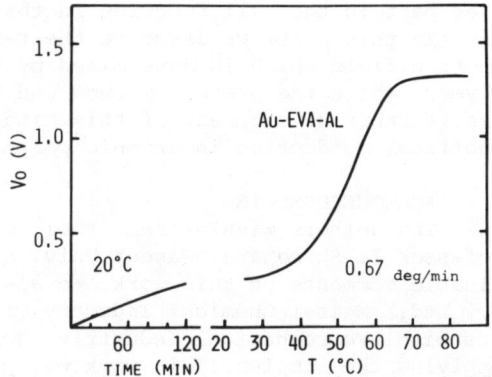

Fig. 5. V_0 as a function of time at 20 °C and temperature for Au-EVA-Al system.

temperature. On the contrary, I is almost independent of temperature for CQ complex salt.[10]

A distinctive feature summarized in Table 2 was obtained from the examination of V_0, I_S and I for the two samples, whose carrier species have been relatively well known. In the following, we take a few polymers tabulated in Table 1, and examine the temperature characteristics for V_0, I_S and I. The values for Au-pl.PVC-Al system and Au-PEO-Al system reach the steady values even at room temperature as shown in Fig.4. For these systems, the short-circuit current I_S and the conduction current I under the externally applied voltage indicate the same temperature characteristics from room temperature.

The value V_0 for Au-EVA-Al system rises with time after the electrometer is connected at room temperature and is almost constant for temperature above about 60 °C as shown in Fig.5. For the Au-EVA-Al system, I_S and I gave the same activation energies above 50 °C.

Supposing that $U_{Ri} \neq U_{Re}$, the temperature region in which the ionic conduction is predominant can be determined from the examination of the temperature characteristics of V_0, I_S and I. The results for various polymers are shown in Fig.6, where the regions of ionic conduction are depicted by the shaded areas. Blanks in Fig.6 indicate the temperature regions in which the dominant carrier could not be discerned as yet, since a very long time was required for establishing the steady value of V_0 in the lower temperature region. The rate of cell reaction is so slow as it takes very long time to reach a thermal equilibrium, or an electron

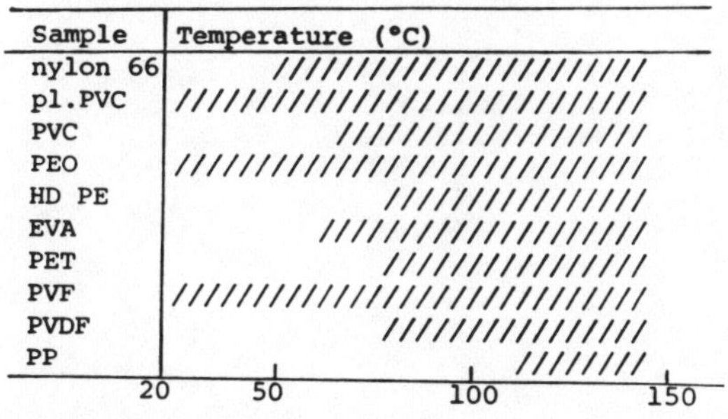

Fig. 6. Temperature regions of ionic conduction in polymers.

takes part in the cell reaction in that temperature region.

In this paper we describe the temperature region of ionic conduction at low electric field which is determined by the galvanic cell system consisting of polymer. While the present method can be successfully applied only to the limited case, further development of this kind of approach may give information on electrical conduction in organic polymers.

ACKNOWLEDGMENTS

The authors wish express their thanks to the members of laboratry of Professor I. Shinohara, Waseda Univ. and Dr. O. Yamamoto, Nagoya Univ. for valuable comments on this work. We also wish to thank Kureha Chemical Industry Co., Ltd., Meisei Chemical Industry Co., Mitsubishi Monsant Chemical Co., Ltd., Mitsubishi Petrochemical Industries Co., Ltd. and Toray Industries, Inc. for supplying the samples. This work was partly supported by the Grant-in-Aid for Scientific Research from the Ministry of Education.

References
1 D. A. Seanor, J. Polymer Sci. A $\underline{2}$ 463 (1968)
2 D. A. Seanor, ibd.   C $\underline{17}$ 196 (1967)
3 E. J. Murphy, Canad. J. Phys. $\underline{41}$ 1022 (1963)
4 G. King and J. A. Medly, J. Colloid. Sci. $\underline{4}$ 1 (1949)
5 S. Saito, H. Sasabe and T. Nakajima, J. Polymer Sci. A $\underline{2}$ 1297 (1968)
6 H. Sasabe and S. Saito, J. Polymer Sci. A $\underline{2}$ 1401 (1968)
7 H. Sasabe, K. Sawamura and S. Saito, Polymer J. $\underline{2}$ 518 (1971)
8 M. Kosaki, M. Yoda and M. Ieda, J. Phys. Soc. Jpn. $\underline{31}$ 1598 (1971)
9 M. Yoda, G, Sawa and M. Ieda, Jpn. J. Appl. Phys. $\underline{12}$ 475 (1973)
10 G. Sawa, T. Inayoshi, Y. Nishio, S. Nakamura and M. Ieda, J. Appl. Phys. $\underline{48}$ 2414 (1977)
11 M. Ieda, G. Sawa, S. Nakamura and Y. Nishio, J. Appl. Phys. $\underline{46}$ 2796 (1975)
12 C. Wagner, Z. Phys. Chem. $\underline{B21}$ 25 (1933)
13 K. Mizoguchi, Y. Kitajima, S. Kajiura, E. Tsuchida and I. Shinohara, J. Chem. Soc. Jpn. Chem. Ind. Chem. 1751 (1974) [in Japanese]

SOME CONSIDERATIONS ON ELECTRIC CONDUCTION AND CHARGE DISTRIBUTION

IN XLPE CABLE INSULATION

> Y. Yamada and S. Miyamoto
>
> Sumitomo Electric Ind., Ltd.
>
> 1-3, Shimaya 1-chome, Konohana-ku, Osaka, 554 Japan

1. Introduction

Many researches have been made on the electric conduction in polyethylene but reports on the conduction in XLPE(cross-linked polyethylene) cable insulation are few. So, mechanisms for the electric conduction in XLPE cable insulation are reported considering the experimental results of the change of (current-time) characteristics with voltage and temperature.

2. Experiments

Two kinds of a experiment on the conduction currents in XLPE cable insulation are made.

Experiment (1) : Leakage current in XLPE cable (thickness of XLPE: 3.5mm; inner diameter of insulation: 10.9mm; outer dia.: 17.9mm; length of electrode: 2 meters) was measured during a 30-minute period under the conditions of (a) voltage: DC 70-350kV (mean stress: 20-100kV/mm), and (b) temperature: 24°C, 60°C, and 90°C.

Experiment (2) : For the same sort of XLPE cable, leakage current during a 30-minute period was measured twice under the same voltage, with the interval between the measurement peirods being 30 minutes, a period during which the inner and other conductors were connected. The applied voltages were 128, 175, and 245KV, and measurement was made at 27°C, 60°C, and 90°C.

3. Results

Current(I) and time(t) characteristics measured for each condition in Experiment (1) and Experiment(2) are shown in Fig. 1 and Fig. 2, respectively. Fig. 1 shows that there are four types of (I-t) characteristics, as is also shown in Fig. 3. Type A appears in low temperature and low stress situations, Type D in high temperature or high stress situations. Type B and C appear in the middle range of stress and temperature between Types A and D. Fig. 2 shows that Type B reappears when voltage is applied twice, with a 30-minute interval, and that for Type D the same (I-t) pattern does not reappear when voltage is applied twice.

4. Discussion

For each of the (I-t) patterns in Fig. 3, it is assumed that there would be the following relationships between the injected current from the electrode(I_i) and the conduction current in XLPE(I_c):

Type A : $I_i = I_c$ (ohmic conduction)
Type B : At the instant of voltage application, I_i is larger than I_c, so space charge is formed in front of the electrode, and the space charge decreases I_i and increases I_c.
Type C : $I_i = I_c$
Type D : I_i would be less than I_c. The decreasing of current with time might be caused by the sweep-out of conduction carriers.

The above considerations are deduced from the following:

(1) Origin of I_i and I_c: It is assumed that I_i is a current by Shottky Emission from the electrode and that I_c is an ionic conduction by ions composed by electron attachment.

(2) Dependency of I_i and I_c on electric stress:
Equation (1) and (2) are assumed.

$$I_i = AT^2 \exp(-\frac{\phi}{kT} + \frac{e}{kT}\sqrt{\frac{e}{4\pi\varepsilon}E}) \propto \exp(\frac{e}{kT}\sqrt{\frac{e}{4\pi\varepsilon}E}) \qquad (1)$$

$$I_c = 2ena\nu \exp(-\frac{U}{kT}) \sinh\frac{eaE}{2kT} \propto \exp(\frac{ea}{2kT}E) \qquad (2)$$

where e: electric charge of an electron; K: Boltzman constant; T: absolute temperature; ε: dielectric constant of XLPE; E: electric stress; and a: jumping distance of ion.

(3) Calculation of I_i and I_e
The dependency of I_i and I_c on stress, calculated from eq.(1) and eq.(2), is shown in Fig. 4(a) and (b), where

$\varepsilon = 2.3 \times \frac{10^{-9}}{4\pi}$ (F/m) and $a = 40 \times 10^{-10}$ (m).

Considering Fig. 4 and the experimental results in Fig. 1, the relation between I_i and I_c at 24°C, 60°C, and 90°C would be as shown in Fig. 5.
At 24°C, I_i and I_c might intersect at two points and in the range of stress where $I_i > I_c$, (I-t) of Type B would appear; at the stress $I_i = I_c$, (I-t) of Type C would appear; in the range of high stresses where $I_i < I_c$, (I-t) of Type D would appear, respectively.
At 60°C, I_i and I_c might intersect at only one point of a stress where (I-t) of Type C would appear and I_c might be larger than I_i except a stress (where $I_c=I_i$), so (I-t) of Type D would appear in the range of experimental stresses.
At 90°C, in the range of experimental stresses I_c might be larger than I_i, and (I-t) of Type D would appear.

Above mentioned relation between I_c and I_i, the charge distribution in the XLPE cable insulation are assumed as is shown in Fig. 6 for each type of (I-t) characteristics and a conduction carrier in (I-t) of Type D must be ionic one because (I-t) of Type D does not reappear when measured twice at a interval of 30 minutes as is shown in Fig. 2. If ionic carriers are assumed in (I-t) of Type D, the decreasing of a current might be caused by the sweep-out of the original carriers in XLPE cable insulation.

5. Conclusion

We wish to report the following conclusions:

(1) Four types of (I-t) characteristics, depending on stress and temperature, appear when DC voltage is applied to XLPE cable.

(2) Each of the four types of (I-t) characteristics may be determined by the relation between injected and conduction currents in XLPE.

(3) Sweep-out of conduction carriers may occur in XLPE insulation when the Type D (I-t) pattern appears.

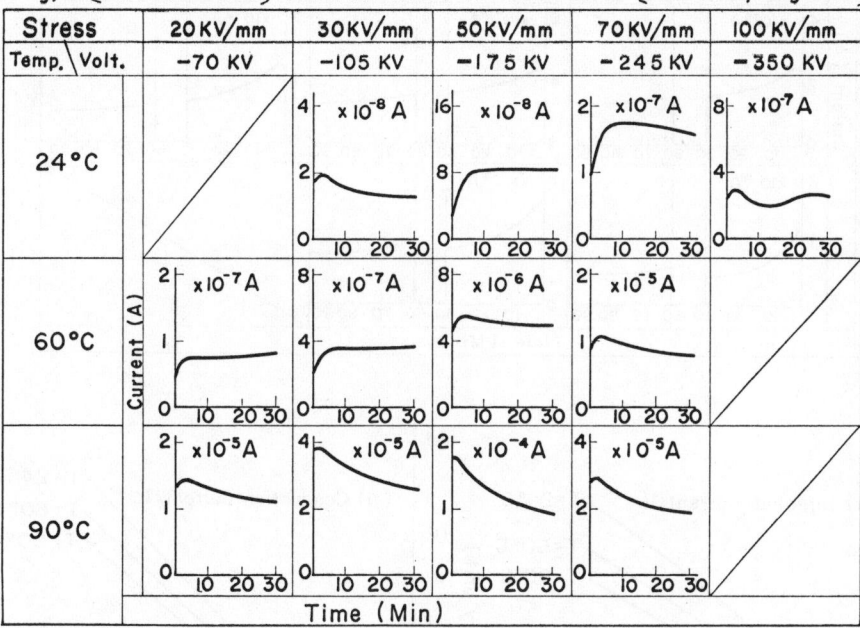

Fig.1 (Current-Time) Characteristics of XLPE Cable [In.Cond.; Negative]

Fig.3. Types of (I-t) characteristic

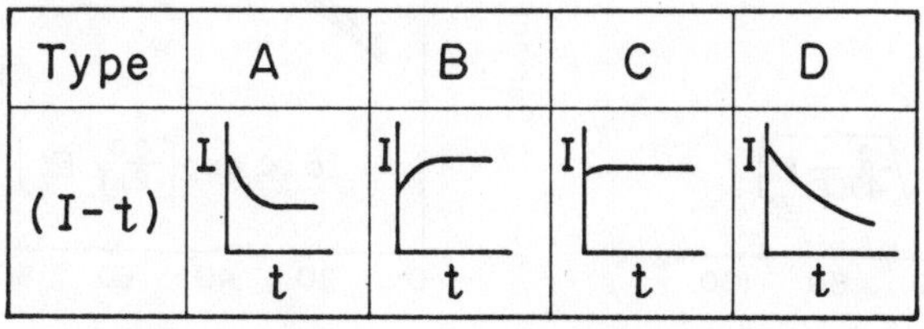

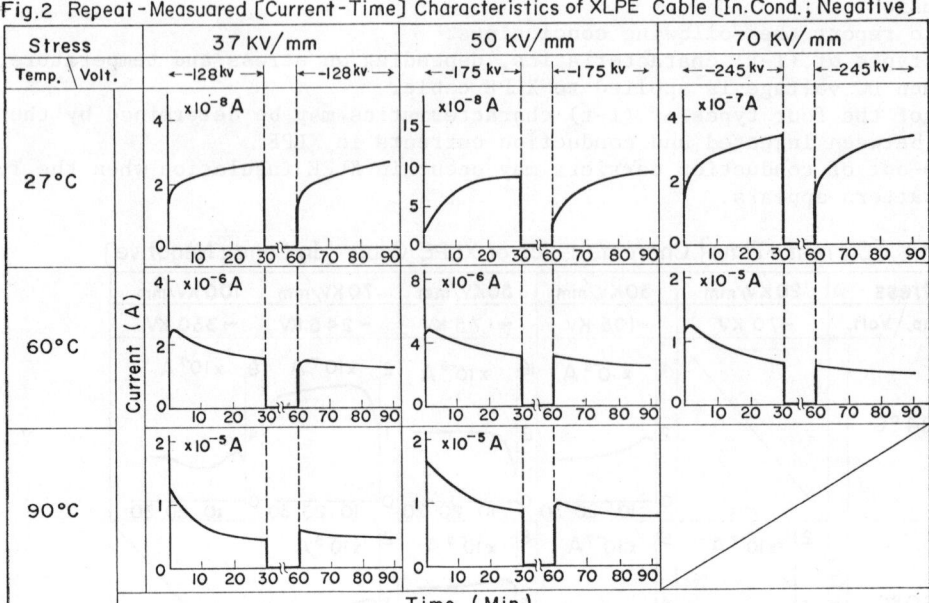

Fig.2 Repeat-Measuared [Current-Time] Characteristics of XLPE Cable [In.Cond.; Negative]

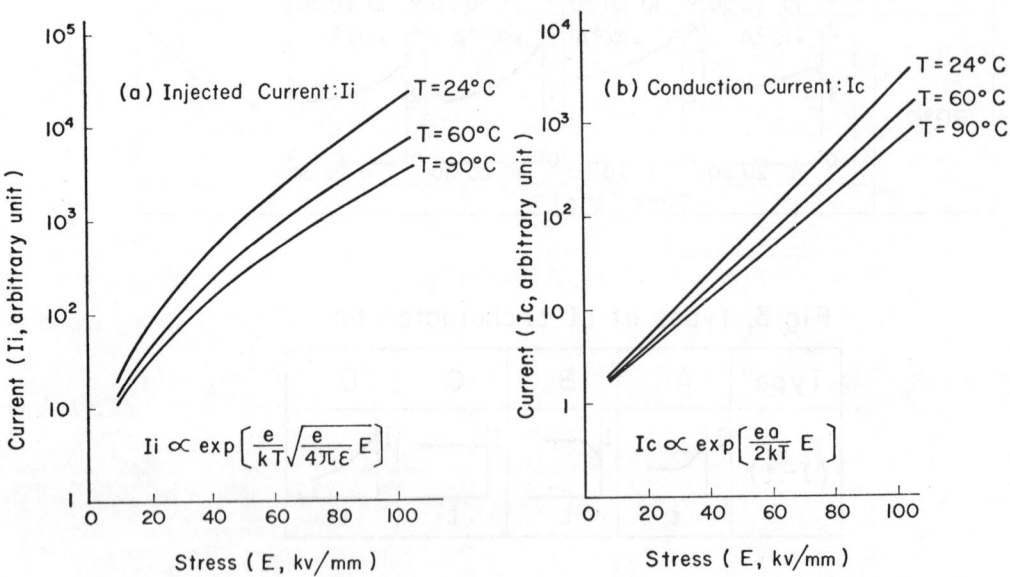

Fig. 4. Dependency of Ii and Ic on Electric stress

(a) Injected Current: Ii

$$I_i \propto \exp\left[\frac{e}{kT}\sqrt{\frac{e}{4\pi\varepsilon}}E\right]$$

(b) Conduction Current: Ic

$$I_c \propto \exp\left[\frac{ea}{2kT}E\right]$$

ELECTRICAL CONDUCTION AND BREAKDOWN

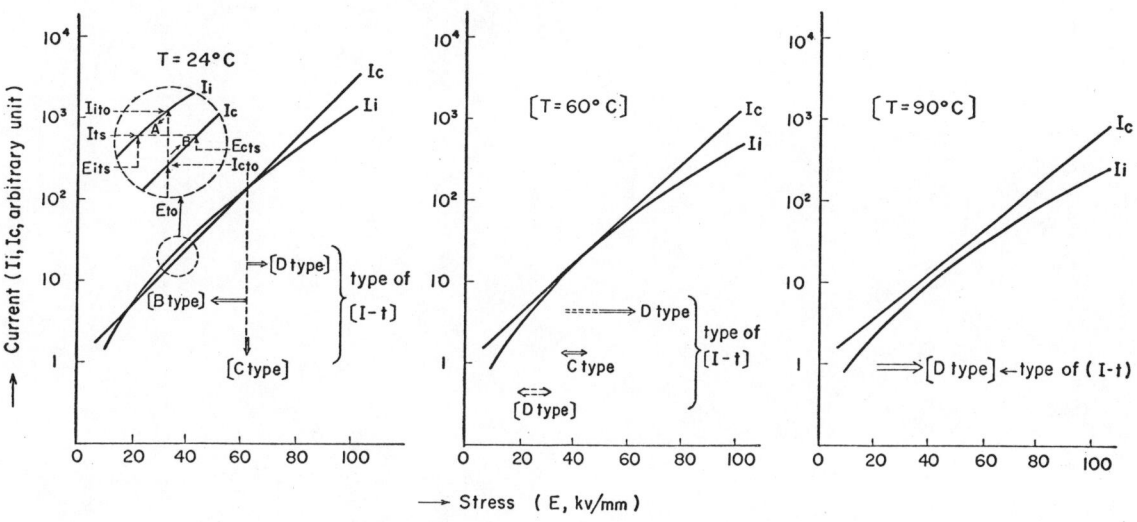

Fig 5. Relation between Ii and Ic, and between type of I-t and stress

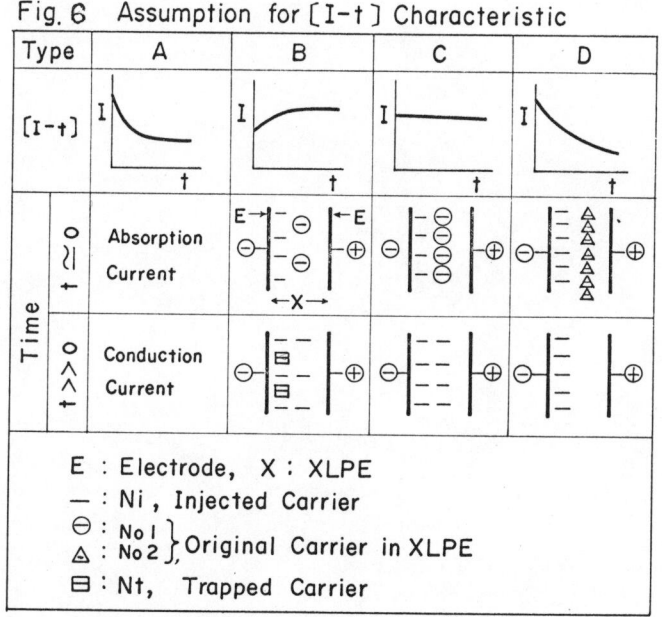

AUTHOR INDEX

A
Adachi, T., 79
AkutsuK., 35
Arii, K., 422
Asakawa, Y., 89

B
Badian, L., 372, 377
Bamji S., 3
Barriere, A. S., 239
Bhatnagar, C. S., 133
Bichon, G., 158
Böhm, M., 261
Bonnafe, J., 256
Broadhurst, M. G., 143

C
Calderwood, J. H., 359
Campos, M., 234
Carr, S. H., 123
Celaschi, S., 97
Chatain, D., 312
Chonan, I., 55
Coelho, R., 412
Collins, R. E., 170

D
Das-Gupta, D. K., 153
Dodelet, J.-P., 228
Donnet, G., 108
Doughty, K., 153

F
Facoetti, H., 158
Felici, N. J., 10
Fillard, J. P., 256
Freeman, G. R., 228

Fujii, H., 218
Fukada, E., 165, 292, 342
Fukushi, N., 307
Furukawa, T., 165

G
Gasiot, J., 251
Greenwood, A., 417
Gubanski, S. M., 372
Guelorget, O., 175
Gupta, N. P., 297
György, I., 228

H
Hashimoto, K., 79
Hatano, Y., 223
Hayashi, K., 218
Hayashi, S., 277
Hino, T., 427
Hirabayashi, S., 277
Horii, K., 213
Hoshino, Y., 322

I
Ieda, M., 322, 327, 397, 432
Inoue, Y., 60
Inuishi, Y., 183, 218
Ishiki, K., 402
Ito, D., 208

J
Jain, K., 297
Jeszka, J., 287
Jimenez-Lopez, J., 251, 256

K
Kaizaki, I., 218

Kamisako, K., 203, 382
Kamiya, T., 342
Kamoto, S, 19
Kanegae, H., 282
Kaneko, F., 427
Kang, D. Y., 427
Kao, K. J., 3
Katsukawa, H., 213
Kawasaki, M., 79
Khare, M. L., 133
Kitani, I., 422
Kobayashi, M., 128
Kobayashi, S., 317
Kodera, Y., 113
Koezuka, H., 282
Kosaki, M., 213
Kryszewski, M., 287
Kubo, U., 218

L
Lacabanne, C., 312
Lachter, A., 239
Lascaud, M., 239
Latour, M., 108, 175
Lemonon, C., 158
Linkens, A., 251

M
Mayeux C., 138
Manifacier, J. C., 251, 256
Mascarenhas, S., 97
Masuda, S., 35, 40
Matsui, M., 337, 387
Mehendru, P. C., 297
Mendiratta, R. G., 368
Micheron, F., 138, 158
Miyairi, K., 327
Miyamoto, S., 437

Miyauchi, M., 213
Mizuno, A., 40
Mizutani, T., 322
Mollah, M., 352
Morita, A., 407
Morita, S., 397
Murakami, I., 244
Murasaki, N., 337
Murashima, S, 24
Murayama, N., 118
Murphy, P. V., 175

N

Nagasaka, H., 45, 387
Nakamura, S., 432
Nakano, M., 70
Namba, H., 223
Narusawa, T., 75

O

Ochiai, H., 244
Odajima, A., 148
Ohara, K., 30
Ohashi, A., 55
Ohtsuka, K., 79
Okada, S., 75
Okuyama, H., 75
Okuyama, T., 342
Ono, H., 282

P

Parot, P., 251
Perlman, M. M., 3
Pillai, C. K., 368
Pillai, P. K. C., 352, 368

Q, R

Quezado, S., 97
Qureshi, M. S., 133
Royer, M., 158

S

Saito, K., 244
Sakai, T., 65
Salardenne, J., 239
Samoc, A., 271
Samoć, M., 271 271
Sasabe, H., 203, 382
Sawa, G., 397, 432
Sawatari, A., 347
Scharmann, A., 261
Schmidt, W. F., 192
Sessler, G. M., 103
Shen, M., 397
Shibayama, K., 277, 282
Shimizu, N., 213
Shibayama, K., 282
Shimozato, M., 322
Shindo, K., 302
Shinohara, I., 342
Shinohara, K., 382
Shinsaka, K., 223
Smycz, E., 377
Stupp, S. I., 123
Sugiyama, K., 332
Suzuoki, Y., 322

T, U

Takada, T., 65
Takahoshi, N., 148
Takahashi, R., 218
Takahashi, S., 266
Takamatsu, T., 292, 342
Takeuchi, M., 45, 50
Tamura, S., 128
Tanaka, T., 277, 417
Tanii, T., 60
Thomae, M., 261
Tokunaga, Y., 322
Togei, R., 70
Toriyama, Y., 65
Tottori, K., 84
Toyoda, T., 113

Ueda, M., 55

V

Vanderschueren, T., 251
Von Seggern, H., 103

W

Washizu, M., 35
Watanabe, S., 128
West, J. E., 392

Y, Z

Yahagi, K., 317, 387, 402
Yamada, Y., 437
Yamashita, K., 427
Yasufuku, S., 60
Yatsuhashi, K., 203
Yoshino, K., 218
Yoshioka, H., 128

Zielinski, M., 287

RAYMOND H. FOGLER LIBRARY
DATE DUE

BOOKS ARE SUBJECT TO
RECALL AFTER TWO WEEKS